MIDLAND MACROMOLECULAR INSTITUTE
1910 WEST ST. ANDREWS DRIVE
MIDLAND, MICHIGAN 48640

HYDROGEN BONDING

HYDROGEN BONDING

Melvin D. Joesten
L. J. Schaad

Department of Chemistry
Vanderbilt University
Nashville, Tennessee

MARCEL DEKKER, INC.   NEW YORK

COPYRIGHT © 1974 by MARCEL DEKKER, INC.
ALL RIGHTS RESERVED

Neither this book nor any part may be reproduced or transmitted in any
form or by any means, electronic or mechanical, including photocopying,
microfilming, and recording, or by any information storage and retrieval system, without permission in writing from the publisher.

MARCEL DEKKER, INC.

270 Madison Avenue, New York, New York   10016

LIBRARY OF CONGRESS CATALOG CARD NUMBER: 74-77104

ISBN: 0-8247-6211-8

Current printing (last digit):
10 9 8 7 6 5 4 3 2 1

PRINTED IN THE UNITED STATES OF AMERICA

CONTENTS

iii

PREFACE

Since the publication of Pimentel and McClellan's book on hydrogen bonding in 1959, thousands of papers on various aspects of hydrogen bonding have appeared. The difficulty presented to persons interested in having an awareness of published results on a particular topic has been reduced by the publication of a number of reviews or monographs. Topics reviewed during the past five years include hydrogen bonding in solids, theory of hydrogen bonding, equilibria and reaction kinetics in hydrogen-bonded solvent systems, and the application of various spectroscopic techniques to hydrogen-bonded systems. The recent book by Vinogradov and Linnell (Van Nostrand-Reinhold, 1971) provides a good introduction to hydrogen bonding for the nonspecialist.

The purpose of the present book is to provide comprehensive coverage of those topics which account for a large percentage of the hydrogen bonding papers published since Pimentel and McClellan's book. The topics chosen include spectroscopic studies and correlations, theory, thermodynamics and kinetics, and general applications. I have written Chapters 1, 3, 4, and 5. These chapters include a brief general introduction with a review of developments during the past decade. Professor Schaad has written Chapter 2, which has an introduction to quantum mechanics and a critical analysis of the present theoretical position of hydrogen bonding.

An extensive annotated bibliography of over three thousand references includes both references related to the topics discussed and a number of key references for topics not discussed. The emphasis was placed on covering the literature from 1960 through 1973. The format of the bibliography is the same as that used by Pimentel and McClellan and both text and bibliography are indexed. It is hoped that the subject and author indices of this book together with those of Pimentel and McClellan's book will provide ready access to the literature on hydrogen bonding.

I wish to acknowledge with thanks the expert secretarial assistance of Miss Brenda Knowles who prepared the final copy for offset photoreproduction. Finally, I am especially grateful to my family for their patience and understanding.

Melvin D. Joesten

HYDROGEN BONDING

I. INTRODUCTION

A. Definition of the Hydrogen Bond

When a covalently bound hydrogen atom forms a second bond to another atom, the second bond is referred to as a hydrogen bond. Although most reviews cite a 1920 paper by Latimer and Rodebush [1418b] as the first definitive reference on hydrogen bonding, the origin of the idea cannot clearly be attributed to any one person. Pimentel and McClellan [1938] have traced the concept from early work on association of liquids through Werner's ideas of complex structure up to the Latimer and Rodebush paper. An interesting sidelight is the role played by Huggins [1103]. Latimer and Rodebush themselves refer to Huggins unpublished work, and in his book on valency G. N. Lewis says "The idea was first suggested by Dr. M. L. Huggins, and was also advanced by Latimer and Rodebush, who showed the great value of the idea in their paper . . ." (p. 109 in Ref. [1452a]).

The hydrogen bond is usually represented as A—H...Y where A and Y are atoms with electronegativity greater than that of hydrogen (C, N, P, O, S, Se, F, Cl, Br, I). We will use the more general representation A—H...B where B is any σ or π electron donor site (Lewis base) and A is one of the atoms listed above. Hydrogen bonds can be either intramolecular or intermolecular. For intramolecular hydrogen bonding to occur A and B of A—H...B must be in a favorable spatial configuration in the same molecule (see Chap. 5). Two types of intermolecular hydrogen bonds will be discussed. Homo-intermolecular hydrogen bonding (self-association) refers to the association of two or more identical molecules while hetero-intermolecular hydrogen bonding refers to the

1

association of two different species (neutral or charged). Discussions
of hetero-intermolecular hydrogen bonding will emphasize 1:1 adducts
although several other stoichiometries are possible.

B. Strength of Hydrogen Bonds

Although the hydrogen bond is much weaker than normal chemical
bonds, its strength varies considerably with A—H and B.   Reported
enthalpy values for the reaction $A—H + B \rightarrow A—H...B$ (tabulated in the
Appendix) range from -0.5 kcal/mole for the thiophenol-benzene hydrogen bond [1623] to -37 kcal/mole for the hydrogen bond in $(CH_3)_4N^+HF_2^-$
[1009]. However, the majority of hydrogen bond energies (i.e., the
energy of the reverse reaction in the previous sentence) range from
3 to 10 kcal/mole. In the present discussion hydrogen bonds will be
designated as weak (<3 kcal/mole), normal (3 to 10 kcal/mole) or
strong (>10 kcal/mole). Examples of these three categories are:
phenol-benzene, ΔH = -1.18 kcal/mole [70]; phenol-triethylamine, ΔH =
-8.85 kcal/mole [70]; and trichloroacetic acid-triphenylphosphine
oxide, ΔH = -16 kcal/mole [973].

C. Methods of Detection

Both nonspectroscopic and spectroscopic techniques have been used
to detect hydrogen bonds. Since the emphasis in the past decade has
been on the use of spectroscopic techniques, we will limit our discussion to these. Comprehensive treatments of nonspectroscopic
techniques are given in Ref. [1938] and [2528].

II.   INFRARED AND RAMAN SPECTROSCOPY

A. Introduction

Historically, ir spectroscopy is the most important spectroscopic
method because of the sensitivity of vibrational modes to the presence
of hydrogen bonds. A majority of the work in this area has been concerned with the stretching frequency of the A—H functional group and
the decrease observed for hydrogen-bonded A—H...B. This example
illustrates an important point about most spectroscopic studies of

hydrogen bonding, namely, that the change in the bond adjacent to the
hydrogen bond is being investigated rather than the hydrogen bond
itself.

Raman and ir spectroscopic methods complement each other and will
be discussed together. The vibrational modes that are of primary
importance in hydrogen bonding are shown in Table 1-1.

TABLE 1-1

A—H...B Vibrations

Region			Description
3500-2500 cm^{-1}	$\leftarrow \; \rightarrow \; \rightarrow$ A—H...B	ν_s	A—H Stretch
1700-1000 cm^{-1}	$\uparrow$ A—H...B $\downarrow$	ν_b	A—H In-plane bend
900-300 cm^{-1}	$\pm$ A—H...B	ν_t	A—H Out-of-plane bend[a]
250-100 cm^{-1}	A—H..$\leftarrow$B$\rightarrow$	ν_σ	H...B Stretch
Below 200 cm^{-1}	$\uparrow$ A—H...B $\downarrow$	ν_β, ν_γ	H...B Bend[b]

[a]$\pm$ Indicates vibrational movement perpendicular to the A—H...B
plane. [b]See footnote, Table 1-4.

B. Weak and Normal Hydrogen Bonds

The major spectral changes that occur when weak and normal hydrogen bonds form are (1) the A—H bending frequency increases, (2) the
A—H stretching frequency decreases, and (3) the band width and band
intensity of the A—H stretching frequency increases.

Ratajczak and Orville-Thomas [2062] have reviewed studies of ν_b,
the in-plane-bending frequency, and ν_t, the out-of-plane bending frequency, for O—H hydrogen-bonded systems. The bending frequencies are
located in the "fingerprint" region, which makes assignments difficult.
Coupling of $\nu_{b(OH)}$ with other vibrational modes (ν_{CO} in carboxylic

acids; rocking CH_3 modes in methanol) further complicates the use of ν_b as a hydrogen bond probe. As a result, A—H bending frequencies have found limited use in hydrogen bonding studies.

Much attention has been given to the use of ν_s as a hydrogen bond probe since it is sensitive to hydrogen bond formation and is easily measured. The following discussion is concerned with changes in ν_s caused by hydrogen bonding. Figure 1-1 shows the change in frequency,

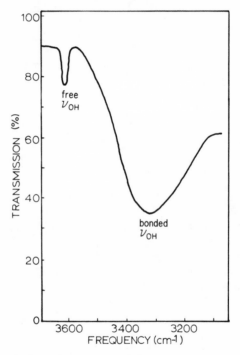

FIG. 1-1. Infrared spectrum of OH band for phenol adduct of N,N-dimethylformamide. 0.14 M phenol + 0.25 M base in CCl_4. Joesten [1180].

band width, and band intensity of ν_{OH} for the phenol-N,N'-dimethylform-amide adduct in carbon tetrachloride. The band width of several hundred wave numbers is typical of solution spectra for normal hydrogen-bonded adducts.

1. Band width of ν_s

The increase in band width of ν_s for the hydrogen-bonded complex has been attributed to (1) coupling of ν_s with ν_σ to give sum-and-difference bands, $\nu_s \pm \nu_\sigma$, (2) anharmonicity changes, (3) Fermi resonance of ν_s with overtone or combination bands, and (4) overlap of bands due to the presence of several different hydrogen-bonded species. The historical development of these explanations of band width is described by Sheppard [2249], Vinogradov and Linnell [2528], and Pimentel and McClellan [1938].

Several gas-phase studies have been carried out in an attempt to gain a better understanding of the cause of band broadening in hydrogen-bonded complexes. These include studies of hydrogen halide-acetonitrile adducts [2443,2445], hydrogen halide-ether adducts [75, 197,219a,221,2444], methanol-ether adducts [1128], alcohol-amine adducts [403,759,1058,1667,2095a], and nitric acid-ether adducts [27, 403,1666]. Results of these studies indicate that $\nu_s \pm \nu_\sigma$ combination bands are an important cause of band broadening. For example, the three bands in the spectrum of the hydrogen chloride-dimethyl ether adduct [73] were assigned initially to $\nu_3 + \nu_1$, ν_3, and $\nu_3 - \nu_1$, where ν_3 is 2570 cm^{-1} and ν_1 is 95 cm^{-1} (Fig. 1-2). However, recent work

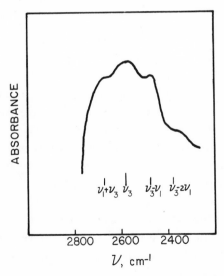

FIG. 1-2. Infrared spectrum of hydrogen chloride adduct of dimethyl ether in ν(HCl) region. Arnold, Berti, and Millen [73].

[219a,1417a] on the temperature dependence of the subband intensities has provided evidence that the $\nu(HCl)$ vibration of the ether adduct is not located at 2570 cm^{-1} but at the subband centered at 2480 cm^{-1}.

Evidence against the importance of a Fermi resonance mechanism in $(CH_3)_2O$—HCl is the observation that the spectrum of deuterated ether with hydrogen chloride is the same as that for the nondeuterated ether [220]. However, Fermi resonance arguments have been used to explain the subband structure of ν_{OH} in spectra of phenol-pyridine adducts [979]. The main problem in assessing the importance of $\nu_s \pm n\nu_\sigma$ for different complexes is the lack of positive identification of ν_σ in the far-infrared region. This is discussed in Section III of this chapter.

The decrease observed in the stretching frequency when hydrogen-bonded liquid-phase systems are cooled has been explained in terms of anharmonic coupling of ν_s with ν_σ modes [1938]. However, Wood and Rice [2095a] found that the stretching frequency for the gaseous tri-fluoroethanol-trimethylamine adduct is independent of temperature. Their calculations employing the Stepanov approximation demonstrate that a small positive temperature coefficient is expected when transitions within 50 cm^{-1} of the band origin contribute, while a larger positive temperature coefficient is expected when all subbands contribute.

The influence of anharmonicity on the ir spectral changes of A—H stretching bands has been examined by Sandorfy and co-workers [776]. Their results indicate that the strength of the hydrogen bond determines the amount of anharmonicity. From a study of the fundamental and overtone bands of ν_{NH} for six secondary amines they found that the anharmonicity decreases when a weak hydrogen bond is formed [219]. However, the anharmonicity increases when medium or strong hydrogen bonds are formed. This has been observed for hydrogen bonds in the phenol dimer and in o-chlorophenol [678,776]. In addition, Fifer and Schiffer [756] have shown that the anharmonicity of O—H...Cl$^-$ is greater than that for O—H...O in crystalline $CuCl_2 \cdot 2H_2O$. Thus, correlations of $\Delta\nu$ with ΔH or with the A...B distance are likely to be complicated because of anharmonicity and its dependence on the strength

of the hydrogen bond. This should be kept in mind when examining the
various correlations in Chap. 4.

The effect of different hydrogen-bonded species on band width has
been examined for hydrogen halide adducts [1108a]. Infrared profiles
for different concentrations of the adducts are interpreted in terms
of chain complexes, A—H...A—H...B, bifurcated complexes, $A—H\overset{\cdot\,\cdot B}{\underset{\cdot\,\cdot B}{\cdots}}$,
and proton-transfer complexes, $(AH)_n A^- \overset{+}{\cdots} H—B; A^- + (B...H...B)^+$.

In summary, factors (1) - (4) have been shown to contribute in
varying degrees in different hydrogen-bonded systems. The most suc-
cessful model calculations are based on adiabatic coupling of ν_s with
ν_σ with a consideration of the anharmonicity of A—H...B motion [1600].

2. *Band intensity of* ν_s

The relative increase in integrated band intensity (B) caused by
hydrogen bonding ranges from 2 to 40 [1938,2528]. As a result, band
intensity measurements are more reliable than frequency shift measurements for detecting weak hydrogen bonds. In 1960 Pimentel and
McClellan [1938] illustrated the potential of band intensity measurements as a probe for weak hydrogen bonds. However, band intensities
have not been reported for very many systems, probably because of the
difficulty in obtaining quantitative data.

Band intensity measurements for adducts of acetylenes [278a],
chloroform-d [1216], and methanol [167] are examples of the use of
band intensity as a probe for weak hydrogen bonds. Iogansen and co-
workers [1135-1137,2053,2055] have recently published several papers
on the measurements of integrated band intensities for ν_{OH} and ν_{NH}
systems. They propose that the correlation of band intensity with
enthalpy is more reliable than the frequency shift-enthalpy correlation
(see Chap. 4, Section IV).

C. Strong Hydrogen Bonds

Potential energy diagrams for hydrogen-bonded complexes have been
used extensively in the interpretation of ir spectra of strong hydrogen
bonds. Most hydrogen bonds are classified as asymmetric, with the
proton remaining in the first potential minimum (Fig. 1-3a). As the
hydrogen bond becomes stronger, the minima approach the same energy

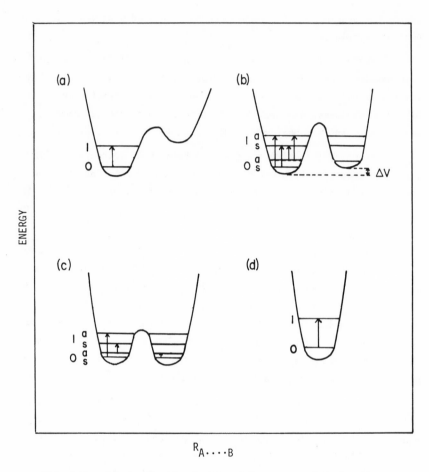

FIG. 1-3. (a) Asymmetric double-minimum potential function;
(b) slightly asymmetric double-minimum potential function; (c) symmetric double-minimum potential function; (d) single-minimum potential
function.

level and the barrier decreases (Fig. 1-3b,c). When the hydrogen atom
reaches the symmetrical position as represented by A...H...B, the
barrier is below the ground-state level and a single-minimum potential
function is obtained (Fig. 1-3d). General discussions of potential
functions of hydrogen-bonded adducts are given in Ref. [989,2528,2700a].
A recent treatment of slightly asymmetric double-minimum potentials is
given in Ref. [2625a].

When the double-minimum potential function is symmetric (Fig. 1-3c), the transitions $\nu_{s,0}$ to $\nu_{a,0}$ (far-infrared), $\nu_{s,0}$ to $\nu_{a,1}$, and $\nu_{a,0}$ to $\nu_{s,1}$ are allowed. Experimental evidence for the first transition is discussed in Section III. The last two transitions should be observable in the ir region with a separation of a few hundred cm^{-1} [2625a]. When the double-minimum potential becomes asymmetric, all four transitions illustrated in Fig. 1-3b are allowed [2321].

The difficulties of interpreting ir spectra in terms of a double-minimum potential have been described by Somorjai and Horning [2321]. They point out that, except for completely symmetric double minima with low barriers (Fig. 1-3c), observation of doublets which can be ascribed to the double-minimum potential is unlikely. They emphasize the importance of temperature and deuteration studies, since their calculations indicate that (1) barriers whose top is in the vicinity of the first excited OH vibrational level would give a low ν_{OH}/ν_{OD} ratio, (2) an asymmetric double-minimum potential would cause intensity inversion or an increase in the doublet spacing for the deuterated form, (3) relative intensities will vary with temperature for a symmetric double-minimum potential, and (4) ir and Raman spectra would give the same position and separation of the doublet if the double minimum were asymmetric. According to point (2) the ν_s band would be expected to be sensitive to changes in asymmetry of the double-minimum function. However, Wood [2625a] has shown that the band is not sensitive to slight changes in asymmetry in double-minimum potential functions. Calculations were applied to the $(B_2H)^+$ cation (see Section II.C.4 in this chapter for description of spectra and references).

1. OHO

Infrared spectra of adducts with strong O—H...O hydrogen bonds have been reviewed by Hadži [968,968a]. The spectra are of two types: type one consists of three main bands (A,B,C) in the 1800-3000 cm^{-1} region caused by OH vibrations; type two has no bands above 1800 cm^{-1}, but rather a strong, broad band around 1500 cm^{-1}.

Examples of systems which give type-one spectra are adducts of mono-, di-, and trichloroacetic acids with sulfoxides and phosphine oxides [973]. The A,B doublet in type-one spectra was interpreted initially [258] in terms of $\nu_{s,0}$ to $\nu_{a,1}$ and $\nu_{a,0}$ to $\nu_{s,1}$ transitions

(Fig. 1-3b). However, the separations in A,B doublets are too large
in compounds such as KH_2PO_4 to be attributed to split vibrational
transitions [968a]. Claydon and Sheppard [469] have proposed that the
extra bands are caused by Fermi resonance of ν_{OH} with $2\nu_b$ (in-plane
deformation) and $2\nu_t$ (out-of-plane deformation). Odinokov and Iogansen
[1819a] have provided additional evidence for the Fermi resonance
mechanism by locating ν_t fundamental and overtone bands in spectra of
adducts of chloroacetic acids with pyridine, dimethylsulfoxide, and
tributylphosphine oxide.

For adducts which give type-one spectra, Hadži and Kolilarov
[973] have estimated that the depths of the double-minima potential
(ΔV, Fig. 1-3b) will differ by more than $1000 \ cm^{-1}$ for the weakest
adducts to nearly zero for the strongest adducts. Odinokov, Maximov,
and Dzizenko [1821] found additional splitting in the A,B doublet in
type-one spectra of the benzoic acid dimer in the solid state and in
carbon tetrachloride solution. The effects which bases have on the
doublet separation was used to calculate the difference in energy (ΔV)
between the minima of the two potential energy curves for adducts of
benzoic and salicylic acids. Estimates of ΔV have also been determined from OH overtone bands for self-associated formic and acetic
acids ([1704a] and Section IV).

Type-two spectra are observed for Type-A acid salts of monocarboxylic acids [968,975a,2323a] and for adducts of trichloroacetic acid
with diphenyl selenoxide and pyridine N-oxide [973]. Speakman [2323a]
has shown that Type-A salts contain symmetrical hydrogen bonds (see
Section VII.A). Therefore, type-two spectra are considered to be
representative of symmetrical hydrogen bonds described by a single-
minimum potential function (Fig. 1-3d).

Other OHO systems which have been analyzed in terms of type-one or
type-two spectra include organophosphorus and phosphinic acids [614],
chromous acid [2303], cobaltous acid [201], and orthophosphates [425].

2. *NHO*

Detoni *et al.* [615] have observed type-one spectra for the
Ph_3PO —HNCS adduct and type-two spectra for the adduct of HNCS with
pyridine N-oxide.

Lindemann and Zundel [1471a] have investigated mixtures of carboxylic acids and cyclic nitrogen bases. Type-one spectra were
observed for pyridine adducts with acetic and monochloroacetic acid;
2-methylpyrazine adducts with di- and tri-chloroacetic acid; and N-
methylimidazole adducts with stearic and acetic acid.

3. XHX$^-$

The bihalide ions are good examples of species which contain
strong hydrogen bonds. Hydrogen bond energies range from 9.1 kcal/mole
for (BrHCl)$^-$ to 37 kcal/mole for (FHF)$^-$. Infrared spectral studies
are described in Ref. [725,726,729,1793,2164].

Evans and Lo [725,726,729] discovered that two different types of
spectra were obtained for different salts of (ClHCl)$^-$ and (BrHBr)$^-$.
For example, the spectrum of $(CH_3)_4N^+HCl_2^-$ (type one) is distinctly
different from that of $(C_2H_5)_4N^+HCl_2^-$ (type two). These differences
were interpreted as evidence for a linear, nonsymmetric anion structure
in $(CH_3)_4N^+HCl_2^-$ and a linear, symmetrical anion structure in
$(C_2H_5)_4N^+HCl_2^-$. A number of compounds of (ClHCl)$^-$ and (BrHBr)$^-$ were
investigated which gave type-one or type-two spectra.

Nibler and Pimentel [1793] have shown that the ν_2 band assignment
of Evans and Lo is incorrect for $Cs^+HCl_2^-$ (type-one spectrum) and
should be assigned as $2\nu_2$. This reassignment builds a stronger case
for the nonsymmetric X...H—X structure for type-one salts, since the
intensity enhancement of $2\nu_2$ requires either $C_{\infty v}$ or C_s symmetry [1793].
If the symmetry were C_s, the bond would be nonlinear. Salts which give
type-two spectra are likely to be examples of linear, symmetric
X...H...X structures, since type-two spectra are observed for salts
of (FHF)$^-$.

4. (NHN)$^+$

Infrared spectra have been obtained for a series of hydrogen-
bonded cations $(B^1HB^2)^+$, where B^1 and B^2 are amines [476-478a,1610b,
1610c,2625b]. When B^1 and B^2 are the same heterocyclic base (*e.g.*,
pyridine) a doublet structure of the H—B stretching band is observed
in the 2000-2500 cm^{-1} region. This observation along with studies of
(BDB)$^+$ and far-infrared spectra are presented in support of a linear,
symmetric complex with a shallow double-minimum potential [477,478a,

2625b]. When B^1 and B^2 are different heterocyclic bases, the structure of the ν_s doublet depends on the difference in pKa values for B^1 and B^2 [476,476a]. If ΔpKa is larger than 6, the lower band disappears and the upper band shifts to a higher frequency. The spectral evidence for a series of cations support a continuous change from a shallow double-minimum potential to an asymmetric single-minimum as the pKa difference between the heterocyclic bases increases.

The ir spectral data for $[(CH_3)_3NHN(CH_3)_3]^+$ are in support of a symmetric single-minimum potential [478,1610c]. There is no evidence of splitting in the ν_s band or in the internal modes of the $(CH_3)_3N$ moities.

Infrared spectra of $(B^1HB^2)^+$ where B^1 is trimethylamine and B^2 is pyridine have been interpreted in terms of an asymmetric single-minimum potential with the proton located nearer trimethylamine [1610b].

D. Complete Proton Transfer

It is convenient to consider strong hydrogen bonding and complete proton transfer together since the transition from a hydrogen-bonded complex A—H...B to a proton-transfer complex $A^-...^+H$—B can be followed by infrared (ir) spectroscopy. For example, Zeegers-Huyskens followed the disappearance of ν_{OH} and the appearance of $\nu_{N^+—H}$ for mixtures of alcohols or phenols with n-propylamine [2684] and aniline [2683] (Table 1-2). The frequency shifts for $\nu_{NH(sym)}$ and $\nu_{NH(asym)}$ are smaller than those for ν_{OH}, but the same trends are observed.

The first two alcohols in Table 1-2 are weak enough to form only hydrogen-bonded complexes with n-propylamine since there is no evidence for $\nu_{N—H^+...O}$. The adducts of p-cresol, phenol, and p-chlorophenol with n-propylamine include both hydrogen-bonded complexes and proton-transfer complexes, as indicated by the two sets of ir bands. For acids stronger than p-chlorophenol the ir data indicate the presence of only the proton-transfer complex.

Zundel and co-workers [2700a,2700b] have studied proton transfer in hydrated polystyrenesulfonic acid. Continuous ir absorption from the O—H stretching frequency down to small wave numbers is observed when more than one molecule per acid group is present. Saturated

TABLE 1-2

Changes in ν_{OH} for Adducts of n-Propylamine[a]

Acid	Solvent	$\nu_{O-H...N}$	$\nu_{N^+-H...O}$	pKa
t-Butanol	CCl$_4$	3322	————	17.62
Methanol	CCl$_4$	3282	————	16.60
p-Cresol	CCl$_4$	3060	2740,2690[b]	10.19
Phenol	CCl$_4$	3050	2740,2690[b]	9.95
p-Chlorophenol	CCl$_4$	3040	2750,2690	9.38
m-Nitrophenol	C$_6$H$_6$	——	2760,2500[c]	8.28
p-Nitrophenol	C$_6$H$_6$	——	2760,2500[c]	7.15
2,6-Dinitrophenol	C$_6$H$_6$	——	2760,2500[c]	3.76
2,4,6-Trinitrophenol	C$_6$H$_6$	——	2760,2500[c]	0.80

[a]Zeegers-Huyskens [2684].   [b]Weak band.   [c]Strong band.

solutions of p-toluenesulfonic acid in dimethylsulfoxide and methanol also exhibit a continuous ir absorption [1227b]. The continuous ir absorption is attributed to proton tunneling across hydrogen bonds of $(H_5O_2)^+$ or $(H_9O_4)^+$. Calculations indicate a high polarizability of these hydrogen bonds [1164,1164a,2703], with contributions to the continuous ir absorption from Fermi resonances and induced dipole interaction of hydrogen bonds with the electric field of solvating molecules or ions.

Lindemann and Zundel [1471a] have used the appearance of continuous ir absorption in spectra of carboxylic acid-nitrogen base mixtures as evidence for proton transfer. Proton transfer was observed in anhydrous systems when the carboxylic acid was more acidic than the nitrogen base by a pKa difference of 4. In aqueous solution proton transfer was observed at ΔpKa of 2. The mechanism for proton transfer in proteins was discussed in terms of the influence of hydration on proton transfer from glutonic acid and aspartic acid residues to basic lysine and arginine residues.

Johnson and Rumon [1183] varied the strength of the hydrogen bond
by using adducts of benzoic acid derivatives with pyridine derivatives
to give systems which could be represented by B...H—A, B...H...A, or
B—H$^+$...A. They reasoned that if the pK values of the protonated base
and acid are similar, the system would be fairly close to symmetrical
B...H...A. The results were interpreted on the basis of the previous
classification scheme of Hadži. Namely, the appearance of two ν_{NH} or
ν_{OH} bands coincides with a double-minimum potential, while a single
ν_{NH} or ν_{OH} band below 1700 cm^{-1} corresponds to a single-minimum potential function. When the difference in pK is about 3.7, proton transfer
takes place. A doubling of ν_{OH} or ν_{NH} is seen on both sides of the
critical pK difference. This was cited as evidence of proton tunneling. Adducts were assigned single- or double-minimum potential
functions on the basis of ir spectral evidence. Examples include the
benzoic acid-pyridine adduct, which is represented as B...H—A with a
double-minimum potential function; the adduct of 3,5-dinitrobenzoic
acid with 3,5-lutidine, which is represented as B...H...A with a
single-minimum potential function; and the 2,4-dinitrobenzoic acid
adduct of 2,6-lutidine, which is represented as B—H$^+$...A$^-$ with a
double-minimum potential function. Sharpening of bands with cooling
rules out the possibility that extra bands are caused by a difference
band or by a tautomeric equilibrium.

Evidence for proton transfer in two adducts of dichloroacetic
acid [973] is the presence of antisymmetric and symmetric bands of the
carboxylate group in the same region as those for metal carboxylates.
The ir spectra of adducts of dichloroacetic acid with trimethylarsine
oxide and trimethylamine oxide have an O—H$^+$ stretching band near
2200 cm^{-1}, deformation bands near 1350 and 950 cm^{-1}, and additional
protonic bands near the antisymmetric band of the CO_2^- group. Car-
bonyl bands at 1640 cm^{-1} ($\nu_{C=O}$) and 1730 cm^{-1} ($\nu_{CO_2^- asym}$) for solutions
of monochloroacetic acid in trioctylamine indicate the presence of both
O—H...N and O$^-$...$^+$H—N species [610a].

Lattice energies are known to stabilize proton transfer. Although
HCl and H$_2$O form an ionic crystal made up of H_3O^+ and Cl$^-$ ions, Ault
and Pimentel [92] have shown that the 1:1 H$_2$O—HCl complex isolated in
a nitrogen matrix at 20°K does not involve proton transfer. The

spectrum is very similar to that for $(CH_3)_2O$—HCl (Fig. 1-2) and indicates the presence of an asymmetric O...H—Cl hydrogen bond. Infrared
spectra of the H_3N—HCl complex in a nitrogen matrix have been interpreted in terms of a symmetrical N...H...Cl hydrogen bond [93].

III. FAR-INFRARED SPECTROSCOPY

Direct spectroscopic observation of hydrogen bonds is possible in
the far-infrared region (10-300 cm^{-1}). Hydrogen bonding modes which
absorb in this region include the hydrogen bond stretching mode ν_σ and
the hydrogen bond bending modes ν_β and ν_γ (Table 1-1). Advances in
instrumentation and experimental techniques [311,760] have resulted in
renewed interest in this area, and the application of far-infrared
spectroscopy to hydrogen bonding has been reviewed by Jakobsen et $al.$
[1159] and by Möller and Rothschild [1681a].

Assignment of hydrogen bond modes is difficult since a variety of
other vibrations occur in the far-infrared region. The use of deuterated derivatives is not much help in making band assignments since
small frequency shifts of a few wave numbers are observed. Three
criteria have been suggested by Hurley et $al.$ [1112] for making positive assignments of hydrogen bond frequencies. First, the bands
associated with hydrogen bond stretching or bending frequencies should
disappear completely when a group incapable of hydrogen bonding is substituted for the hydrogen atom ($e.g.$, —OCH_3 for —OH). Second, the
hydrogen bond frequencies should be very weak in the Raman spectrum
since the polarizability change during the formation of the hydrogen
bond will be small. Third, far-infrared absorptions due to hydrogen
bonding should be broad and asymmetric. These criteria were used to
make ν_σ assignments for self-associated acids (Table 1-3).

A. Far-Infrared Spectra of Self-Associated Acids

A number of techniques have been used to improve the quality of
far-infrared spectra of self-associated HA. These include low temperature studies of crystalline films [1897], spectra of samples between
polyethylene plates [1158], or samples absorbed in a polyethylene
matrix [309,1156]. Examples of spectra obtained under various physical

TABLE 1-3

ν_σ Assignments for Self-Associated Acids[a]

Compound	ν_σ, cm^{-1}	Reference	Compound	ν_σ, cm^{-1}	Reference
A. Phenols					
Phenol	162	[1156]	2-Methyl-6-*t*-butylphenol	104	[1156]
	150	[859]	*o*-Chlorophenol (intra)	84	[1112]
	175	[2334]	*o*-Nitrophenol (intra)	95	[2334]
Phenol-OD	143	[859]	*m*-Chlorophenol	130	[1112]
o-Cresol	121	[1156]	*p*-Chlorophenol	132	[1112]
	124	[1112]	*m*-Nitrophenol	125	[2334]
m-Cresol	143	[1156]	*p*-Nitrophenol	99	[2334]
	146	[1112]	2,6-Dinitrophenol (intra)	100	[2334]
p-Cresol	126	[1156]	Hydroquinone	172	[2334]
	124	[1112]	Resorcinol	158	[2334]
o-Isopropylphenol	134	[1156]			
	106	[1156]			
o-*t*-Butylphenol	127	[1156]			
2,6-Dimethylphenol	150	[1156]			
2,6-Diisopropylphenol	145	[1156]			
	106	[1156]			
B. Alcohols					
Ethanol	110	[1404]	*n*-Hexanol	145	[1404]
n-propanol	145	[1404]	*n*-Heptanol	165	[1404]
n-Butanol	148	[1404]	*sec*-Butanol	180	[1404]
n-Pentanol	150	[1404]	*t*-Butanol	150	[1404]

C. Acetic acids

1. Monosubstituted-RCH_2COOH

Acetic	176	[1159]	Undecanoic	175	[1404]
Propanoic	151	[1159]	Decanoic	170	[1404]
Pentanoic	185	[1159]	Nonanoic	167	[1158]
Decanoic	170	[1159]	Heptanoic	167	[1158]
Butanoic	163	[1158]	Fluoroacetic	115	[1159]
			Chloroacetic	147	[1159]
			Phenylacetic	151	[1159]
			3-Methyl pentanoic	154	[1159]
			Hexanoic	168	[1158]

2. Disubstituted-$R_1R_2CHCOOH$

2-Methyl propanoic	113	[1159]	2-Methyl butanoic	121	[1159]
Difluoroacetic	105	[1159]	2-Ethyl butanoic	110	[1159]
Dichloroacetic	95	[1159]	2-Ethyl hexanoic	115	[1159]
			2-Methyl pentanoic	123	[1159]

3. Trisubstituted-$R_1R_2R_3COOH$

2,2-Dimethyl propanoic	98	[1159]	2,2-Dimethyl butanoic	96	[1159]
Trifluoroacetic	93	[1159]	2,2-Dimethyl pentanoic	105	[1159]
Trichloroacetic	79	[1159]			

aSpectra run at 25°C for neat liquids or liquids absorbed on a polyethylene matrix.

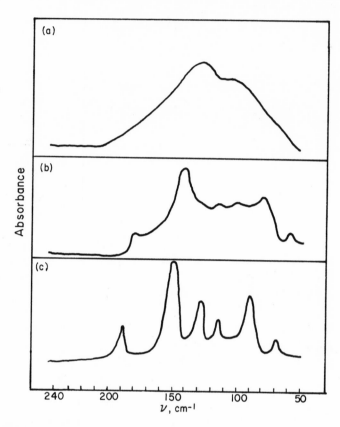

FIG. 1-4.   Far-infrared spectrum of imidazole. (a) 0.5 M solution in CHCl$_3$; (b) crystalline film at room temperature; (c) crystalline film at liquid nitrogen temperature. Perchard and Novak [1897].

conditions are shown in Fig. 1-4 for imidazole [1897]. In the poly-ethylene matrix method the hydrogen-bonded system is absorbed initially as the polymer. By heating the polyethylene matrix, the polymer can be converted to the monomer. Comparison of changes in the ν_{OH} and ν_{σ} region can aid in the assignment of ν_{σ}. An example of how this method was used for 2,6-dimethylphenol is shown in Fig. 1-5. The spectrum of 2,6-dimethylphenol absorbed into a polyethylene film is shown in Fig. 1-5a. The spectrum obtained after the polyethylene film had been heated to 100°C and cooled to 25°C is illustrated in Fig. 1-5b.

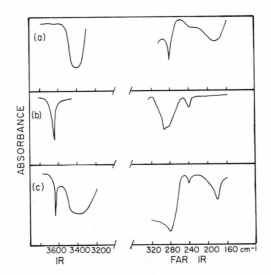

FIG. 1-5. Infrared spectra of 2,6-dimethylphenol in a polyethylene matrix. (a) 25°C; (b) heated to 100°C and cooled to 25°C; (c)
-190°C. Jakobsen and Brasch [1156].

The spectrum in Fig. 1-5c was obtained at -190°C. The disappearance
of the bands at 3400 cm^{-1} and 160 cm^{-1} when the polyethylene film is
heated, and the reappearance of these bands at -190°C provide evidence
for assigning the 160 cm^{-1} band as ν_σ (O—H...O).

A dramatic example of the effect of temperature on the quality of
far-infrared spectra is shown in Fig. 1-6 for hexanoic acid [1158].
However, use of solids makes the assignment of ν_σ more difficult because of the presence of lattice modes.

1. *Hydrogen bond stretch*

A summary of ν_σ assignments for self-associated acids is given in
Table 1-3. Assignments for solids have also been made and may be found
in the references cited.

The frequency of ν_σ for self-associated carboxylic acids has been
shown to fall into three ranges which depend on the α-substituent
[1157b,1159,2179,2340]. The three ranges are 100-185 cm^{-1} for monosubstituted, 95-125 cm^{-1} for disubstituted, and 80-105 cm^{-1} for trisubsti-
tuted acetic acids. Jakobsen and Katon [1157b] have shown that the

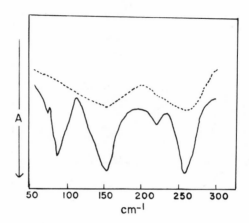

FIG. 1-6. Low-frequency spectra of hexanoc acid. (-) solid at
-150°C; (---) liquid at 25°C. Reprinted from Jakobsen, Mikawa, and
Brasch [1158], p. 841, by courtesy of Pergamon Press.

frequencies of the three ranges are related to the mass of the substituent on the alpha carbon while frequencies within a given range are
related to the total mass of the molecule. In general the decrease in
ν_{OH} for compounds listed in Table 1-3 does not correlate with the magnitude of ν_σ [1156,1112].

2. Hydrogen bond deformations

Assignments of hydrogen bond deformations have been limited to
cyclic dimers of carboxylic acids primarily because of the difficulty
in observing these bands. They usually occur at lower frequencies and
have lower intensities than the hydrogen bond stretching vibration.
Normal coordinate analysis of formic acid dimers and acetic acid dimers
has been carried out, and the results of these calculations have been
used to aid in band assignments [1276]. Previous electron diffraction
studies in the vapor phase had shown formic acid and acetic acid to
have planar dimer ring structures with C_{2h} symmetry [1233]. Six of the
twenty-four vibrational degrees of freedom involve motions of hydrogen
bonds. Three are ir-active (Au, Bu symmetry) and three are Raman-
active (Ag, Bg symmetry).

The calculated and observed ir frequencies for cyclic dimers of
formic and acetic acid are given in Table 1-4. Raman spectral data are
not available for these acids, but low frequency Raman and ir spectral
data have been reported for single crystals of benzoic acid (Table 1-5).

TABLE 1-4

Infrared Hydrogen Bond Frequencies

Mode[a]	$(HCOOH)_2$ Calculated	$(HCOOH)_2$ Observed vapor[b]	$(HCOOH)_2$ Observed hexane[c]	$(HCOOD)_2$ Calculated	$(HCOOD)_2$ Observed vapor[d]	$(HCOOD)_2$ Observed hexane[c]	$(CH_3COOH)_2$ Calculated	$(CH_3COOH)_2$ Observed vapor[d]	$(CH_3COOH)_2$ Observed hexane[c]
$\nu_\sigma + \nu_\beta$ (Bu)	249	248	248	243	240	240	187	190	176
ν_γ (Au)	167	164	173	158	158	167	79	--	69
ν_{twist} (Au)	68	68	--	68	68	68	54	50	--

[a] ν_σ = hydrogen bond stretching mode; ν_β = in-plane hydrogen bond bending mode; ν_γ = out-of-plane hydrogen bond bending mode; ν_{twist} = ring torsion along hydrogen bonds. [b] Kishida and Nakamoto [1276]. [c] Jakobsen, Mikawa, and Brasch [1161]. [d] Carlson, Witkowski, and Fateley [404].

TABLE 1-5

Single-Crystal Polarized Spectral Data: Benzoic Acid Dimer

Infrared-Active[a]

Mode[b]	Au, Calculated	Au, Observed	Bu, Calculated	Bu, Observed
$\nu_\sigma + \nu_\beta$	108	104	110	106
ν_γ	63	62	58	65
ν_{twist}	84	86	74	76

Raman-Active[c]

Mode	Ag, Calculated	Ag, Observed	Bg, Calculated	Bg, Observed
ν_σ	126	122	122	126
ν_β	113	114	113	114
ν_γ	100	93	100	96

[a]Meshitsuka, et al. [1639a].   [b]See footnote, Table 1-4, for descriptions.
[c]Colombo and Furic [495].

The far-infrared spectrum of imidazole at room temperature has six bands at 179, 142, 114, 103, 89, and 62 cm^{-1} (Fig. 1-4b). At liquid nitrogen temperature the peaks are sharpened and shifted to higher frequencies (Fig. 1-4c). All six bands are associated with hydrogen bond motion, with the first two associated with hydrogen bond stretching, the next three with hydrogen bond bending, and the last with the torsion mode [1897]. Raman studies of imidazole [494] are in agreement with the ir assignments.

B. Heteroassociated Systems

Positive identifications of ν_σ for several intermolecular hydrogen-bonded adducts have been reported. Ginn and Wood [859] have assigned the hydrogen bond stretching vibration for the phenol adducts of tri-methylamine, triethylamine, and pyridine as 143, 123, and 134 cm^{-1}, respectively. For phenol-OD adducts the bands shift to 141, 120, and 130 cm^{-1}, respectively. Normal coordinate calculations resulted in stretching force constants of 0.27, 0.19, and 0.23 for the trimethyl-amine, triethylamine, and pyridine adducts. The force constants cal-culated for the corresponding phenol-OD adducts were 0.27, 0.18, and 0.23. Thus, no change in the stretching force constant occurred as a result of deuteration.

A systematic study of twenty-four 1:1 adducts of p-substituted phenols with p-substituted pyridines has been carried out by Hall and Wood [978a]. The band maxima are sharply defined and can be located within ±1 cm^{-1}. Values of ν_σ for the adducts range from 101.0 to 130.5 cm^{-1}.

Values of ν_σ for phenol adducts of substituted pyridine N-oxides range from 144 cm^{-1} for 2-methyl-pyridine N-oxide to 218 cm^{-1} for 4-methoxy-pyridine N-oxide [845a]. A 1:1 mixture of carbon tetrachloride and chloroform was used as the solvent. Far-infrared peaks for the phenol adduct of 4-methyl-pyridine N-oxide dissolved in carbon disul-fide have been observed at 78, 135, 194, and 324 cm^{-1} [1109]. These have been assigned as in-plane bending, in-plane twisting, stretching, and out-of-plane deformation vibrations, respectively. The ν_σ assign-ment of 194 cm^{-1} is in good agreement with the value of 205 cm^{-1} observed for the same adduct dissolved in a 1:1 carbon tetrachloride/chloroform mixture [845a].

Carlson, Witkowski, and Fately [403] assigned the band at 175 cm^{-1} in the far-infrared spectrum of the $(CH_3)_2O—HNO_3$ adduct to ν_σ. This value is in good agreement with the value of 185 cm^{-1} predicted by Al-Adhami and Millen [27] from an examination of the O—H band structure for gaseous $(CD_3)_2O—HNO_3$.

Thomas [2444] observed a far-infrared band at about 180 cm^{-1} for gaseous hydrogen fluoride-ether adducts. This is consistent with the predictions of Arnold and Millen [75] in their study of band broadening in the H—F stretching region.

C. Testing Band Width Theories and the Double-Minimum Potential Function

Far-infrared studies are important for testing theories proposed to explain band width. In II.C. evidence was presented in support of the theory that the width of ν_{AH} is primarily due to sum-and-difference bands $\nu_{AH} \pm n\nu_\sigma$. However, several studies [311,477,1808,1897,2340] have shown that the band shape and frequency of ν_{AH} do not change in the same way as ν_σ. For example, the hydrogen bond frequencies of imidazole increase when the temperature is lowered (Fig. 1-4), but a corresponding trend in the ν_{NH} submaxima is not observed [1897]. The width of the ν_{NH} fundamental in imidazole has been attributed to Fermi resonance of ν_{NH} with various combinations and overtones of internal vibrations. Another example is the lack of agreement between changes in ν_σ for different $(B_2H)^+$ species and changes in the doublet structure of ν_s [477]. Therefore, it is likely that combination bands are not the only important cause of band widths.

Difficulties are often encountered in assignment of ν_σ. One illustration of this is the controversy over the assignment of ν_σ for the trimethylamine-methanol adduct. Ginn and Wood [861] assigned a band at 142 cm^{-1} to ν_σ on the basis of a predicted value of 145 cm^{-1} [1668]. However, Carlson, Witkowski, and Fately [403] believe the band at 142 cm^{-1} to be a methanol band shifted to higher frequencies by pressure, since they observed a shift of a weak doublet at 139 cm^{-1} for methanol vapor in the presence of nitrogen or argon just as in the presence of trimethylamine.

The proposed double-minimum potential function described in II.C. has resulted in attempts to observe the transitions between the split levels in the far-infrared region. Hadži [970] assigned bands observed in the 121-151 cm^{-1} region for ferroelectric crystals such as KH_2PO_4 to the $\nu_{s,0}$ to $\nu_{a,0}$ transition (Fig. 1-3c). However, Ginn and Wood [857] have presented arguments against the use of the proton tunneling mechanism to account for the doublet structure (163 and 187 cm^{-1}) in the far-infrared spectrum of the acetic acid dimer. Witkowski [2608] proposed that the doublet structure in the far-infrared spectrum of the acetic acid dimer is a result of nonadiabatic coupling of the hydrogen bond vibrations with the O—H stretching modes. Better agreement between the calculated and experimental spectra is obtained by assuming an anharmonic Morse potential for the hydrogen bond vibration [1449]. The experimental value for ν_σ for the acetic acid dimer is 170-180 cm^{-1}, and the calculated values are 110 cm^{-1} when anharmonicity is absent and 160 cm^{-1} when anharmonicity is present.

D. Summary

In summary, the results of far-infrared spectral studies are not definitive in elucidating the causes of broad ν_s bands for hydrogen-bonded systems. Certainly anharmonic coupling of ν_s with ν_σ is an important factor. The importance of other factors, such as Fermi resonance, appears to vary from adduct to adduct.

Additional work is needed to determine the degree of localization of the A—H...B stretching mode and the validity of using ν_σ as a measure of hydrogen bond strength. At present, ν_σ is less reliable than ν_{OH} for predictions of hydrogen bond strength.

IV. NEAR-INFRARED SPECTROSCOPY

Near-infrared spectroscopy (0.7-3.0 μ) is quite useful for hydrogen bonding studies since overtones of O—H, N—H, and C—H stretching vibrations occur in this region with little interference from other bands. Whetsel [2582] has reviewed the equipment, techniques, and general applications of near-infrared spectroscopy. Near-infrared instruments can also be used to study the fundamental O—H stretching vibration which absorbs near 2.75 μ.   This use has not been extensive because of the tendency to use conventional ir instruments to study the fundamental region.   However, workers contemplating quantitative measurements of the O—H stretching vibration should consider the use of a near-infrared instrument.

The intensities of overtone bands are quite low, with molar absorptivity values of less than 2.   This requires the use of sample cells with long path lengths, usually 10 cm.   As a result, the amount of solution required (about 30 ml) is one of the main disadvantages of this technique.

The ranges for bands which can be examined with near-infrared spectroscopic methods are given in Table 1-6.   The first overtone bands

TABLE 1-6

Bands in Near-Infrared Region[a]

Group	Band type	Range (μ)
C—H	Fundamental	3.0-3.6
	Combination	2.0-2.4
	First overtone	1.6-1.8
	Second overtone	1.1-1.2
N—H	Fundamental	2.8-3.0
	First overtone	1.5
	Second overtone	1.0
O—H	Fundamental	2.8
	Combination	2.0
	First overtone	1.4
	Second overtone	1.0
P—H	First overtone	1.89

[a]Whetsel [2582].

of O—H and N—H groups have been used in studies of self-association, intermolecular bonding, and the structure of water.

Overtones of ν_{AH} for self-associated O—H [289bc,290], N—H [212abc], and S—H [239d] groups have been examined to determine the effect of hydrogen bonding on the anharmonicity of the A—H vibration. Such studies have also provided valuable information about the nature of self-associated species as a function of temperature and concentration.

The O—H overtone bands for the chain polymer of formic acid and the cyclic dimer of formic acid or acetic acid have been used to calculate the barrier height in the double-minimum potential curves for these species [1704a]. The calculated values for the cyclic dimer and chain polymer of formic acid are 6400-6900 cm^{-1} and 6500-7000 cm^{-1}, respectively. The values are in agreement with the estimated value of greater than 6000 cm^{-1} from microwave spectral data for gaseous CF_3COOH—HCOOH [510].

The overtone region has provided evidence for the presence of simultaneous vibrational transitions ("double excitation" theory) in hydrogen-bonded adducts. Ron and Horning [2111] attributed extra bands in the overtone region of crystalline HCl to "double excitation." Asselin and Sandorfy [88] observed two bands in the first overtone region of ν_{OH} in low temperature spectra of self-associated alcohols. At -180°C the fundamental band is at 3230 cm^{-1} and the bands in the first overtone region are at 6220 and 6428 cm^{-1}. The band at 6220 cm^{-1} is considered the true overtone band, while the higher frequency band is attributed to the simultaneous excitation of the fundamental ν_{OH} in two neighbor O—H groups coupled by a hydrogen bond. Results of deuteration studies were used to discount the Fermi resonance mechanism and the proton tunneling mechanism.

A detailed examination of the problems encountered in sorting out overtone bands, simultaneous excitations, ν_{CH} + ν_{OH} combination bands, and effects of anharmonicity has been carried out by Bourdéron and Sandorfy [290] for self-associated alcohols.

Recently, evidence for "double excitation" in heteroassociated systems has been obtained by Rossarie et al. [2113a]. Overtone bands for adducts of alcohols with acetonitrile correspond to simultaneous excitation of transitions for the O—H group of the alcohol and the CH_3 group of acetonitrile. Burneau and Corset [371] observed extra bands at 5287 cm^{-1}, 4720 cm^{-1}, and 4004 cm^{-1} for chloroform adducts of CD_3CN, $(CD_3)_2CO$, and $(CD_3)_2SO$, respectively, which they assigned to

$\nu_{CH} + \nu_{AB}$ where AB is CN, CO, and SO. In an extension of this work
Burneau and Corset [372] have assigned extra bands in the near-infrared
spectra of adducts of chloroform, water, pentachlorophenol, and methanol with acetonitrile, acetone, dimethyl sulfoxide, and pyridine to
simultaneous vibrational transitions. They also observed that the
intensity of simultaneous transitions in RXH...R'AB is sometimes
greater than that for $2\nu_{XH}$.

V. ELECTRONIC ABSORPTION SPECTROSCOPY

Ultraviolet and visible spectroscopy may be used to study hydrogen bonding if the chromophoric portion of one of the molecules (acid
or base) is perturbed by the hydrogen bond [1432,1938,2528]. Generally, hydrogen bonding causes a red (bathochromic) shift for chromophores
acting as proton donors and a blue (hypsochromic) shift for chromophores
acting as proton acceptors. For example, studies of shifts for aniline
and 4-nitroaniline in hydroxylic solvents indicate that aniline acts as
a proton acceptor and 4-nitroaniline acts as a proton donor [1889a,
2501a].

Kamlet and co-workers [690,1226-1227a,1671a] have made extensive
use of chromophore electronic shifts in studies of solvated species and
proton affinity of solvents. For example, shifts of 4-nitroaniline
have been compared to those for N,N-diethyl-4-nitroaniline in a series
of solvents. The shift for N,N-diethyl-4-nitroaniline represents a
correction for the shift caused by solvent polarity. Table 1-7 lists
red shifts for these compounds in several hydroxylic solvents. The
enhanced red shift ($-\Delta\Delta\nu$) correlates with both σ^* of R in ROH and pK_{HB}
of ROH (see Chap. 4, Section II). These correlations provide support
for the use of enhanced red shifts as a measure of solvent proton
affinity.

The electronic shifts are of the same order of magnitude as hydrogen bond energies (250-7000 cm^{-1}). Pimentel [1934] has discussed the
significance of these shifts in terms of the Franck-Condon principle.
The electronic frequency shift is represented as

$$\Delta\nu_e = \nu_H - \nu_0 = W_0 - W_1 + \omega_1,$$
                                                              (1-1)

TABLE 1-7

Electronic Spectral Shifts[a]

Solvent ROH, R =	σ^*	$4\text{-}NO_2\text{-}C_6H_5NH_2$ $-\Delta\nu_{max}$, kK	$4\text{-}NO_2\text{-}C_6H_5N(C_2H_5)_2$ $-\Delta\nu_{max}$, kK	$-\Delta\Delta\nu_{max}$ kK
$(CH_3)_3C-$	-0.30	4.52[b]	1.79[b]	2.73[c]
$n\text{-}C_4H_9-$	-0.13	4.34	1.89	2.45
C_2H_5-	-0.10	4.16	1.92	2.24
CH_3-	0.00	4.06	2.24	1.82
$C_6H_5CH_2-$	+0.22	4.76	3.07	1.69
$H-$	+0.49	4.73	4.17	0.56

[a]Minesinger, Kayser, and Kamlet [1671a].   [b]Red shift relative to ν_{max} in cyclohexane.   [c]Enhanced red shift for $4\text{-}NO_2\text{-}C_6H_5NH_2$ compared with $4\text{-}NO_2\text{-}C_6H_5N(C_2H_5)_2$.

where ν_H and ν_0 are transition energies for the adduct and acid, W_0 and W_1 are the hydrogen bond energies of the ground state and excited state, and ω_1 is the extra transition energy received by the Franck-Condon principle (Fig. 1-7).

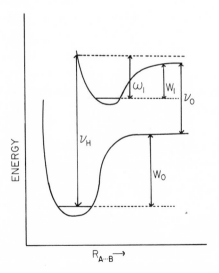

FIG. 1-7. Potential energy curve for the formation of A—H...B when the hydrogen bond in the excited state is weaker than in the ground state.

For 1:1 adducts $W_0 \simeq \Delta H$. The relationship between W_1 and ω_1 depends on the shapes of the potential curves in the ground and excited states. When the hydrogen bond is weakened by the transition, $\omega_1 > W_1$ and $\Delta\nu_e > W_0$. If the hydrogen bond is only slightly weakened or is strengthened by excitation, then $\omega_1 < W_1$ and $\Delta\nu_e < W_0$. The use of Eq. (1-1) for estimating hydrogen bond energies for the excited state is important since it provides one of the few ways to obtain information on hydrogen bond energies of molecules in the excited state.

Gramstad and Sandstrom [915] have provided support for Pimentel's assessment of the contribution of the Franck-Condon principle to blue shifts. They examined shifts in $\pi \rightarrow \pi^*$ and $n \rightarrow \pi^*$ transitions for a series of thioamide adducts of ethanol. Table 1-8 lists $\Delta\nu_e$, energies

TABLE 1-8

Electronic Frequency Shifts for Ethanol Adducts of Thioamides[a]

Thioamide	$\Delta\nu_e$, $n \rightarrow \pi^*$		$\Delta\nu_e$, $\pi \rightarrow \pi^*$		$-\Delta H$
	cm^{-1}	kcal/mole	cm^{-1}	kcal/mole	kcal/mole
$CH_3C(S)N(CH_3)_2$	2900	8.3	410	1.2	2.2
$CH_3SC(S)N(CH_3)_2$	1150	3.3	260	0.7	1.2
$(CH_3)_2NC(S)N(CH_3)_2$	1750	5.0	900	2.6	2.4
$C_6H_5C(S)N(CH_3)_2$	2000	5.7	380	1.1	1.6
$NCC(S)N(CH_3)_2$	890	2.5	-360	-1.0	2.3

[a]Gramstad and Sandstrom [915].

calculated from $\Delta\nu_e$ using $\Delta E = hc\Delta\nu$, and experimentally determined $-\Delta H°$ values. The energy calculated from $\Delta\nu_e$ for the $n \rightarrow \pi^*$ transition is larger than $-\Delta H$, while the energy calculated from $\Delta\nu_e$ for the $\pi \rightarrow \pi^*$ is smaller than $-\Delta H$. This suggests that the hydrogen bond is weakened much more by the $n \rightarrow \pi^*$ transition than by the $\pi \rightarrow \pi^*$ transition. This is as expected, since the $n \rightarrow \pi^*$ transition is due to the excitation of one of the sulfur lone-pair electrons engaged in the hydrogen bond with ethanol. The $\pi \rightarrow \pi^*$ transition shifts some electron density away from the sulfur atom but does not reduce the basicity of the sulfur atom as much as the $n \rightarrow \pi^*$ transition. The deviations shown by $NCC(S)N(CH_3)_2$ may be caused by the formation of a 1:2 complex and/or hydrogen bonding with the NC group.

Interest in the effect of hydrogen bonding on electronic transitions was stimulated by Mulliken's discussion of hydrogen bonding in terms of charge-transfer theory [1731-1733]. The familiar charge-transfer model

$$\Psi = a\Psi_0 + b\Psi_1 \qquad\qquad\qquad (1\text{-}2)$$

can be used to represent hydrogen bonding by considering HA as an electron acceptor and B as an electron donor. Ψ_0 is the wavefunction of the nonbonded structure B...H—A (includes classical electrostatic and van der Waals terms). Ψ_1 is the wavefunction for the "dative" form B^+—H...A^-, which represents electron transfer from B to A. In weak and normal hydrogen bonds, a >> b.

Mulliken and Person [1733] suggest that the term "charge-transfer" is probably not appropriate for hydrogen-bonded complexes since Ψ_1 is important only for strong hydrogen bonds. If the charge-transfer theory were applicable to hydrogen-bonded adducts, we would expect to observe charge-transfer bands similar to those observed for iodine adducts. However, charge-transfer bands for hydrogen-bonded adducts would occur at much higher energies, usually in the vacuum ultraviolet region. As a result, they are much more difficult to observe experimentally. The only published example is a charge-transfer band assignment for the hydrogen maleate ion [1763].

Fluorescence spectral measurements [1614] for the β-naphthol adduct of triethylamine have been interpreted in terms of a strong charge transfer from the nitrogen nonbonding orbital to the OH antibonding orbital in the Franck-Condon excited state, which is followed by essentially complete proton transfer from oxygen to nitrogen in the excited equilibrium state.

Infrared spectra of hydrogen-bonded systems have also been discussed in terms of charge-transfer theory [147,1763,2021,2399,2661]. An example is the study of the variation of the ir spectrum for phenol adducts of aromatic hydrocarbons in terms of the relative contributions of Ψ_0 and Ψ_1 [2399]. However, the hydrogen bonds in these adducts are too weak to have any degree of charge transfer.

Ratajczak and Orville-Thomas [2061b,2061c] have extended Mulliken's charge-transfer theory to explain the correlation between hydrogen bond enthalpy and the enhancement of the dipole moment in hydrogen-bonded adducts. Interpretation of this correlation in terms of charge-transfer theory leads to the conclusion that electron delocalization and *not* proton transfer is responsible for the enhanced polarity of hydrogen-bonded adducts. Spectroscopic studies of carboxylic acid-tertiary amine systems in inert solvents support the importance of charge transfer. For example, spectra of solutions of acetic acid and triethylamine in carbon tetrachloride have been interpreted in terms of hydrogen bonding without ionization [613].

A correlation of intensity of charge-transfer band with increase in the amount of charge transfer had been predicted by Mulliken in 1964 [1731]. Bhowmik [236] found that electronic spectral shifts ($\Delta\nu_e$) and oscillator strengths (f) increase with increase in hydrogen bond energy in the case of β-naphthol adducts of azaaromatics. However, just the opposite trend was observed for β-naphthol adducts of aromatic hydrocarbons. This agrees with the results for adducts of n-donors and π-donors with other Lewis acids [1733], since azaaromatics would be expected to be n-donors. Mulliken and Person [1733] attribute this reversal to the inapplicability of the intensity theory to π-donors. Kubota [1353] had shown earlier that the oscillator strength of the absorption band of hydrogen-bonded complexes varies directly with hydrogen bond enthalpy for trimethylamine oxide adducts of phenol, β-naphthol, and α-naphthol.

Vinogradov and Linnell [2528] have described the application of electronic spectroscopic techniques to the study of the equilibrium between hydrogen-bonded complexes and proton-transfer complexes

$$B\ldots H\!\!-\!\!A \rightleftarrows B\!\!-\!\!H^+\ldots A^-. \tag{1-3}$$

Evidence in support of equilibrium (1-3) has been obtained from ultraviolet spectra for adducts of p-nitrophenol with amines in various solvents [2218,2219]. The absorption bands for the undissociated and dissociated forms of p-nitrophenol are at 317.5 and 402.5 nm, respec-

tively. Spectrophotometric titration of p-nitrophenol with amines in
various solvents provided data for estimating the importance of solvent polarity in proton-transfer reactions. The absorbances measured
in the 400 nm and 320 nm region were plotted as log [A$^-$]/[HA] versus
log [B]. Values of K$_{PT}$, where K$_{PT}$ is the equilibrium constant for the
formation of a proton-transfer complex, were estimated from the intercepts of the above plots. The agreement of log K$_{PT}$ values obtained
from absorbance data at both the 320 and 400 nm region and the near-
unity slopes of the plots support the proposed 1:1 adduct. Table 1-9
summarizes values of log K$_{PT}$ in three different solvents.

TABLE 1-9

Estimates of Log K$_{PT}$ for Adducts of p-Nitrophenol[a]

Solvent	Base	Log K$_{PT}$, 25° 400 nm region	Log K$_{PT}$, 25° 310 nm region
(CH$_3$)$_2$SO	(C$_2$H$_5$)$_3$N	2.61	2.52
	n-C$_4$H$_9$NH$_2$	5.73	5.51
HCON(CH$_3$)$_2$	(C$_2$H$_5$)$_3$N	1.90	1.78
	n-C$_4$H$_9$NH$_2$	3.03	2.81
n-C$_4$H$_9$OH	(C$_2$H$_5$)$_3$N	1.57	1.58
	n-C$_4$H$_9$NH$_2$	2.45	2.34

[a]Scott, DePalma, and Vinogradov [2218].

The data in Table 1-9 can be used to illustrate how the magnitude
of K$_{PT}$ is affected by the basicity of the amine and the polarity and
basicity of the solvent. n-Butylamine favors proton transfer to a
greater extent than triethylamine, which does not agree with the
measured basicities in water (pKa = 10.65 for triethylamine and 10.59
for n-butylamine). Scott et $al.$ [2218] propose that this reversal is
caused by enhancement of proton transfer, either by solvation of NH$_2$
groups (dimethyl sulfoxide and dimethylformamide) or by solvation of
the phenolic oxygen (n-butyl alcohol). However, water would solvate
both acid and base species, and the differences in basicity in water
are probably a result of differences in solvation energies of the
various species produced.

The importance of solvent polarity is further illustrated by the
effect of triethylamine added (dielectric constant = 2.4) to a solution
of p-nitrophenol in dimethylformamide (dielectric constant = 37.2).
Proton transfer is enhanced up to 8 vol% amine. Further addition of
amine depresses the ionization of p-nitrophenol until 80 vol% of amine
(dielectric constant $\simeq$ 13), where the absorption band for the dissociated p-nitrophenylate anion disappears. Experiments in butanol-
cyclohexane mixtures of known dielectric constant indicate that
ionization of p-nitrophenol occurs at about 50 vol% butanol (dielectric
constant = 7.84).

Results of spectrophotometric titrations of p-nitrophenol with
amines in aqueous mixtures of dioxane, t-butyl alcohol, and acetone
are given in Table 1-10. The same trends are observed in log K_{PT} as

TABLE 1-10

Estimates of K_{PT} for Adducts of p-Nitrophenol
in Aqueous Solvent Mixtures[a]

Solvent	Base	Log K_{PT}
Dioxane/H_2O 7:3	n-$C_4H_9NH_2$	3.78
Dioxane/H_2O 8:2	n-$C_4H_9NH_2$	2.21
t-Butyl alcohol/H_2O 8:2	$(C_2H_5)_3N$	2.30
	n-$C_4H_9NH_2$	5.03
t-Butyl alcohol/H_2O 9:1	$(C_2H_5)_3N$	1.71
	n-$C_4H_9NH_2$	3.28
Acetone/H_2O 9:1	$(C_2H_5)_3N$	2.80
	n-$C_4H_9NH_2$	2.21
Acetone/H_2O 19:1	$(C_2H_5)_3N$	2.21
	n-$C_4H_9NH_2$	2.37

[a]Scott and Vinogradov [2219].

those in Table 1-9. The assumption is made in these studies that the
spectrum of the proton-transfer complex is the same as that of the p-
nitrophenylate anion. This means that ultraviolet studies cannot
distinguish between an ion pair and a hydrogen-bonded ion pair.

VI. ION CYCLOTRON RESONANCE SPECTROSCOPY

Ion cyclotron resonance spectroscopy (ICR) is a fairly new technique for studying ion-molecule reactions in the gas phase. Beauchamp
[162] has reviewed the techniques and applications of ICR. Strong
hydrogen bonds in a variety of ion-molecule reactions have been investigated by ICR. For example, Kebarle and co-workers have determined
experimental dissociation energies for hydrogen-bonded clusters, such
as $H^+(H_2O)_n$ [1250], $H^+(CH_3OH)_n$ [937b], and $H^+(CH_3OCH_3)_n$ [937b].
Thermodynamic data determined by ICR techniques are given in Chapter
3 (Table 3-5).

The role of hydrogen bonding in nucleophilic displacement reactions [162], acid-catalyzed elimination reactions [162], decomposition
reactions [2096], and remote functional group interactions [1716] have
also been examined by ICR.

VII. DIFFRACTION METHODS

A. X-Ray Diffraction

The use of x-ray diffraction methods to investigate hydrogen bonding in solids has been reviewed by Speakman [2323a], Hamilton and
Ibers [989], Pimentel and McClellan [1937,1938], and Vinogradov and
Linnell [2528]. The position of hydrogen atoms cannot be located
accurately by x-ray diffraction. As a result, the total length
$R(A...B)$ in A—H...B is used rather than the actual hydrogen bond
length H...B. When the sum of van der Waals radii [277] for A and B
is greater than $R(A...B)$, a hydrogen bond is assumed to be present.
Average A...B lengths are listed in Table 1-11. Values of the van der
Waals sum vary from 0.3 Å larger than $R(A...B)$ for strong hydrogen
bonds ($e.g.$, O—H...O) to 0.04 Å smaller than $R(A...B)$ for weak hydrogen bonds ($e.g.$, N—H...S). Comparisons of H...B distances obtained
from neutron diffraction methods with the sum of van der Waals radii
for H...B are more appropriate for predicting the presence of weak
hydrogen bonds (see VII.B).

Evidence for symmetrical O...H...O bonds in a number of Type-A
acid salts of monocarboxylic acid has been obtained by Speakman and

TABLE 1-11

Comparison of A...B Distances for Nonsymmetrical Hydrogen Bonds with van der Waals Contact Distances

A—H...B	R(A...B)[a] calc.	observed	Ref.	A—H...B	R(A...B)[a] calc.	observed	Ref.
O—H...O	3.04	2.72 ± 0.04	[2547a]	N—H...O	3.07	2.93 ± 0.11	[2547a]
O—H...O$^-$	3.04	2.75 ± 0.13	[2547a]	N—H...O$^-$	3.07	2.98 ± 0.15	[2547a]
O—H...N	3.07	2.78 ± 0.10	[2547a]	N—H...N	3.10	3.07 ± 0.11	[2547a]
O—H...S	3.32	3.31 ± 0.08	[1937]	N—H...S	3.35	3.39 ± 0.12	[1937]
O—H...F$^-$	2.92	2.72 ± 0.09	[1937]	N—H$^+$...F$^-$	2.95	2.92 ± 0.11	[1937]
O—H...Cl$^-$	3.28	3.12 ± 0.13	[1937]	N—H$^+$...Cl$^-$	3.31	3.23 ± 0.12	[1937]
O—H...Br$^-$	3.37	3.28 ± 0.10	[1937]	N—H$^+$...Br$^-$	3.40	3.37 ± 0.15	[1937]
O—H...I$^-$	3.48	3.53 ± 0.06	[1937]	N—H$^+$...I$^-$	3.51	3.66 ± 0.11	[1937]

[a]Bondi [277]. van der Waals contact distances calculated with the following van der Waals radii: O, 1.52; N, 1.55; C, 1.70; S, 1.80; F, 1.40; Cl, 1.76; Br, 1.85; I, 1.96 Å.

co-workers [2323a]. The two acidic groups in Type-A salts are crystal-
lographically equivalent. The symmetry element is usually a center of
inversion, $\bar{1}$, which places the hydrogen atom at the midpoint of the
O...H...O bond. Table 1-12 lists O...O distances for eleven Type-A

TABLE 1-12

Hydrogen Bonds in Type-A Acid Salts of Monocarboxylic Acids[a]

MHX$_2$		Symmetry	
HX	M	of bond	Distance, Å
Acetic	Na	2	2.444(10)[b]
Trifluoroacetic	Cs	$\bar{1}$	2.38(3)
	K	$\bar{1}$	2.435(7)
Phenylacetic	K	$\bar{1}$	2.443(4)
Benzoic	K	$\bar{1}$	2.51(4)
p-Cl-Benzoic	K	$\bar{1}$	2.457(13)
m-Cl-Benzoic	K	$\bar{1}$	2.437(7)
p-OH-Benzoic	K	$\bar{1}$	2.458(6)
o-NO$_2$-Benzoic	Rb	$\bar{1}$	2.43(6)
Cinnamic	NH$_4$	$\bar{1}$	2.51(3)
Aspirin	K	$\bar{1}$	2.455(5)
Anisic	K	2	2.476(18)

[a]Speakman [2323a]. [b]Numbers in parentheses are the standard
deviations in the least significant digits.

salts. The O...H...O bonds are very short, with a weighted mean distance of 2.445(2) Å for the nine most precise values. These results
provide support for Hadži's interpretation of type-two infrared spectra
(II.C.) as evidence for symmetrical hydrogen bonds, since Type-A salts
give type-two spectra. However, Speakman [2323a] has pointed out a
number of technical difficulties in locating the hydrogen atom at a
center of inversion. For example, in many Type-A acid salts the choice
of space group is ambiguous. Where a choice exists (e.g., potassium
hydrogen bis-phenylacetate), the evidence for the more symmetrical
space group is whether the use of this space group led to a satisfactory refinement. X-ray methods cannot exclude the possibility of a

disordered structure with the hydrogen atoms occupying positions up to
0.1 Å on either side of the midpoint. Comparison of x-ray and neutron
diffraction studies [551a,1336-1338,1571a] for potassium hydrogen *meso*-
tartrate also illustrates some of the problems involved in accurate
location of the hydrogen atom.

B. Neutron Diffraction

The position of hydrogen atoms can be located accurately by neutron diffraction techniques, and a number of structures have been
investigated by this method. Values for A...B, A—H, and H...B distances and the A—H—B angle have been tabulated by Hamilton and Ibers
[989]. The van der Waals sum for H...B is always much larger than the
average H...B distance determined by neutron diffraction. Values for
selected hydrogen bonds are listed in Table 1-13. This illustrates

TABLE 1-13

Comparison of Average H...B Distances with van der
Waals Contact Distances[a]

A—H...B	$R(H...B)_{exp}$	$R(H...B)_{calc}$
O—H...O	1.7 Å	2.6 Å
O—H...N	1.9	2.7
N—H...O	2.0	2.6
N—H...N	2.2	2.7
N—H...S	2.4	3.1
C—H...O	2.3	2.6

[a]Hamilton and Ibers [989].

the greater reliability of comparisons involving H...B rather than
A...B.

The recent neutron diffraction study of potassium hydrogen
bis(trifluoroacetate) and its deuterium analogue [1571a] is an excellent example of the applicability of neutron diffraction analysis to
Type-A acid salts. The hydrogen bond, lying across a center of inver-

sion, has identical O...H...O and O...D...O distances of 2.437 Å. The
absence of an isotope effect is evidence for a symmetrical hydrogen
bond with the hydrogen atom vibrating anharmonically in a single potential minimum [2283]. The ir spectra are characteristic of Type-A acid
salts, with a ν_{OH}/ν_{OD} ratio close to 4/3 and a strong absorption at
800 cm^{-1} and 600 cm^{-1} for the hydrogen and deuterium compounds, respectively.

In summary, a survey [1937] of R(O...O) distances determined by
x-ray and neutron diffraction methods indicate that symmetrical hydrogen bonds are probably present if R is less than 2.47 Å. If R is
greater than 2.54 Å, the hydrogen bond is nonsymmetric. Between 2.47
Å and 2.54 Å both symmetrical and nonsymmetrical hydrogen bonds may
be present.

VIII. NEUTRON SCATTERING SPECTROSCOPY

Neutron scattering spectroscopy can serve as a complementary
technique to ir and Raman spectroscopy. Collins, Haywood, and Stirling
[493] have applied this technique to hydrogen-bonded systems. The
advantages are (1) neutron scattering is primarily from the motions of
the hydrogen nuclei, with small contributions from other nuclei; and
(2) optical selection rules do not apply to neutron scattering, so
there are no forbidden transitions. The main disadvantage is the need
for a reactor to produce the neutrons. As a result, neutron scattering
spectroscopy has found only limited use. A comparison of the neutron
scattering data with ir data is given in Table 1-14.

IX. NUCLEAR MAGNETIC RESONANCE SPECTROSCOPY

The principles and techniques of nuclear magnetic resonance (NMR)
spectroscopy have been discussed in detail by Pople, Schneider, and
Bernstein [1971] and more recently by Emsley, Feeney, and Sutcliffe
[711]. Application of NMR to hydrogen-bonding studies has been reviewed by Davis and Deb [579]. In this section we will limit our
discussion to those references which illustrate the scope of NMR

TABLE 1-14

Neutron Scattering Data[a]

Compound	Observed ir frequency, cm^{-1}	Optical value	Assignment
KHF_2	1240	1233	$\nu_t{}^b$, $\nu_b{}^c$
	589	600	$\nu_\sigma{}^d$
$CsHCl_2$	1136	1170	ν_t, ν_b
	652	631	ν_t, ν_b
$CsDCl_2$	438	449	ν_t, ν_b
$NaHCO_3$	1357	1395	$\nu_b + \nu_5(CO_3^=)$
		1300	
	1164		$2\nu_b + \nu_3(CO_3^=)$
	998	998	ν_t
	646	650	$\nu_b + \nu_3(CO_3^=)$
$KH(CF_3COO)_2$	1053		ν_b
	802		ν_t
	444		
	399		
$KH(CCl_3COO)_2$	1145	1200	ν_b
	988	990	ν_t
	423		

[a]Collins, Haywood, and Stirling [493]. [b]ν_t hydrogen atom in-plane bend. [c]ν_b hydrogen atom out-of-plane bend. [d]ν_σ hydrogen bond stretch.

studies of hydrogen bonding. The specific application of NMR spectroscopy to thermodynamic measurements of hetero-association and self-association is discussed in Chapters 3 and 5, respectively.

A. Proton

The effect of hydrogen bonding on the proton resonance signal was
first noted for the hydroxyl proton resonance of ethanol [77,1460].   A
shift to higher magnetic fields was observed with either an increase
in temperature or a dilution of ethanol with carbon tetrachloride.
This chemical shift in the resonance was attributed to breaking the
hydrogen bonds in ethanol either by dilution or by heating [1460].
Studies by Schneider [2202] on gaseous and liquid ethanol illustrate
the magnitude of the chemical shift for the hydroxyl proton resonance
(Fig. 1-8).

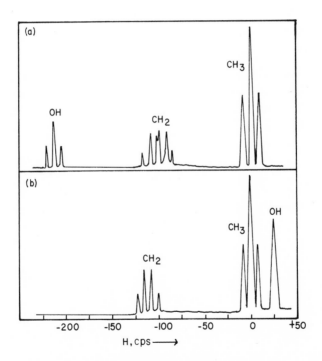

FIG. 1-8. Proton NMR spectrum of ethanol at 60 MHz.   (a) Liquid
ethanol at -114°C. The complex pattern of the CH_2 multiplet is caused
by coupling between OH and CH_2 protons.   (b) Gaseous ethanol at 195°C.
Schneider [2202].

Comparison of the spectra in Fig. 1-8 emphasize several important
points about the study of hydrogen bonds by proton magnetic resonance.

First, the proton resonance signal of the associated OH hydrogen is
shifted to a lower magnetic field relative to its position in the unassociated molecule. Shifts are generally to lower field except when
aromatic donor sites are involved in hydrogen bonding. Second, only
one OH proton resonance line is usually observed. Third, spin-spin
coupling of the OH group may be observed.

1. Downfield shift

The reason for the downfield shift is not clear. Intuitively,
one would predict an increase in the electron density at the proton
when a hydrogen bond forms. However, theoretical calculations show
that the proton loses electron density on forming a hydrogen bond
(see Chap. 2). Pople, Schneider, and Bernstein [1971] have explained
the observed shifts by considering the relative influence of (1) anisotropic magnetic currents in B and (2) polarization of the AH bond by B.

The effect of (1) may lead to upfield or downfield shifts. If B
is an aromatic molecule, (1) becomes more important than (2), and the
induced ring current creates a secondary magnetic field at the proton
which causes an upfield shift. As an example of this effect, the
chloroform-proton resonance for a mixture of chloroform and benzene
gives an upfield shift of 1.35 ppm at infinite dilution [2075].

If the primary function of B is to produce a strong electric field
in the vicinity of the AH bond, (2) will be more important and shifts
to lower field will be observed. The electron distribution about the
proton in the hydrogen bond is deformed by the polarizing effect of
the electric field of B. This decreases the electron density around
the proton and increases the asymmetry of the electron density. Both
effects decrease the shielding of the proton.

The contribution of electric field effects to hydrogen bonding
chemical shifts has been calculated for chloroform adducts [208] and
for quinuclidine adducts [2290]. The quinuclidine study also included
calculations of the contribution to chemical shift from polarizability
of the acid along the hydrogen bond [2290].

A large number of hydrogen bond shifts have been reported in the
literature. Two methods have been used to obtain the shifts. First,
the hydrogen resonance for the gaseous state (unassociated) can be
subtracted from the hydrogen resonance for the liquid just above the

melting point (associated). This method is illustrated in Fig. 1-8
for ethanol. The second and more common method is to obtain the unassociated value by successively diluting the solute with an inert
solvent and extrapolating the results to infinite dilution. Values
for the associated state are commonly taken at room temperature. The
use of the second method undoubtedly introduces some error in the hydrogen bond shifts. However, comparisons can be made for data collected under similar conditions. For example, comparison of trends in
hydrogen bond chemical shifts with other parameters, such as ir frequency shifts, have been made (Chap. 4, Section III).

2. *Resonance line characteristics*

The observed shape of the proton NMR signal depends upon (1) the
rate of chemical exchange, (2) dipole-dipole interactions, and (3)
paramagnetic impurities. Chemical exchange is the most important
factor in hydrogen bonding studies.

Usually, one AH proton resonance signal is observed in solutions
containing both associated and free acid. This represents a major
difference from ir spectroscopy, where both associated and free acid
can be observed simultaneously (Fig. 1-1). The slower time scale of
NMR spectroscopy generally prevents observation of resonances for both
associated and free states. To have an AH proton signal for both
states would require the lifetime of each state to be longer than the
reciprocal of the hydrogen bond chemical shift (10^{-2} to 10^{-3} sec).
Normal hydrogen-bonded systems have much faster exchange rates than
this at room temperature. As a result, a single, sharp line is observed that is a weighted average of the resonances for the various
environments.

The rate of AH proton exchange can be reduced either through
steric hindrance of the hydrogen-bonding site or by lowering the temperature. For example, OH proton resonance signals of both 2,6-
diisopropylphenol and ethanol are observed for an equimolar mixture
of the two (Fig. 1-9). The only reported example of the use of cooling
to observe both free and associated signals is for the adduct of
$(CF_3)_2CHOH$ with triethylamine [799a]. Both coordinated and free OH
resonances are observed by cooling the system to -125°C.

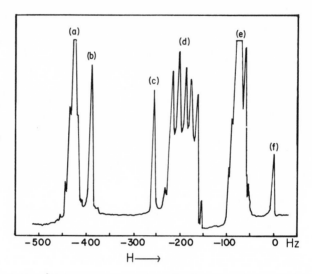

FIG. 1-9. ^{1}H NMR spectrum of equimolecular mixture of 2,6-
diisopropylphenol and ethanol at 25°; 60 MHz. (a) phenyl proton lines,
(b) phenolic hydroxyl line, (c) ethanolic hydroxyl line, (d) methylene
group lines of ethanol superimposed on septet due to lone proton of
isopropyl group, (e) methyl group of ethanol, (f) TMS. Somers and
Gutowsky [2320].

The rate of proton exchange also determines whether spin-spin
coupling will be observed. For gaseous ethanol at ten atmospheres
pressure and 195°C the exchange is fast enough to give only a singlet
for the OH proton (Fig. 1-8b). However, the OH signal for liquid
ethanol at -114°C is a multiplet since the proton exchange is slower
than the frequency separation of the components of the spin-spin
multiplet.

Literature reports are not in agreement with regard to the concentration dependence of spin-spin coupling for alcohols dissolved in
inert solvents [170,344,396,2293]. There is general agreement that the
observation of the hydroxyl proton multiplet in solutions of alcohols
in hydrogen-bonding solvents such as acetone [344] and dimethyl sulfoxide [428,1177,1563,2293] is evidence for the importance of hydrogen
bonding to the retention of the multiplet structure.

The disagreement over the concentration dependence of spin-spin
multiplets for alcohols dissolved in inert solvents is probably a

result of not using rigorously dried alcohol and solvent. Becker, Liddel, and Shoolery [170] reported collapse of the OH triplet for ethanol at 94 mole % in carbon tetrachloride, while Slocum and Jennings [1177,2293] were able to observe the OH triplet at 5 mole % in carbon tetrachloride and, with repeated effort, as low as 1 mole % in benzene. Therefore, the multiplet structure can be retained over a wider concentration range in nonhydrogen-bonding solvents if special efforts are made to purify the alcohol and the solvent. This means that the rate of making and breaking hydrogen bonds in 1:1 adducts with the solvent or in polymeric self-associated species must be slower than 10^{-2}-10^{-3} sec.

The influence of intramolecular hydrogen bonding on the multiplicity of the hydroxyl proton has been investigated for β-chloroethanols [2293]. Two opposing effects on proton exchange must be considered. Proton exchange will be increased by the increased acidity of the hydroxyl proton and decreased by the formation of intramolecular hydrogen bonds. At lower concentrations intramolecular hydrogen bonding is predominant as shown by the observation of the OH proton multiplet structure down to 0.2 mole % for β-chloroethanols in carbon tetrachloride. The influence of proton acidity becomes important at higher concentrations, particularly for solutions of β-chloroethanols in dimethyl sulfoxide [2293].

B. Other Nuclei

Only a limited amount of NMR work has been done on the effect of hydrogen bonding on nuclei other than hydrogen. Laszlo [1418a] has reviewed the research published before 1967. As a rule, the resonance of atom A in A—H...B shifts downfield and the resonance of atom B shifts upfield when the hydrogen bond forms. For example, the ^{14}N resonance of pyrrole shifts downfield in the presence of acetonitrile [2152] while the ^{14}N resonances of pyridine [2153], acetonitrile [1511], and methyl isocyanide [1511] shift upfield in the presence of methanol.

1. Deuterium

Deuterium magnetic resonance spectroscopy has proven very useful in the study of hydrogen-bonded ferroelectric crystals [257a]. This

application of NMR to solid-state studies is a result of the sensitivity of deuteron quadrupole coupling constants to small changes in
$R(O...O)$ distances in ferroelectrics such as KD_2PO_4. Deuteron quadrupole coupling constants increase with decreasing $R(O...O)$ distance
(increasing O_1—D distance in O_1—D$...O_2$ [440a]). Comparison of
deuteron coupling constants in the region of short hydrogen bonds
$(R(O...O) = 2.4$-2.6 Å) can help to differentiate between symmetrical
and nonsymmetrical hydrogen bonds. For example, the $R(O...O)$ distance
for triglycine sulfate is the same as that for potassium hydrogen
maleate, which has a symmetrical hydrogen bond. Since deuterated triglycine sulfate has a higher deuteron quadrupole coupling constant than
potassium deuterium maleate (78.8 versus 56 kc/sec), the O_1—D distance
is shorter and the hydrogen bond is not symmetrical [259a].

2. Carbon

Lichter and Roberts [1457] have measured the ^{13}C and ^1H chemical
shifts of chloroform in a variety of solvents. Downfield shifts from
the ^{13}C and ^1H chemical shifts for chloroform in cyclohexane were observed. Evans [723] had noted previously that the ^{13}C—H coupling
constant for chloroform adducts increased with increase in hydrogen
bond strength. The ^{13}C shifts correlate with ^{13}C—H coupling constants
(Fig. 1-10) and with ^1H shifts (Fig. 1-11).

3. Nitrogen

Saito and Nukada [2152] used the heteronuclear double resonance
technique to obtain ^{14}N chemical shifts for hydrogen-bonded complexes
of pyrroles and indole. The ^{14}N shift was determined from the maximum
in a plot of N—H proton signal versus ^{14}N irradiation frequency.
Using the chemical shifts in carbon tetrachloride as the reference, the
^{14}N hydrogen bond shift for self-associated pyrroles is -5.0 ppm compared to a proton shift of +0.55 ppm. This is an example of the greater sensitivity of the ^{14}N shift to N—H$...\pi$ interactions. The stronger
the base, the larger the downfield shift in ^{14}N. Indole adducts give
larger shifts, which indicates that indole forms stronger bonds than
pyrrole. Saito *et al.* [2155] used the same techniques to examine ^{14}N
chemical shifts for the self-association of formamide and N-methylacetamide.

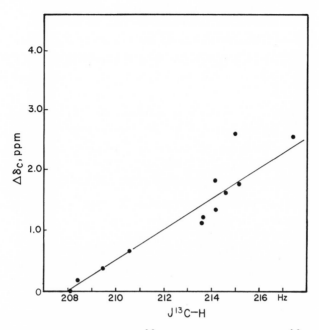

FIG. 1-10. Correlation of ^{13}C chemical shifts with ^{13}C—H coupling constants for chloroform-solvent mixtures. Lichter and Roberts [1457].

The ^{15}N resonance of liquid ammonia is 15.9 ppm downfield from that of the gas [1488]. A decrease in temperature or dilution of $^{15}NH_3$ in organic solvents also causes a downfield shift in the ^{15}N resonance of $^{15}NH_3$ [1488,1489]. These shifts are given in Table 1-15. The proposed model assumes that the observed shifts are the sum of contributions from (1) the interaction of the lone-pair electrons of ammonia with solvent-molecule groups, and (2) hydrogen bonding between unshared electron pairs in solvent-molecule groups with protons of ammonia. The results of statistical analysis of the various possible contributions using this model indicate that interaction (1) causes large paramagnetic shifts while (2) causes small diamagnetic shifts.

4. *Oxygen*

A downfield shift of 36 ppm is observed for the ^{17}O NMR resonance in going from gaseous to liquid water at room temperature [773]. The

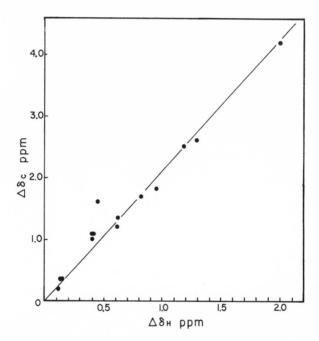

FIG. 1-11. Correlation of ^{13}C chemical shifts with ^{1}H chemical
shifts for chloroform-solvent mixtures. Lichter and Roberts [1457].

effect of solvents on [17]O chemical shifts of water has also been
examined [2093]. The values at infinite dilution are given in Table
1-16. Each water molecule in the liquid at room temperature is assumed
to have four hydrogen bonds with two *via* the hydrogen atoms and two
bonding to the oxygen atom lone pairs. The contributions from proton
donation and proton acceptance are estimated to be -12 and -6 ppm,
respectively.

A single [17]O resonance is observed for acetic acid and solutions
of acetic acid. The resonance is approximately the average of the
chemical shifts of the carbonyl and hydroxyl oxygen atoms [449]. Figure 1-12 illustrates the dilution curves of [17]O chemical shifts for
acetic acid dissolved in various solvents [2093]. The results in the
basic solvents acetone and acetonitrile may be interpreted in terms of
the equilibrium

$$1/2 \ (CH_3COOH)_2 \ + \ B \ \rightleftarrows \ CH_3COOH...B. \qquad\qquad (1-4)$$

TABLE 1-15

^{15}N Shifts for $^{15}NH_3$[a]

Solvent	^{15}N Shift, ppm, relative to $^{15}NH_3$ (gas)
H_2O	-19.4
$(CH_3)_2O$	- 7.7
$(CH_3)_3N$	-10.0
$(CH_3)_4C$	-11.1
CH_3NH_2	-12.3
CH_3OH	-13.4
C_2H_5OH	-16.6
$(CH_3)_2NH$	-10.1
$C_2H_5NH_2$	-13.7
$(C_2H_5)_2O$	-10.8
$(C_2H_5)_3N$	-12.4
$(C_2H_5)_2NH$	-12.7
NH_3	-15.9

[a]Litchman, Alei, and Florin [1488].

TABLE 1-16

Oxygen-17 Chemical Shifts of Water[a]

Solvent	Shift, ppm
Vapor (175-215°)	36
p-Dioxane	18.3
CH_3OH	13.6
$(CH_3)_2CO$	12.1
C_5H_5N	8.3
CH_3COOH	2.1
$H_2{}^{17}O$ (24°)	(0)

[a]Reuben [2093].

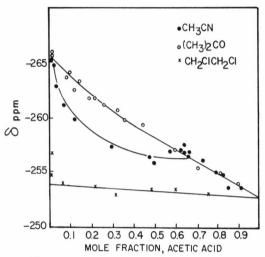

FIG. 1-12. ^{17}O chemical shift of acetic acid (relative to water) versus mole fraction in acetonitrile, acetone, and 1,2-dichloroethane. Reuben [2093].

Carbon-13 chemical shifts of these systems have been studied by Maciel and Traficante [1578].

The ^{17}O dilution shift of $(CH_3)_2C^{17}O$ in H_2O and D_2O is shown in Fig. 1-13. The shift toward higher field when the hydrogen bond forms has been attributed to the increased importance of the resonance structure $>\overset{+}{C}-O^-$ [2093]. No deuterium isotope effect was observed.

Alei and Florin [29] have measured ^{17}O shifts for $H_2^{17}O$—NH_3, $H_2^{17}O$—$(CH_3)_2CO$, and $H_2^{17}O$—$(CH_3)_3N$ adducts. The dilution curves are shown in Fig. 1-14. The similarity in the ^{17}O shift for the adducts of water with trimethylamine and acetone suggests that shielding of the oxygen nucleus in water is dependent upon the extent of involvement of lone pair electrons on the oxygen atom in hydrogen bonding. Other conclusions were that (1) the hydrogen bond in H_2O—NH_3 is stronger than in $(H_2O)_2$, as evidenced by the downfield shift of the ^{17}O resonance for water in liquid ammonia, (2) the hydrogen bond in H_2O—$(CH_3)_2CO$ could be 0.6 times as strong as the hydrogen bond in liquid water since the ^{17}O resonance for the water-acetone adduct is 21 ppm downfield from the resonance for water vapor. This assumes complete hydrogen bonding in liquid water and a linear relationship between hydrogen bond strength and ^{17}O shifts.

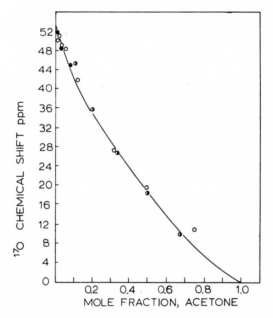

FIG. 1-13. ^{17}O chemical shift of acetone as a function of concentration in H_2O and D_2O. O, H_2O; ●, D_2O; ◐, data from [449a]. Reprinted from Reuben [2093], p. 5729, by courtesy of the American Chemical Society.

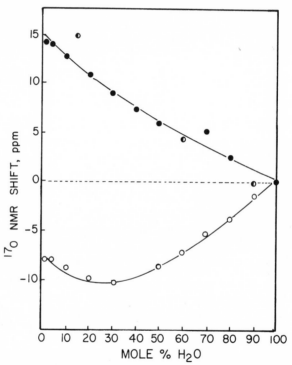

FIG. 1-14. $H_2{}^{17}O$ shifts in mixed liquid systems: O, H_2O-NH_3; ●, H_2O-Me_2CO; ◐, H_2O-Me_3N. Reprinted from Alei and Florin [29], p. 864, by courtesy of the American Chemical Society.

CHAPTER 2

THEORY OF THE HYDROGEN BOND

L. J. Schaad

I. THEORETICAL BACKGROUND

A. Introduction

Within the last five years *ab initio* calculations, especially those on the water dimer, have brought the theory of the hydrogen bond to a greater, but not yet final, level of sophistication. At the same time, new semiempirical methods have made possible the theoretical examination of larger systems. This chapter is an attempt to describe the current theoretical picture of the hydrogen bond, and to give some guidance in how far one may trust the various theoretical results. Excellent reviews of this subject already exist, each with a somewhat different emphasis. A well known paper by Coulson [516] summarizes work to 1957 in a very readable way. Bratož [313] and Lin [1465] give more mathematical details of the older work. Murthy and Rao [1748] are especially complete on potential functions for proton motion in the hydrogen bond. Kollman and Allen [1292] give an authoritative summary of recent work. This chapter is more concerned with explaining and pointing out the limitations of those theoretical methods which have been applied to the study of the hydrogen bond.

The phrase "theory predicts" might refer to anything from the crudest Hückel model to variational calculations of high accuracy. At the one extreme, the work by Kołos and Wolniewicz [1300b] on H_2 shows what can be done if the system is small enough and if one works hard enough. These authors obtained a calculated energy in agreement with

the experimental to 3 cm$^{-1} \approx$ 0.01 kcal/mole.[*] This is certainly well
within the accuracy required for most purposes, but Kołos and Wolniewicz felt that their errors were not as large as this disagreement. A
reexamination of the experimental data followed [1051a], with the result that theory and experiment now agree to a fraction of a reciprocal
centimeter. At the other extreme, many embarrassing false predictions
of approximate theories can be cited. To see where in this range the
theory of the hydrogen bond now stands, we begin with a survey of the
methods of theoretical chemistry.

B. The Schrödinger Equation

There is no reason to doubt that the answers to all chemical problems can in principle be obtained by solution of the Schrödinger
equation

$$i\hbar \frac{\partial}{\partial t} \psi(\underline{r},t) = H\psi(\underline{r},t), \tag{2-1}$$

where $|\psi(\underline{r},t)|^2$ gives the probability of finding the system at point $\underline{r}$
at time t. It is not usually difficult to write down the proper Hamil-
tonian operator H for a given system, but it is almost always extremely
difficult to solve the resulting equation. If H is not a function of
time t, the space and time variables separate in Eq. (2-1). The time
equation can then be solved to give an imaginary phase factor which is
usually of no interest to us,

$$\psi(\underline{r},t) = \psi(\underline{r}) \exp\left(\frac{-iE}{\hbar} t\right), \tag{2-2}$$

---

[*] We shall most commonly express energies in kcal/mole, but other
units also occur. Theoretical results are usually stated in atomic
units (au). In these units the Schrödinger equation simplifies slightly, and, more important, results so expressed are independent of current values of the fundamental constants. Electron volts (eV) will
also be used, and for accurate spectroscopic data reciprocal centi-
meters (cm^{-1}) are the usual unit. Wherever possible the conversion
factors recommended by Cohen and DuMond [482a] will be used with the
modifications of Taylor, Parker, and Langenberg [2421a]:
   1 au of energy = 1 Hartree = 27.2117 eV = 627.52 kcal/mole =
      219,475 cm^{-1}
   1 au of length = 1 Bohr = 0.529177 Å

with the spatial part of the wavefunction given by the time-independent
Schrödinger equation

$$H\psi(\underline{r}) = E\psi(\underline{r}). \tag{2-3}$$

For isolated molecules, or molecules in a constant external electromagnetic field, we can therefore make this separation. In a time-
dependent field, such as those of electromagnetic radiation, we cannot.
Although this separation simplifies the problem considerably, it still
leaves the enormously difficult solution of Eq. (2-3).

The Hamiltonian operator in Eq. (2-3) can be written as

H = T + V(electrostatic) + V(spin-orbit) + V(spin-spin)

+ H(other relativistic) + H(hyperfine) + H(nuclear spin). (2-4)

Here T is the kinetic energy operator of all the particles, and in
Cartesian coordinates is given by

$$T = -\frac{\hbar^2}{2m_e}\sum_i \left(\frac{\partial^2}{\partial x_i^2} + \frac{\partial^2}{\partial y_i^2} + \frac{\partial^2}{\partial z_i^2}\right)$$

$$-\frac{\hbar^2}{2}\sum_\alpha \frac{1}{m_\alpha}\left(\frac{\partial^2}{\partial x_\alpha^2} + \frac{\partial^2}{\partial y_\alpha^2} + \frac{\partial^2}{\partial z_\alpha^2}\right). \tag{2-5}$$

The first sum in Eq. (2-5) is taken over all electrons of the system
and the second over all nuclei. The mass of the αth nucleus is m_α, and
m_e is the mass of an electron; $\hbar$ is the usual symbol for Planck's constant divided by 2π. The second term in H gives the coulombic interaction between particles

$$V(\text{electrostatic}) = \sum_{i>j} e^2/r_{ij} + \sum_{\alpha>\beta} \frac{Z_\alpha Z_\beta e^2}{r_{\alpha\beta}}$$

$$- \sum_{i,\alpha} Z_\alpha e^2/r_{i\alpha} . \tag{2-6}$$

Indices i and j refer to electrons; α and β to nuclei; e is the electronic charge; and Z_α is the charge, in units of electronic charge, of
the αth nucleus. The distance between electrons i and j is r_{ij}. The
first sum in Eq. (2-6) gives the electrostatic energy of repulsion
between electrons; the second, that between nuclei; and the third gives
the electron-nuclear attractions.

As in most chemical problems, as long as we are concerned with
accuracy no greater than about 0.1 kcal/mole, and as long as we treat
first-row elements, we may safely disregard the remaining terms in the
Hamiltonian. On the other hand, if we want spectroscopic accuracy of,
say, 1 cm^{-1}, we may not disregard these terms.

The first of these terms to be discarded is V(spin-orbit), the
spin-orbit energy. It arises from the interaction of the moving magnetic field of the electrons with the electric field of each other particle. The resulting energy increases with nuclear charge. The ^2P
state of Na, emission from which gives the yellow D-line, is split by
spin-orbit interaction into two levels 17 cm^{-1} = 0.05 kcal/mole apart;
while in the corresponding state of Li, the splitting is 0.001
kcal/mole. It is true that these splittings represent only *differences*
in spin-orbit energy between the two substates, but the total spin-
orbit energy is of the same order of magnitude [265a]. Accordingly we
shall neglect V(spin-orbit).

We shall also discard V(spin-spin), the interactions between electron spins. This runs contrary to a picture of the chemical bond that
one sometimes meets, and one that was in fact suggested by G. N. Lewis
[1452a] in the early days of quantum theory. This plausible but incorrect view runs as follows. Electrons tend to form pairs, one with spin
up and one with spin down. Such a pair is like a pair of magnets
aligned in opposite directions. A pair of bar magnets is most stable
when the two magnets are aligned in opposition, just like the electrons.
Therefore electron pairs, and hence chemical bonds, are formed because
of the alignment of electronic magnetic dipoles. There are two things
wrong here. First, the magnetic dipole-dipole interaction is only part
of V(spin-spin), for there is also a second term (the "Fermi contact
term"), and, depending upon the system, the one or the other may domi-
nate. But what is more significant to us is that neither of these

terms is ever large relative to a chemical bond. In H_2 the spin-spin interaction energy amounts to only 0.004 kcal/mole [1300a]. The true cause of electron pair formation is more subtle, as we shall see.

The term H(hyperfine) is due to the fact that nuclei are not point charges, but have finite sizes. It is too small to be important in the theory of the hydrogen bond, as are terms due to nuclear spin, H(nuclear spin).

There is as yet no proper quantum relativistic treatment of a system for more than one electron, but relativistic effects may be taken into account approximately by treating the particles as independent. The spin terms above come out of such a treatment, as do a series of other terms which we have included in H(other relativistic). Relativistic terms may not be small compared to bond energies, but they are most important for inner electrons, and these do not vary much in going from atoms to molecules. It has therefore been assumed that relativistic terms do not make a significant contribution to the bond energy. A recent paper by Matcha [1615a] examines this question directly. He finds that total relativistic effects, including the spin energies above, account for from 0.6% of the bond energy in the case of LiH to 7.8% in the case of LiBr. The effect increases with atomic weight and should be minimal for hydrogen bonds.

No gravitational effects have been included. Presumably atomic particles exert gravitational forces just as do larger bodies, but we can easily dispose of these. Both gravitational and electrostatic forces have a $1/r^2$ dependence. The corresponding energies then will simply be in the ratio of the gravitational to the Coulomb's law constants. It follows at once that for the hydrogen molecule, gravitational forces contribute less than one part in 10^{35} of the bond energy.

The last of what might be called the fundamental approximations, that is, those approximations that will be implicit in all the following treatments of the hydrogen bond, is the Born-Oppenheimer separation. The assumption here is that electrons move rapidly compared to nuclei, so that for each position of the nuclei the electronic wavefunction is the same as that for a system with nuclei fixed at those positions. As a result, the kinetic energy of the nuclei may be dropped from Eq. (2-5), leaving only the first sum. The second sum in Eq. (2-6) is a

constant for each nuclear configuration and may be ignored until com-
pletion of the electronic calculation, after which it is simply added
to the electronic energy. This leaves for the time-independent
Schrödinger equation

$$\left[\frac{-\hbar^2}{2m_e}\sum_i \left(\frac{\partial^2}{\partial x_i^2} + \frac{\partial^2}{\partial y_i^2} + \frac{\partial^2}{\partial z_i^2}\right) + \sum_{i>j} e^2/r_{ij}\right.$$

$$\left. - \sum_{i,\alpha} z_{i\alpha} e^2/r_{i\alpha}\right] \psi = E\psi \ . \tag{2-7}$$

Finally, one might ask the following. We have thrown away spin-
orbit and other energy terms and have left only kinetic energy and
Coulomb's law electrostatic terms in Eq. (2-7), but in discussions of
the hydrogen bond one sees mention of van der Waals, ionic and covalent, ion-dipole and dipole-dipole, dispersion, London, and resonance
energies. Why are there no terms in the Schrödinger equation to
account for these?

The answer is that these result from particular interpretations
of various approximate solutions of Eq. (2-7). They do not correspond
to fundamental physical interactions in the sense that electrostatic
and spin-spin interactions do. They all come out of Eq. (2-7), which
is the quantum mechanical equation of motion for a set of point
charges moving in the field of each other. The ultimate cause of the
hydrogen bond, like that of all other chemical bonds, is therefore the
coulombic attraction of oppositely charged particles, with particle
motion described by quantum, rather than classical, mechanics.

C. Solution of the Schrödinger Equation

1. *The indistinguishability of electrons*

Before looking at the methods for solving the Schrödinger equation
[Eq. (2-7)], it is worth considering some general properties of the
wavefunction ψ. This function will depend upon all N electrons in the
system. We may indicate this by writing $\psi = \psi(1,2...N)$. All electrons
are identical, and in a quantum mechanical system there is no way of

following individual electron paths in time. Thus, there is no way to
tell which electron is which, and therefore the interchange of any pair
can produce no observable physical result. Let us define an operator
P_{12} which interchanges the coordinates of electrons 1 and 2 and apply
it to Eq. (2-3).

$$P_{12}(H\psi) = (P_{12}H)(P_{12}\psi) = H(P_{12}\psi) = E(P_{12}\psi) \qquad (2-8)$$

The first equality in Eq. (2-8) simply says that interchanging electrons 1 and 2 in the quantity $H\psi$ is the same as interchanging them
separately in each factor. But since all electrons appear in a completely symmetric way in H [see Eq. (2-7)], H is unchanged by this
interchange, giving the second equality in Eq. (2-8). The third equality comes from the right-hand side of Eq. (2-3), and it shows that if
ψ is a solution of Eq. (2-3), with energy E, then $(P_{12}\psi)$ is also a
solution with the same energy. Now suppose we are interested in a non-
degenerate state of our system. This means that only ψ, or an arbitrary constant times ψ, is a solution of Eq. (2-3) with energy E.
Hence it follows from Eq. (2-8) that

$$P_{12}\psi = \lambda\psi, \qquad (2-9)$$

where λ is some constant. If P_{12} is applied again to ψ, the electrons
are reexchanged giving back the original ψ. Combining this with Eq.
(2-9) gives

$$\psi = P_{12}^2\psi = P_{12}(P_{12}\psi) = P_{12}(\lambda\psi) = \lambda(P_{12}\psi) = \lambda^2\psi . \qquad (2-10)$$

Therefore $\lambda^2 = 1$, $\lambda = \pm 1$. The same result follows in a more elaborate
way for degenerate wavefunctions. This means that if the coordinates
of two identical particles are interchanged, the wavefunction must
either remain unchanged or else simply change sign. Further, by considering various routes from an original to some multiply interchanged
function, one can show that λ must be the same for all particles in an
identical set. It is impossible, then, for a wavefunction to change
sign upon interchanging electrons 1 and 2, but to remain unchanged upon
interchanging electrons 2 and 3. There are theories that relate the

sign of λ to the spin of the particle, but it will be sufficient for
us to take the fact that $\lambda = -1$ for electrons as an experimental
observation.

Now return to Eq. (2-3) and suppose that the Hamiltonian can be
written as the sum of two terms, with the first depending upon electrons 1 through i and the second upon the remaining i + 1 through N.

$$H(1,2...N) = H_1(1,2...i) + H_2(i + 1,...N) \qquad (2-11)$$

This says physically that the first i particles do not interact with
the others. Substituting Eq. (2-11) into Eq. (2-3) and trying a solution of the form

$$\psi(1,2..N) = \psi_1(1,2..i)\ \psi_2(i + 1,...N) \qquad (2-12)$$

gives

$$(H_1 + H_2)\psi_1\psi_2 = E\psi_1\psi_2 \ . \qquad (2-13)$$

Divide by $\psi_1\psi_2$ to obtain

$$\frac{H_1\psi_1}{\psi_1} + \frac{H_2\psi_2}{\psi_2} = E \qquad (2-14)$$

or

$$\frac{H_1\psi_1}{\psi_1} = E - \frac{H_2\psi_2}{\psi_2} \ . \qquad (2-15)$$

The left side of Eq. (2-15) is independent of variables
(i + 1,...N), but it equals the right side which is independent of
variables (1...i). Therefore, neither side can depend upon any of the
N variables, and hence both must equal a constant. The result is that
the original N-particle equation separates into two simpler equations

$$H_1\psi_1 = \varepsilon_1\psi_1 \quad \text{and} \quad H_2\psi_2 = \varepsilon_2\psi_2$$

with $E = \varepsilon_1 + \varepsilon_2$.   (2-16)

This result generalizes, so that if all particles are noninteracting

$$H(1,2...N) = H_1(1) + H_2(2) \ldots + H_N(N) \tag{2-17}$$

then

$$\psi(1,2...N) = \psi_1(1)\psi_2(2) \ldots \psi_N(N) . \tag{2-18}$$

If it were not for the electron-electron repulsion terms (e^2/r_{ij}) in Eq. (2-7), the molecular Hamiltonian would be of the separable form [Eq. (2-17)]. It is therefore common to take the product form, Eq. (2-18), as the starting point for approximate wavefunctions. However, the simple product, Eq. (2-18), has a serious flaw in that it does not change sign upon electron exchange. Consider the two-electron case

$$P_{12}[\psi_1(1)\psi_2(2)] = \psi_1(2)\psi_2(1) \neq - \psi_1(1)\psi_2(2) . \tag{2-19}$$

But the simple product $\psi_1(2)\psi_2(1)$ is a solution with the same energy as $\psi_1(1)\psi_2(2)$. So, therefore, is any linear combination of the two, and the particular linear combination

$$\psi(1,2) = \psi_1(1)\psi_2(2) - \psi_1(2)\psi_2(1) \tag{2-20}$$

does have the proper antisymmetry under exchange. Equation (2-20) can be recognized as the expanded form of the determinant

$$\begin{vmatrix} \psi_1(1) & \psi_2(1) \\ \psi_1(2) & \psi_2(2) \end{vmatrix} . \tag{2-21}$$

This correctly suggests the proper form for the wavefunction in the case of N independent particles

$$\psi(1,2...N) = \det \left| \psi_1(1) \; \psi_2(2) \; ... \; \psi_N(N) \right|$$

$$= \begin{vmatrix} \psi_1(1) & \psi_2(1) & ... & \psi_N(1) \\ \psi_1(2) & \psi_2(2) & ... & \psi_N(2) \\ & \cdot \; \cdot \; \cdot & \\ \psi_1(N) & \psi_2(N) & ... & \psi_N(N) \end{vmatrix} . \tag{2-22}$$

Exchanging particles i and j in this form, which is known as a Slater determinant, exchanges rows i and j of the determinant and hence changes its sign as required.

The Pauli Principle follows at once for a system with wavefunction of form Eq. (2-22). The one-electron functions ψ_i in Eq. (2-22) depend upon all the coordinates, spin as well as space, of electron i, and are called spinorbitals. Suppose two of these are made the same in Eq. (2-22); that is, suppose one puts two electrons into the same spinorbital. Then the determinant will have two identical columns and hence will vanish. There is thus zero probability of finding the system in such a state. This is the Pauli Principle. It can be put into more familiar form if we take explicit account of the spin variable. Since the original Hamiltonian, Eq. (2-4), separated into terms independent of spin plus terms dependent upon spin, and since all the latter were discarded, it follows that just as we separated $\psi(1...N)$ into a product of spinorbitals, each spinorbital separates into space and spin factors. It happens that the spin factor can be only one of two (α or β)

$$\psi_i = \phi_i\alpha \quad \text{or} \quad \psi_i = \phi_i\beta . \tag{2-23}$$

The orbital ϕ_i depends only upon the space coordinates of one electron. Each such space orbital can be used at most twice in Eq. (2-22), once with spin α and once with spin β, otherwise there will be two identical spinorbitals and the determinant will vanish.

We are now in a position to understand the cause of electron pairing. For two independent particles Eq. (2-16) shows that the total energy of the system is the sum of ε_1, the energy of the orbital ψ_1, plus ε_2, that of ψ_2. A similar situation holds for N independent particles, and use of the determinantal wavefunction, Eq. (2-22), in place of the simple product does not change this result. The orbital energy depends only upon the space factor, so that $\phi_i\alpha$ and $\phi_i\beta$ both have the same energy. The total energy of the system $E = \Sigma_i\varepsilon_i$ will be lowest if the individual orbital energies are as low as possible. The ground state is obtained by putting two electrons, one with spin α and one β, into each space orbital starting with that of lowest orbital energy and working up.

Thus electron pairing is caused by the indistinguishability of electrons which leads to exchange antisymmetry of the wavefunction, and then to the impossibility of more than one electron occupying any spinorbital. Then, because the Hamiltonian is independent of spin, the spinorbitals occur in degenerate pairs with a common space factor times one of the two possible spin factors. The observed electron pairing is a manifestation of the ability of each space orbital to contain two electrons. We can imagine a macroscopic analogy. Suppose marbles of diameter R are put into a box of rectangular cross section R x 2R. The marbles will fill the box in pairs from the bottom up. The cause of the marble (electron) pairing is not a marble-marble (electron-electron) attraction, but is due to the fact that the marbles (electrons) seek a position of lowest gravitational (electronic) energy. Each available gravitational (electronic) level happens to be able to hold two marbles (electrons).

Thus electron pairing occurs in a model in which all electron-electron interaction is neglected. However, we must include the important electrostatic repulsion between electron pairs (but not the spin-spin effects). This modifies the picture above so that the total energy is no longer simply a sum of orbital energy terms but also contains integrals due to the electron repulsions. Nevertheless, the orbital energies will often be the major factor for systems in their ground state, and the discussion above on electron pairing will still apply. There are, however, cases——the ground state of O_2, for example——where, because of the importance of electron repulsions, the electrons are not all paired.

We shall next consider in outline how one solves the Schrödinger equation [Eq. (2-7)]. The solution can be carried out exactly only in very simple cases, the least simple of which is the hydrogen atom. We shall not consider these exact solutions, but will go at once to approximation methods that must be used for all other atoms and molecules. This does not mean that the exact solutions are unimportant; in fact, many of the approximate methods are based upon the exact hydrogen atom results.

Although there are borderline cases, by and large atomic and molecular calculations may be classed as *ab initio* or semiempirical. In an *ab initio* calculation one solves the Schrödinger equation without the introduction of further experimental data (except that one might use the experimental molecular geometry). Simplifications such as neglect of, or approximate evaluation of, integrals were common in earlier work, but are now usually avoided. In the limit, these calculations converge to the exact solution of Eq. (2-7). These methods are similar in spirit to a Taylor series evaluation of a function. If one takes enough terms, one can come as close as one wishes to the exact result. The approximations occur because for reasons of time or money one must truncate the calculation somewhere short of perfection.

Semiempirical methods start with the exact Hamiltonian and make sweeping approximations. At the mildest these consist in wholesale neglect of difficult integrals. There may also be hand-waving sorts of manipulations whose justification is difficult to pin down. Certain remaining theoretical parameters are then determined by reference to experimental quantities, such as ionization potentials and dipole moments.

It will be convenient to discuss the *ab initio* methods first since semiempirical schemes are usually patterned after them.

2. Ab initio methods

These are almost always of either the variation or perturbation type, though there are others; also, the two can be used in combination. In calculations of molecular energies and geometries of the kind we shall discuss, variation calculations are more common than perturbation calculations by a factor of perhaps 10 to 1.

a. *Variation methods.* These are all based upon the important *variation theorem* which fortunately is easy to state and easy to prove, given the orthogonality of the wavefunctions of various states of a system. Suppose H is the Hamiltonian of the system under consideration

whose allowed states ψ_1, ψ_2, ... are arranged in order of increasing
energy $E_1 \leq E_2 \leq ...$, where

$$H\psi_i = E_i\psi_i .$$ (2-24)

Now suppose we pick some arbitrary function χ as an approximation to
the ground state ψ_1. Let us define the quantity E_χ, called the energy
of the trial function χ, by

$$E_\chi \equiv \frac{\int \chi^* H\chi d\tau}{\int \chi^* \chi d\tau} ,$$ (2-25)

where $\int$ ... $d\tau$ indicates integration over all coordinates of the system. If by luck we had happened to pick χ as one of the exact solutions (ψ_i, say), E_χ would be the corresponding E_i

$$E_\chi = \frac{\int \psi_i^* H\psi_i d\tau}{\int \psi_i^* \psi_i d\tau} = \frac{\int \psi_i^* E_i\psi_i d\tau}{\int \psi_i^* \psi_i d\tau} = \frac{E_i \int \psi_i^* \psi_i d\tau}{\int \psi_i^* \psi_i d\tau} = E_i .$$ (2-26)

The variation theorem says that no matter how we pick the approximation
χ, its energy E_χ will not be lower than the true ground state energy
E_1. To prove this, expand the trial function χ in a series of the ψ's

$$\chi = \sum_i a_i\psi_i$$ (2-27)

and substitute into Eq. (2-26)

$$E_\chi = \frac{\sum_{i,j} a_i^* a_j \int \psi_i^* H\psi_j d\tau}{\sum_{i,j} a_i^* a_j \int \psi_i^* \psi_j d\tau} = \frac{\sum_{i,j} a_i^* a_j E_j \int \psi_i \psi_j d\tau}{\sum_{i,j} a_i^* a_j \int \psi_i \psi_j d\tau}$$

$$= \frac{\sum_i |a_i|^2 E_i}{\sum_i |a_i|^2} \geqslant \frac{\sum_i |a_i|^2 E_1}{\sum_i |a_i|^2} = E_1 . \qquad (2\text{-}28)$$

Comparing the first and last expressions in Eq. (2-28) gives the variation theorem

$$E_\chi \geqslant E_1 . \qquad (2\text{-}29)$$

The third equality in Eq. (2-28) which goes from double to single summations uses the orthogonality of the exact wavefunctions

$$\int \psi_i^* \, \psi_j d\tau = \begin{cases} 1 & i = j \\ 0 & \text{otherwise} \end{cases} . \qquad (2\text{-}30)$$

A variation calculation consists in choosing a χ that is a function of a number of parameters (orbital size and degree of hybridization, for example) and varying them to get the lowest possible energy, knowing from the variation theorem that one can never go too low. The self-consistent field (SCF) method is a particular type of variation calculation that is very frequently used. We have seen that if electron-electron repulsion is neglected, the exact solution can be written as a determinant of spinorbitals [Eq. (2-22)]. This is no longer an exact solution when electron-electron repulsion is included, but we may take Eq. (2-22) as an approximate solution and ask for the best possible ϕ_i's. These best ϕ_i's, or molecular orbitals (MOs), satisfy eigenvalue equations, the Hartree-Fock self-consistent-field equations, whose eigenvalues are the orbital energies. If a simple product of ϕ's is used instead of the anti-symmetrized product, Eq. (2-22), one obtains the Hartree self-consistent-field equations. These are now of only historic interest. The Hartree-Fock SCF equations have been solved numerically for many atoms, but a further simplication is almost always made for molecules. If one approximates the unknown MOs as finite sums of some n given basis function f_j

$$\phi_i = \sum_{j=1}^{n} a_{ij} f_j \ , \tag{2-31}$$

then the problem reduces to finding the best coefficients a_{ij}. The f_j's are usually taken to be atomic orbitals, that is, one-electron functions localized on the various atoms in the molecule, and the resulting equations are known as Roothaan's equations [2111a], although C. A. Coulson [516a] had applied them to the hydrogen molecule as early as 1938. They are of a matrix eigenvalue form, though the matrix elements themselves involve the unknown eigenfunctions. Consequently, the Roothaan equations must be solved iteratively to give the best "linear combination of atomic orbitals, self-consistent-field molecular orbitals" (LCAO SCF MOs). As the number of basis functions increases, the LCAO SCF result approaches the Hartree-Fock limit.

The basis functions f_j may be chosen as one pleases. Most of the earlier work done in the 1950s and early 1960s on diatomics used Slater orbitals which are like the exact hydrogen atom wavefunctions. The most time-consuming part of an LCAO SCF calculation is the evaluation of the integrals representing electron-electron repulsion. These are both numerous and difficult. In 1950 S. F. Boys [296a] showed how the use of Gaussian instead of Slater orbitals greatly simplifies the integral evaluation. These differ in that Gaussians contain the factor $\exp(-\alpha r^2)$ in place of the $\exp(-\alpha r)$ in the Slater orbitals. It requires two to three times as many Gaussian as Slater orbitals to get an equivalent energy result; and the number of electron-electron integrals increases as the fourth power of the number of AOs used. It was not at once clear, therefore, which basis functions would be better in a large calculation. As it turned out, the Gaussians have grown steadily in popularity, so that by the end of the 1960s most polyatomic SCF calculations—and these now number in the hundreds—are done with a Gaussian basis. Even many which nominally use a Slater basis first expand each Slater orbital in a series of Gaussians so that in fact the integrals are done over Gaussians. As an example of the size of the job involved, eight million integrals were evaluated by Clementi

[470] for *each* of many geometric configurations in his study of the
hydrogen bond in NH_4Cl.

Even in the Hartree-Fock limit with an infinite number of basis
functions one of course still does not have the exact wavefunction and
energy. The difference between the exact and limiting Hartree-Fock
energies is called the *correlation energy*, and the study of this correlation energy is currently the subject of extensive research.
Perhaps the most straightforward way to compute correlation energy
involves the virtual orbitals from the ground state SCF calculation.
If one uses n basis functions f_i in the SCF calculation, one gets n
MOs ϕ_j. The ground state is described approximately by putting the
electrons into the lowest of these, leaving empty ϕ's of higher energy.
Determinants constructed by replacing some of the ground state ϕ's by
some of these form approximations to excited states. Call the SCF
approximate ground state ψ_0 and the approximate excited states ψ_1,
ψ_2 ... . If we write

$$\Xi \; = \; C_0\psi_0 \; + \; C_1\psi_1 \; + \; ... \tag{2-32}$$

and vary the coefficients C_i to lower the energy, then Ξ will be a
better approximation to the ground state than ψ_0. This process is
known as configuration interaction and Ξ is described as an LCAO SCF
MO + CI function. Boys [296a] has shown that if the set of basis functions f_i is complete, that is, if the expansion [Eq. (2-31)] has con-
verged to the best MOs, then Eq. (2-32) will also converge to the exact
wavefunction. At present very few hydrogen bond calculations have been
carried from the SCF to the CI stage [36,1638b,1642a,1925,2644].

A slightly different approach is to say that since even the best
single-determinant wavefunction is not exact, one might avoid the
difficult SCF procedure and go directly to the CI step [Eq. (2-32)].
One now doesn't have the MOs Eq. (2-31) from which to construct the ψ_i,
but one might use the individual AO's, f_i, instead. The series, Eq.
(2-32), would not be expected to converge as fast, but with the time
saved from the SCF step one can afford more effort here. This is the
valence-bond (VB) method in contrast to the molecular-orbital (MO)
method above. In the 1930s and 1940s there was much discussion on
whether the MO or the VB method is the better. Of course, in the limit

the two are both expected to converge to the same exact result, but Coulson and Fischer [518a] demonstrated how they may be identical even for finite and small basis sets. In more recent years the MO method has become the more common, and there are no VB studies of the hydrogen bond of a quality to compare with the current MO work.

b. *Perturbation methods.* The spirit here is quite different from, and in many ways more attractive than, that of the brute force variational methods. Suppose in the Schrödinger equation [Eq. (2-3)] the Hamiltonian can be written

$$H = H_0 + H_1 \, , \qquad (2-33)$$

where H_1 is small compared to H_0 and solutions of

$$H_0 \psi^0 = E^0 \psi^0 \qquad (2-34)$$

are known. The solutions of Eq. (2-34) would be expected to be similar to those of Eq. (2-3). The perturbation method consists in starting with the unperturbed energies and wavefunctions of Eq. (2-34) and correcting them using the perturbing Hamiltonian H_1 to get the exact solutions of Eq. (2-3) in the limit.

Molecules are not very different from collections of isolated atoms. For example, the total bond energy in a molecule is on the order of 1/100 of the total electronic energy in first-row molecules. Further, molecular structure determinations are done using atomic scattering factors, and only the most precise work reveals a difference between these and the true molecular scattering. Since hydrogen bonds are among the weakest bonds, they would appear ideal for a perturbation description based upon the nonbonded fragments.

Many of the terms used in discussing the hydrogen bond arise from a perturbation description. Let us write the Hamiltonian of the hydrogen-bonded system,

$$H = H_0 + \lambda H_1 \, , \qquad (2-35)$$

where λ is a parameter that will be set equal to 1 at the end of the calculation; its purpose is to allow keeping track of the various

orders of correction to the energy and wavefunction. H_0 is the Hamiltonian for the fragments AH and B at infinite separation. H_1 gives
the additional terms between the two for finite separation. The energy
of the combined system is written as a power series in λ

$$E = E^{(0)} + \lambda E^{(1)} + \lambda^2 E^{(2)} + \ldots . \qquad (2\text{-}36)$$

Such a treatment has long been used to represent the long-range forces
between molecules, and overlap integrals between the molecules are
usually neglected. Then $E^{(0)}$ is the energy of separated fragments.
$E^{(1)}$ is the *classical electrostatic energy* of interaction between the
two molecules. $E^{(2)}$ is the *polarization energy* and contains two kinds
of terms, *induction* and *dispersion*. Higher terms are not usually considered. At closer distances overlap is more important, and a perturbation treatment has been developed [1742,1742a,2600a] as a double
power series of the interaction energy and the overlap. The terms
zero-order in overlap are named as above. In addition, there are terms
first-order in interaction energy and second-order in overlap (*exchange
energy*); terms second-order in both interaction and overlap (*exchange
polarization energy* and *charge-transfer energy*), as well as higher
terms which have not yet been examined. There have been recent applications of this method to the hydrogen bond, but they contain a large
number of approximations. One difficulty in the application of perturbation methods is that to improve beyond the first in the infinite
series of corrections requires knowing all the excited states of the
unperturbed system.

3. Semiempirical methods

   a. *Hückel and extended Hückel methods.* The Hückel method is one
of the oldest and simplest of the semiempirical techniques. It was
developed originally for the treatment of π electrons in conjugated
organic molecules, and has been used extensively in this application.
One starts by postulating that each electron may be described as moving
in the electrostatic field of all the nuclei plus a field due to
averaged position of all other electrons. However, this assumption
does not really uncouple the Schrödinger equation of one electron from
all others. That is, although each individual orbital satisfies an
equation that looks like Eq. (2-3), the Hamiltonian is actually depen-

dent upon all others. So far, this is exactly as in the *ab initio*
methods. There one takes proper account of the electron coupling, but
here one ignores it. Since, as it turns out, one never needs to write
out the explicit form of the Hückel Hamiltonian, this neglect does not
make obvious difficulties.

The molecular orbitals are approximated as in Eq. (2-31) by a sum
of atomic orbitals, using only the $2p\pi$ AOs from each atom in the conjugated
system, and constants are found by the variation method. Because
of the simplified form assumed for H, these turn out to be roots of a
simple matrix eigenvalue problem. Approximations are then made in
matrix elements leaving finally only two integrals in the entire scheme.
These are evaluated empirically by comparing calculated results with
experimental data (usually heat of hydrogenation or the energy of the
first excited state).

The Hückel method is well known to most experimentalists, and
details of its application are readily available (J. D. Roberts,
[2102a]; A. Streitwieser, Jr., [2357a]), but it is primarily for unsaturated systems and not especially suited to investigations of the
hydrogen bond, though it has been used.

The extended Hückel theory developed by Hoffmann [1067a] is similar, but σ valence electrons as well as π's are included. Not quite as
many integrals are discarded as in the original Hückel method, and the
distance dependence of those that are kept is expressed in terms of
rigorously computed overlap integrals. There have been calculations
of this kind on the hydrogen bond.

b. Pariser-Parr-Pople method. The first widely used semiempirical method to go beyond the Hückel was that of Pariser and Parr [1873a]
and Pople [1970b]. Like the original Hückel method, this also focuses
on the π electrons only, but now the coupling of the orbital equations
through electron-electron repulsion is taken into explicit account.
This leads to difficult integrals. The worst of these are dropped, and
those remaining are determined semiempirically.

In the original versions, the Pariser-Parr procedure differed from
the Pople in two ways. The Pople version was based on an SCF calculation, while Pariser and Parr's used configuration interaction. The
integral approximations were also different, with Pople using a point-

charge model and Pariser and Parr a fit to experimental data. In the
1950s the Pariser-Parr version was more common, but as larger computers arrived the SCF problem became less formidable, and the common
version now combines Pople's SCF approach and the Pariser-Parr
integral approximations.

Again, the original version of this method was for π systems, and
hence was not appropriate to the hydrogen bond.

   c. *CNDO and related methods.* In 1965 Pople, Santry, and Segal
[1971a] published a semiempirical method that treats all valence electrons and includes electron-electron repulsion. They started with the
exact SCF equations and discarded any integral involving an electron
spread between two atomic orbitals. This approximation is like that
in the original Pariser-Parr method and gives this particular technique
its name——CNDO: complete neglect of differential overlap. Remaining
integrals are evaluated according to one of two sets of recipes, CNDO/1
or CNDO/2, with the second being the preferred version. In a sense the
CNDO method stands in relation to the Pariser-Parr-Pople as the extended Hückel does to the Hückel.

   Related methods, including INDO (intermediate neglect of differential overlap), MINDO (modified INDO), NNDO (neglect of diatomic
differential overlap), and PNNDO (partial neglect of diatomic differential overlap), have also been developed to retain more integrals than
the CNDO. These methods have become very popular since 1965, there
being perhaps several hundred published applications. Details are
given in a new book by Pople and Beveridge [1970c] and computer programs for applying the CNDO method are available from the Indiana
University Quantum Chemistry Program Exchange. CNDO methods have been
used in many hydrogen bond calculations.

   d. *Other semiempirical methods.* The methods above are the most
popular, but there are many others. In the early days a semiempirical
method based upon the valence bond approach, in the way that the Hückel
method is based upon the SCF MO, was popular. The method of "Atoms in
Molecules" published by Moffitt [1678a] is interesting in that, unlike
most of the others, it is derived from a perturbation rather than a
variation technique. Daudel and Sandorfy [569a] give a useful review
of the current state of semiempirical calculations.

A difficulty in semiempirical work is that one has no estimate of
the error in any result. One is on safest ground if the molecule of
interest is similar to a set of other molecules where the particular
semiempirical method to be used has been shown to give good results.

Figure 2-1 indicates in a rough way the quality of various *ab*

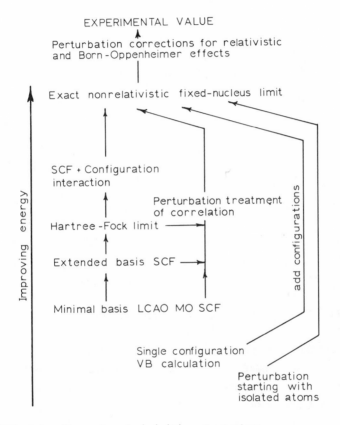

FIG. 2-1. Hierarchy of *ab initio* calculations.

initio methods, and Fig. 2-2 does the same for the semiempirical.
Going up the scale in Fig. 2-1 lowers the computed energy. As the
energy improves, usually other calculated properties do also [693a].
This is not always true, and especially for small changes an improved
energy can result in a worse dipole moment, for example. But for big
energy changes it does hold in the main. The vertical scale must be

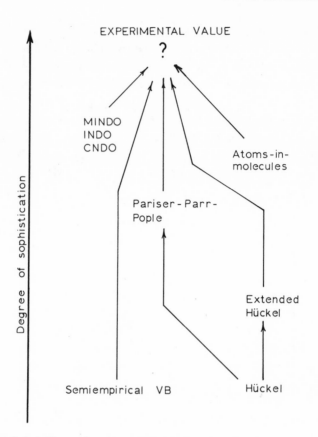

FIG. 2-2. Hierarchy of semiempirical calculations.

left less definite for Fig. 2-2. It is labeled "Degree of Sophistication," but one is never certain that more sophistication will give
better results. It is equally impossible to say which semiempirical
method will give results comparable with any particular *ab initio*
method.

II. THE THEORY OF THE HYDROGEN BOND

A. The Situation Before 1957

Coulson's [516] influential 1957 review makes a convenient mark
to separate older from newer theoretical treatments of the hydrogen
bond. In 1957 the IBM 704 had recently been built, but most results

then in the literature had been carried out by hand or with smaller
computers. SCF calculations with small basis sets were being done on
some first-row diatomics, but such treatment of even the water dimer
was out of the question. Pariser-Parr calculations were being published for aromatics, but the CNDO method for σ systems was still
eight years in the future.

Theoretical studies of the hydrogen bond were made using either
classical electrostatic models or approximate quantum mechanical treatments of fragments of the hydrogen-bonded system.

1. Electrostatic models

In their 1920 paper proposing the hydrogen bond, Latimer and
Rodebush [1418b] wrote the water dimer as

$$
\begin{array}{ccccccc}
 & & \overset{\cdot\cdot}{} & & & H & \\
H & : & \overset{\cdot\cdot}{\underset{\cdot\cdot}{O}} & : & H & : & \overset{\cdot\cdot}{\underset{\cdot\cdot}{O}} & : \\
 & & & & & H &
\end{array}
$$

(1)

They stated that a free pair of electrons on the one water might exert
a force on the hydrogen of the other. The physical basis of the chemical bond was not clear in 1920 so it is not certain exactly what force
was meant, except that Latimer and Rodebush did not feel it was simply
an electrostatic force between the two dipolar water molecules. A more
common view of the time was that since, according to the Lewis-Langmuir
theory, hydrogen can form only one covalent bond, any attraction between two water molecules must be electrostatic. A number of model
calculations were made, of which that of Lennard-Jones and Pople
[1443a,1970a] is typical. To represent the hydrogen bond, the O and H
of the proton donor and the O of the proton acceptor were drawn in a
line as in Structure (2). The O—H bond length b and the O—O distance

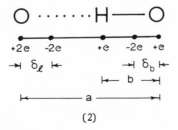

(2)

a were taken equal to their experimental values, and the nuclei were
represented by point charges of +e for hydrogen, +2e for the proton-
accepting oxygen, and +e for the proton donor. A point charge of -2e
to represent the electron lone pair was placed a distance δ_ℓ from the
proton-accepting oxygen, and another δ_b from the proton donor to represent the O—H bonding electrons. The distances δ_ℓ and δ_b were chosen
to make the lone pair and the O—H bond have correct dipole moments.
The coulomb energy of the set of charges was then easily found to equal
5.95 kcal/mole, in good agreement with the experimental value for the
hydrogen-bond energy in water.

   Since this and other similar calculations gave results in agreement with experiment, the problem of the hydrogen bond at first sight
appeared to be solved. But a closer look showed several difficulties.
The first concerns the placing of the bond and lone pair charges. The
dipole moment of an isolated water molecule is definite enough, but
partitioning it into various bond and lone pair components is not.
There is, therefore, considerable arbitrariness in the distances δ_ℓ
and δ_b. The computed energy turns out to depend rather sensitively
upon the arbitrary placing of the point charges. Perhaps, too, other
oxygen electrons ought to be included.

   A more fundamental difficulty of simple electrostatic models is
that they cannot give correct predictions of the hydrogen-bond length.
An attempt to do so by varying a in Structure (2) to give minimum
energy causes the proton to collapse upon the point charge representing
the oxygen lone pair. At the same time the computed hydrogen-bond
energy becomes infinite. This behavior is connected to Earnshaw's
theorem, which derives from Coulomb's law, stating that no set of
electrostatic charges can be placed in equilibrium under the influence
of only their mutual forces.

   A way out is to postulate an additional force to prevent a too-
close approach of the two oxygens. A hydrogen bond would then form at
the length where the repulsive and attractive forces are in equilibrium,
and the bond energy would be the difference in area under the repulsive
and attractive curves (Fig. 2-3). But specifying the repulsive curve
introduces additional arbitrary parameters in a model already containing awkwardly many. Furthermore, unless it can be shown that repulsive

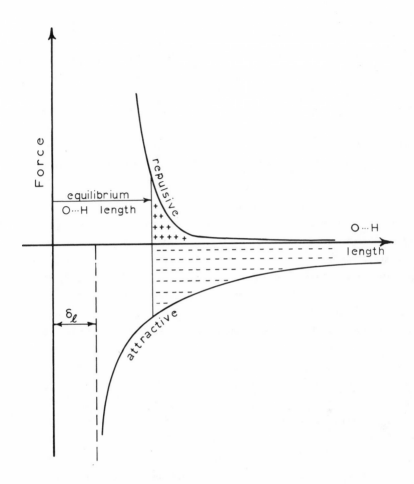

FIG. 2-3. Balance of attractive and repulsive forces in modified
point-charge model.

force is small at larger distances and increases sharply only near the
equilibrium O...H length, the work done against this force in forming
the hydrogen bond is significant. Adding this to the electrostatic
energy ruins the previous agreement between computed and experimental
bond energy. The point charges might be readjusted to restore agreement, but by this time one has pretty much lost faith in the simple
point-charge model.

The electrons involved in the hydrogen bond are not, of course, fixed at points, but are distributed in a diffuse way as specified by the wavefunction. If one had exact wavefunctions for AH and B separately, one could compute the exact value of the electrostatic contribution to the hydrogen bond by keeping the wavefunctions fixed, but allowing AH and B to approach to their hydrogen-bonded distance. This requires a calculation much more elaborate than the point-charge model, and could not be done before 1957 since good wavefunctions for AH and B were not available. As will be described below, this calculation has now been carried out for water dimer.

2. *Approximate quantum mechanical treatments*

The exact electrostatic calculation in the last paragraph fails to be a complete description of the hydrogen bond because it does not take into account two essential effects. The first may be called polarization. As two molecules join to form a hydrogen bond, the charge cloud of each is distorted by the other. The second effect is due to the identity of electrons which, as we have seen, requires the wavefunction of the hydrogen-bonded system to be antisymmetric with respect to exchanges of electrons between bonded molecules. The resulting exchange integrals modify the total energy in a way that might be expected to be important. As Heitler and London showed in the early days of quantum mechanics [1045a], this exchange energy is the most important contributor to the bond energy in H_2. But before 1957 computers were not large enough to take all these effects into complete account for any real system. A typical calculation then ignored all but three nuclei (the proton, the proton-donating, and the proton-accepting atoms) and four electrons (the two in the A—H bond and the lone pair on the proton acceptor). This is a reasonable start, since if there are any general properties of the hydrogen bond one would expect them to depend upon this region rather than upon peripheral parts of the system. A look at Coulson and Danielsson's approximate valence-bond calculation [517] and at Pimentel's approximate MO treatment [1934a] will give the spirit of those times, and show what could be said and what problems remained.

Coulson and Danielsson considered three valence-bond structures:

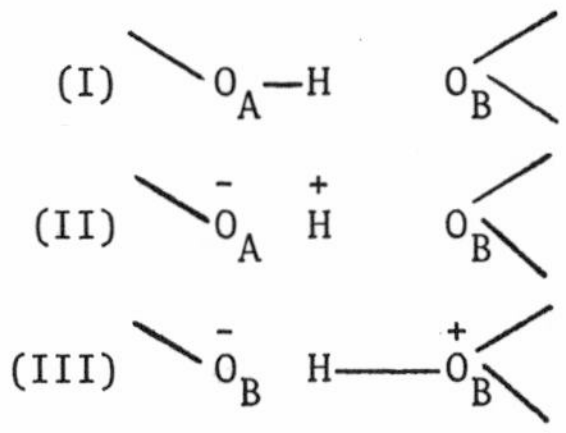

$$(I) \quad \diagdown O_A\text{—}H \quad O_B\diagup$$

$$(II) \quad \diagdown \overset{-}{O}_A \quad \overset{+}{H} \quad O_B\diagup$$

$$(III) \quad \diagdown \overset{-}{O}_B \quad H\text{——}\overset{+}{O}_B\diagup$$

$$(3)$$

Let h be a 1s orbital on hydrogen with spin α and $\bar{h}$ be the same space orbital but with spin β. In the same way define a and $\bar{a}$ as orbitals on O_A and b and $\bar{b}$ as orbitals on O_B. A range of hybrid a's and b's was studied from pure p to tetrahedral sp^3. The wavefunction of the ionic form (II) is easiest to write down as

$$\psi_{II} = N_{II} \det \mid a\bar{a} \, b\bar{b} \mid . \tag{2-37}$$

This represents a structure in which the electrons from the O—H bond are both moved onto the O atom; the unshared pair on the other oxygen is left undisturbed in orbital b. N_{II}, Eq. (2-37), is a normalizing constant chosen so that the denominator in the variational energy [Eq. (2-25)] will equal 1. We may ignore this detail. There are two possible ways to pair the electrons in the bonds of (I) and (III), Structure (3). Both must be taken into account, making their valence-bond functions a little more complicated

$$\psi_I = N_I \, [\det \mid a\bar{h} \, b\bar{b} \mid + \det \mid h\bar{a} \, b\bar{b} \mid] , \tag{2-38}$$

$$\psi_{III} = N_{III} \, [\det \mid a\bar{a} \, h\bar{b} \mid + \det \mid a\bar{a} \, b\bar{h} \mid] . \tag{2-39}$$

(I) is a structure with a covalent bond between O_A and H with a lone pair on O_B. (III) is a charge-transfer structure with the O_A—H bonding electrons moved onto O_A and a long covalent bond to H formed to O_B using what were the lone-pair electrons. The exact wavefunction of the system was then written as the best possible linear combination of (I)-(III) as in Eq. (2-40),

$$\Xi = C_I \, \psi_I + C_{II} \, \psi_{II} + C_{III} \, \psi_{III} \tag{2-40}$$

with the C's determined variationally. Note that all this does tally
with our description of a valence-bond calculation in I.C.2a above.
The individual determinants are constructed from simple AOs rather than
MOs, and the work goes into the configuration interaction part of the
calculation.

A number of approximations in the resulting integrals had to be
made to complete the calculation, but with these Coulson and Danielsson
studied the hydrogen bond energy and the varying contribution of structures (I)-(III) as a function of bond length and type of hybridization
of AOs a and b. Some of their results are in agreement with experiment
(as the O...O distance is shortened, the O—H length increases and the
bond energy increases), but others are in serious error (an equilibrium
O...O distance is not found; the energy of the system continues to decrease as this length shortens).

Tsubomura [2483] carried out a similar calculation at about the
same time. He included two further valence-bond structures, and made

$$
\text{(IV)} \quad \diagdown \overset{+}{\text{O}}_A \quad \bar{\text{H}} \quad \text{O}_B \diagup \quad , \text{ and}
$$

$$
\text{(V)} \quad \diagdown \text{O}_A \quad \bar{\text{H}} \quad \overset{+}{\text{O}}_B \diagup
$$

$$(4)$$

fewer approximations in the integral evaluations. This increased the
computational labor, and accordingly only one geometry was considered.

Both Coulson and Danielsson [517] and Tsubomura [2483] examined
the question of electrostatic versus covalent contributions to the
hydrogen bond. Coulson and Danielsson conclude, "the long bond (2.8 Å)
is essentially electrostatic, the covalent contribution . . . amounting
to only a few percent." But according to Tsubomura, who uses "delocalization" in place of Coulson's "covalent," "the delocalization energy
is of the same order of magnitude as, or rather larger than, the electrostatic energy." This discrepancy is due to differences in interpretation rather than differences in computed results. Of the structures
(I)-(III), only (III) shows a covalent bond between H and O_B. Since
C_{III} is found to be small and hence (III) to make only a few percent
contribution to the total wavefunction Ξ, the system may be accurately

represented in terms of (I) and (II) only. Since these do not have a
covalent bond between H and O_B, Coulson and Danielsson conclude that
the hydrogen bond is essentially electrostatic. Tsubomura calculates
the total energy of the system, first with structures (I), (II), and
(IV), and then including (III) and (V) with a covalent bond to O_B.
The addition of these two lowered the energy of the system by 8.1
kcal/mole. This is larger than the 6 kcal computed by Pople [1970a]
for the strength of the hydrogen bond using the electrostatic model.
Hence, Tsubomura concludes that the covalent contribution to the hydrogen bond is at least as large as the electrostatic.

One might wish to modify both these approaches. Although ψ_{III}
makes only a few percent contribution to the total wavefunction, the
wavefunction describes the complete O—H...O system, not just the
hydrogen bond. Since the energy of the hydrogen bond is only a small
part of the energy of the system (5 to 10 parts in 120), it is not
impossible that the small contribution of ψ_{III} to Ξ is actually a
major part of the hydrogen bond. Tsubomura found that (III) and (V)
lowered the total energy by 8.1 kcal, but the hydrogen bond energy is
not necessarily lowered by this amount. Of the two results, that of
Tsubomura seems closer to what one wants—in fact, it is Tsubomura's
result that Coulson [516] quotes in his review as the delocalization
energy of the hydrogen bond—but it would be more interesting to have
the change in hydrogen bond energy, not total energy, due to (III) and
(IV).

There is a second feature in both calculations that might be misleading. How one defines the electrostatic contribution to the hydrogen bond is more arbitrary than one might like, but the method in the
last paragraph of the previous section is in complete agreement with
Coulson [515] who says, "By the word 'electrostatic' . . . we mean only
such forces as would arise if, in some hypothetical fashion, we could
bring the interacting species together, without deformation of either
charge cloud, or any electron exchange." Others might prefer to include the charge-cloud deformation since this would be a classical
electrostatic effect, but reasons for including exchange are not appar-
ent. As we have seen, exchange terms arise from the indistinguishability of electrons; they have no classical analogue. Nevertheless, since

ψ_I-ψ_V are all completely antisymmetric, they do include exchange between the interacting fragments O_A—H and O_B. This must be kept in
mind when comparing the calculations above with others using a
different definition of "electrostatic."

Early molecular-orbital treatments of the hydrogen bond are
typified by Pimentel's 1951 paper [1934a] on HF_2^-. The symmetrical
hydrogen bond simplifies the work, but the method is similar in other
cases, and some of the HF_2^- results are true in general. Molecular
orbitals for A—H...B were formed from $2p\sigma$ orbitals (p_A and p_B) on A
and B plus a 1s (h) on hydrogen, as shown in Structure (5). Hybrid

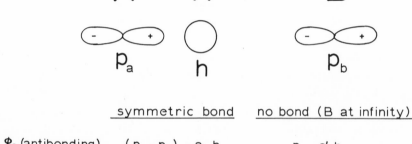

	symmetric bond	no bond (B at infinity)
Φ_3 (antibonding)	$(p_a - p_b) - a_2 h$	$p_a - a_2' h$
Φ_2 (nonbonding)	$p_a + p_b$	p_b
Φ_1 (bonding)	$(p_a - p_b) + a_1 h$	$p_a + a_1' h$

(5)

orbitals might be preferred on A and B, but Coulson and Danielsson
[517] note that hybridization did not affect their results to any great
degree. Symmetry and orthogonality fix the molecular orbitals almost
entirely, leaving only the coefficients a_1 and a_2 for the symmetric
bond to be determined by the variation calculation.

The attractive feature of this MO treatment is that it appears to
explain both the usual hydrogen bond between electronegative atoms and
also the bond in the electron-deficient boron hydrides. The four electrons of a hydrogen bond fill orbitals ϕ_1 and ϕ_2. ϕ_1 is nonbonding and
places electrons in the p orbitals of A and B. Hence, the more electro-

negative are A and B, the lower the expected orbital energy of ϕ_2. In
the boron hydrides there are only enough electrons to fill ϕ_1, leaving
ϕ_2 empty. ϕ_1 is a bonding orbital between H and A and B. Its energy
depends upon the bonding interactions rather than solely upon the
electronegativity of A and B. Thus, while the usual hydrogen bond
requires that A and B be electronegative, the electron-deficient
hydrogen bond does not. However, total energies are not simply the
sum of orbital energies, so that arguments based only upon orbital
energies can be misleading [2186a]. It is, therefore, not clear whether this plausible explanation is correct.

From these and similar calculations Coulson [516] concluded that
electrostatic forces contribute +6 kcal/mole to the hydrogen bond in
ice; delocalization, +8 kcal/mole; repulsive forces, -8.4 kcal/mole;
and dispersion forces, +3 kcal/mole. The sum of these is +8.6
kcal/mole, in fair agreement with the experimental value of +6.1. As
Coulson pointed out, these figures must be regarded as tentative, and
that the safest conclusion is that all four contributions are important
and must be included. In addition to the uncertainties caused by the
approximate nature of the calculations, and the ambiguities in the
definitions of the four terms, there is the difficulty that each of
the four is obtained from a different kind of calculation, rather than
from the partitioning of a single calculation. It is not obvious that
the four are independent energies that should add to give the total
hydrogen bond energy.

As of 1957, then, one could say that theory had found the hydrogen
bond to be describable in terms of no single effect.

B.   Advances since 1957

Today's fastest computers have several hundred times the speed of
the best existing in 1957. There have also been advances in the technique of evaluating the electron repulsion integrals necessary in
molecular calculations, but it is mainly the availability of larger
and faster computers that now makes it possible to treat small polyatomics such as the water dimer as accurately as one could first-row
diatomics in 1957. On the semiempirical side, the development and
application of all-valence-electron methods such as CNDO has also
depended upon these computers.

Three systems will be considered in detail. The first of these
is the water dimer, which might be considered the prototype for the
typical hydrogen bond. Extensive calculations have been published for
the water dimer, and the results are more accurate and trustworthy
than for any other hydrogen-bonded system. We shall also examine
theoretical results for the bifluoride ion (FHF)$^-$ as representative of
the smaller class of strong hydrogen bonds. Accurate calculations are
also available here. Third studied will be the guanine-cytosine dimer.
The methods in this third case are typical of those that must now be
used for most organic hydrogen-bonded systems.

1. The water dimer

One approach would be to compare theoretical results, at various
levels of approximation, to experimental, and in this way decide how
complex a calculation must be done to get accurate theoretical estimates of the various experimental properties. This cannot be done in
the case of the water dimer, a species that is not easily studied
experimentally. As will be seen, current theoretical and experimental
studies of the water dimer are very much tied together, and constitute
a nice case of the two advancing hand-in-hand. Other hydrogen-bonded
systems are more available to measurement but much less so to accurate
theoretical treatment. It is therefore necessary to go back a step to
the water monomer, where the required experimental data are well known
and where a range of theoretical treatments is possible, to make our
comparison.

A high resolution infrared study of water vapor by Benedict,
Gailar, and Plyler [198a] has established the equilibrium O—H bond
length in water to be 0.9572 ± 0.0003 Å and the H—O—H angle as
104.52° ± 0.05°. Nelson, Lide, and Maryott [1783a] list the dipole
moment of water in the gas phase as 1.85 D and estimate an accuracy of
1%. The total bond strength in water, $i.e.$, the enthalpy of the
reaction

$$O(g) + 2 H(g) \longrightarrow H_2O(g) \qquad\qquad\qquad (2\text{-}41)$$

at 0°K is -219.3 kcal/mole [2542b]. At 0°K the water molecule has
neither translational energy nor rotational energy, but it does
have vibrational energy of ½ hν for each of its three normal modes.

The fundamental frequencies ν are listed by Benedict, Gailar, and Plyler [198a] as 3656.65, 1594.59, and 3755.79 cm^{-1}. The sum of the corresponding zero-point vibrational energies is 4503.52 cm^{-1} = 12.88 kcal/mole. This must be subtracted from the experimental bond energy to obtain a vibration-free energy of -232.2 kcal/mole to be compared with the quantum mechanical results.

The quantum mechanical bond energy is obtained by subtracting the computed energy of O + 2 H from that of H_2O. The variation theorem applies only to these individual energies, which are computed relative to separated electrons and nuclei, and not to the bond energy. To obtain the corresponding experimental total energy for H_2O it is necessary to add the -232.2 kcal bond energy to the energy of

Oxygen nucleus + 2 protons + 10 electrons $\longrightarrow$ O + 2 H.   (2-42)

The energy of this process is very high and can be obtained by summing all the ionization potentials of O and H [1683a]. Adding the resulting -72.109 au to the -0.370 au bond energy gives -72.479 au as the experimental total energy of H_2O in an imaginary vibrationless state at 0°K. Here as always, bond energies are a small fraction of total energies, thus requiring very accurate calculations of total energies to assure reliable bond energies.

Three recent very high-quality calculations on the water molecule are an SCF calculation using Gaussian basis functions by Neumann and Moskowitz [1789a], a configuration interaction calculation by Schaefer and Bender [2186b] using 1027 configurations with a Gaussian basis, and an elaborate perturbation calculation by Miller and Kelly [1670a]. This last is unusual for a polyatomic. Each of the three is an extensive work, and each was carried out only for a single geometry. Only Schaefer and Bender did their calculation at the experimental geometry of R_{OH} = 0.9572 Å, $\angle HOH$ = 104.52°. Miller and Kelly used older experimental values of 0.9584 Å and 104.45°, and Neumann and Moskowitz used 0.9526 Å and 105°. Table 2-1 shows a comparison of the results of these three calculations.

The "experimental" total energy in Table 2-1 is the -76.479 au above with Miller and Kelly's estimated relativistic correction of

TABLE 2-1

Calculated and Experimental Data for H_2O Molecule

	Total energy (au)	Bond energy (kcal/mole)	Dipole moment (D)
Neumann and Moskowitz [1789a] SCF	-76.059	156	1.995
Schaefer and Bender [2186b] CI	-76.2418	189	---
Miller and Kelly [1670a] Perturbation	-76.48	263	---
Experiment	-76.430[a]	232[b]	1.85

[a]Estimated nonrelativistic part of total energy.   [b]The energy required to split the molecule into three neutral atoms.

0.049 au.   The SCF and CI numbers both come out of variational calculations, and therefore cannot be below the experimental number.   The perturbation method has no such restriction, and in fact does lie below experiment.   All three computed total energies look good in that none is more than 0.5% in error.   But it is sobering to note that this small percent error still corresponds to 233 kcal, which is not small compared to the bond energy.   The total energy of Miller and Kelly is an extrapolated value obtained by estimating energy contributions higher than third order.   Their best actually computed value was -76.233 au.

The binding energies were not computed in comparable ways.   Miller and Kelly subtracted the experimental values of O + 2 H from their calculated H_2O energy.   A more common method is to calculate the energy of O + 2 H in a way analogous to that used for H_2O and take the difference between calculated numbers as the binding energy.   This was done by Neumann and Moskowitz and Schaefer and Bender.   The SCF bonding energy is in error by 76 kcal/mole, or 33%.   This is not surprising since an SCF calculation, by definition, cannot include correlation energy.   Correlation between electrons in the same orbital is thought to be the most important; and in the formation of H_2O from atoms, two extra pairs are created.   The other two calculated bond energies include correlation and are in better agreement with experiment (13% and 19% errors).

Neumann and Moskowitz's calculated dipole moment is within 8% of the experimental. They list six previous calculations giving from 1.42 D to 2.57 D with an average of 1.77 D. Neither Miller and Kelly nor Schaefer and Bender compute the dipole moment.

To calculate a predicted molecular geometry it is necessary to compute the energy at each of many geometries and interpolate to that of lowest energy. This multiplies the work by a factor of 10 to 100 and makes it impossible for the three calculations above. Moskowitz and Harrison [1716a] have done a geometry search with an SCF calculation and a smaller Gaussian basis of 5 s and 3 p (*i.e.*, 3 each of p_x, p_y and p_z) orbitals on O plus 3 s orbitals on each H. The resulting energy was poorer by about 0.5 au than the Neumann and Moskowitz calculation with 10 s, 6 p and 2 d orbitals on O plus 4 s and 2 p on each H. Their computed geometry was R_{OH} = 0.98 Å, $\underline{/HOH}$ = 113°. In a more elaborate calculation Diercksen [621] searched for the minimum energy geometry with 5 s, 4 p and 1 d functions on O and 3 s and 1 p on H. His functions were what are called "contracted Gaussians," *i.e.*, each was a fixed linear combination of several simple Gaussians of the kind used above. A fairly common shorthand notation for this is to say that Diercksen used a [5,4,1/3,1] set contracted from a (11,7,1/6,1) set. Square brackets are used for contracted sets; round for uncontracted. The comma's are often omitted, and one must decide whether (1171/61) means 11 s, 7 p and 1 d (as it does) or perhaps 1 s, 17 p and 1 d (which would not be a very sensible choice). Note again that 7 p and 1 d means a total of 7 x 3 p's and 1 x 5 d's. With this set Diercksen found a minimum SCF energy of -76.053 au at a bond length of 0.9443 Å (a 1% error) and an angle equal 105.33° (a 0.8% error). Thus, except for the bond energy, agreement between theory and experiment is satisfactory, and even the bond energy results are not wildly in error.

For the water dimer, and this is the most favorable case, there are no theoretical results comparable to the CI work of Schaefer and Bender on water. The best will approach the SCF level of Neumann and Moskowitz, which because of missing correlation energy gave the worst bonding energy in Table 2-1. However, in hydrogen bond formation no new electron pairs are formed and the correlation contribution may be expected to be less significant. A demonstration of this point can be seen in the H_3^+ system as shown below:

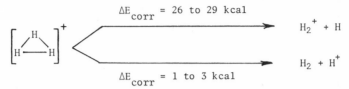

The three hydrogen nuclei in this ion are at the corners of an equilat-
eral triangle with each bonded to the other two. It requires 97 kcal
to remove a proton to form $H_2 + H^+$ [2212]. No unpaired electrons are
left, and the change in correlation energy is only 1 to 3 kcal [2212,
1300a]. On the other hand, pulling off a hydrogen atom to give $H_2^+ + H$
does form two unpaired electrons and involves a 26-29 kcal change in
correlation energy.

In contrast with the monomer results the properties of the water
dimer are not as well known experimentally. Gebbie, Burroughs, Chamberlain, Harries, and Jones [833] studied the temperature variation of the
30 cm^{-1} absorption of water dimer in the vapor. Plotting ln(absorbance)
versus 1/T they found the standard enthalpy of dimerization at 300°K to
be -5.2 ± 1.5 kcal per mole of dimer. This use of ln(absorbance) in
place of ln K_p is valid only if the monomer concentration does not
change significantly over the temperature range studied. Minton [1672]
has estimated the entropy of water dimerization to be -32 cal K^{-1} $mole^{-1}$
at 300°K. Combining this with Gebbie's ΔH gives an equilibrium constant 6.2 x 10^{-4} and hence only 4 x 10^{-4}% dimer formation at the highest pressure used. There may be considerable error in this number
which is quite sensitive to the computed entropy, but it does seem that
the monomer concentration is essentially constant as required.

The experimental ΔH°_{300} = -5.2 kcal/mole cannot be compared directly with the result of a variational calculation which gives only the
electronic contribution ΔE°(electronic) to this enthalpy. The two are
related by

$$\Delta H^\circ_{300} = \Delta E^\circ(electronic) + \Delta E^\circ_{300}(trans) + \Delta E^\circ_{300}(rot)$$

$$+ \Delta E^\circ_{300}(vib) + \Delta(PV). \qquad (2\text{-}43)$$

At 300°K, $\Delta(PV) = -RT = -0.596$ kcal. In going from two monomer molecules to one dimer, three degrees of translational and three degrees of
rotational freedom are lost, each of which contribute $\frac{1}{2}RT$ to the enthalpy. Thus $\Delta E_{300}^{\circ}(trans) + \Delta E_{300}^{\circ}(rot) = -3RT$. Each vibration contributes

$$h\nu \left[\frac{1}{2} + \left(e^{h\nu/RT} - 1 \right)^{-1} \right] \qquad (2-44)$$

to the internal energy. The first term in Eq. (2-44) is the zero-point
energy of the vibration, and the second is due to the population of
excited levels. The water dimer is fairly weakly bound so that of the
$3n-6 = 12$ dimer fundamental frequencies, three should be similar to
the three fundamentals of the proton-donor monomer and three to those
of the proton-acceptor. Tursi and Nixon [2492] list these six vibration frequencies together with the three of the monomers. Using Eq.
(2-44) these vibrations are computed to contribute -0.15 kcal/mole to
the enthalpy of the dimer bond. The remaining six vibrations of the
dimer correspond to motion of one monomer unit relative to the other
and are expected to occur at low frequencies. These six frequencies
have not been identified for the water dimer, but they almost certainly
all lie below 200 cm^{-1} (see Chap. 1, III). In the limit of low frequencies, the internal energy of a quantum harmonic oscillator given
by Eq. (2-44) goes to the classical limit, RT. For a frequency of
200 cm^{-1} the difference between quantum and classical energy is only
0.01 kcal/mole. We shall not be in serious error, then, if we allow
the classical energy RT for each of the six weak dimer modes. These
various contributions to H_{300}° are summarized in Table 2-2. They give
$\Delta E^{\circ}(electronic) = -6.2 \pm 1.5$ kcal/mole as the quantity to be compared
with theoretical bond strengths.

The geometry of the dimer is not well fixed. Experimental discussions have been concerned chiefly with deciding, from the number of
observed ir fundamental frequencies, whether or not the dimer has a
center of symmetry. If a center of symmetry exists, half of the 12
vibrations of the dimer are symmetric with respect to it (A_g) and half
are antisymmetric (A_u). Motion in an A_u mode changes the dipole moment
of the dimer, but motion in an A_g mode does not. Therefore, only the

TABLE 2-2

Experimental Bond Energy of Water Dimer

ΔH_{300}° (experimental)[a]	-5.2 ± 1.5 kcal/mole
$-\Delta E_{300}^{\circ}$ (translational)	+0.894
$-\Delta E_{300}^{\circ}$ (rotational)	+0.894
$-\Delta E_{300}^{\circ}$ (6 strong dimer vibrations - corresponding monomer vibrations)	+0.154
$-\Delta E_{300}^{\circ}$ (6 weak dimer vibrations)	-3.577
$-\Delta(PV)$	+0.596
ΔE° (electronic) =	-6.2 ± 1.5 kcal/mole

[a]Gebbie, Burroughs, Chamberlain, Harries, and Jones [833].

A_u modes will be observed in the ir spectrum. Hence, if more than six fundamental modes are observed, the dimer cannot have a center of symmetry. But, as mentioned above, six of the normal modes are expected to have weak force constants and to occur at very long wavelengths. The other six modes are sums and differences for each of two pairs of O—H stretches and one pair of HOH bends of the monomers. The three sums give A_u ir active modes; the three differences give A_g ir inactive modes. Hence, if one observes more than three normal modes (two stretches and one bend) in the normal ir region, then the molecule cannot have a center of symmetry.

Likely dimer structures are shown in Structure (6), and only the cyclic has a center of symmetry. Solid bonds are in the plane of the paper, dotted, behind; and wedge-shaped, in front. Hydrogen bonds are represented by dashes and are in the plane of the paper. In 1957 Van Thiel, Becker, and Pimentel [2513] studied the ir spectrum of water frozen in a solid N_2 matrix at 20°K and attributed three peaks to fundamental vibrations of the dimer, which they concluded had a cyclic structure. Tursi and Nixon [2492] extended this work in 1970 using higher resolution, but again at 20°K in a nitrogen matrix. They assigned five dimer fundamentals with confidence (the sixth, used above in the computation of ΔE° (electronic), was identified with less cer-

linear

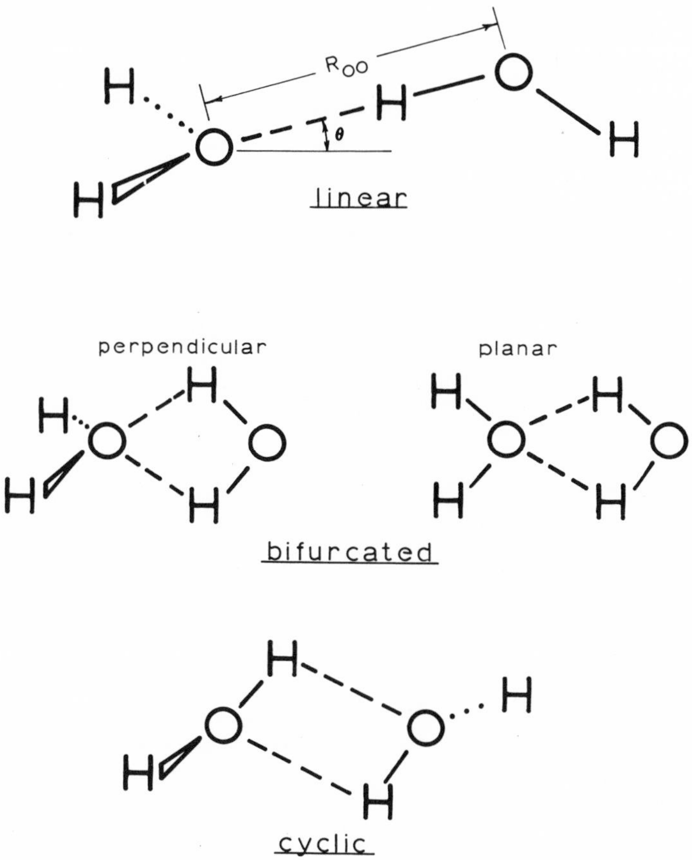

perpendicular   planar

bifurcated

cyclic

(6)

tainty), and hence ruled out the cyclic structure. On the basis of
frequency shifts relative to the monomer, they postulated the linear
rather than the bifurcated structure. Magnusson at the same time
[1585] measured the ir spectrum of water in CCl_4 at 25°C using a
double-beam technique to cancel out the monomer absorption. He found
only two stretching frequencies (he did not look at the bending region)
and therefore concluded that the dimer is cyclic. Kollman and Buckingham [1299] suggested that the differencing method might have removed a
dimer peak, but Magnusson [1584] thought not. An interesting alternative interpretation of Magnusson's spectrum has been suggested very
recently by Atkins and Symons [89a]. They propose a rapid interchange

between two equivalent linear dimers with a resulting coalescence of
otherwise distinct dimer peaks. This effect is, of course, well known
in NMR and ESR spectroscopy, but in order to appear in an ir spectrum
the interchange must be extremely fast. Atkins and Symons estimate
10^{13} interchanges per second, $i.e.$, about one every collision, are
required.

No rotational analysis has been done on the dimer spectra, hence
bond lengths and angles are not known.

The first complete ab $initio$ calculation on water dimer was as
recent as 1968 by Morokuma and Pedersen [1713]. Thirteen others, all
of the LCAO SCF type, have been reported since. Morokuma and Pedersen
found a total energy of -151.118 au and a hydrogen bond energy of 12.6
kcal/mole for their best geometry with their (5,3/3) basis. The next
year Kollman and Allen [1298] published similar calculations but with
a [3,1/1] Gaussian basis contracted from (10,5/5). Their total energy
was improved to -151.957 au and the hydrogen bond energy to 5.27
kcal/mole. In 1970 Morokuma [1714] bettered his first calculations
using a (2,1/1) Slater basis and obtained a total energy of -151.420 au
and a 6.55 kcal bond energy. In the same year Del Bene and Pople [603]
also used a (2,1/1) Slater basis, but to simplify integral evaluation
they expanded each Slater function as a sum of four Gaussians. Their
total energy was -151.010 au, slightly worse than Morokuma's; and their
bond energy was 6.09 kcal/mole. Also in 1970, Hankins, Moskowitz, and
Stillinger [1001] used a [5,3,1/2,1] Gaussian basis to compute a total
energy of -152.091 au and a bond energy of 4.73 kcal/mole. In 1971
Del Bene [601] repeated her earlier calculations, but using three
Gaussians instead of four to describe each Slater function, the aim
being to decrease the number of required integrals in order to extend
the method to larger systems. This raised the total energy to -149.94
au, but the hydrogen bond energy was changed only to 5.88 kcal/mole.
The water monomer work of Diercksen described above was only preliminary to his even more extensive calculations on the water dimer reported
in the same paper [621]. Using the same [5,4,1/3,1] Gaussian basis as
in the monomer, Diercksen obtained the lowest dimer energy to that
date, -152.112 au, and a hydrogen bond energy of 4.84 kcal/mole.

Several LCAO SCF calculations with Slater orbitals expanded in Gaussians [600,1182b,1182e], a bond orbital analysis [294a], a study of the formation of $H_3O^+ + OH^-$ from the water dimer [738], and of NMR proton shifts in the dimer [1172a,b] followed.

These were capped by Popkie, Kistenmacher, and Clementi [1969a], who used a [8,5,2,1/4,2,1] contracted Gaussian basis (note the use of f orbitals) to give a dimer energy of -152.1378 au, the best present result. Their hydrogen bond energy of 3.67 kcal/mole is rather low. As they point out, a poor basis set might be expected to give larger hydrogen bond energies than a better set because when monomer units are brought together, basis functions on the one can make up for deficiencies in those on the other. This favors the dimer over the separated monomers, and leads to bond energies that are too large. Popkie, Kistenmacher, and Clementi estimate that correlation effects would increase their 3.67 kcal bond energy by from 0.3 to 0.5 kcal.

A paper by Meunier, Lévy, and Berthier [1638b] makes similar points for the water-ammonia dimer. They found that a minimal Slater basis gave a hydrogen bond energy of 7.66 kcal. Extending each monomer basis by using the orbitals of the other monomer, positioned as in the dimer, lowered the bond energy to 3.96 kcal. Including correlation energy by a direct perturbation calculation then raised this to 4.63 kcal.

The energies just quoted all refer to geometry of lowest computed energy. Extensive geometry searches were done involving variations of the geometries shown in Structure (6). In the "linear" dimer, the two monomer units lie in perpendicular planes with the two O atoms and the H involved in the hydrogen bond all in a line. The O—O distance and the angle θ were varied. Variations in which H was not in the O—O line and in which the two monomer planes were not at right angles were also examined but were found to be unfavorable. Two bifurcated forms were considered. In the one, all atoms are in a plane; in the other, one monomer unit is at right angles to the other. In either case, both hydrogens of the proton donor are bonded to the proton-accepting oxygen. The two oxygens and two hydrogens of the two hydrogen bonds in the cyclic structure are all in a single plane, with the two remaining

hydrogen atoms one in front and the other behind that plane. As shown
in Fig. 2-4, Morokuma and Pedersen calculated the total energy of the

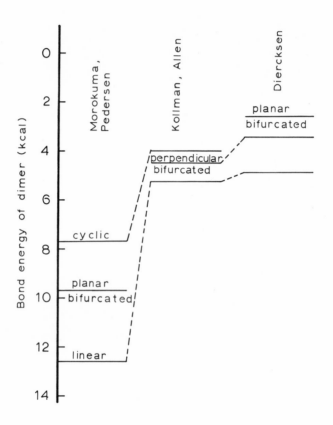

FIG. 2-4. Calculated electronic energies of water dimer.

linear structure to be lowest, and therefore the hydrogen bond energy
to be of greatest magnitude. The planar bifurcated structure was next
most favorable, and the cyclic structure least. Kollman and Allen
found the same order; Diercksen, who did not examine the cyclic form,
also found the linear to be better than the perpendicular bifurcated
form, which was better than the planar bifurcated structure. Because
of the agreement of Kollman and Allen with Morokuma and Pedersen, most
later authors considered only variations of the linear form.

One might imagine that the best value of the angle θ in the
linear dimer either would be approximately 55° corresponding to protonation of an oxygen lone pair, or alternatively that the proton might
be attracted to both lone pairs simultaneously giving θ = 0°. The
energy turns out not to be very sensitive to θ, and best computed
values lie from 12° to 58°. Morokuma and Pedersen are sometimes quoted
as predicting θ = 0°. Of the structures whose energy they compute, one
with θ = 0° does give the lowest energy. But they do not examine a
very fine grid of θ values, and interpolating their data with a cubic
polynomial predicts lowest energy at θ = 17°. Popkie, Kistenmacher,
and Clementi [1969a] were concerned with mapping the entire monomer-
monomer interaction surface rather than with finding the best dimer
geometry. Their calculation of lowest energy was for a linear dimer
with θ = 0°. But their potential surface, computed with a smaller
basis set, indicates an optimum θ of 30-45°. They also mention that
the dimer energy is not very sensitive to moving the hydrogen off the
oxygen-oxygen line. As can be seen from Fig. 2-5, there is no observable trend in computed θ; the lower energy calculations do not agree
with each other better than with those of higher energy. A similar

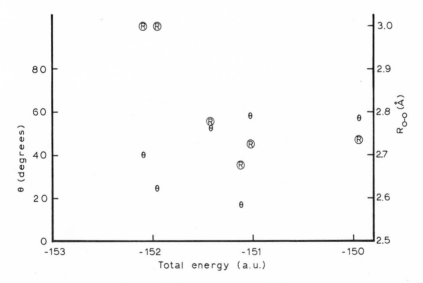

FIG. 2-5. Bond angle θ [see Structure (6)] and oxygen-oxygen
distance R versus calculated total energy.

scatter is seen in the computed O—O distances which range from 2.68 Å to 3.00 Å.

All of these calculations have been of the SCF type. Therefore, ignoring relativistic effects, they must all lie above the true energy by an amount at least equal to the correlation energy. Adding the total energy of two water molecules (-2 x 76.479 au) to the experimental energy of dimerization (-5.2 kcal/mole) gives -152.966 au as a close approximation to the total dimer energy. Subtracting twice the estimated relativistic energy of the monomer gives approximately -152.868 au as the total nonrelativistic energy. Schaefer and Bender [2186b] estimate the SCF limit to be -76.07 au for the water molecule, corresponding to a correlation energy of (-76.430 + 76.07) = -0.36 au. That of the dimer must be roughly double this figure, giving a nonrelativistic Hartree-Fock limit of -152.148 au for the dimer. These quantities are shown in Fig. 2-6 on a plot of computed hydrogen bond

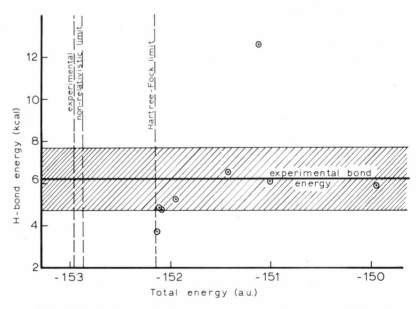

FIG. 2-6. Hydrogen bond energy versus total energy of the water dimer. The shaded band is centered about the experimental value and gives the estimated error range.

energy versus total energy. As expected, none of the computed total
energies lie to the left of the Hartree-Fock limit, but Popkie, Kistenmacher, and Clementi's result does approach it very closely. However,
the horizontal and vertical axes of Fig. 2-6 are drawn to very different scales, and even this last result is 6 kcal/mole above the Hartree-
Fock limit which itself is over 500 kcal/mole above the exact value.
Most of the calculated bond energies do fall within the experimental
range; but Popkie, Kistenmacher, and Clementi's, the most rigorous of
all, does not. In their dimer calculations, the monomer units were
held rigid. Allowing them to adjust to optimum geometry in the dimer
might well increase the hydrogen bond energy by the required amount.

All of this points up to the extreme difficulty of obtaining an
accurate theoretical estimate of an energy as small as that of the
hydrogen bond.

Geometry and energy have been the primary objectives of these
calculations, but other properties have been considered in less detail.
Several of the authors above examined the change of charge distributions caused by hydrogen bond formation. Diercksen shows contour
plots of these charge differences, and Kollman and Allen give less
detailed drawings which are easier to understand at a glance. Notice
that Morokuma and Pedersen and Kollman and Allen use charge densities;
hence, their signs are opposite to those of the other authors who use
electron densities. Two interesting points come out of these charge
differences. First, the proton in the hydrogen bond becomes more
positive. One might have expected the opposite effect, since the proton in the hydrogen bond is imbedded in the electrons of both oxygens.
The calculated result is in qualitative agreement with the observed
shift to lower field of the NMR resonance of a hydrogen-bonded proton.
The second point is that there are significant charge effects on the
protons not in the hydrogen bond. These are, of course, missed by
model calculations using only the O—H...O fragment.

Kollman and Allen and also Diercksen have made some preliminary
calculations on the infrared spectrum of the water dimer. The main
effects to be explained are: (1) the decreased force constant for O—H
stretching; (2) the increased force constant for bending; and (3) the
factor-of-12 increase in the stretching intensity. The difficulty in

examining these effects theoretically is that one needs to examine
motion of the dimer in its normal modes. These depend upon the force
constants, and obtaining those requires the computation of extensive
potential surfaces. This has not yet been done for the water dimer.
Instead, force constants and dipole moment changes were computed for
motion of the proton along the O...O distance, keeping all other nuclei
fixed. Kollman and Allen [1298] and Diercksen [621] give this potential curve for an O—O distance of 3.0 Å and Morokuma and Winick [1714]
for 2.78 Å. All agree that the curve shows only a single minimum with
no hint of a second minimum or even a shoulder. So far as predictions
about the infrared spectrum can be made, they seem to be in qualitative
agreement with experiment. The water dimer is perhaps one of the least
favorable cases for this sort of treatment because the remaining protons must also contribute to the normal modes. The sensitivity of the
charge distribution on these peripheral atoms to hydrogen bond formation shows that their motion might make a significant contribution to
the dipole moment change and hence to the infrared intensity.

Morokuma, van Duijneveldt, and Kollman and Allen have partitioned
the hydrogen bond energy of the water dimer into components similar to
Coulson's [516]. Morokuma's partitioning [1710] of the Morokuma and
Winick wavefunction is the most complete and rigorous of the three.
Suppose ψ_A^0 is a single product function composed of SCF MOs for one
isolated water molecule, and let A be an antisymmetrizing operator.
Then the antisymmetrized single-determinant SCF wavefunction can be
written as $A\psi_A^0$, and similarly $A\psi_B^0$ is such a wavefunction for a second
water molecule. Let the sum of energies of these two at infinite separation be E_0. Bring the monomers to bonding distance and let E_1 be
the energy of the product $(A\psi_A^0)(A\psi_B^0)$, a function allowing exchange
within each monomer, but not between monomers. The difference,
$E_0-E_1 = E_{es}$, is the electrostatic contribution to the hydrogen bond.
Insofar as $A\psi_A^0$ is an exact wavefunction for the water monomer ($i.e.$,
the SCF calculation has approached the Hartree-Fock limit and correlation energy can be neglected), E_{es} corresponds exactly to Coulson's
definition of the electrostatic contribution (see this chapter, II.A.2).
Next, allow each monomer function to respond to the field of the other,
but still prohibit electron exchange between the two. The wavefunction
may be written as $(A\psi_A)(A\psi_B)$ with energy E_2. Morokuma defines E_1-E_2 to

be E_{pd}, the polarization plus dispersion energy. This is not exactly
the same as the polarization plus dispersion defined as the second
term of a perturbation expansion, but it appears to be similar. Suppose that, instead of polarization, exchange is allowed between the
original monomer functions. The wavefunction is then $A(\psi_A^0 \psi_B^0)$ with
energy E_3, and exchange energy may be defined $E_{ex} = E_1 - E_3$. Finally,
consider the complete SCF wavefunction for the dimer $A\psi_{AB}$ with energy
E_4. In going from $A(\psi_A^0 \psi_B^0)$ to $A\psi_{AB}$, polarization of the individual
monomers is allowed and so is transfer of charge from one monomer to
the other. Hence, $E_3 - E_4 = E_{pd} + E_{ct}$, giving for the charge-transfer
energy $E_{ct} = E_2 + E_3 - E_1 - E_4$. Again this use of "charge-transfer"
and "exchange" is similar, but not identical, to that in perturbation
theory. Note that Morokuma's partitioned terms do sum to give the
hydrogen bond energy, $E_0 - E_4 = E_{es} + E_{pd} + E_{ex} + E_{ct}$.

The partitioned energies in Table 2-3 agree with Coulson's conclusion that the hydrogen bond results from the combination of several
factors, with no one predominating. The electrostatic contribution is
on the same order as the bond energy itself, but so are the exchange
and charge transfer terms. Coulson's estimated energies are also
shown in Table 2-3, each aligned, as Morokuma has done, with Morokuma's
term to which it most closely corresponds. Considering the differences
in definition and the approximations in Coulson's estimates, the correspondence is surprisingly good.

Unfortunately, as a result of work by Kollman and Allen [1297],
it appears that this agreement is in part accidental. These authors,
using a better wavefunction (E = -152.01787 au) than in their first
calculations [1298], also partitioned the hydrogen bond energy of water
dimer. Their "electrostatic energy" includes exchange and corresponds
exactly to Morokuma's $E_{es} + E_{ex}$, and their "delocalization energy" is
Morokuma's $E_{pd} + E_{ct}$. They compute values of 4.50 and 3.05 kcal/mole
for these quantities, to be compared to -1.86 and 8.41 from Morokuma's
data. Thus, the partitioning appears to be very sensitive to the
wavefunction.

Van Duijneveldt [2507] has analyzed the hydrogen bond of water
dimer using Murrell, Randić, and Williams' [1742] perturbation method.
However, he used an approximate model of the dimer roughly on the level

TABLE 2-3

Partitioning of the Hydrogen Bond Energy in Water Dimer[a]

	Morokuma[b]	Kollman and Allen[c]	Coulson[d]	van Duijneveldt[e]
Electrostatic	8.00	4.50	6 (electrostatic)	7.2 (electrostatic)
Exchange	-9.86		-8.4 (repulsive)	-5.0 (exchange repulsion)
Polarization and dispersion	0.25	3.05	3 (dispersion)	1.0 (dispersion)
Charge transfer	8.16		8 (delocalization)	1.1 (delocalization)
Total	6.55	7.55	8.6	4.3

[a]All energies are in kcal/mole. Positive terms are stabilizing; negative are destabilizing. [b]Morokuma [1710]. [c]Kollman and Allen [1297]. [d][516]. [e]van Duijneveldt - van de Rijdt and van Duijneveldt [2507]. The names in parentheses are those used by Coulson.

of Coulson and Danielsson's [517]. As shown in Table 2-3, his results
are also in approximate agreement with Coulson's.

   In contrast with these *ab initio* data, which are on the whole
fairly accurate, the semiempirical results are somewhat erratic.
Murthy and Rao [1750] used the Extended Hückel method to compute the
potential for motion of the hydrogen-bonded proton in the water dimer.
In disagreement with the *ab initio* results they found a double minimum
potential with the $H\bar{O}...\overset{+}{H}OH_2$ minimum at lower energy than the $HOH...OH_2$
minimum. This appears to be a doubtful result. Murthy and Rao do not
report a hydrogen bond energy from their Extended Hückel work; but
Rein, Clarke, and Harris [2080], using an iterative version of the
Extended Hückel theory, compute the water dimer to be unstable with
respect to separated monomer units. Minton [1672] examined these results and concluded that the Extended Hückel method is unable to take
proper account of the interaction between distant charge distributions.
He therefore repeated the calculations, computed charges on the atoms
of the dimer using a Mulliken population analysis, and then added the
electrostatic energy of this set of point charges to the Extended
Hückel energy. In this way a total energy minimum, corresponding to a
hydrogen bond strength of about 2 kcal/mole, was found at an O—O
separation of about 3.1 Å.

   There have been at least eleven CNDO calculations on the water
dimer, see Refs. [444,879,1020,1093,1290a,1294,1705a,1750,2047,2208,
2380d]. The results are in reasonable agreement with *ab initio* work
and give computed hydrogen bond energies ranging from 5 kcal/mole [444]
to 8.7 [2208]. The order of dimer stability is linear > bifurcated >
cyclic [1093,1294] as in the *ab initio* calculations. The potential
curve for motion of the proton along the hydrogen bond shows only a
single minimum, but with a shoulder that becomes more pronounced as the
O—O distance is lengthened from 2.5 to 2.9 Å [444,1750].

   The satisfaction of seeing the somewhat more sophisticated CNDO
method overcome the faults of the Extended Hückel results is spoiled by
Kollman and Allen's NDDO calculation [1294]. The NDDO method is one of
the most elaborate of its kind in that it discards fewest integrals.
However, it predicts the absurd value of 76 kcal/mole for the hydrogen-
bond strength in the water dimer.

INDO [444,2208] and MINDO [444] calculations have also been carried
out for the water dimer. INDO and CNDO results are similar, but the
potential for proton motion seems incorrectly given by the MINDO
method.

2. *The bifluoride ion*

The hydrogen bond in bifluoride ion is the strongest known
(unless one wishes to count the bonds in H_3^+ [2212] as hydrogen bonds).
A number of experimental values have been reported for the heat of
reaction of

$$HF~(g) + F^-~(g) \longrightarrow FHF^-~(g). \qquad (2\text{-}45)$$

Waddington's value of -58 ± 5 kcal/mole [2542a] is often quoted, but it
is not entirely experimental since it requires the calculation of
lattice energies of solid metal bifluorides. Pimentel and McClellan
[1937] take Harrell and McDaniel's [1009] 37 kcal/mole as the currently
accepted value of the bifluoride ion hydrogen bond strength. These
authors measured the heat of reaction of

$$N(CH_3)_4^+F^-~(s) + HF~(g) \longrightarrow N(CH_3)_4^+(HF_2)^-~(s) \qquad (2\text{-}46)$$

and found it to be -37 kcal/mole at 100°C. They felt that the lattice
energies of tetramethylammonium fluoride and bifluoride would not be
very different and that therefore the 37 kcal/mole represents, to
within 1 or 2 kcal/mole, the hydrogen bond strength in bifluoride ion.
The Harrell and McDaniel value might appear to be freer of theoretical
corrections, and therefore a more purely experimental result, than
Waddington's, but in fact both require a theoretical estimate of the
same quantity. Waddington's detailed calculation of the difference in
lattice energies might be more accurate than Harrell and McDaniel's
assumption that this difference vanishes.

Neckel, Kuzmany, and Vinek [1777a], and very recently Dixon,
Jenkins, and Waddington [628a] have reexamined this question. Consider
the reactions

$$MF~(s) \xrightarrow{\Delta H_1} M^+~(g) + F^-~(g), \qquad (2\text{-}47)$$

$$MHF_2 \ (s) \xrightarrow{\Delta H_2} M^+ \ (g) + HF_2^- \ (g), \qquad\qquad (2\text{-}48)$$

$$MF \ (s) + HF \ (g) \xrightarrow{\Delta H_3} MHF_2 \ (s), \qquad\qquad (2\text{-}49)$$

where M is an alkali metal or tetramethylammonium group. In all four papers under discussion the heat of reaction of Eq. (2-45) was obtained as

$$\Delta H = -\Delta H_3 + (\Delta H_1 - \Delta H_2). \qquad\qquad (2\text{-}50)$$

The heat of reaction between solid MF and HF gas was measured experimentally; but $(\Delta H_1 - \Delta H_2)$, the difference between fluoride and bifluoride lattice energies, had to be estimated theoretically. The simplest, and perhaps least likely, estimate was that of Harrell and McDaniel, who assume the difference to be negligible. In both of Waddington's papers [2542a,628a] crystal energies were computed in the usual way by summing point-charge electrostatic contributions and adding dispersion and higher-order corrections using classical empirical formulas. The more recent paper [628a] obtains an H-bond strength of 60.2 kcal/mole, in good agreement with the earlier result. Neckel, Kuzmany, and Vinek treat the bifluoride ion in a somewhat more sophisticated way. They take Clementi and McLean's [474] bifluoride ion wavefunction and make a multipole expansion of the charge distribution. The crystal energy was then computed in terms of the multipole moments. From the potassium, rubidium and cesium compounds these authors find an average hydrogen bond strength in bifluoride ion to be 55.9 kcal/mole.

But the calculated quantity $(\Delta H_1 - \Delta H_2)$ is about 30 to 40 kcal/mole. This is not small compared either to the experimental ΔH_3 or to the bifluoride hydrogen bond strength. The calculations of $(\Delta H_1 - \Delta H_2)$ are much less rigorous than current direct calculations of the bifluoride ion hydrogen bond strength, and there appears to be no reason to suppose them more reliable. It seems, therefore, that the numbers usually quoted as experimental strengths of the bifluoride hydrogen bond are no less theoretical than the theoretical results with which they are compared.

The F—F distance is not known for the isolated bifluoride ion, but it has been measured in solid salts, and has been found by Ibers

[1117a] to be 2.277 ± 0.006 Å in KHF$_2$ and by McGaw and Ibers [1562a]
to be 2.264 ± 0.003 Å in NaHF$_2$. Solid NH$_4$HF$_2$ contains two bifluoride
ions that differ in crystal environment, but McDonald [1560a] has
found the two F—F distances to be practically the same (2.275 Å and
2.269 Å).

Ibers [1116,1117a] has stressed the difficulty of distinguishing
a hydrogen bond with a symmetric double minimum potential from one with
a symmetric single minimum, but combining ir and neutron diffraction
data [1117a] he concluded that the single minimum is more likely in the
case of HF$_2^-$. The NMR study of KHF$_2$ by Waugh, Humphrey, and Yost
[2563a] is in accord with this in showing that the proton lies within
± 0.06 Å of the center of the F—F distance, and that hence a single
potential minimum is very likely. Very recently Williams and Schneemeyer [2602a] have reported a neutron diffraction study of solid
p-toluidinium bifluoride in which the H-bond is found to be asymmetric.
One H—F length is 1.025 Å, and the other is 1.235 Å. The total F—F
length of 2.260 Å is very close to that in the simpler bifluorides.
This system was chosen because of the asymmetric environment in which
the bifluoride ion lies, and the authors ascribed the asymmetry of the
hydrogen bond to the asymmetry of the environment. These results then
do not contradict the assumption of a symmetric bond in the free
bifluoride ion.

Ab initio SCF calculations on HF$_2^-$ were carried out by Hamano [984a]
as early as 1957, but many of the multicenter integrals in this work
were approximated. Hamano examined only the experimental F—F distance
of 2.26 Å; he reported neither the calculated total energy nor the
hydrogen bond energy. In 1960 an *ab initio* valence bond calculation
was reported by Bessis and Bratož [230] giving a total energy of
-199.5742 au. But because integrals were again approximated, this
cannot be compared easily with later exact calculations, none of which
give so low an energy. Bessis and Bratož did not calculate the energy
of the hydrogen bond.

Clementi [474a] in 1961 and Clementi and McLean [474] in 1962 published SCF studies on HF$_2^-$ using Slater orbitals and exact evaluation of
all integrals. The F—F distance was fixed at 2.26 Å, and the total
energies of -198.28264 au and -199.39266 au were calculated with an

estimated Hartree-Fock limit of -199.616 au. Again, the hydrogen bond
energy was not computed.

In 1967 McLean and Yoshimine [1568a,2661a] examined 21 geometries
of HF_2^- using an SCF calculation and an extended Slater basis. The
lowest energy of these was -199.57296 au and had the hydrogen atom in
the center of a 2.22 Å F—F bond. The hydrogen bond energy was not
investigated.

It was not until 1970 that *ab initio* calculations of the hydrogen
bond strength in HF_2^- were made. In that year Gaussian basis SCF
studies were published by Kollman and Allen [1295] and by Noble and
Kortzeborn [1805]. Kollman and Allen's lowest energy was -199.52303
au for a symmetric hydrogen bond and an F—F length of 2.285 Å. The
H-bond energy for this structure was computed to be 52 kcal/mole. They
found that if the F—F length is stretched to 2.38 Å, the potential for
motion of the proton along the bond changes from a symmetric single to
a symmetric double minimum curve. Noble and Kortzeborn obtained a
somewhat lower energy of -199.55454 au for an F—F length of 2.25 Å.
Their computed hydrogen bond energy of 40 kcal is in good agreement
with the experimental value of 36.

There have been two recent semiempirical studies of the HF_2^- system.
In 1969 Koller, Borstnik, and Ažman [1290] examined the potential for
proton motion using both CNDO/2 and INDO methods. Allen and Kollman
[35] in 1970 obtained the rather poor value of 103 kcal for the hydrogen bond energy using CNDO/2.

3. The guanine-cytosine pair
Finally we examine the guanine-cytosine (G—C) base pair shown in
Structure (7) to see how well theory can predict properties for a

Guanine   Cytosine

(7)

hydrogen-bonded system of average size. However, the G—C pair is not really a typical example of systems of this size. For a typical system one would find at most one or two semiempirical calculations published; but because of the crucial biochemical role of the G—C pair, more and better calculations are available here.

For the most part, calculations on the G—C system have investigated one of two questions. The first concerns the stability of G—C compared to other nucleotide base pairs; this is related to the specificity of DNA replication. The second is the shape of the potential curve for interbase proton exchange; this is involved in a possible mechanism of gene mutation. Here we shall attempt only to assess the reliability of the theoretical calculations, and shall concentrate on the shape of the potential curve.

Most discussions of either single proton motion or simultaneous exchange of two protons within a DNA base pair assume a double well potential. The only experimental evidence for this seems to be in a short paper by Shulman [2262]. From pK_a's of adenine and thymine he estimates the energy required to move a proton from NH of the thymine ring to the adenine ring nitrogen to be no more than 7.6 kcal/mole = 2650 cm^{-1}. This is below the energy of the first excited vibrational level of the thymine N—H bond which is at about 3400 cm^{-1}. If in the base pair the potential increases monotonically as the proton is moved from thymine to adenine, then the portion of the curve below 3400 cm^{-1} is very different from that of free thymine, and the first excited vibrational level of free thymine should lie at a very different energy from that of the base pair (see Fig. 2-7). If, on the other hand, a potential maximum exists between two minima for the complex, the potential near the thymine equilibrium position might rise to 3400 cm^{-1} without differing much in shape from the free thymine curve. In fact, base pair formation shifts the NH vibration frequency rather slightly (about 200 cm^{-1}). Shulman therefore feels that the double minimum curve is the more likely.

The earliest of the theoretical papers was a Hückel calculation by Pullman and Pullman [2019] in which the N—H...N and N—H...O groups were each treated as a single conjugated link with $\beta_{H-bond} = 0.2\ \beta_{CC}$. This approach will not yield details of the hydrogen bond itself, but

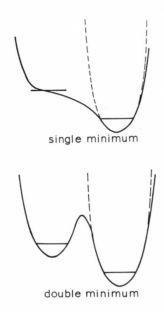

single minimum

double minimum

FIG. 2-7. Possible potential curves for proton transfer from thymine to adenine. Dashed line represents N—H potential curve for free thymine.

might indicate its effect on the remainder of the conjugated system. The guanine-cytosine pair was computed to be more stable than adenine-thymine. The potential for proton motion could, of course, not be investigated.

A number of papers by the Pullmans and their co-workers followed, see Refs. [216,386,2016,2017], using more elaborate semiempirical techniques. These were concerned primarily with nucleotide configurations and not with hydrogen bond proton potentials.

First studies of the proton potentials were dominated by a series of papers by Rein and Harris [2084-2088], who used a Pariser-Parr-Pople calculation which included the π electrons plus four electrons in each hydrogen bond treated. Both single proton motion and simultaneous exchange of a proton pair were considered. Rein and Harris found double minimum potentials, and later workers, see Refs. [240-242,1528, 2038,2088a,2091] obtain similar results.

Few systems of this size have been subjected to such a range of semiempirical treatments. None can be claimed to be better understood theoretically. Since these various methods gave results in at least qualitative agreement, the theoretical description of the G—C pair seemed to be fairly satisfactory, but a recent paper by Clementi, Mehl, and von Niessen [475] has upset the situation. As these authors point out, it would be more accurate to say that Rein and Harris *assume* a double minimum potential than to say that this potential is derived from their calculations. Rein and Harris [2086] consider the effect of various approximations used in the Pariser-Parr-Pople calculation. In particular, they study variations in the Wolfsberg-Helmholtz parameter K for the calculation of off-diagonal matrix elements. Some values of K lead to double minima, but others do not. They choose one that does.

To examine the problem more rigorously, Clementi, Mehl, and von Niessen [475] have carried out *ab initio* calculations on an IBM 360/195 computer for the G—C pair using a set of 334 Gaussians contracted to 105. The work required 70 billion two-electron integrals, computed at the rate of 100,000 per sec, and must be the largest molecular calculation carried out to date. Guanine and cytosine were placed in their experimentally observed configurations, and protons were allowed to move singly along each of the three hydrogen bonds. None of the resulting potentials for single proton motion showed double minima.

In spite of the heroic size of this work, the basis used is not a large one (per atom) compared with those of the water dimer calculations. It would perhaps correspond most nearly to Morokuma's (5,3/3) basis, and that was one of the poorest. Increasing basis size or studying all concerted motions of the protons was beyond even the IBM 360/195. To consider these problems Clementi, Mehl, and von Niessen went to a smaller system, the formic acid dimer. As protons were moved along the hydrogen bonds, the remaining atoms were held fixed in their equilibrium positions; otherwise, symmetry alone would necessitate a double minimum potential. Expanding the basis set in the single proton motion calculations and examination of coupled motion with the smaller basis both gave inflexion points, hinting of incipient double minima, for formic acid dimer. When coupled motion was studied with the larger basis, definite double minima appeared.

In view of all this, the shape of the guanine-cytosine curves must still be considered unresolved, especially for the case of coupled proton transfer.

C. Summary of the Present Theoretical Position

1. *Energy and geometry*

The total energy of a system appears as the eigenvalue of the Hamiltonian operator in the Schrödinger equation. It is the parameter that a theoretical calculation yields first and most easily, and it formed the main goal of the work described above. A bond energy is obtained as the difference between total energies of separated and combined systems. Ideally, one would like errors in total energy calculations to be small in comparison to bond energies (which unfortunately are themselves small compared to total energies). This is never the case for hydrogen-bonded systems. Even the best water dimer calculations give total energies that are in error by 100 times the hydrogen bond energy. Thus, in spite of huge advances in computing ability during the last fifteen years, one still must hope, much as one did fifteen years ago, that errors will cancel on taking the energy difference between bonded structure and separated fragments and so give accurate bond energies. The theoretical results since 1960 collected in Table 2-4 (see p. 113) show that this seems to work reasonably well, especially for thorough *ab initio* calculations. But one can never be really certain ahead of time of obtaining even approximately correct bond energies.

The problem of computing molecular geometries is an order of magnitude more difficult than computing the energy of a given geometry, since one must search the energy-geometry surface for the point of minimum energy. There are no hydrogen-bonded cases where this has yet been done completely. Even in the water dimer, the monomer units were held in their best monomer geometries.

2. *Vibrational spectra*

This is a problem still more difficult. After determining the complete potential surface one must find the normal modes and energy levels for vibration of the nuclei in this potential. Only very recently is there an example of carrying out this program completely

for a hydrogen-bonded system. Almlöf [44a] has made an *ab initio*
Gaussian basis calculation of the HF_2^- potential surface. He then computed the energy of eight vibrational transitions ranging from 660
cm^{-1} to 5157 cm^{-1}. All were in good agreement with experiment.

However, HF_2^- has a uniquely small number of vibrational modes for
a hydrogen-bonded system. Further, the vibrational frequencies themselves were calculated. One is usually concerned with a shift in
vibrational frequency due to hydrogen bonding. These differences are
smaller numbers and may be more difficult to compute to good accuracy.

It is more usual to find rough estimates of frequency shifts and
intensity changes in approximate normal modes. Even when potential
curves are computed, they are for proton motion only and may or may
not resemble those for motion in a normal mode. The review of Kollman
and Allen [1292], and more especially that of Murthy and Rao [1748],
give details of this kind of calculation.

3. Electronic spectra

To make a thorough prediction of an electronic spectrum, all the
work for an accurate ground state calculation must be repeated for each
excited state. Estimates of the effect of peripheral hydrogen bonds
upon the electronic spectra of conjugated chromophores have been made,
but none compare in accuracy to the work on the water dimer ground
state above.

4. NMR spectra

In hydrogen bond formation a proton originally attached to one
electronegative atom becomes bonded in addition to an unshared electron
pair of a second such atom. At first thought, this would appear to
lead to an increase in electron density around the proton. Experiment,
however, shows that hydrogen bond formation leads to NMR proton chemical shifts in the direction of *lower* magnetic fields, indicating less
shielding and hence lower electron density. One satisfying theoretical
result is that all good calculations show that hydrogen bond formation
does lower electron density near the proton, in agreement with experiment and contrary to first thoughts.

However, the correlation between charge density and chemical
shift is itself questionable. Attempts to go beyond this and make

rigorous direct calculations of chemical shift are not easy. The only
work of this type on a hydrogen-bonded system appears in a recent
paper by Jaszuński and Sadlej [1172a,b] who treated Diercksen's accurate water dimer wavefunction [621] by the Karplus-Kolker method to
give a chemical shift of 27.61 ppm for the hydrogen bond proton, in
reasonable agreement with the observed value of 25.62 ppm. But the
difference between this value and that of water monomer, $i.e.$, the
change due to H-bond formation, is small and temperature-dependent.
The calculated value of -0.69 ppm agrees only in sign with the observed.

Calculations of spin-spin coupling constants are also difficult.
Such a calculation on H_2 [708a] with a very accurate wavefunction gave
a calculated coupling constant of 163 Hz, in poor agreement with the
experimental 278 Hz. There is no comparable work on hydrogen-bonded
systems.

5. *The origin and nature of the hydrogen bond*

Perhaps even more than for reliable predictions of physical properties of hydrogen-bonded systems the chemist looks to theory for
answers to questions like, "What is the cause of a hydrogen bond,"
"How does it differ from other bonds," and "Why does hydrogen appear
to be unique in its ability to form such a bond?" As already mentioned,
it can be said with certainty that the ultimate cause of the hydrogen
bond is the Coulomb attraction between electrons and positive nuclei.
No other fundamental forces are important. Of course, this does not
distinguish the hydrogen bond from any other, and this is mirrored in
the theoretical treatments above which are the same as one would apply
to any chemical bond. There is nothing in them particular to a
hydrogen bond.

One may ask how the Coulomb attraction energy of the hydrogen
bond can be divided into various subcategories. Commonly one asks
about electrostatic, covalent, repulsive, and dispersion contributions.
Not everyone means the same thing by these terms, and care is needed in
comparing results of different authors. The electrostatic contribution
is especially tricky since in a fundamental sense the entire energy is
electrostatic. In his 1957 review Coulson concluded that the hydrogen
bond could not be described adequately in terms of any single one of

these four quantities. Each contributed significantly, some to stabilize and some to destabilize the bond, and each was roughly of the
same magnitude as the total bond energy itself. These conclusions
were based upon rough calculations of not very precisely defined
quantities. Today, with precise definitions and accurate calculations,
it is found that this partitioning is very sensitive to the wavefunction used, and that hence we cannot really say more than Coulson did
fifteen years ago.

Kollman, Liebman, and Allen [1300] have reported interesting calculations in which they examine the effect of replacing hydrogen by
lithium, with the aim of looking for those things that are unique to
the hydrogen bond. So far, however, the hydrogen bond looks much like
any other; theory has not revealed with certainty any striking features
that are peculiar to it.

Theoreticians are well aware of the limitations of their methods,
but it is natural that reports of theoretical work emphasize successes
and go softly on the problem of reliability; it would be tedious repe-
tition to give the same warnings in every paper. As a balance, the
present chapter goes to the other extreme and concentrates upon the
dangers and cautions. But it would be wrong to go too far in that
direction. It is a tempting trap for the experimentalist to conclude
that since few theoretical predictions are completely certain one can
ignore all theoretical results. The difficult alternative of weighing
each separately is probably more useful.

ACKNOWLEDGMENT

I am grateful to Professors Kollman and Morokuma for helpful
correspondence on several points.

Theoretical Studies on Hydrogen-Bonded Systems from 1960 through 1973*

Key to Table 2-4

In order of increasing molecular weight. Certain systems that might be expected in this section are instead below. For example, HF_2^- is classified as a dimer with HF as the proton donor and F^- as proton acceptor.

In order of increasing molecular weight of proton donor. Systems with a common proton donor are in order of increasing molecular weight of proton acceptor. The systems are listed as: proton donor...proton acceptor. In cases such as the acetic acid dimer where both members donate and accept protons, the system is listed under the member of lower molecular weight.

In order of increasing molecular weight of monomer unit. Systems with a common monomer unit are in order of increasing number of monomer units.

In order of increasing total molecular weight of system.

---

*The literature search was concluded on December 31, 1973. Papers appearing later but with a nominal publication date in late 1973 are missing from the table.

TABLE 2-4

Theoretical Studies on Hydrogen-Bonded Systems

1. MONOMERS

System	Method	Comments	Authors (reference in brackets)
HCOOH	LCAO SCF + CI, Gaussian basis	Possibility of hydrogen bond in monomer	Peyerimhoff and Buenker, 1969 [1925]
HCOOH	INDO		Gordon and Tallman, 1972 [889b]
Malondialdehyde	CNDO/2	Model for acetylacetone	Schuster, 1969 [2210]
Allyl alcohol	Extended Hückel	Hydrogen bond to π system	Morokuma, Kato, Yonezawa, and Fukui, 1965 [1712]
Glycine	LCAO SCF Gaussian basis	Electron density	Almlöf, Kvick, and Thomas, 1973 [44b]
$CH_2OHCH_2CH_2OH$	LCAO SCF Slater 3-G basis and CNDO	3 Conformations	Johansson, Kollman, and Rothenberg, 1973 [1182d]
$CH_3COCOOH$	INDO		Gordon and Tallman, 1972 [889b]
Acetylacetone	INDO	Geometry search	Gordon and Koob, 1973 [889a]
Acetylacetone	MINDO/1	Keto-enol equilibrium	Dewar and Shanshal, 1971 [617]

TABLE 2-4, continued

1. MONOMERS

System	Method	Comments	Authors (reference in brackets)
Acetylacetone	Extended Hückel and CNDO/2		Murthy, Bhat, and Rao, 1970 [1746]
Acetylacetone	Extended Hückel		Morokuma, Kato, Yonezawa, and Fukui, 1965 [1712]
Benzyl alcohol	Extended Hückel	Hydrogen bond to π system	Morokuma, Kato, Yonezawa, and Fukui, 1965 [1712]
3-Methylpentane-2,4-dione	MINDO/1	Keto-enol equilibrium	Dewar and Shanshal, 1971 [617]
Maleate ion	CNDO/2		Morita and Nagakura, 1973 [1705a]
Maleate ion	Extended Hückel and CNDO/2		Murthy, Bhat, and Rao, 1970 [1746]
Methyl acetylacetate	MINDO/1	Keto-enol equilibrium	Dewar and Shanshal, 1971 [617]
Salicylaldehyde	Extended Hückel and CNDO/2		Murthy, Bhat, and Rao, 1970 [1746]
Salicylaldehyde	Hückel	Effect of hydrogen bond on C=O vibration	Grinter, 1965 [938]

TABLE 2-4, continued

1. MONOMERS

System	Method	Comments	Authors (reference in brackets)
Salicyaldehyde	Extended Hückel		Morokuma, Kato, Yonezawa, and Fukui, 1965 [1712]
3-Ethylpentane-2,4-dione	MINDO/1	Keto-enol equilibrium	Dewar and Shanshal, 1971 [617]
Methyl α-acetylpropionate	MINDO/1	Keto-enol equilibrium	Dewar and Shanshal, 1971 [617]
Ethyl acetylacetate	MINDO/1	Keto-enol equilibrium	Dewar and Shanshal, 1971 [617]
KH_2PO_4	Semiempirical VB	Delocalization of hydrogen bond	Hidaka, 1972 [1052a]
o-Nitrophenol	Extended Hückel and CNDO/2		Murthy, Bhat, and Rao, 1970 [1746]
Methyl α-acetylbutyrate	MINDO/1	Keto-enol equilibrium	Dewar and Shanshal, 1971 [617]
Ethyl α-acetylpropionate	MINDO/1	Keto-enol equilibrium	Dewar and Shanshal, 1971 [617]
Trifluoroacetylacetone	INDO	Geometry search	Gordon and Koob, 1973 [889a]

TABLE 2-4, continued

1. MONOMERS

System	Method	Comments	Authors (reference in brackets)
Ethyl α-acetylbutyrate	MINDO/1	Keto-enol equilibrium	Dewar and Shanshal, 1971 [617]
1-Hydroxy-2-naphthaldehyde	Hückel	Effect of hydrogen bond on C=O vibration	Grinter, 1965 [938]
2-Hydroxy1-1-naphthaldehyde	Hückel	Effect of hydrogen bond on C=O vibration	Grinter, 1965 [938]
3-Hydroxy1-2-naphthaldehyde	Hückel	Effect of hydrogen bond on C=O vibration	Grinter, 1965 [938]
Ethyl α-acetylvalerate	MINDO/1	Keto-enol equilibrium	Dewar and Shanshal, 1971 [617]
2-Hydroxy1-1,4-naphthoquinone	Hückel	Effect of hydrogen bond on C=O vibration	Grinter, 1965 [938]
Ethyl α-acetylhexanoate	MINDO/1	Keto-enol equilibrium	Dewar and Shanshal, 1971 [617]
2,3-Dihydroxy-1,4-naphthoquinone	Hückel	Effect of hydrogen bond on C=O vibration	Grinter, 1965 [938]
5,8-Dihydroxy-1,4-naphthoquinone	Hückel	Effect of hydrogen bond on C=O vibration	Grinter, 1965 [938]

TABLE 2-4, continued

1. MONOMERS

System	Method	Comments	Authors (reference in brackets)
1-Hydroxybenzophenone	Hückel	Effect of hydrogen bond on C=O vibration	Grinter, 1965 [938]
10-Hydroxy-9-phenanthrenealdehyde	Hückel	Effect of hydrogen bond on C=O vibration	Grinter, 1965 [938]
1-Hydroxy-9,10-anthroquinone	Hückel	Effect of hydrogen bond on C=O vibration	Grinter, 1965 [938]
1-Hydroxy-9,10-phenanthraquinone	Hückel	Effect of hydrogen bond on C=O vibration	Grinter, 1965 [938]
1,2-Dihydroxy-9,10-anthroquinone	Hückel	Effect of hydrogen bond on C=O vibration	Grinter, 1965 [938]
1,3-Dihydroxy-9,10-anthroquinone	Hückel	Effect of hydrogen bond on C=O vibration	Grinter, 1965 [938]
1,4-Dihydroxy-9,10-anthroquinone	Hückel	Effect of hydrogen bond on C=O vibration	Grinter, 1965 [938]
1,5-Dihydroxy-9,10-anthroquinone	Hückel	Effect of hydrogen bond on C=O vibration	Grinter, 1965 [938]
1,2,4-Trihydroxy-9,10-anthroquinone	Hückel	Effect of hydrogen bond on C=O vibration	Grinter, 1965 [938]

TABLE 2-4, continued

2. DIMERS

System	Method	Comments	Authors (reference in brackets)
LiH...Li+	Double perturbation expansion		Shaw, 1969 [2243]
NH3...NH3	LCAO SCF Gaussian basis	Linear, cyclic, and bifurcated	Kollman and Allen, 1971 [1293]
NH3...NH3	CNDO	Linear, cyclic, and bifurcated	Kollman and Allen, 1970 [1294]
NH3...NH3	CNDO/2		Schuster, 1970 [2208]
NH3...NH3	CNDO/2	Two geometries	Hoyland and Kier, 1969 [1093]
NH3...H2O	LCAO SCF Gaussian basis		Diercksen, Kraemer, and von Niessen, 1972 [624a]
NH3...H2O	LCAO SCF Slater basis expanded in Gaussians		Johansson and Kollman, 1972 [1182b]
NH3...H2O	LCAO SCF Gaussian basis	Linear, cyclic, and bifurcated	Kollman and Allen, 1971 [1293]
NH3...H2O	CNDO		Kollman and Allen, 1970 [1294]

TABLE 2-4, continued

2. DIMERS

System	Method	Comments	Authors (reference in brackets)
$NH_3 \cdots H_2O$	CNDO/2		Schuster, 1970 [2208]
$NH_3 \cdots HF$	LCAO SCF Gaussian basis	Bifurcated, cyclic and linear	Kollman and Allen, 1971 [1293]
$NH_3 \cdots HF$	CNDO		Kollman and Allen, 1970 [1294]
$NH_3 \cdots CH_2O$	CNDO/2		Schuster, 1970 [2208]
$NH_3 \cdots (CH_3)_2NO$	INDO		Morishima, Endo, and Yonezawa, 1973 [1703a]
$NH_4^+ \cdots NH_3$	LCAO SCF Gaussian basis		Merlet, Peyerimhoff, and Buenker, 1972 [1638a]
$H_2O \cdots OH^-$	LCAO SCF Gaussian basis		Kraemer and Diercksen, 1972 [1327]
$H_2O \cdots OH^-$	CNDO/2		De Paz, Ehrenson, and Friedman, 1970 [611]
$H_2O \cdots OH^-$	CNDO/2		Schuster, 1970 [2208]

TABLE 2-4, continued

2. DIMERS

System	Method	Comments	Authors (reference in brackets)
H_2O...NH_3	LCAO SCF STO 3-G basis		Del Bene, 1973 [599a]
H_2O...NH_3	LCAO SCF minimal Slater basis	Correlation energy	Meunier, Lévy, and Berthier, 1973 [1638b]
H_2O...NH_3	LCAO SCF Gaussian basis		Diercksen, Kraemer, and von Niessen, 1973 [624a]
H_2O...NH_3	LCAO SCF Slater basis		Piela, 1972 [1929b]
H_2O...NH_3	LCAO SCF Gaussian basis	Bifurcated, cyclic, and linear	Kollman and Allen, 1971 [1293]
H_2O...NH_3	CNDO		Kollman and Allen, 1970 [1294]
H_2O...NH_3	CNDO/2		Schuster, 1970 [2208]
H_2O...H_2O	LCAO SCF minimal Slater basis	Bond orbital analysis	Bowers and Pitzer, 1973 [294a]
H_2O...H_2O	INDO		Gordon and Tallman, 1972 [889a]

TABLE 2-4, continued

2. DIMERS

System	Method	Comments	Authors (reference in brackets)
$H_2O\ldots H_2O$	LCAO SCF minimal STO 3-G basis		Johansson, Kollman, and Rothenberg, 1973 [1182e]
$H_2O\ldots H_2O$	LCAO SCF extended Gaussian basis	Lowest energy to date	Popkie, Kistenmacher, and Clementi, 1973 [1969a]
$H_2O\ldots H_2O$	CNDO/2	Perturbation treatment	Sustmann and Varenholt, 1973 [2380d]
$H_2O\ldots H_2O$	CNDO/2		Morita and Nagakura, 1973 [1705a]
$H_2O\ldots H_2O$	CNDO	Study of charge transfer	Koller, Kaiser, and Ažman, 1972 [1290a]
$H_2O\ldots H_2O$		Localized orbitals	Koller and Ažman, 1972 [1289a]
$H_2O\ldots H_2O$	LCAO SCF Slater basis expanded in Gaussians		Del Bene, 1972 [600]
$H_2O\ldots H_2O$	LCAO SCF Gaussian basis	Calculation of NMR proton shift	Jaszuński and Sadlej, 1972, 1973 [1172a,b]

TABLE 2-4, continued

2. DIMERS

System	Method	Comments	Authors (reference in brackets)
$H_2O...H_2O$	LCAO SCF Slater basis expanded in Gaussians		Johansson and Kollman, 1972 [1182b]
$H_2O...H_2O$	Electrostatic model		Bonaccorsi, Petrongolo, Scrocco, and Tomasi, 1971 [276]
$H_2O...H_2O$	CNDO/2, INDO, MINDO/1		Chojnacki, 1971 [444]
$H_2O...H_2O$	LCAO SCF with Slater orbitals expanded in Gaussians		Del Bene, 1971 [601]
$H_2O...H_2O$	LCAO SCF Gaussian basis		Diercksen, 1971 [621]
$H_2O...H_2O$	CNDO/2		Goel, Murthy, and Rao, 1971 [879]
$H_2O...H_2O$	Iterative Extended Hückel (Modified)		Minton, 1971 [1672]
$H_2O...H_2O$	Empirical	Comparison of empirical potential functions and M.O. results	Moore and O'Connell, 1971 [1684]
$H_2O...H_2O$	CNDO/2		Rao, Goel, and Murthy, 1971 [2047]

TABLE 2-4, continued

2. DIMERS

System	Method	Comments	Authors (reference in brackets)
$H_2O...H_2O$	Extended Hückel		Zhogolev and Matyash, 1971 [2687]
$H_2O...H_2O$	LCAO SCF minimal Slater basis expanded in Gaussians	Extensive geometry search	Del Bene and Pople, 1970 [603]
$H_2O...H_2O$	LCAO SCF minimal Slater basis expanded in Gaussians	Formation of H_3O^+ + OH^-	Fang and de la Vega, 1970 [738]
$H_2O...H_2O$	LCAO SCF Gaussian basis		Hankins, Moskowitz, and Stillinger, 1970 [1001]
$H_2O...H_2O$	CNDO/2		Hasegawa, 1970 [1020]
$H_2O...H_2O$	NDDO		Kollman and Allen, 1970 [1294]
$H_2O...H_2O$	CNDO		Kollman and Allen, 1970 [1294]
$H_2O...H_2O$	LCAO SCF Gaussian basis	Study of energy partitioning	Kollman and Allen, 1970 [1297]
$H_2O...H_2O$	CNDO	Study of energy partitioning	Kollman and Allen, 1970 [1297]

TABLE 2-4, continued

2. DIMERS

System	Method	Comments	Authors (reference in brackets)
$H_2O...H_2O$	LCAO SCF minimal Slater basis		Morokuma and Winick, 1970 [1714]
$H_2O...H_2O$	CNDO/2		Schuster, 1970 [2208]
$H_2O...H_2O$	Approximate perturbation calculation	Study of energy partitioning	van Duijneveldt-van de Rijdt and van Duijneveldt, 1970 [2507]
$H_2O...H_2O$	CNDO/2	Linear, bifurcated, and cyclic	Hoyland and Kier, 1969 [1093]
$H_2O...H_2O$	LCAO SCF Gaussian basis	Linear geometry best of several studied	Kollman and Allen, 1969 [1298]
$H_2O...H_2O$	LCAO SCF Gaussian basis	Linear geometry best	Morokuma and Pedersen, 1968 [1713]
$H_2O...H_2O$	CNDO/2	Potential for proton motion	Murthy and Rao, 1968 [1750]
$H_2O...H_2O$	Extended Hückel	Potential for proton motion	Murthy and Rao, 1968 [1750]
$H_2O...H_2O$	Iterative Extended Hückel	Predicted to be unstable	Rein, Clarke, and Harris, 1968 [2080]

TABLE 2-4, continued

2. DIMERS

System	Method	Comments	Authors (reference in brackets)
$H_2O...H_2O$	Approximate M.O.	Treated as four-electron problem	Bowen and Linnett, 1966 [294]
$H_2O...F^-$	LCAO SCF Gaussian basis	Geometry search	Kistenmacher, Popkie, and Clementi, 1973 [1276a]
$H_2O...F^-$	LCAO SCF Gaussian basis		Diercksen and Kraemer, 1970 [623]
$H_2O...HF$	CNDO		Kollman and Allen, 1970 [1294]
$H_2O...HF$	LCAO SCF Gaussian basis	Linear, bifurcated, and cyclic	Kollman and Allen, 1970 [1296]
$H_2O...CH_2O$	LCAO SCF minimal STO 3-G basis		Del Bene, 1973 [599b,c]
$H_2O...CH_2O$	LCAO SCF minimal Slater basis	Energy partitioning in excited states	Iwata and Morokuma, 1973 [1153a,b]
$H_2O...CH_2O$	LCAO SCF Slater basis expanded in Gaussians		Johansson and Kollman, 1972 [1182b]
$H_2O...CH_2O$	LCAO SCF minimal Slater basis		Morokuma, 1971 [1710]

TABLE 2-4, continued

2. DIMERS

System	Method	Comments	Authors (reference in brackets)
$H_2O \cdots CH_2O$	Extended Hückel		Rao, Goel, Rao, and Murthy, 1971 [2048]
$H_2O \cdots CH_2O$	CNDO/2		Rao, Goel, Rao, and Murthy, 1971 [2048]
$H_2O \cdots CH_2O$	CNDO/2	Effect on electronic levels	Rao and Murthy, 1971 [2049]
$H_2O \cdots CH_2O$	CNDO/2		Schuster, 1970 [2208]
$H_2O \cdots CH_2O$	CNDO/2	Several geometries considered	Schuster, 1969 [2209]
$H_2O \cdots CH_3OH$	LCAO SCF Slater orbitals expanded in Gaussians		Del Bene, 1971 [601]
$H_2O \cdots NH_2OH$	LCAO SCF Slater basis expanded in Gaussians		Del Bene, 1972 [600a]
$H_2O \cdots H_2O_2$	LCAO SCF Slater orbitals expanded in Gaussians		Del Bene, 1972 [600]
$H_2O \cdots Cl^-$	LCAO SCF Slater basis		Piela, 1973 [1929a]
$H_2O \cdots Cl^-$	LCAO SCF Gaussian basis	Geometry search	Kistenmacher, Popkie, and Clementi, 1973 [1276a]

TABLE 2-4, continued

2. DIMERS

System	Method	Comments	Authors (reference in brackets)
$H_2O...HOF$	LCAO SCF Slater basis expanded in Gaussians		Del Bene, 1972 [600a]
$H_2O...cyclopropene$	Electrostatic model		Alagona, Cimiraglia, Scrocco, and Tomasi, 1972 [27b]
$H_2O...ethylene imine$	Electrostatic model		Alagona, Cimiraglia, Scrocco, and Tomasi, 1972 [27b]
$H_2O...ethylene oxide$	Electrostatic model		Alagona, Cimiraglia, Scrocco, and Tomasi, 1972 [27b]
$H_2O...CH_2ONH$ (oxaziridine)	Electrostatic model		Alagona, Cimiraglia, Scrocco, and Tomasi, 1972 [27b]
$H_2O...NH_2CHO$	LCAO SCF Slater basis expanded in Gaussians		Johansson and Kollman, 1972 [1182b]
$H_2O...(CH_3)_2CO$	CNDO/2	Effect on electronic levels	Rao and Murthy, 1971 [2049]
$H_2O...Cl_2$	CNDO/2	Hydrogen-bonded complex not most stable	Fredin and Nelander, 1973 [799b]

TABLE 2-4, continued

2. DIMERS

System	Method	Comments	Authors (reference in brackets)
$H_2O...pyridine$	Extended Hückel		Adam, Grimison, Hoffmann, and Ortiz, 1968 [7]
$H_3O^+...H_2O$		Potential function fit to *ab initio* results	Anderson and Jiang, 1973 [54a]
$H_3O^+...H_2O$	LCAO SCF, and INDO	Correlation study	Meyer, Jakubetz, and Schuster, 1973 [1642a]
$H_3O^+...H_2O$	LCAO SCF double zeta Slater basis		Alagona, Cimiraglia, and Lamanna, 1973 [27a]
$H_3O^+...H_2O$	CNDO	Proton tunneling	Flanigan and de la Vega, 1973 [768a]
$H_3O^+...H_2O$	LCAO SCF	vibrations	Janoschek, Weidemann, and Zundel, 1973 [1164a]
$H_3O^+...H_2O$	LCAO SCF Gaussian basis		Janoschek, Weidemann, Pfeiffer, and Zundel, 1972 [1164]
$H_3O^+...H_2O$	CNDO/2		De Paz, Ehrenson, and Friedman, 1970 [611]
$H_3O^+...H_2O$	LCAO SCF Gaussian basis		Kollman and Allen, 1970 [1295]

TABLE 2-4, continued

2. DIMERS

System	Method	Comments	Authors (reference in brackets)
$H_3O^+...H_2O$	LCAO SCF Gaussian basis		Kraemer and Diercksen, 1970 [1329]
$H_3O^+...H_2O$	CNDO/2		Schuster, 1970 [2208]
HF...Li	LCAO SCF Gaussian basis	Potential curves	Lester and Krauss, 1970 [1446]
HF...NH$_3$	LCAO SCF minimal STO 3-G basis	Solvent effect on formation of NH$_4$F	Yambe, Kato, Fujimoto, and Fukui, 1973 [2639a]
HF...NH$_3$	LCAO SCF Gaussian basis		Kollman and Allen, 1971 [1293]
HF...NH$_3$	CNDO		Kollman and Allen, 1970 [1294]
HF...NH$_3$	CNDO/2		Schuster, 1970 [2208]
HF...H$_2$O	CNDO		Kollman and Allen, 1970 [1294]
HF...H$_2$O	LCAO SCF Gaussian basis	Linear and cyclic	Kollman and Allen, 1970 [1296]
HF...H$_2$O	CNDO/2		Schuster, 1970 [2208]

TABLE 2-4, continued

2. DIMERS

System	Method	Comments	Authors (reference in brackets)
HF...F⁻	LCAO SCF Gaussian basis		Noble and Kortzeborn, 1970 [1805]
HF...F⁻		Fit of potential function to *ab initio* results	Jiang and Anderson, 1973 [1179a]
HF...F⁻		Localized orbitals	Koller and Ažman, 1972 [1289a]
HF...F⁻	LCAO SCF Gaussian basis	Potential surfaces and vibrational levels	Almlöf, 1972 [44a]
HF...F⁻	LCAO SCF Gaussian basis		Kollman and Allen, 1970 [1295]
HF...F⁻	Extended Hückel and CNDO/2		Murthy, Bhat, and Rao, 1970 [1746]
HF...F⁻	LCAO SCF Gaussian basis		Noble and Kortzeborn, 1970 [1805]
HF...F⁻	CNDO/2		Schuster, 1970 [2208]
HF...F⁻	CNDO/2 and INDO		Koller, Borstnik, and Ažman, 1969 [1290]
HF...F⁻	LCAO SCF		McLean and Yoshimine, 1967 [1568a, 2661a]

TABLE 2-4, continued

2.   DIMERS

System	Method	Comments	Authors (reference in brackets)
HF...F⁻	Approximate M.O.	Treated as four-electron problem	Bowen and Linnett, 1966 [294]
HF...F⁻	LCAO SCF Slater basis		Clementi and McLean, 1962 [474]
HF...F⁻	Approximate VB		Bessis, 1961 [229]
HF...F⁻	LCAO SCF Slater basis		Clementi, 1961 [474a]
HF...F⁻	Approximate VB		Bessis and Bratož, 1960 [230]
HF...HF	LCAO SCF minimal STO 3-G basis		Johansson, Kollman, and Rothenberg, 1973 [1182e]
HF...HF	Perturbation CNDO		Bacon and Santry, 1971 [104]
HF...HF	SCF + CI Gaussian basis		Allen and Kollman, 1970 [36]
HF...HF	LCAO SCF Gaussian basis		Diercksen and Kraemer, 1970 [624]
HF...HF	LCAO SCF Gaussian basis	Linear and cyclic	Kollman and Allen, 1970 [1296]

TABLE 2-4, continued

2. DIMERS

System	Method	Comments	Authors (reference in brackets)
HF...HF	CNDO	Linear and cyclic	Kollman and Allen, 1970 [1294]
HF...HF	LCAO SCF Gaussian basis	Study of energy partitioning	Kollman and Allen, 1970 [1297]
HF...HF	CNDO	Study of energy partitioning	Kollman and Allen, 1970 [1297]
HF...HF	CNDO/2		Schuster, 1970 [2208]
HF...HF	CNDO/2		Hoyland and Kier, 1969 [1093]
HF...HF	LCAO SCF Slater basis with integral approximations	Calculation of electrostatic contribution	Howard, 1963 [1088]
HF...HCN	LCAO SCF minimal STO 3-G basis	Comparison of bond to lone pair and π system	Del Bene and Marchese, 1973 [601a]
HF...HCN	LCAO SCF Gaussian basis		Johansson, Kollman, and Rothenberg, 1972 [1182c]
HF...CH$_2$O	CNDO/2		Schuster, 1970 [2208]
HF...CH$_2$O	INDO		Schuster, 1970 [2208]

TABLE 2-4, continued

2. DIMERS

System	Method	Comments	Authors (reference in brackets)
HF...N$_2$H$_2$	CNDO/2	Hydrogen bond to π system	Sabin, 1972 [2138a]
HF...benzene	CNDO/2		Jakubetz and Schuster, 1971 [1162]
HF...pyridine	CNDO/2		Jakubetz and Schuster, 1971 [1162]
HF...pyridine	CNDO/2		Schuster, 1970 [2208]
DF...F$^-$	LCAO SCF Gaussian basis	Potential surfaces and vibrational levels	Almlöf, 1972 [44a]
H$_2$F$^+$...HF	LCAO SCF Gaussian basis		Diercksen, von Niessen, and Kraemer, 1973 [624b]
CH≡CH...NH$_3$	CNDO/2		Goel and Rao, 1971 [881]
CH≡CH...CH≡CH	CNDO/2		Goel and Rao, 1971 [881]
CH≡CH...CH$_2$O	CNDO/2		Goel and Rao, 1971 [881]
CH≡CH...CH$_3$CN	CNDO/2		Goel and Rao, 1971 [881]
CH≡CH...(CH$_3$)$_2$NO	INDO		Morishima, Endo, and Yonezawa, 1973 [1703a]
CH≡CH...pyridine	CNDO/2		Goel and Rao, 1971 [881]

TABLE 2-4, continued

2. DIMERS

System	Method	Comments	Authors (reference in brackets)
HCN...NH$_3$	CNDO/2		Goel and Rao, 1971 [881]
HCN...HF	LCAO SCF Gaussian basis		Johansson, Kollman, and Rothenberg, 1972 [1182c]
HCN...HCN	LCAO SCF minimal STO 3-G basis		Johansson, Kollman, and Rothenberg, 1972 [1182e]
HCN...HCN	LCAO SCF Slater basis expanded in Gaussians	Linear and cyclic	Johansson, Kollman, and Rothenberg, 1972 [1182a]
HCN...HCN	Perturbation CNDO		Bacon and Santry, 1971 [104]
HCN...HCN	CNDO/2	Considers both proton donation to N and to triple bond	Goel and Rao, 1971 [881]
HCN...HCN	CNDO/2		Hoyland and Kier, 1969 [1093]
CH$_3$OH...NH$_3$	LCAO SCF STO 3-G basis		Del Bene, 1973 [599a]
CH$_3$OH...H$_2$O	LCAO SCF Slater orbitals expanded in Gaussians		Del Bene, 1971 [601]
CH$_3$OH...CH$_2$O	LCAO SCF minimal STO 3-G basis		Del Bene, 1973 [599b,c]

TABLE 2-4, continued

2. DIMERS

System	Method	Comments	Authors (reference in brackets)
$CH_3OH...CH_2O$	CNDO/2	Effect on electronic levels	Rao and Murthy, 1971 [2049]
$CH_3OH...CH_3OH$	CNDO/2		Morita and Nagakura, 1973 [1705a]
$CH_3OH...CH_3OH$	LCAO SCF Slater orbitals expanded in Gaussians		Del Bene, 1971 [601]
$CH_3OH...CH_3OH$	Extended Hückel and CNDO/2	Linear	Murthy, Bhat, and Rao, 1970 [1746]
$CH_3OH...CH_3OH$	CNDO/2	Linear and cyclic	Hoyland and Kier, 1969 [1093]
$CH_3OH...CH_3OH$	Extended Hückel and CNDO/2	Linear and cyclic	Murthy, Davis, and Rao, 1969 [1747]
$CH_3OH...CH_3CN$	Extended Hückel and CNDO/2		Murthy, Bhat, and Rao, 1970 [1746]
$CH_3OH...(CH_3)_2CO$	CNDO/2	Effect on electronic levels	Rao and Murthy, 1971 [2049]
$CH_3OH...(CH_3)_2CO$	Extended Hückel and CNDO/2		Murthy, Bhat, and Rao, 1970 [1746]

TABLE 2-4, continued

2. DIMERS

System	Method	Comments	Authors (reference in brackets)
CH$_3$OH...(CH$_3$)$_2$NO	INDO	Geometry search	Morishima, Endo, and Yonezawa, 1973 [1703a]
CH$_3$OH...(CH$_3$)$_2$NO	INDO	Spin densities and hyperfine coupling constants	Morishima, Endo, and Yonezawa, 1971 [1702]
CH$_3$OH...benzene	Extended Hückel		Murthy, Bhat, and Rao, 1970 [1746]
CH$_3$OH...pyridine	Extended Hückel and CNDO/2		Murthy, Bhat, and Rao, 1970 [1746]
CH$_3$OH...pyridine	Extended Hückel		Adam, Grimison, Hoffmann, and Ortiz, 1968 [7]
NH$_2$OH...NH$_3$	LCAO SCF STO 3-G basis		Del Bene, 1973 [599a]
NH$_2$OH...H$_2$O	LCAO SCF Slater basis expanded in Gaussians		Del Bene, 1972 [600a]
NH$_2$OH...CH$_2$O	LCAO SCF minimal STO 3-G basis		Del Bene, 1973 [599b,c]
NH$_2$OH...NH$_2$OH	LCAO SCF Slater basis expanded in Gaussians		Del Bene, 1972 [600a]
NH$_2$OH...NH$_2$OH	Point-charge model		Salaj, 1970 [2162]

TABLE 2-4, continued

2. DIMERS

System	Method	Comments	Authors (reference in brackets)
$H_2O_2...NH_3$	LCAO SCF STO 3-G basis		Del Bene, 1973 [599a]
$H_2O_2...H_2O$	LCAO SCF Slater orbitals expanded in Gaussians		Del Bene, 1972 [600]
$H_2O_2...CH_2O$	LCAO SCF minimal Slater basis	Ground and excited states	Iwata and Morokuma, 1973 [1153b]
$H_2O_2...CH_2O$	LCAO SCF minimal STO 3-G basis		Del Bene, 1973 [599b,c]
$H_2O_2...H_2O_2$	LCAO SCF Slater orbitals expanded in Gaussians	Cyclic structure best	Del Bene, 1972 [600]
$H_2S...SH^-$	LCAO SCF Gaussian basis	Symmetric linear form best	Sabin, 1971 [2139]
$H_2S...H_2S$	CNDO	Results do not agree well with ab $initio$	Sabin, 1971 [2140]
$H_2S...H_2S$	LCAO SCF Gaussian basis	Linear form best	Sabin, 1971 [2140]
$HOF...NH_3$	LCAO SCF STO 3-G basis		Del Bene,1973, [599a]
$HOF...H_2O$	LCAO SCF Slater basis expanded in Gaussians		Del Bene, 1972 [600a]
$HOF...CH_2O$	LCAO SCF minimal STO 3-G basis		Del Bene, 1973 [599b,c]

TABLE 2-4, continued

2. DIMERS

System	Method	Comments	Authors (reference in brackets)
HOF...HOF	LCAO SCF Slater basis expanded in Gaussians		Del Bene, 1972 [600a]
HCl...NH$_3$	LCAO SCF Gaussian basis		Clementi, 1967 [471]
HCl...NH$_3$	LCAO SCF Gaussian basis	The question of inner and outer complexes	Clementi and Gayles, 1967 [473]
HCl...Cl$^-$	LCAO SCF Gaussian basis	Infrared and Raman	Janoschek, 1973 [1163a]
Ethylene imine...H$_2$O	Electrostatic model		Alagona, Cimiraglia, Scrocco, and Tomasi, 1972 [27b]
Oxaziridine...H$_2$O	Electrostatic model		Alagona, Cimiraglia, Scrocco, and Tomasi, 1972 [27b]
NH$_2$CHO...H$_2$O	LCAO SCF Slater basis expanded in Gaussians		Johansson and Kollman, 1972 [1182b]
NH$_2$CHO...NH$_2$CHO	LCAO SCF Gaussian basis	Model for peptide conformations	Berthod and Pullman, 1972 [216a]
NH$_2$CHO...NH$_2$CHO	LCAO SCF Slater basis expanded in Gaussians	Linear	Johansson and Kollman, 1972 [1182b]

TABLE 2-4, continued

2. DIMERS

System	Method	Comments	Authors (reference in brackets)
$NH_2CHO...NH_2CHO$	Iterative Extended Hückel		Almlöf and Martensson, 1971 [48]
$NH_2CHO...NH_2CHO$	CNDO/2	Linear	Rao, Goel, Rao, and Murthy, 1971 [2048]
$NH_2CHO...NH_2CHO$	LCAO SCF Gaussian basis	Cyclic	Dreyfus, Maigret, and Pullman, 1970 [658]
$NH_2CHO...NH_2CHO$	LCAO SCF Gaussian basis	Partitioning of energy	Dreyfus and Pullman, 1970 [659]
$NH_2CHO...NH_2CHO$	LCAO SCF Gaussian basis		Dreyfus and Pullman, 1970 [660]
$NH_2CHO...NH_2CHO$	CNDO/2		Momany, McGuire, Yan, and Scheraga, 1970 [1682]
$NH_2CHO...NH_2CHO$	Extended Hückel and CNDO	Cyclic	Murthy, Bhat, and Rao, 1970 [1746]
$NH_2CHO...NH_2CHO$	Extended Hückel and CNDO/2	Linear and cyclic	Murthy, Rao, and Rao, 1970 [1753]
$NH_2CHO...NH_2CHO$	CNDO/2		Pullman and Berthod, 1968 [2004]

TABLE 2-4, continued

2. DIMERS

System	Method	Comments	Authors (reference in brackets)
$(CH_3)_2NH\cdots(CH_3)_2NO$	INDO		Moroshima, Endo, and Yonezawa, 1973 [1703a]
$HCOOH\cdots Cl^-$	CNDO/2		Rode, 1973 [2105a,b]
$HCOOH\cdots HCOOH$	CNDO/2		Morita and Nagakura, 1973 [1705a]
$HCOOH\cdots HCOOH$		Localized orbitals	Koller and Ažman, 1972 [1289a]
$HCOOH\cdots HCOOH$	Iterative Extended Hückel		Almlöf and Mårtensson, 1971 [47]
$HCOOH\cdots HCOOH$	LCAO SCF Gaussian basis	Potential curves for proton motion	Clementi, Mehl, and von Niessen, 1971 [475]
$HCOOH\cdots HCOOH$	Extended Hückel and CNDO/2	Cyclic	Murthy, Bhat, and Rao, 1970 [1746]
$HCOOH\cdots HCOOH$	CNDO/2		Hoyland and Kier, 1969 [1093]
$HCOOH\cdots HCOOH$	Extended Hückel and CNDO/2	Linear and cyclic	Murthy, Davis, and Rao, 1969 [1747]
$HCOOH\cdots HCOOH$	CNDO/2	Potential for proton motion	Schuster, 1969 [2209]

TABLE 2-4, continued

2. DIMERS

System	Method	Comments	Authors (reference in brackets)
HCOOH...HCOOH	CNDO/2	Several ring and linear structures	Schuster and Funck, 1968 [2211]
HCOOH...HCOOH	Extended Hückel		Morokuma, Kato, Yonezawa, and Fukui, 1965 [1712]
C_2H_5OH...crotonaldehyde	Pariser-Parr-Pople	Effect of hydrogen bond on $\pi \rightarrow \pi^*$ transition	Besnainou, Prat, and Bratož, 1964 [228]
C_2H_5OH...mesityl oxide	Pariser-Parr-Pople	Effect of hydrogen bond on $\pi \rightarrow \pi^*$ transition	Besnainou, Prat, and Bratož, 1964 [228]
C_2H_5OH...p-benzoquinone	Pariser-Parr-Pople	Effect of hydrogen bond on $\pi \rightarrow \pi^*$ transition	Besnainou, Prat, and Bratož, 1964 [228]
C_2H_5OH...acetophenone	Pariser-Parr-Pople	Effect of hydrogen bond on $\pi \rightarrow \pi^*$ transition	Besnainou, Prat, and Bratož, 1964 [228]
C_2H_5OH...p-methylacetophenone	Pariser-Parr-Pople	Effect of hydrogen bond on $\pi \rightarrow \pi^*$ transition	Besnainou, Prat, and Bratož, 1964 [228]
CH_3COOH...CH_3COOH	CNDO/2		Rode, Engelbrecht, and Jakubetz, 1973 [2105c]
CH_3COOH...CH_3COOH	Iterative Extended Hückel		Almlöf and Mårtensson, 1971 [47]

TABLE 2-4, continued

2. DIMERS

System	Method	Comments	Authors (reference in brackets)
$CH_3COOH...CH_3COOH$	CNDO/2		Hoyland and Kier, 1969 [1093]
$CH_3COOH...CF_3COOH$	CNDO/2		Rode, Engelbrecht, and Jakubetz, 1973 [2105c]
$CH_3COOH...CCl_3COOH$	CNDO/2		Rode, Engelbrecht, and Jakubetz, 1973 [2105c]
Pyrrole...pyridine	Iterative Extended Hückel		Almlöf and Mårtensson, 1972 [46]
Pyrrole...pyridine	Extended Hückel and CNDO/2		Murthy, Bhat, and Rao, 1970 [1746]
Pyrrole...pyridine	Pariser-Parr-Pople		Sabin, 1968 [2141]
Imidazole...imidazole	Pariser-Parr-Pople	Effect of hydrogen bond on $\pi \rightarrow \pi^*$ transition	Chojnacki, 1968 [446]
$F_3CH...CH_2O$	CNDO/2		Goel and Rao, 1971 [881]
$F_3CH...F_3CH$	CNDO/2	Considers both proton donation to F and to C	Goel and Rao, 1971 [881]
N-methylacetamide... N-methylacetamide	CNDO/2		Momany, McGuire, Yan, and Scheraga, 1970 [1682]

TABLE 2-4, continued

2. DIMERS

System	Method	Comments	Authors (reference in brackets)
N-methylacetamide... N-methylacetamide	Extended Hückel and CNDO/2		Murthy, Bhat, and Rao, 1970 [1746]
N-methylacetamide... N-methylacetamide	Extended Hückel and CNDO/2		Murthy, Rao, and Rao, 1970 [1753]
[Pyridine H]$^+$...pyridine	Pariser-Parr-Pople		Sabin, 1968 [2142]
Phenol...dioxane	Pariser-Parr-Pople	Effect of hydrogen bond on $\pi \to \pi^*$ transition	Besnainou, Prat, and Bratož, 1964 [228]
Phenol...NR$_3$	Modified Hückel	Effect of hydrogen bond on electronic levels of phenol	Julg and Bonnet, 1962 [1213]
2-Pyridone...glucose	CNDO/2		Gold, 1971 [882]
CF$_3$COOH...CF$_3$COO$^-$	CNDO/2		Ocvirk, Ažman, and Hadži, 1968 [1819]
CF$_3$COOH...CF$_3$COOH	CNDO/2		Rode, Engelbrecht, and Jakubetz, 1973 [2105c]
7-Azaindole...7-azaindole	Pariser-Parr-Pople	Potential curves for proton motion	Pechenaya and Danilov, 1971 [1890]
CHCl$_3$...(CH$_3$)$_2$NO	INDO		Morishima, Endo, and Yonezawa, 1973 [1703a]

TABLE 2-4, continued

2. DIMERS

System	Method	Comments	Authors (reference in brackets)
α-Naphthol...dioxane	Pariser-Parr-Pople	Effect of hydrogen bond on π→π* transition	Besnainou, Prat, and Bratož, 1964 [228]
β-Naphthol...dioxane	Pariser-Parr-Pople	Effect of hydrogen bond on π→π* transition	Besnainou, Prat, and Bratož, 1964 [228]
Guanine...cytosine	LCAO SCF Gaussian basis	Study of proton motion	Clementi, Mehl, and von Niessen, 1971 [475]
Guanine...cytosine	Approximate VB		Hasegawa, Daiyasu, and Yomosa, 1970 [1021]
Guanine...cytosine	CNDO/2		Blizzard and Santry, 1969 [265]
Guanine...cytosine	Pariser-Parr-Pople	Four potential curves	Rai and Ladik, 1968 [2038]
Guanine...cytosine	Semiempirical perturbation method		Rein and Pollak, 1967 [2089]
Guanine...cytosine	Pariser-Parr-Pople	Proton vibrational levels	Rein and Svetina, 1967 [2091]
Guanine...cytosine	Pariser-Parr-Pople		Rein and Harris, 1966 [2084]; 1965 [2085,2087]; 1964 [2086,2088]

TABLE 2-4, continued

2. DIMERS

System	Method	Comments	Authors (reference in brackets)
Guanine...cytosine	Pariser-Parr-Pople	Mutagenic effect of radiation	Rein and Ladik, 1964 [2088a]
DNA base pairs	Calculation of vibrational levels	Proton tunneling	Tolpygo and Ol'khovskaya, 1972 [2457a]
DNA base pairs	Pariser-Parr-Pople		Ladik and Sundaram, 1969 [1400]
DNA base pairs	Empirical potential function	Calculation of equilibrium conformation	Caillet, 1968 [386]
DNA base pairs	Pariser-Parr-Pople	Potentials for proton motion	Lunell and Sperber, 1967 [1528]
DNA base pairs	Empirical potential functions	Proton tunneling frequency	Biczó, Ladik, and Gergely, 1966 [240]
DNA base pairs	Electrostatic		Nash and Bradley, 1966 [1775]
DNA base pairs	Empirical		Pollak and Rein, 1966 [1967]
DNA base pairs	Semiempirical		Pullman, Claverie, and Caillet, 1966 [2016]

TABLE 2-4, continued

2. DIMERS

System	Method	Comments	Authors (reference in brackets)
DNA base pairs	Empirical		Pullman, Claverie, and Caillet, 1966 [2017]
DNA base pairs	Hückel (π)-Del Re (σ)	Influence of hydrogen bond on dipole moment	Berthod and Pullman, 1964 [216]
DNA base pairs	Empirical potential function	Proton tunneling frequency	Biczó, Ladik, and Gergely, 1964 [241]
DNA base pairs	Hückel		Pullman, 1963 [2003]
$H_3Co(CN)_6 \cdots H_3Co(CN)_6$	LCAO SCF Gaussian basis	Calculations on a fragment of this system with unusually short hydrogen bonds	Güdel, 1972 [947]

TABLE 2-4, continued

3. HOMOGENEOUS POLYMERS

System	Method	Comments	Authors (reference in brackets)
$(NH_3)_n$	INDO		Goren, 1972 [893a]
$(H_2O)_3$	LCAO SCF Gaussian basis		Lentz and Scheraga, 1973 [1443c]
$(H_2O)_3$	CNDO/2, INDO, MINDO/1		Chojnacki, 1971 [444]
$(H_2O)_3$	Modified iterative Extended Hückel		Minton, 1971 [1672]
$(H_2O)_3$	CNDO/2	On structure of "polywater"	Goel, Murthy, and Rao, 1970 [880]
$(H_2O)_3$	LCAO SCF Gaussian basis	Hydrogen bond stronger than in dimer	Hankins, Moskowitz, and Stillinger, 1970 [1001]
$(H_2O)_3$	CNDO/2	Two linear structures	Hoyland and Kier, 1969 [1093]
$(H_2O)_4$	LCAO SCF Gaussian basis		Lentz and Scheraga, 1973 [1443c]
$(H_2O)_4$	Modified iterative Extended Hückel		Minton, 1971 [1672]
$(H_2O)_4$	CNDO	Cyclic model for "polywater"	Ažman, Koller, and Hadži, 1970 [98]

TABLE 2-4, continued

3. HOMOGENEOUS POLYMERS

System	Method	Comments	Authors (reference in brackets)
$(H_2O)_5$	LCAO SCF Slater basis	Tetrahedral geometry	Guidotti, Lamanna, and Maestro, 1972 [948b]
$(H_2O)_5$	CNDO	Tetrahedral geometry	Kollman and Allen, 1970 [1294]
$(H_2O)_5$	CNDO/2	Tetrahedral geometry	Hoyland and Kier, 1969 [1093]
$(H_2O)_6$	CNDO	Cyclic model for "polywater"	Ažman, Koller, and Hadži, 1970 [98]
$(H_2O)_6$	CNDO/2	On structure of "polywater"	Goel, Murthy, and Rao, 1970 [880]
$(H_2O)_6$	LCAO SCF Gaussian basis	A model for "polywater"	Sabin, Harris, Archibald, Kollman, and Allen, 1970 [2144]
$(H_2O)_8$	Modified iterative Extended Hückel		Minton, 1971 [1672]
$(H_2O)_n$		Analysis of recent *ab initio* results	Del Bene and Pople, 1973 [601b]
$(H_2O)_n$ n = 4...12	CNDO/2	Linear chains	Cignitti and Paoloni, 1971 [457]

TABLE 2-4, continued

3. HOMOGENEOUS POLYMERS

System	Method	Comments	Authors (reference in brackets)
$(H_2O)_n$	CNDO/2		Rao, Goel, and Murthy, 1971 [2046]
$(H_2O)_n$ n = 3,5,6	Extended Hückel		Zhogolev and Matyash, 1971 [2687]
$(H_2O)_n$	CNDO/2	Study of "polywater"	Allen and Kollman, 1970 [35]
$(H_2O)_n$ n = 3...6	LCAO SCF minimal Slater basis expanded in Gaussians	Linear and cyclic polymers	Del Bene and Pople, 1969 [604]; 1970 [603]
$(H_2O)_n$ n = 2...14	INDO		Pedersen, 1969 [1891]
$(H_2O)_n^-$	CNDO/2	Hydrated electron	Weissmann and Cohan, 1973 [2570a]
$(HF)_3$	CNDO	Linear	Kollman and Allen, 1970 [1294]
$(HF)_3$	CNDO/2	Linear	Hoyland and Kier, 1969 [1093]
$(HF)_4$	SCF + CI Gaussian basis		Allen and Kollman, 1970 [36]

TABLE 2-4, continued

3. HOMOGENEOUS POLYMERS

System	Method	Comments	Authors (reference in brackets)
$(HF)_4$	CNDO	Linear and square	Kollman and Allen, 1970 [1294]
$(HF)_4$	CNDO/2	Linear	Hoyland and Kier, 1969 [1093]
$(HF)_5$	CNDO	Linear and pentagonal	Kollman and Allen, 1970 [1294]
$(HF)_6$	CNDO/2	Study of an ionic defect in the chain	Sabin, 1972 [2138]
$(HF)_6$	CNDO	Linear and hexagonal	Kollman and Allen, 1970 [1294]
$(HF)_6$	CNDO/2	Linear and cyclic	Hoyland and Kier, 1969 [1093]
$(HF)_7$	CNDO	Linear	Kollman and Allen, 1970 [1294]
$(HF)_8$	CNDO	Linear and octagonal	Kollman and Allen, 1970 [1294]
$(HF)_n$ $n = 1...16$	LCAO SCF Slater orbitals expanded in Gaussians		Del Bene and Pople, 1971. [602]

TABLE 2-4, continued

3.   HOMOGENEOUS POLYMERS

System	Method	Comments	Authors (reference in brackets)
$(HCN)_3$	LCAO SCF Slater basis expanded in Gaussians	Linear	Johansson, Kollman, and Rothenberg, 1972 [1182a]
$(HCN)_4$	Extended Hückel		Loew and Chang, 1972 [1494]
$(CH_3OH)_3$	CNDO/2		Morita and Nagakura, 1973 [1705a]
$(CH_3OH)_3$	CNDO/2	Linear	Murthy, Davis, and Rao, 1969 [1747]
$(CH_3OH)_4$	CNDO/2	Cyclic	Hoyland and Kier, 1969 [1093]
$(NH_2OH)_3$	Point charge model		Salaj, 1970 [2162]
$(NH_2CHO)_3$	CNDO/2		Momany, McGuire, Yan, and Scheraga, 1970 [1682]
$(NH_2CHO)_n$	CNDO/2, MINDO/2		Suhai and Ladik, 1972 [2371]
$(HCOOH)_3$	Extended Hückel and CNDO/2	Two linear forms	Murthy, Davis, and Rao, 1969 [1747]
$(Imidazole)_3$	Pariser-Parr-Pople	Effect of hydrogen bond on $\pi \rightarrow \pi^*$ transition	Chojnacki, 1968 [446]

TABLE 2-4, continued

3. HOMOGENEOUS POLYMERS

System	Method	Comments	Authors (reference in brackets)
$(Glycine)_n$ $n = 1...5$	LCAO SCF floating spherical Gaussian basis		Shipman and Christoffersen, 1973 [2259a]

4. HETEROGENEOUS POLYMERS

System	Method	Comments	Authors (reference in brackets)
$(H_2O)_2 \cdots OH^-$	CNDO/2		De Paz, Ehrenson, and Friedman, 1970 [611]
$(H_2O)_2 \cdots F^-$	LCAO SCF Gaussian basis		Kraemer and Diercksen, 1973 [1326a]
$H_3O^+ \cdots (H_2O)_2$	CNDO/2		De Paz, Ehrenson, and Friedman, 1970 [611]
$(H_2O)_3 \cdots OH^-$	CNDO/2		De Paz, Ehrenson, and Friedman, 1970 [611]
$H_3O^+ \cdots (H_2O)_3$	CNDO/2	H_3O^+ surrounded by three water molecules	Daly and Burton, 1970 [562]
$H_3O^+ \cdots (H_2O)_3$	CNDO/2		De Paz, Ehrenson, and Friedman, 1970 [611]
$H_3O^+ \cdots (H_2O)_3$	Point charge model	Planar	Salaj, 1969 [2163]

TABLE 2-4, continued

4. HETEROGENEOUS POLYMERS

System	Method	Comments	Authors (reference in brackets)
$H_3O^+\cdots(H_2O)_3$	Approximate LCAO SCF		Grahn, 1962 [902]
$(H_2O)_n\cdots OH^-$ n = 0...4	LCAO SCF Gaussian basis		Newton and Ehrenson, 1971 [1791]
$H_3O^+\cdots(H_2O)_n$ n = 0...4	LCAO SCF Gaussian basis		Newton and Ehrenson, 1971 [1791]
$CH_3NH_3^+\cdots(H_2O)_3$	CNDO/2		Hoyland and Kier, 1969 [1093]
$NH_4^+\cdots(H_2O)_4$	CNDO/2		Hoyland and Kier, 1969 [1093]
(DNA base)$_3$	Empirical		Pullman, Claverie, and Caillet, 1967 [2015]
(DNA base)$_4$	Electrostatic	Study of replication plane	Bass and Schaad, 1971 [149]
Trinucleotide...tri-nucleotide	Electrostatic		Nash and Bradley, 1965 [1776]
Polypeptides	Empirical	Hydrogen bond and van der Waals potentials to predict conformation	Brant, 1968 [308]
Polypeptides	Empirical potential functions	Calculation of best configuration	Ramachandran, Venkatacha-lam, and Krimm, 1966 [2041]

CHAPTER 3

THERMODYNAMICS AND
KINETICS OF
HYDROGEN BONDING

I. THERMODYNAMIC PARAMETERS FOR 1:1 ADDUCTS

A number of spectroscopic methods have been used to obtain thermodynamic parameters for hetero-intermolecular hydrogen bonding. Since
1960, emphasis has been placed on improving the accuracy of the
methods. The order of reliability of methods for measuring hydrogen
bond enthalpies is calorimetric > near-infrared > NMR $\sim$ ir > uv [70,
652,1937]. This should be kept in mind when selecting a method for
obtaining quantitative thermodynamic data for hydrogen-bonded complexes. Key references will be given for each method.

The general procedure followed in obtaining thermodynamic parameters by spectroscopic measurements is to measure equilibrium constants
at several different temperatures and then to calculate ΔH and ΔS from
a plot of ln K versus 1/T. As a result, the description of spectroscopic methods emphasizes determination of equilibrium constants.
Regardless of what spectroscopic method is used, careful attention
must be paid to temperature control and measurement.

A. Equilibrium Constant Determinations

1. *Methods*
The various spectroscopic methods which have been used to obtain
thermodynamic data for hydrogen-bonded systems are based upon an examination of changes in the spectrum of the equilibrium mixture when
compared to the spectrum of the acid and base. The reaction of HA with
B to form B...HA is assumed to occur rapidly to give a 1:1 adduct.

$$B + HA \rightleftarrows B...HA \qquad (3-1)$$

Studies are usually carried out in an inert solvent to minimize self-association and to approximate ideal conditions. The assumption is also made that the activity coefficients for HA, B, and B...HA are unity.

$$K = \frac{[B...HA]}{[HA][B]} \qquad (3-2)$$

Most of the equations discussed here were derived initially for Lewis acid-base reactions with iodine as the reference acid. However, the symbolism will be given for hydrogen bonding reactions.

Benesi and Hildebrand [199] derived an equation which can be used to determine equilibrium constants if the initial concentration of the base, C_B^0, is much larger than the equilibrium concentration of the complex C_C. The equation which they derived is

$$\frac{\ell C_{HA}^0}{A} = \frac{1}{C_B^0} \cdot \frac{1}{\varepsilon_C K} + \frac{1}{\varepsilon_C}, \qquad (3-3)$$

where C_{HA}^0 and C_B^0 are initial molar concentrations of acid and base, A is the absorbance due to the complex, ℓ is the path length in cm, and ε_C is the molar absorptivity of the complex. It is important to keep in mind the assumptions made by Benesi and Hildebrand—namely, that $C_B^0 \gg C_A^0 > C_C$, and that the complex is the only absorbing species at the wavelength being used. If C_{HA}^0/A is plotted versus $1/C_B^0$, K and ε_C can be calculated from the slope and intercept.

The Scott modification [2220] of the Benesi-Hildebrand procedure uses Eq. (3-3) in the form

$$\frac{\ell C_{HA}^0 C_B^0}{A} = \frac{1}{K\varepsilon_C} + \frac{C_B^0}{\varepsilon_C}, \qquad (3-4)$$

and K and ε_C are calculated from the intercept and slope of a plot of $C_{HA}^0 C_B^0/A$ versus C_B^0.

Equations derived by Rose and Drago [2112a] are more general in that they do not assume that the concentration of the base is much larger than the equilibrium concentration of the complex. Equations will be derived here which can be used when (1) all three species, HA, B, B...HA, absorb at a given wavelength, (2) two of the three species absorb, HA, B...HA, or (3) only B...HA absorbs. These equations can be used for any system with proper spectral characteristics and an equilibrium involving the formation of a 1:1 adduct.

The equilibrium concentrations of [HA] and [B] in Eq. (3-2) can be expressed in terms of the initial concentrations, C_{HA}^0 and C_B^0:

$$[HA] = C_{HA}^0 - C_C , \tag{3-5}$$

$$[B] = C_B^0 - C_C . \tag{3-6}$$

If Beer's law holds for the absorbing species, HA, HA...B, C_C, and B, the total absorbance at a given wavelength is

$$A = \varepsilon_{HA}[HA] + \varepsilon_C C_C + \varepsilon_B[B] , \tag{3-7}$$

where ε_{HA}, ε_C, ε_B are molar absorptivities of the absorbing species and the path length is 1 cm. The absorbance of the free acid is

$$A^0 = \varepsilon_{HA} C_{HA}^0 . \tag{3-8}$$

Rose and Drago [2112a] combined Eqs. (3-5)-(3-8) to give

$$K^{-1} = \frac{A-A^0-A_B}{\varepsilon_C - \varepsilon_{HA} - \varepsilon_B} - C_{HA}^0 - C_B^0 + \frac{C_{HA}^0 C_B^0 (\varepsilon_C - \varepsilon_{HA} - \varepsilon_B)}{A-A^0-A_B} . \tag{3-9}$$

For systems where the base shows no absorption in the region of interest, A_B and ε_B are zero.

$$K^{-1} = \frac{A-A^0}{\varepsilon_C - \varepsilon_{HA}} - C_{HA}^0 - C_B^0 + \frac{C_{HA}^0 C_B^0 (\varepsilon_C - \varepsilon_{HA})}{A-A^0} \tag{3-10}$$

If only the complex absorbs then $\varepsilon_{HA} = 0$ and

$$K^{-1} = \frac{A}{\varepsilon_C} - C_{HA}^0 - C_B^0 + \frac{C_B^0 C_{HA}^0 \varepsilon_C}{A}.$$ (3-11)

Equations (3-9)-(3-11) contain two unknowns, K^{-1} and ε_C. When solving two simultaneous equations for these unknowns, the entire molar absorptivity term is treated as an unknown, $i.e.$, $(\varepsilon_C - \varepsilon_{HA} - \varepsilon_B)$ in Eq. 9 and $(\varepsilon_C - \varepsilon_{HA})$ in Eq. 10. For example, measurement of $A-A^0$ for two different combinations of C_B^0 and C_{HA}^0 would provide sufficient experimental data to solve two simultaneous equations for K^{-1} and ε_C in Eq. (3-10). However, this would provide no measure of the precision of the method, so $A-A^0$ should be measured for at least three combinations of C_B^0 and C_{HA}^0. Usually the same concentration of acid is mixed with different concentrations of base in sufficient excess to prevent self-association of the acid.

During preliminary studies of a new system a graphical method can be used to solve for K^{-1} and $\varepsilon_C - \varepsilon_{HA}$. Application of the graphical method to the phenol-trimethylphosphine oxide adduct is illustrated in Fig. 3-1. Arbitrary values for $\varepsilon_C - \varepsilon_{HA}$ and the measured values for $A-A^0$ are used with C_{HA}^0 and C_B^0 to obtain a series of solutions to Eq. (3-10). The intersection of the lines gives the values of K^{-1} and $\varepsilon_C - \varepsilon_{HA}$. Usually the three lines cross at three different points to form a triangle, and the size of this triangle is a measure of the precision of the experiments. If the lines do not intersect, either a side reaction is occurring or a simple 1:1 adduct is not formed.

Several conditions must be met in order to determine equilibrium constants accurately by Eqs. (3-9)-(3-11). The slopes of the plots of K^{-1} versus $\varepsilon_C - \varepsilon_{HA}$ should differ as much as possible; the contribution of concentration errors to slope errors must be small compared to variations between slopes; and a wavelength should be chosen where the contribution of the complex to the observed absorbance is large. The best way to meet these conditions is to make a preliminary run and use the data to calculate rough values of K and $\varepsilon_C - \varepsilon_{HA}$. Once the best conditions have been established, a series of runs should be made with variation in the base concentration for each run. Computer programs

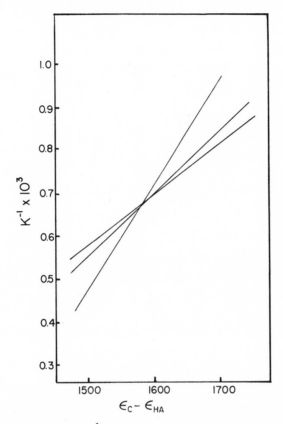

FIG. 3-1. Plot of K^{-1} versus $\varepsilon_c - \varepsilon_{HA}$ for adduct of phenol with trimethylphosphine oxide. Joesten [1180].

can be used for calculations of K^{-1} and $\varepsilon_c - \varepsilon_{HA}$ which include error analysis [949].

All of the discussion on the use of Eqs. (3-9)-(3-11) is based on the assumption that only a 1:1 adduct is present. Maier and Drago [1586a] have reported a procedure for calculating equilibrium constants when both 1:1 and 2:1 adducts are present in solution. The Liptay method [1486] is also well suited for checking the assumption of 1:1 adduct formation. Briegleb [325] and Mulliken and Person [1733] have discussed the application of this method to charge-transfer complexes. Here the method will be described briefly for hydrogen bonding complexes.

The absorbance at frequency i (A_i) for a solution of C_{HA}^0 and C_B^0 is compared with the absorbance at a reference frequency m (A_m). The reference frequency is usually at the absorption maximum. The ratio A_i/A_m, referred to as ζ, is obtained for all values of C_B^0 (and/or C_{HA}^0). Values of ζ are then averaged to give $\bar{\zeta}$ ($\bar{\zeta} = \bar{A}_i/\bar{A}_m$). If a 1:1 complex is present, the value of ζ of each solution must be the same as that for any other solutions made with different concentrations of C_B^0 (and/or C_{HA}^0). Scatter or trends in absorbance data are checked by computing $A_m(\zeta)$ for a given solution from A_i and $\bar{\zeta}$:

$$A_m(\zeta) = A_i/\bar{\zeta} . \tag{3-12}$$

If the data were perfect, all computed values of $A_m(\zeta)$ would agree with the measured absorbance A_m for the particular solution. The values of $A_m(\zeta)$ for each solution are then averaged over all frequencies i to obtain $\bar{A}_m$, the average absorbance for a given solution at frequency m. These averaged values of $\bar{A}_m$ for several different solutions are then used in Eq. (3-4) to obtain K and ε.

2. Error analysis

Error analysis is essential since the accuracy of a particular spectroscopic method varies considerably for different hydrogen-bonded systems, particularly those which involve weak hydrogen bonds. Person [1913] has proposed a reliability criterion for Benesi-Hildebrand determinations. In general, reliable equilibrium constants are obtained for weak complexes only when the equilibrium concentration of the complex is the same order of magnitude as the equilibrium concentration of the most dilute component. The best concentration range is 0.1 K^{-1} to 9.0 K^{-1}. If it is less than 0.1 K^{-1}, a Benesi-Hildebrand plot will give zero intercept and a Scott plot will give zero slope. Deranleau [612] has also discussed the reliability of spectral data in terms of base concentration and magnitude of K (see p. 176 for an application of his method).

The concentration scale is an important consideration since several studies have demonstrated that equilibrium constants expressed in molarity units (Kc) show less variation than equilibrium constants expressed in molality or mole fraction units (Kn). Scott [2220] was the first to point out the concentration scale dependence of parameters

obtained by using the Benesi-Hildebrand method. He found that different values of ε_c were obtained by plotting the same data in molar and
mole fraction units. Additional work by Trotter and Hanna [2470] has
shown that it is not possible to use a Benesi-Hildebrand plot to determine which concentration scale is best since straight lines are
obtained for all scales.

One way to test the constancy of Kc and Kn is to measure both in
inert solvents which have different molar volumes. Two examples of
this are cited by Kuntz *et al*. [1370]. The first compares values of
Kc and Kn for the heptyne-1 adduct with acetone [354]. Values of Kc
are 0.44, 0.45, 0.43, 0.43 ℓ/mole for the solvents cyclohexane, hexane,
decane, and tetradecane, respectively. Values of Kn for the same
series of solvents are 4.0, 3.4, 2.2, and 1.6. The second example is
a comparison of NMR chemical shifts for the chloroform-benzene complex.
At constant molarity (5.11 M benzene) the values of δ_{obsd} were 35.7
and 36.1 cps for the complex in heptane and hexadecane, respectively.
At constant mole fraction (0.490, 0.478 benzene) the values of δ_{obsd}
were 31.3 and 22.9 cps. The results of a near-infrared study of
di-*t*-butyl carbinol adducts [2095b] are also in support of the use of
molarity rather than molality or mole fraction.

If the equilibrium constant is larger than 100, the choice of
concentration scale is not critical. However, many of the equilibrium
constants reported in the Appendix are less than 10. In these cases it
is necessary to use a large excess of base, and the selection of the
concentration scale becomes an important consideration.

Conrow *et al*. [501] discuss error analysis for Eq. (3-9). They
point out that the slopes will differ the most when the concentration
of the absorbing species is held constant and the concentration of the
nonabsorbing species is varied over a wide range. Usually a variation
of base concentration over a 10:1 range gives satisfactory results.
The wavelength should be selected to keep $\varepsilon_c C_c$ large compared to
$\varepsilon_{HA}[HA] + \varepsilon_B[B]$. Adamek and Ksandr [9,10] have examined the optimum
analytical concentration of reacting species and the effect which
experimental errors have on the value of the equilibrium constant.

Wentworth *et al*. [2572] have derived a rigorous least-squares
adjustment of spectrophotometric data for 1:1 complexes that is appli-

cable to hydrogen-bonded systems. In a similar statistical analysis of literature data Rosseinsky and Kellawi [2114] found discrepancies between the reported value and the least-squares value for a number of association constants. The importance of using a wider range of concentrations as well as a larger number of solutions is emphasized. Rosseinsky and Kellawi [2114] question the validity of using the "sharpness of fit" parameter [501] as a basis for rejecting data since this parameter does not reflect improvement in statistical errors caused by an increase in the number of measurements.

3. Determination of enthalpy and entropy
   The van't Hoff equation

$$\log K = \frac{-\Delta H^0}{2.3RT} + \frac{\Delta S^0}{2.3R} \tag{3-13}$$

can be used to obtain ΔH^0 from measurements of K at more than one temperature. If the Rose-Drago method is used to determine accurate values of K^{-1} and $\varepsilon_c - \varepsilon_{HA}$ at one temperature, $A-A^0$ can be measured for a single acid-base solution over a wide range of temperature (minimum of five temperatures over a 30° range). Then, the equilibrium constants can be calculated at the various temperatures by using Eq. (3-10). The effect of temperature change on the density of the solvent must be considered in calculating molar concentrations of C_{HA}^0 and C_B^0 at the various temperatures. Since $\varepsilon_c - \varepsilon_{HA}$ is known at one temperature, Eq. (3-10) can be solved for K^{-1} by assuming that $\varepsilon_c - \varepsilon_{HA}$ is independent of temperature. The temperature dependence of ε for HA, B...HA, and B should always be checked before making this assumption.

   If ΔH^0 is constant over the temperature range involved, a plot of log K versus 1/T will give a straight line, and ΔH^0 can be calculated from the slope, b:

$$-\Delta H \text{ (kcal/mole)} = \frac{4.58b}{1000} . \tag{3-14}$$

The entropy can be calculated from

$$\Delta G^0 = \Delta H^0 - T\Delta S^0 . \tag{3-15}$$

The importance of statistical treatment of experimental data has already been emphasized. Error analysis should be included in the calculations of K, ΔH, and ΔS [999]. Computer programs are available which can be used to compute the thermodynamic parameters and their error limits from experimental data [949].

B. Hydrogen Bond Energies

Most thermodynamic data for hydrogen bonding have been obtained in an "inert" solvent such as carbon tetrachloride. A Born-Haber energy cycle for hydrogen bonding illustrates the effect which solvents may have on hydrogen bond energies, as shown in Scheme 1.

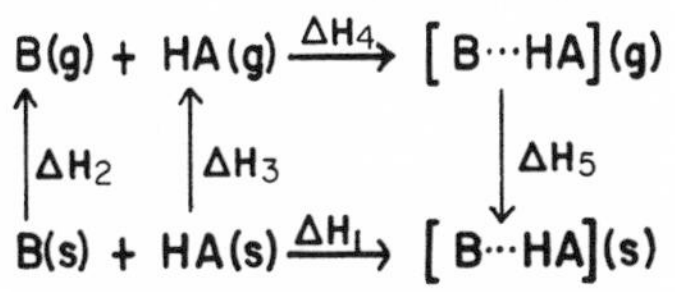

$$B(g) + HA(g) \xrightarrow{\Delta H_4} \left[B \cdots HA \right](g)$$

$$\uparrow \Delta H_2 \quad \uparrow \Delta H_3 \qquad\qquad \downarrow \Delta H_5$$

$$B(s) + HA(s) \xrightarrow{\Delta H_1} \left[B \cdots HA \right](s)$$

Scheme 1. Enthalpy cycle: g = gas; s = solvent.

The enthalpy measured in solution can be represented as the sum of the contributions in the Born-Haber cycle:

$$\Delta H_1 = \Delta H_2 + \Delta H_3 + \Delta H_4 + \Delta H_5 \ . \tag{3-16}$$

The assumption is that $\Delta H_2 + \Delta H_3$ will be approximately equal to ΔH_5 for "inert" solvents. The gas-phase thermodynamic data that are available indicate that this is not a good assumption. However, more gas-phase thermodynamic data for hydrogen-bonded adducts are needed in order to determine the extent of solvent effects. Methods used to obtain thermodynamic data in the gas phase will be discussed after a survey of the solution methods.

1. Solution studies

The above discussion emphasizes the importance of considering solvent effects on hydrogen bonding systems. An examination of the thermodynamic data in the Appendix reveals that carbon tetrachloride has been used as the solvent for the vast majority of hydrogen bonding

solution studies. However, carbon tetrachloride is not always inert,
and at least two types of bases, nitrogen and sulfur donors, react
with it, see Refs. [650,1811,1811a,1879,2533]. In these cases, a
saturated hydrocarbon, such as cyclohexane, should be used as the solvent.

Although existing evidence suggests that carbon tetrachloride
does not solvate reactants and products equally, one can still assume
that the variation in solvation energies for different hydrogen-bonded
systems in carbon tetrachloride is negligible. The same assumption
appears to be valid for saturated hydrocarbons. One should avoid
making quantitative comparisons of thermodynamic data measured in
polar solvents with those measured in nonpolar solvents until the
effects of solvation (Scheme 1) have been examined.

Several studies of solvation effects have been carried out. One
of these [666] is discussed in 1.b in connection with an assessment of
Arnett's pure-base calorimetric method [70]. Another study involves
an evaluation of solvent effects so that thermodynamic data measured
in polar solvents can be compared with data measured in nonpolar solvents [651,1811a]. The procedure is based on the investigation of a
displacement reaction (A—H...B + B' $\rightleftarrows$ A—H...B' + B) for a given acid
with a series of bases. If the enthalpies measured in a polar solvent
differ by a constant amount from those measured in an inert solvent or
in the gas phase, the assumption is made that the nonspecific solvation
contributions in the displacement reaction cancel. The procedure has
been shown to be applicable to m-fluorophenol adducts.

    a. *Calorimetry procedures*. Most of the work in calorimetry has
occurred since Arnett *et al*. [69] described a calorimeter suitable for
measuring small heat changes (5-20 calories). Some modifications of
Arnett's apparatus should be given serious consideration if research
in this area is being contemplated. One modification consists of using
a reference calorimeter that is identical to the reaction flask to provide a reference zero by compensating for heating of stirring, heating
effect of the thermistor, etc. [1046,2255]. The use of a reference
calorimeter also allows the slope of the base line on the recorder to
be changed simply by adjusting the relative stirring rates in the two
calorimeters. A base-line compensator has also been used to control

the slope of the base line. A precision power supply that provides a known, reproducible voltage to the heater is available from Luminon (Irvington, New Jersey) as an accessory for their calorimeter, and from CEA, division of Berkleonics, Monrovia, California.

Henrickson *et al.* [1046] give detailed instructions on the experimental procedure. The problems which cause the greatest difficulty in the use of the calorimetric method are (1) proper stirring, (2) base-line drift, (3) uniform sample injection, and (4) accurate voltage source.

Arnett *et al.* [70] have used two methods for obtaining hydrogen bond energies calorimetrically. The first method is referred to as the high dilution method. In this case, a small quantity of base (50-300 μl) is injected into a dilute solution of HA. The observed heat of interaction, ΔH_{obsd}, is related to ΔH_f, the hydrogen bond enthalpy according to

$$\Delta H_{obsd} = \Delta H_F C_C V , \qquad (3-17)$$

where V is the volume in liters of solution in the calorimeter and C_C is the equilibrium concentration of [A—H...B]. The value of ΔH_{obsd} is determined experimentally by subtracting the heat of solution for a given quantity of base in pure carbon tetrachloride from its heat of solution in the same amount of carbon tetrachloride containing the reference acid at initial molar concentration C_{HA}^0.

The expression for the equilibrium constant is

$$K = \frac{C_C}{[C_{HA}^0 - C_C][C_B^0 - C_C]} , \qquad (3-18)$$

where C_{HA}^0 and C_B^0 are initial concentrations of proton donor and acceptor, respectively. If Eq. (3-18) is written as

$$K(C_{HA}^0 C_B^0 - C_{HA}^0 C_C - C_B^0 C_C + C_C^2) = C_C \qquad (3-19)$$

and solved for C_C by using the quadratic

$$C_c = \frac{-(C_{HA}^0 + C_B^0 + K^{-1}) \pm \sqrt{(C_{HA}^0 + C_B^0 + K^{-1})^2 - 4C_{HA}^0 C_B^0}}{2} , \quad (3\text{-}20)$$

only one of the roots will be reasonable.   The equation is solved for C_c by using the value of K determined from ir spectroscopic studies [70].

A value for ΔH_F is obtained from the slope of a plot of ΔH_{obsd} versus $C_c V$.   Figure 3-2 is a plot of ΔH_{obsd} versus $C_c V$ for the

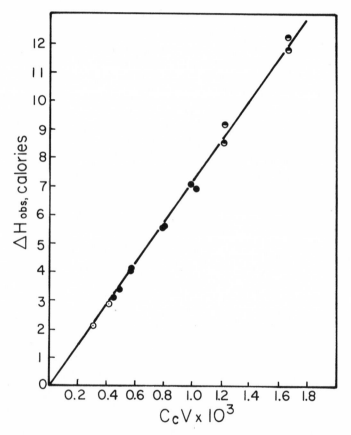

FIG.   3-2. ΔH_{obsd} versus $C_c V$ for p-fluorophenol adduct of pyridine. Concentrations of p-fluorophenol are: ◑ = 0.02 M,   ● = 0.01 M, 0 = 0.005 M. ΔH_F is calculated from the slope according to Eq. (3-17). Reprinted from Arnett et al. [71a], p. 5956, by courtesy of the American Chemical Society.

p-fluorophenol-pyridine complex. The equilibrium constant determined
by ir measurements was 75 at 298°K. This is used in Eq. (3-20) to calculate C_C. The value of ΔH_F from the slope of Fig. 3-2 is -7.2
kcal/mole. Values of ΔH_F determined in this way depend upon the availability and reliability of K.

Bolles and Drago [275] determine K and ΔH simultaneously by using
a modified version of the Rose-Drago equation [2112a]. If the symbolism from the above equations is used,

$$C_C = \frac{\Delta H_{obsd}}{(\Delta H_F)(V)} . \tag{3-21}$$

Combining Eqs. (3-18) and (3-21) gives

$$K = \frac{\dfrac{\Delta H_{obsd}}{(\Delta H_F)V}}{C_{HA}^0 - \left[\dfrac{\Delta H_{obsd}}{(\Delta H_F)V}\right]\left[C_B^0 - \dfrac{\Delta H_{obsd}}{(\Delta H_F)V}\right]} . \tag{3-22}$$

Equation (3-22) may be rearranged to

$$K^{-1} = \frac{\Delta H_{obsd}}{(\Delta H_F)(V)} + \frac{C_{HA}^0 C_B^0 (\Delta H_F)(V)}{\Delta H_{obsd}} - C_{HA}^0 + C_B^0 . \tag{3-23}$$

Equation (3-23) has two unknowns, K^{-1} and ΔH_F. The procedure described
for solving Eq. (3-10) is followed. Values of ΔH_{obsd} are obtained for
several different acid and/or base concentrations. Using arbitrary
values of ΔH_F, measured values of ΔH_{obsd}, and the known initial acid
and base concentrations, a series of solutions to Eq. (3-23) are obtained. In plotting K^{-1} versus ΔH_F a curve is obtained for each set
of concentrations, and the intersections represent graphical solutions
of the simultaneous equations. Figure 3-3 is a plot of $-\Delta H_F$ versus K^{-1}
for phenol-acetonitrile in carbon tetrachloride [719] . A computer
program for Eq. (3-23) is given in Ref. [949].

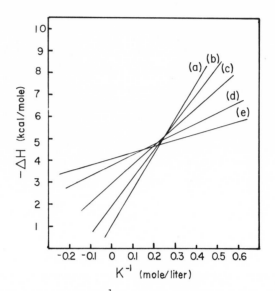

FIG. 3-3. $-\Delta H_F$ versus K^{-1} for phenol-acetonitrile adduct.
Molarity of CH_3CN: (a) 0.0806, (b) 0.1716, (c) 0.3249, (d) 0.7873,
(e) 1.4652. Reprinted from Epley and Drago [719], p. 5771, by courtesy of the American Chemical Society.

Systematic errors in ΔH and K may occur when the method of Bolles
and Drago is used [275]. Cabani and Gianni [384] discuss ways to avoid
systematic errors which include (1) a variation in the concentration of
both reagents, (2) a change in the order of addition of reagents, (3)
the use of separate experiments to obtain data for the curve, (4)
weighting experiments carried out under different conditions, and (5)
a check on whether the pair of best values of K and ΔH give a good fit
both when the ratio of concentrations and the concentrations themselves
are varied.

Purcell, Stikeleather, and Brunk [2026] have used a modification
of Eq. (3-23) which is obtained by making approximations similar to
those made in the Scott equation [2220]. The resulting expression is

$$\frac{C^0_{HA} C^0_B V}{\Delta H_{obsd}} = \frac{C^0_B}{\Delta H_F} + \frac{1}{\Delta H_F} . \tag{3-24}$$

Values for ΔH_F can be obtained from a plot of $C^0_{HA} C^0_B V/\Delta H_{obsd}$ versus C^0_B.

The pitfalls of the Scott method for spectrophotometric determination of equilibrium constants should be kept in mind if this method is used.

Equation (3-23) is the most reliable approach if K and ΔH are both obtained from a calorimetric method. Arnett *et al.* [70] have proposed an alternative method, which they refer to as the pure-base method, to avoid the need for accurate K values at high dilution. In this method the base is used as the solvent, so complete association of the proton donor can be assumed. The two heat terms involved are (1) the heat due to hydrogen bonding interactions, and (2) the heat produced in the absence of hydrogen bonding. Estimates of contributions to the heat term from nonhydrogen bonding interactions are obtained by using model compounds which resemble the acids in all ways except in the ability to hydrogen bond.

Arnett *et al.* [70] use anisole and *p*-fluoroanisole as model compounds for phenol and *p*-fluorophenol. The value of ΔH_F is obtained by measuring heats of solution for the acid (ΔH_S^a) and the model compound (ΔH_S^m) in pure base and in reference solvent (r.s.). The equation

$$\Delta H_F = (\Delta H_S^a - \Delta H_S^m)_{base} - (\Delta H_S^a - \Delta H_S^m)_{r.s.} \qquad (3-25)$$

is used to calculate ΔH_F directly, since one assumes that all acid molecules are converted to a 1:1 complex. ΔH_F values obtained by both the high dilution and the pure-base methods (Appendix, phenol, *p*-fluorophenol) are generally in good agreement. The pure-base method offers the advantages of larger and more easily measured heats, and the reduced need for careful temperature control.

b. *Calorimetry reliability.* Duer and Bertrand [666] have tested the reliability of both of Arnett's methods (dilution method and pure-base method). The equation

$$HA(I) + B(I) \longrightarrow BHA(I) \qquad (3-26)$$

is used to represent the high dilution method where I is an inert solvent. For the pure-base method

$$HA(I) + B(B) + M(B) \longrightarrow BHA(B) + M(I) \qquad (3-27)$$

is used where M represents the model compound for the acid. Parentheses indicate that the preceding species is very dilute in the indicated solvent.

The heats of solution of phenol and its model compounds, anisole and toluene, were measured in the inert solvents carbon tetrachloride, n-heptane, and cyclohexane and in the basic solvents pyridine, tetrahydrofuran, p-dioxane, and acetone. These heats were extrapolated to infinite dilution and combined by Arnett's pure-base method for Eq. (3-27) to give hydrogen bond energies. The results (Table 3-1) indicate

TABLE 3-1

Hydrogen Bond Energies for Phenol Adducts

Calculated by Pure-Base Method[a]

Base	Inert solvent	$-\Delta H$[b] kcal/mole	$-\Delta H$[c] kcal/mole
C_5H_5N	C_6H_{12}	7.5	8.5
	n-C_6H_{14}	7.7	8.6
	CCl_4	7.3	7.7
$(CH_2)_4O$	C_6H_{12}	5.9	6.9
	n-C_6H_{14}	6.1	7.0
	CCl_4	5.7	6.1
$(CH_3)_2CO$	C_6H_{12}	5.5	6.6
	n-C_6H_{14}	5.7	6.7
	CCl_4	5.3	5.8
p-Dioxane	C_6H_{12}	5.3	6.3
	n-C_6H_{14}	5.5	6.4
	CCl_4	5.0	5.6

[a]Duer and Bertrand, [666]. [b]Model compound is anisole. [c]Model compound is toluene.

that the inert solvent used for the acid and the model compound does affect the values obtained for hydrogen bond energies. The lower enthalpy values for carbon tetrachloride solutions have been interpreted in terms of a reaction of carbon tetrachloride with phenol [666]

or with base [1811a]. The use of anisole as a model compound for
phenol has also been criticized [1811a] since anisole forms both σ- and
π-donor complexes with phenol [2564].

    c. Near-infrared spectroscopy. Thermodynamic data for a large
number of phenol-base systems have been obtained by measurements in
the near-infrared region. Since the first overtone of the free O—H
stretching mode of phenol occurs at 7220 cm^{-1}, the equilibrium concentration of free phenol can be measured if one assumes that only the
O—H group of free phenol molecules contributes to the peak at 7220
cm^{-1}.

    Powell and West [1981] have described a precise method for deter-
mining equilibrium constants for phenol adducts by measuring the peak
height of the overtone absorption at 7220 cm^{-1} as a function of base
concentration and temperature. The method of Liddel and Becker [1459]
was used to make a correction for dimerization of phenol. The initial
concentrations of phenol and base should be corrected for density
changes whenever temperatures other than the preparation temperature
are used. In most of the work reported thus far, carbon tetrachloride
is used as the solvent.

    The equilibrium concentration of phenol, [HA], is obtained by
using the equation

$$A = \varepsilon \ell [HA] \, , \tag{3-28}$$

where A is the absorbance, ε is the molar absorptivity of phenol, ℓ is
the path length, and [HA] is the equilibrium concentration of phenol.
Absorbance data at several temperatures for carbon tetrachloride solutions of phenol were used to calculate the functions

$$\varepsilon = 3.4054 - 0.000434 t (°C) \quad \text{and} \tag{3-29}$$

$$\log K_2 = \frac{798.9}{T(°A)} - 2.836 \, , \tag{3-30}$$

where K_2 is the dimerization constant for phenol.

An equation was derived for calculating equilibrium constants for
1:1 adducts of phenol which is based upon C_{HA}^0, C_B^0, K_2, and [HA]:

$$K = \frac{C_{HA}^0 - [HA] - 2K_2[HA]^2}{[HA]\{C_B^0 - C_{HA}^0 + [HA] + 2K_2[HA]^2\}} . \tag{3-31}$$

The enthalpy of association was calculated from a plot of ln K versus
1/T fitted to the best straight line by the method of least squares.

A similar approach has been used to obtain thermodynamic parameters for 1:1 adducts of aniline and N-methylaniline [1401,1803]. The
first overtone N—H stretching bands for cyclohexane solutions of aniline and N-methylaniline occur at 6696 and 6734 cm^{-1}, respectively.

The main disadvantage of the near-infrared method is the small
magnitude of ε (phenol, 3.4; aniline, 1.9; N-methylaniline, 1.2), which
necessitates the use of long cell paths and/or higher concentrations.

d. Infrared spectroscopy. It is first necessary to use concentrations which cause the dimerization of the acid and the base to be
negligible. This varies with the acid and base and must be investigated before making measurements for the acid-base system. If a Beer's
law plot of absorbance versus initial concentration of acid is linear,
the dimerization can be assumed to be negligible over the range studied.
The equilibrium concentration of the acid, [HA], is determined from the
free O—H peak. Arnett *et al.* [70] describe the procedure for determining equilibrium constants by ir spectroscopy.

Since measurement of the temperature dependence of ir absorption
bands has been used to determine ΔH, it is important to emphasize the
effect of solvents on the intensity of ir bands. The intensity of ν_{OH}
for phenol varies from 2.1 cm^2 molecules^{-1} sec^{-1} in the vapor phase to
5.8 cm^2 molecules^{-1} sec^{-1} in carbon tetrachloride.

e. Ultraviolet spectroscopy. Ultraviolet spectroscopic methods
should not be used to obtain thermodynamic data unless one of the other
methods is not applicable. References [100], [1181], and [2347] describe the application of uv spectroscopy to measurement of equilibrium
constants.

f. NMR spectroscopy——symbols. Some confusion exists in the literature over the use of ν and δ to represent frequencies and chemical shifts. Pople, Schneider, and Bernstein [1971] encouraged the use of ν for the resonant frequency in cycles per second (cps), and the use of δ for chemical shifts in dimensionless units $[\delta_{ppm} \equiv (\nu_S - \nu_R)/\nu_R \times 10^6]$.

Recently, both ν and δ have been used in proton NMR hydrogen bonding studies to represent chemical shifts in cps or Hz. Actually, $\Delta\nu$ should have been used instead of ν for chemical shifts in cps, since ν implies a frequency and not a chemical shift. Most workers, see Refs. [461,579,883,2593], have switched to the use of δ to present chemical shifts either in Hz (cps) or ppm. This practice will be followed here. However, it is important for the reader to keep in mind the various uses of ν and δ. If chemical shifts measured at different reference frequencies (*e.g.*, 60 MHz and 100 MHz) are being compared, the dimensionless δ_{ppm} must be used. The symbol Δ will be used to represent the difference between the chemical shift for the adduct and that for the free acid $(\Delta = \delta_{AH...B} - \delta_{AH})$.

All of the equations developed in this section can be used for chemical shifts $(\delta_{Hz}, \delta_{ppm}, \Delta\nu_{cps})$ or frequencies (ν_{cps}) provided the user pays careful attention to internal consistency in the units (Hz, ppm) and the reference materials (TMS, cyclohexane) for the chemical shift data.

g. PMR spectroscopy. Most of the thermodynamic data obtained by PMR methods has been for weak complexes. Examples include adducts of chloroform, thiols, and phenylacetylene. Factors which must be considered in PMR studies of weak complexes are (1) the importance of anisotropic contributions to the chemical shift, (2) the self-association of the acid, and (3) the suitability of the Benesi-Hildebrand equation. Recent reports, see Refs. [883,1096a,2290,2291,2593], have included a consideration of these factors in assessing the reliability of published thermodynamic data. The present treatment is based on these references along with earlier reports, see Refs. [790,1004,1623,1772].

Since the formation of 1:1 hydrogen-bonded adducts is rapid, the observed chemical shift of the hydrogen bonding proton δ_{obsd} will be a time-weighted average of the chemical shift of the protons in the adduct, δ_C, and the chemical shift of the uncomplexed HA, δ_M:

$$\delta_{obsd} = \left(\frac{C_{HA}^0 - C_C}{C_{HA}^0}\right)\delta_M + \frac{C_C}{C_{HA}^0}\delta_C \, . \tag{3-32}$$

Rearranging Eq. (3-32) gives

$$C_C = \frac{\delta_{obsd} - \delta_M}{\delta_C - \delta_M} C_{HA}^0 \, . \tag{3-33}$$

Rearranging the equilibrium expression gives

$$KC_{HA}^0 C_B^0 - KC_C(C_{HA}^0 + C_B^0 - C_C) = C_C \, . \tag{3-34}$$

Combining Eqs. (3-33) and (3-34) gives

$$\frac{C_B^0}{\Delta_{obsd}} = \frac{1}{\delta_C - \delta_M}(C_{HA}^0 + C_B^0 - C_C) + \frac{1}{K(\delta_C - \delta_M)} \, , \tag{3-35}$$

where $\Delta_{obsd} = \delta_{obsd} - \delta_M$. Equation (3-35) contains two unknowns, C_C and δ_C, which can be calculated by means of an iterative procedure [1772].

A plot of C_B^0/Δ_{obsd} versus $(C_{HA}^0 + C_B^0)$ gives a line with a slope which is approximately equal to $1/(\delta_C - \delta_M)$. This is substituted into Eq. (3-33) to calculate an approximate value of C_C. Then this value of C_C is used in Eq. (3-35) to calculate an improved value of the slope. The procedure is repeated until two successive cycles yield essentially identical values for the slope. The final value of the equilibrium constant is calculated from the limiting slope and intercept values. The value of δ_C can also be obtained from the final slope.

Nakano *et al.* [1772] checked Eq. (3-35) by computing equilibrium constants for the phenol adduct of dimethylacetamide. The concentration of phenol was fixed at 5.04×10^{-3} M and the base concentration varied from 7.21 to 36.1×10^{-3} M. As a first approximation, C_B^0/Δ_{obsd} was plotted versus $(C_{HA}^0 + C_B^0)$ with C_C set equal to zero. The slope of this line was used to calculate an approximate value of C_C as described above. Continuing plots of C_B^0/Δ_{obsd} versus $(C_{HA}^0 + C_B^0 - C_C)$ were made and the slopes used to calculate more refined values of C_C. The number of iterations necessary for convergence depends upon the concentration ratio and the strength of the hydrogen bond. For phenol-DMA one iterative cycle is sufficient. The values of K obtained by this method for phenol adducts with DMA and acetone are in good agreement with those obtained by other spectroscopic methods (Appendix).

If $C_B^0 >> C_{HA}^0$, Eq. (3-35) reduces to

$$\frac{1}{\Delta_{obsd}} = \frac{1}{K(\delta_C - \delta_M)} \frac{1}{C_B^0} + \frac{1}{\delta_C - \delta_M} \qquad (3\text{-}36)$$

or the rearranged

$$\frac{\Delta_{obsd}}{C_B^0} = K(\Delta - \Delta_{obsd}) , \qquad (3\text{-}37)$$

where $\Delta \equiv \delta_C - \delta_M$. A number of workers have used Eqs. (3-36) and (3-37), see Refs. [790, 1004, 1623, 1624, 2593].

Nishimura, Ke, and Li [1802] used Eq. (3-36) to study 1:1 adducts of chloroform with bases in cyclohexane. The concentration of chloroform was held at 0.05 M and the concentration of the bases were kept in large excess. The results for the 1:1 adduct of chloroform with tri-n-butylamine are given in Fig. (3-4). A value of 96 cps for $\delta_C - \delta_M$ is obtained from the intercept and values of 0.45, 0.36, 0.28, 0.22 ℓ/mole for K at 2°, 10°, 22°, 36.5° are obtained from the slopes. An enthalpy change of -3.6 kcal/mole is obtained from a plot of log K versus 1/T.

The equation used by Wiley and Miller [2593] can be obtained from Eq. (3-35) by substituting Eq. (3-33) for C_C. The solution of the resulting quadratic equation for $\delta_{obsd} - \delta_M$ gives

$$\delta_{obsd} = \delta_M + \frac{\Delta\left[C_{HA}^0 + C_B^0 + K^{-1} - \sqrt{(C_{HA}^0 + C_B^0 + K^{-1})^2 - 4C_{HA}^0 C_B^0}\right]}{2C_{HA}^0},$$

$$(3\text{-}38)$$

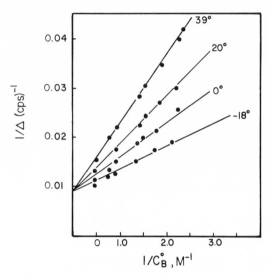

FIG. 3-4.   Plots of $1/\Delta$ versus $1/C_B^0$ for the 1:1 adduct of chloro-form with tri-n-butylamine.   Nishimura, Ke, and Li [1802].

where $\Delta \equiv \delta_C - \delta_M$.   Equation (3-38) simplifies to Eq. (3-37) when $C_B^0 \gg C_{HA}^0$.   Estimates of K and Δ are obtained from plots of Δ_{obsd}/C_B^0 versus Δ_{obsd} according to Eq. (3-37).   These estimates are used as initial input for an iterative solution of Eq. (3-38).

Wiley and Miller [2593] use the saturation factor developed by Deranleau [612] to analyze the data.   The saturation factor, s, is defined as

$$s \equiv C_C/C_{HA}^0 = \Delta_{obsd}/\Delta \ . \tag{3-39}$$

The most reliable equilibrium constants are obtained for an s range of 0.2-0.8.   The value of K must be between 0.1 and 10 ℓ/mole and a broad concentration range must be accessible in order to obtain a saturation factor in this range.   Most of the published work on adducts of chloro-form, see Refs. [762,1802,2244,2588], and phenylacetylene [883] was obtained with s less than 0.2.   Therefore, the reliability of these data is questionable.

Wiley and Miller [2593] also investigated the temperature depen-dence of $\delta_C - \delta_M$ or Δ.   They found that Δ decreases as the temperature increases. The average slope for twelve adducts of chloroform is -0.06 Hz/°C. The temperature dependence of Δ was attributed to changes in the hydrogen bond length [1729]. The hydrogen bond becomes longer at higher temperatures which causes δ_C to approach δ_M.

Equation (3-9) has been modified for use with PMR data [2291]. The resulting equation is

$$K = \frac{\Delta_{obsd}}{(\Delta - \Delta_{obsd})\left\{C_B^0 - \left[\frac{(\Delta_{obsd})C_{HA}^0}{\Delta}\right]\right\}} \cdot \qquad (3-40)$$

Experimental values of Δ_{obsd} are measured for n different combinations of C_{HA}^0 and C_B^0 to give n equations with two unknowns, K and Δ. The computer program described for Eq. (3-23) [949] has been modified for use with Eq. (3-40), and the error analysis procedure is described in reference [2291].

The effect of solvent on the values of K was also examined by the iterative method [1772]. Equilibrium constants for the phenol-isophorone adduct were measured in a variety of solvents (Table 3-2). Of interest is the solvent dependence of δ_M and K.

TABLE 3-2

K for Phenol-Isophorone in Various Solvents[a]

Solvent	K, 25°C, ℓ/mole	Solvent	K, 25°C, ℓ/mole
CCl$_4$	29.6	Cyclopentane	63.2
C$_2$Cl$_4$	39.7	Cyclohexane	69.8
CS$_2$	39.8	Methylcyclohexane	71.0
n-C$_5$H$_{12}$	67.9	Isooctane	81.0
n-C$_6$H$_{14}$	65.1	Decalin	78.3
n-C$_7$H$_{16}$	67.2		

[a]Nakano, Nakano, and Higuchi [1772].

h. *Fluorine NMR spectroscopy*. Gurka and Taft [955] have used
^{19}F NMR data to determine equilibrium constants for the association of
p-fluorophenol with bases in carbon tetrachloride. p-Fluoroanisole is
used as the internal reference to represent intramolecular screening
effects similar to p-fluorophenol so that the chemical shifts observed
for p-fluorophenol would be due entirely to hydrogen bond formation.
Equilibrium constants determined by ^{19}F NMR are in good agreement with
those determined by ir and calorimetric methods (Table 3-3).

TABLE 3-3

Comparison of K Values at 25°C for p-Fluorophenol Adducts in CCl_4[a]

Base	ir, K, ℓ/mole	^{19}F NMR, K, ℓ/mole	Calorimetric, K, ℓ/mole
$(CH_3)_2S$	346 ± 8	338 ± 7	—
$CH_3C(O)N(CH_3)_2$	260 ± 12	242 ± 6	—
$HC(O)N(CH_3)_2$	116 ± 3	115 ± 2	122 ± 9
$(C_2H_5)_3N$	85.2 ± 1.9	82 ± 2	—
C_5H_5N	76.2 ± 1.1	76 ± 1	74 ± 5
$(CH_2)_4O$	17.7 ± 0.5	18.4 ± 0.5	19.4 ± 1.0
$CH_3C(O)OC_2H_5$	12.3 ± 0.3	12.0 ± 0.2	13.0 ± 0.7

[a]Gurka, Taft [955].

A comparison of enthalpy values obtained by NMR and calorimetric
[70] methods is given in Table 3-4. The NMR enthalpy values were calculated using the van't Hoff equation and K values at -20° and 25°C.
The good agreement of thermodynamic values obtained by these methods
is support for their use for other hydrogen-bonded systems.

i. *Gas-Liquid chromatography*. In recent years gas-liquid chromatography (GLC) has been used to obtain thermodynamic data for hydrogen-
bonded adducts, see Refs. [385,1491,1608,2030,2251,2530]. Examples
include alcohol adducts with di-n-octyl ether [1608], di-n-octyl
ketone [1608], and didecyl sebacate [385]; bicyclic alcohol adducts

TABLE 3-4

Comparison of Enthalpy Values of p-Fluorophenol Adducts in CCl_4[a]

Base	$-\Delta H°$ (NMR) kcal/mole	$-\Delta H°$ (calorimetric) kcal/mole
$(CH_2)_4O$	5.7	5.8 ± 0.1
2-Butanone	4.8	5.2 ± 0.1
$CH_3C(O)OC_2H_5$	4.7	4.7 ± 0.1
p-Dioxane	4.6	5.1 ± 0.1
$(C_6H_5CH_2)_2O$	4.7	4.6 ± 0.1
$C_6H_5N(CH_3)_2$	4.2	4.0 ± 0.1

[a]Arnett *et al.* [70].

with tris(p-*tert*-butylphenyl) phosphate [2530]; and haloalkane adducts with di-n-octyl ether and di-n-octyl thioether [2251].

The advantages of GLC are (1) measurements are made at infinite dilution of the solute, (2) purification is not required, and (3) the method is simple and rapid. References [1608] and [385] provide a detailed description of the experimental procedure. Thermodynamic data obtained by the GLC method appear to be as reliable as that obtained by spectroscopic methods (see Appendix).

2. *Gas-phase studies*

   a. *Infrared spectroscopy.* Fild, Swiniarski, and Holmes [759] have used ir methods to determine the heat of formation for the methanol-trimethylamine adduct. The value of -7.1 kcal/mole is in good agreement with the value obtained by vapor pressure measurements (-7.5 kcal/mole).

   Thomas [2444] used the Benesi-Hildebrand method to determine equilibrium constants for hydrogen fluoride-ether complexes in the gas phase. Measurements were made on the absorption band of the H—F stretching vibration for the complex.

   b. *PMR spectroscopy.* Clague, Govil, and Bernstein [461] used a PMR spectroscopic method to determine thermodynamic data for the self-

association of methanol and the association of methanol with trimethylamine. In the presence of trimethylamine the dimerization of methanol
was considered as a competing reaction, and equations were derived
which can be used to calculate the enthalpy for a hetero-association
$[(CH_3OH...N(CH_3)_3]$ with allowance for the competing homo-association
$[(CH_3OH)_2]$. The value obtained for the enthalpy of dimerization is
-4.15 kcal/mole, which is in good agreement with values obtained by
using P-V-T curves and heat capacity measurements (-4.0 kcal/mole
[1333]) and an ir method (-3.5 kcal/mole [1129]). The amount of
tetramer is assumed to be negligible under the conditions of the experiment. Their enthalpy value for the hetero-association of methanol-
trimethylamine is -5.8 kcal/mole [461], which is substantially lower
than the value obtained by vapor pressure and ir methods (ΔH = -7.5
kcal/mole).

An enthalpy value of -7.1 kcal/mole was obtained from a PMR study
of the hydrogen chloride-dimethyl ether adduct [896]. This agrees well
with the enthalpy value of -7.6 kcal/mole from vapor pressure measurements [871] but not with results of an ir study (-5.6 kcal/mole) [197].
Much more gas-phase enthalpy data are needed to determine the cause of
these discrepancies and the reliability of the various methods.

  c. *Ion cyclotron resonance spectroscopy (ICR)*. This relatively
new technique can be used to obtain hydrogen bond energies for a
variety of ion-molecule reactions. For example, Yamdagni and Kebarle
[2640] have investigated the gas-phase equilibria $B^- —HR(g) \rightleftarrows B^-(g)$ +
HR(g) for B^- = Cl^- and HR = CH_3OH, $(CH_3)_3COH$, $CHCl_3$, C_6H_5OH, CH_3COOH,
and HCOOH. The enthalpy and entropy changes were calculated from the
temperature dependence of equilibrium constants. The thermodynamic
data are summarized in Table 3-5. Kebarle and co-workers [551,937b,
1250] have also used ICR to determine dissociation energies of protonated clusters such as $H^+(H_2O)_n$ and $H^+(CH_3OH)_n$. Stepwise dissociation energies for $H^+(H_2O)_n$ range from 10.3 kcal/mole for n = 8 to 31.6
kcal/mole for n = 2.

3. Applications

  a. *Solvent effects*. The data in the Appendix illustrate the
effect of solvent on hydrogen bond enthalpies. Gas-phase enthalpy
values for methanol adducts are at least 1 kcal/mole more negative

TABLE 3-5

Thermodynamic Data for $Cl^- - HR \rightleftarrows Cl^- + HR$[a]

RH	ΔG, kcal/mole[b]	ΔH, kcal/mole	ΔS, eu[c]
H_2O	8.2	13.1	16.5
CH_3OH	9.7	14.1	14.8
$(CH_3)_3COH$	11.1	14.2	10.3
CH_3Cl	10.8	15.2	14.8
C_6H_5OH	14.8	19.4	15.5
CH_3COOH	15.8	21.6	19.3
HCOOH	25.4	37.2	39.6

[a]Yamdagni and Kebarle [2640].   [b]298°K, standard state 1 atm.
[c]Standard state 1 atm.

than those for carbon tetrachloride solutions of the adducts [759, 1058]. An increase in polarity of the solvent causes a further decrease in enthalpy. For example, hydrogen bond enthalpies for the m-fluorophenol adduct of ethyl acetate are -6.7, -5.2, -4.7, -4.0, and -3.7 kcal/mole in cyclohexane, carbon tetrachloride, o-dichlorobenzene, benzene, and 1,2-dichloroethane, respectively [1811a].

Procedures for converting thermodynamic data obtained in polar solvents to that expected in an "inert" solvent have been proposed [452a,651,1811a]. These methods are based on the assumption that specific interactions of the solvent with the acid, base, or adduct are absent. In the first method the enthalpy (or free energy) of transfer of the adduct from one solvent to another is assumed to be proportional to the enthalpies (or free energies) of transfer of the acid and base for the same solvent pair [452a]:

$$\Delta H^0_{AB} = \alpha(\Delta H^0_A + \Delta H^0_B) .$$
$$I \rightarrow II \qquad I \rightarrow II \quad I \rightarrow II$$

$$(3-41)$$

The quantity α is a constant for a given acid-base pair and is assumed to be independent of temperature and solvent.

The second method is based on the assumption that the solvation energy for transferring any adduct from one solvent to another minus that for transferring the base is a constant [1811a]. The assumption was tested by showing that the displacement reaction, BA + B' → B'A + B, is independent of solvent for a reference acid. Once the assumption is tested for a given acid, enthalpies can be predicted from

-ΔH (poorly solvating media) = -ΔH (weakly polar solvent) + S .

$$(3-42)$$

In the case of m-fluorophenol adducts S = 0.5 for o-dichlorobenzene and 1.2 for benzene when the poorly solvating medium is carbon tetrachloride.

Both methods have been analyzed for m-fluorophenol adducts [1811a], but testing of other acid-base systems is needed before the utility of these methods can be assessed.

    b. Correlations. The thermodynamic data in the Appendix can be used to test Δν-ΔH, Δν-ΔG, and ΔH-ΔS correlations. A description of these correlations is given in Chapter 4.

    c. Deuterium versus hydrogen bonds. Enthalpy data for a number of A—H...B and analogous A—D...B systems indicate that the relative strength of H...B versus D...B depends on the system. In intramolecular hydrogen bonds the experimental data available indicate that deuterium bonds are weaker than hydrogen bonds. For example, the gasphase enthalpy values for o-Cl, o-Br, and o-I phenols are 3.4, 3.1, and 2.75, while the values for the corresponding o-phenol-d are 2.8, 2.65, and 2.65, respectively [1467].

In hetero-associated systems the deuterium bonds are stronger for weak bases while the hydrogen bonds are stronger when the base strength is increased. For example, adducts of phenol-d with triethylamine and pyridine have much smaller enthalpies than those of phenol (see Appendix). However, the enthalpy for the phenol-d adduct of n-heptyl fluoride is larger than that for phenol (2.5 versus 2.1, [1190]). Another example is $CDCl_3$-acetone, which has a larger enthalpy than $CHCl_3$-acetone [1216,1249]. Reversals in relative strengths of deuterium and hydrogen bonds have also been observed for self-associated systems. Enthalpy values in kcal/mole are: 5.3, 7.9 for $(C_6H_5COOH)_2$, $(C_6H_5COOD)_2$ [873]; 4.5, 4.2 for $(C_6H_5OH)_2$, $(C_6H_5OD)_2$ [1955]; and .6.0, 4.9 for $[(CH_3)_3COOH]_2$, $[(CH_3)_3COOD]_2$ [2547].

II. KINETICS

A. Methods

The relaxation methods developed by Eigen and co-workers [698-700] for determining rates of very fast reactions are suitable for investigating the dynamics of hydrogen bonding. Relaxation methods are based on the measurement of the length of time (relaxation time) required for equilibrium to be re-established after a sudden change in an external parameter, such as temperature or pressure. The relaxation time is related to the rate constants of the equilibrium being disturbed. Detailed descriptions of the theory and experimental methods have been published [390,700]. The present discussion will be limited to the application of these methods to selected hydrogen-bonded systems.

The principal methods used for hydrogen bonding studies are the temperature-jump and the ultrasonic methods. The time ranges of these two methods are 1 to 10^{-6} seconds and 10^{-5} to 5 x 10^{-10} seconds, respectively. In the temperature-jump method developed by Eigen and co-workers [700] the temperature is raised 5-10°C in a few microseconds by discharging a large voltage through the solution. Caldin and Crooks [391] obtain a much smaller temperature jump (0.5°) by using microwaves for a few microseconds. The main advantage of this method is its suitability for solvents of low dielectric constant.

The ultrasonic relaxation method, see Refs. [390,700,993,994,2403, 2404], is based on the use of sound waves for periodic variation of temperature and pressure. The temperature changes are of the order of 0.001°C or less. When the relaxation time for an equilibrium is comparable with the periodic time of the sound wave, a time lag will occur and energy will be absorbed. The absorption will be a maximum at a certain frequency and the relationship between absorption coefficient per wavelength μ, angular frequency in radians per second ω, and relaxation time τ is [390]

$$\mu = 2\mu_{MAX}\omega\tau/(1 + \omega^2\tau^2) \ . \qquad\qquad (3\text{-}43)$$

For a single relaxation time μ is a maximum value when $\omega\tau = 1$.

When Eq. (3-43) is written in terms of frequency f (= $\omega/2\pi$) and frequency for maximum absorption f_c (= $\omega_{MAX}/2\pi = 1/2\pi\tau$),

$$\mu = 2\mu_{MAX}(f/f_c)/[1 + (f/f_c)^2] \ . \qquad\qquad (3\text{-}44)$$

The quantity measured is generally not μ but α, the coefficient of absorption per centimeter. The relationship between μ and α is

$$\alpha = \mu/\lambda = \mu f/c \ , \qquad\qquad (3\text{-}45)$$

where c is the velocity of the sound wave. Therefore

$$\alpha/f^2 = \mu/cf = (2\mu_{MAX}/cf_c)/[1 + (f/f_c)^2] \ . \qquad\qquad (3\text{-}46)$$

The velocity c is usually not measured but changes are generally less than 2%. Therefore, the numerator on the right side of Eq. (3-46) is independent of frequency and may be represented as A:

$$\alpha/f^2 = A/[1 + (f/f_c)^2] \ . \qquad\qquad (3\text{-}47)$$

An additional term, B, must be added to Eq. (3-47) to take care of background absorption:

$$\alpha/f^2 = A/[1 + (f/f_c)^2] + B \ . \qquad\qquad (3\text{-}48)$$

The value of A depends on the concentrations and the thermodynamics of the reaction while B is characteristic of the solvent.

The relaxation time is determined as follows. Values of α/f^2 are obtained at the lowest and highest frequency available. If the values of α at the two ends of the frequency range are different, it is probably due to relaxation, and sufficient α/f^2 measurements are made to develop the changeover region shown in Fig. (3-5). Equation (3-48) is assumed to apply to the data and it is used to obtain the best values

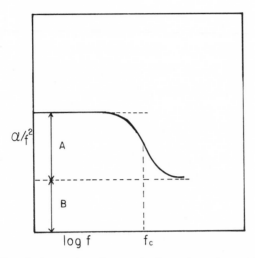

FIG. 3-5. Sound absorption for a single relaxation with background.

for A, B, and f_c. If the experimental values are reproduced over a
sufficient range of frequencies, it is assumed that only a single
relaxation is taking place and that

$$\tau^{-1} = 2\pi f_c . \tag{3-49}$$

The relaxation time, τ, is determined for a series of concentrations
and plots are made of τ versus $(\bar{a} + \bar{b})$ where $\bar{a}$ and $\bar{b}$ are final equilibrium concentrations of reactants [390]. The slope gives k_F and the
intercept gives k_B.

If Eq. (3-48) does not fit the data, more than one relaxation is
probable and measurements over a wider range of frequency and concentration are necessary. The main disadvantages of the ultrasonic method
are the need for high concentrations ($\geq 10^{-2}$ M) and large volumes (100
ml at 10 Hz and a few liters at 1 MHz).

B. Applications

1. Self-association

Hammes and co-workers [993,994] have studied the dynamics of a
number of self-associated systems. Specific examples of their work
will be discussed to illustrate the applicability of relaxation methods.

Equation (3-50),

$$\alpha/f^2 \;=\; A\tau/(1 + \omega^2\tau^2) + B + \Delta B \;, \tag{3-50}$$

where ΔB accounts for possible changes in solvent absorption due to
the presence of solute [998], was used to obtain kinetic data for the
self-association of 2-pyridone, shown in Structure (1).

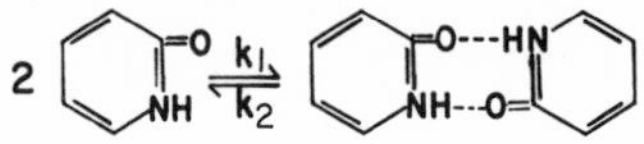

(1)

Rearrangement gives

$$1/\Delta \;\equiv\; [(\alpha/f^2) - B - \Delta B]^{-1} \;=\; 1/A\tau + (\tau/A)\omega^2 \;, \tag{3-51}$$

which permits convenient analysis of the data using $1/\Delta$ and ω as
variables. Weighted least-squares calculations were carried out to
obtain values for A and τ with standard deviations.

If the mechanism is assumed to proceed according to Structure (1),
the relaxation time is

$$1/\tau \;=\; k_2 + 4k_1 C_M \;, \tag{3-52}$$

where C_M is the molarity of monomeric pyridone. Rearrangement gives

$$1/\tau^2 \;=\; (k_2)^2 + 8k_1 k_2 C_0 \;, \tag{3-53}$$

where C_0 is the total concentration of 2-pyridone. Plots of $1/\tau^2$ versus C_0 are used to evaluate k and k_2.

Rate data are summarized in Table 3-6 for dimerization of 2-
pyridone in 50 wt% dioxane/carbon tetrachloride and in pure dioxane.
The magnitude of k_1 indicates that hydrogen bond formation is primarily
a diffusion-controlled process. The activation energy of 3.4 kcal/mole
also supports this [998].

TABLE 3-6

Rate Data for Dimerization of 2-Pyridone[a]

Solvent	$10^{-8}k_1$, $M^{-1}sec^{-1}$	$10^{-8}k_2$, sec^{-1}	k/k_2, M^{-1}	Temp, °C
Dioxane	17	0.9	19	13
	21	1.3	16	25
	31	2.2	15	40
50 wt% Dioxane/CCl$_4$	17	0.14	120	3
	21	0.29	73	25
	31	0.46	67	40

[a]Hammes and Spivey [998].

Sometimes a combination of methods is necessary in order to determine accurate rate constants in fast reactions. The work of Hammes and Lillford [995] on the 2-pyridone dimerization in solvent mixtures is an illustration of the combined use of ultrasonic absorption, uv, ir, and PMR measurements. Their results indicate that the rate of dimerization of 2-pyridone is diffusion-controlled in chloroform/dimethylsulfoxide mixtures, but dissociation of the 2-pyridone-dimethylsulfoxide hydrogen bond becomes rate-determining in carbon tetrachloride/dimethylsulfoxide mixtures [995].

Kinetic data for the self-association of benzoic acid, ε-caprolactam, 1-cyclohexyluracil, 2-pyridone, thiopental, mephobarbital, and 2-thiopyridine are summarized in Table 3-7. All association rate constants have values characteristic of a diffusion-controlled process (10^9 $M^{-1}sec^{-1}$) so the dissociation rate constant is a direct measure of the stability of hydrogen bonds.

Rates for the dissociation of benzoic acid dimers in N,N-dimethylformamide have been calculated from ultrasonic absorption coefficients [2060,2645]. Results of the most recent study of reaction (3-54) are given in Table 3-8. The forward and reverse activation

$$D \underset{k_{21}}{\overset{k_{12}}{\rightleftarrows}} 2M \ . \tag{3-54}$$

enthalpies for benzoic acid are 8.4 and 3.3 kcal/mole, respectively. The rate constants correlate with Hammett substituent constants and values calculated for p were 1.2 for the dissociation and -0.33 for the association.

Corsaro and Atkinson [507] have used ultrasonic absorption techniques to test the importance of the cyclic dimer (Dc) and the open dimer (D_O) of acetic acid in acetone. The equilibrium between D_C and D_O was shown to be more important than the dissociation to a monomer. The rate constants determined for $D_C \rightleftarrows D_O$ at 25° are $k_F = 3.47 \times 10^5$ sec^{-1} and $k_B = 3.37 \times 10^6$ sec^{-1}. The stability of D_C increases with increase in the size of the R group in RCOOH [2169a]. Kinetic data are summarized in Table 3-9.

TABLE 3-7

Kinetic Data for Self-Association

Reactant	Solvent	$10^{-9}k_F$ M⁻¹sec⁻¹	$10^{-7}k_R$ sec⁻¹	T, °C	Ref.
Benzoic acid	CCl₄	5	0.073	25°	[1588]
	CHCl₃	4.7	0.75	20°	[285]
	C₆H₁₂	8.1	0.022	20°	[285]
ε-Caprolactam	CCl₄	5.5	4.6	22°	[206]
	C₆H₆	6.5	26.0	22°	[206]
2-Pyridone	CHCl₃	2.2	2.2	25°	[997]
2-Thiopyridone	CHCl₃	1.5	4.7	10°	[997]
1-Cyclohexyluracil	CHCl₃	1.5	25.0	25°	[996]
Mephobarbital	CHCl₃	1.1	48.0	25°	[997]
Thiopental	CHCl₃	0.57	23.0	10°	[997]

TABLE 3-8

Rate of Dissociation of Benzoic Acid Derivatives

at 25°C in N,N-Dimethylformamide[a]

Acid	$k_{12} \times 10^{-8}$ sec^{-1}	$k_{21} \times 10^{-8}$ sec^{-1}M^{-1}
p-Toluic	0.67	0.28
m-Toluic	0.90	0.19
Benzoic	1.6	0.15
p-Cl-Benzoic	2.1	0.24
m-Cl-Benzoic	3.3	0.16

[a]Yasunga, Nishikawa, Tatsumoto [2645].

2. *Intermolecular association and proton transfer*

A combination of spectroscopic, thermodynamic, and kinetic evidence has been presented in support of a hydrogen-bonded intermediate
in proton-transfer reactions of Bromophenol Blue adducts [544,544a,545]
and Bromophthalein Magenta E adducts [391a]. The kinetics have been
analyzed on the basis of a two-stage mechanism with hydrogen bond
formation followed by proton transfer [Reaction (3-55)]. A temperature-
jump apparatus [391] was used to determine rate constants for the
reaction scheme

$$\text{ROH} + \text{B} \underset{k_{21}}{\overset{k_{12}}{\rightleftarrows}} \text{ROH...B} \underset{k_{32}}{\overset{k_{23}}{\rightleftarrows}} \text{RO}^-\text{...HB}^+ \,. \qquad (3\text{-}55)$$

Since Bromophenol Blue (Structure 2-I) is a diprotic acid, ROH in

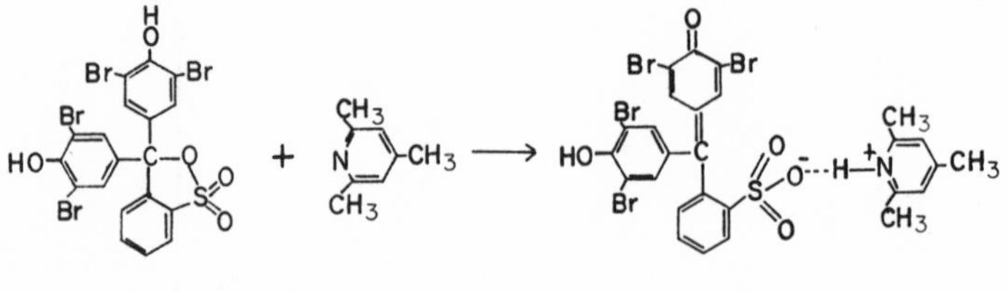

I   II

(2)

TABLE 3-9

Kinetic Data at 30° for $D_C \rightleftharpoons D_O$ [a]

Acid	$k_B \times 10^{-7}$ sec^{-1}	$\Delta H_B^{\ddagger}$ kcal/mole	$k_F \times 10^{-5}$ sec^{-1}	$\Delta H_F^{\ddagger}$ kcal/mole
Acetic	0.47	8.9	10.3	12
Propionic	1.7	6.8	5.9	11
Butyric	1.5	5.3	4.6	10
Valeric	2.7	5.5	3.9	10
Caproic	2.3	6.0	2.3	11
Isobutyric	5.5	5.0	1.9	12
Isovaleric	3.0	5.8	2.2	13

[a]Sano *et al.* [2169a].

reaction (3-55) may represent either Structure (2)-I or (2)-II. Results of measurements for the reaction of Bromophenol Blue with substituted pyridines in chlorobenzene indicate that the hydrogen-bonded
intermediate is present in both proton transfer reactions [544a,545].
Kinetic and thermodynamic data for the first and second proton transfer reactions of Bromophenol Blue with 2,4,6-trimethylpyridine are
given in Table 3-10.

TABLE 3-10

Thermodynamic and Kinetic Data for Reaction (3-55) at 25°C

for Bromophenol Blue with 2,4,6-Trimethylpyridine in Chlorobenzene

	ROH = Structure (2)-I[a]	ROH = Structure (2)-II[b]
K_{12}	$\sim$11 ℓ/mole	53 ℓ/mole
K_{23}	9.2×10^4 ℓ/mole	1.32 ℓ/mole
$-\Delta H_{12}$	$\sim$5.6 kcal/mole	6.8 kcal/mole
$-\Delta H_{23}$	9 kcal/mole	4.45 kcal/mole
$-\Delta S_{12}$	$\sim$14 cal deg^{-1} mole^{-1}	15 cal deg^{-1} mole^{-1}
$-\Delta S_{23}$	7 cal deg^{-1} mole^{-1}	14.4 cal deg^{-1} mole^{-1}
k_{23}	8.9×10^6 sec^{-1}	6×10^5 sec^{-1}
k_{32}	98 sec^{-1}	4×10^5 sec^{-1}
$\Delta H_{23}^{\ddagger}$	2 kcal/mole	$\sim$0 kcal/mole
$\Delta H_{32}^{\ddagger}$	10.8 kcal/mole	3.8 kcal/mole
$-\Delta S_{23}^{\ddagger}$	20 cal deg^{-1} mole^{-1}	32 cal deg^{-1} mole^{-1}
$-\Delta S_{32}^{\ddagger}$	13.3 cal deg^{-1} mole^{-1}	20 cal deg^{-1} mole^{-1}

[a]Crooks and Robinson [544a]. [b]Crooks and Robinson [545].

Grunwald and Ralph [946] have reviewed kinetic studies of hydrogen-
bonded solvation complexes of amines. The solvation process is analyzed in terms of the variation of k_H, the rate constant for breaking

the hydrogen bond in R_3N—HOH, with the structure of the amine and with the solvent medium.

Hine [1055a] has presented arguments in support of the formation of hydrogen-bonded intermediates in proton-exchange reactions of phenol and substituted benzoic acids in hydroxylic solvents. Equations are derived which relate the equilibrium constants for hydrogen bonding between oxygen acids and bases in water or methanol to pK values.

3. Intramolecular

The effect of intramolecular hydrogen bonding on the rate of proton transfer has been investigated by Rose and Stuehr [2112]. The rate of hydrolysis of a number of indicator systems was measured by the temperature-jump method. The rates at 25°C for Alizarin Yellow, Structure (3)-I, Alizarin Yellow G, Structure (3)-II, and Tropaeolin O, Structure (3)-III, are 4.0×10^7, 1.7×10^7, and 9.0×10^5 M^{-1} sec^{-1}, respectively. These rates are several orders of magnitude lower than diffusion-controlled reactions and are attributed to a finite chemical

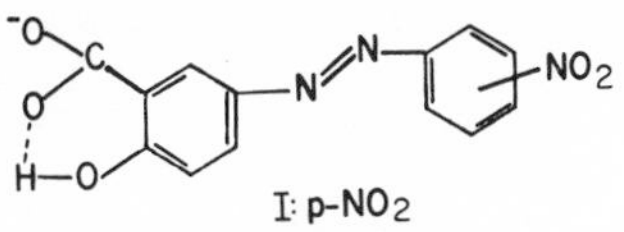

I: p-NO$_2$
II: m-NO$_2$

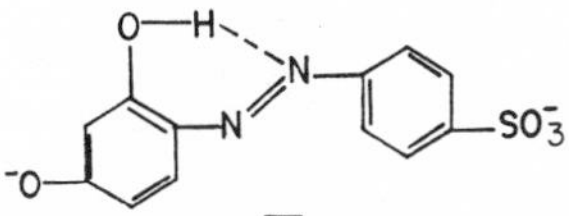

III

(3)

barrier caused by intramolecular hydrogen bonding. The smaller rate constant of Structure (3)-III, relative to Structure (3)-I and II, is presumably due to the formation of a stronger hydrogen bond which is in agreement with the strength of intramolecular hydrogen bonds proposed by Friedman [800] (O—H...N > O—H...O). In similar rate studies on the hydrolysis of Tropaeolin O, 2,4-dihydroxy-4'-nitroazobenzene, N,N-dimethylanthranilic acid, N-methyl-N-ethyl-anthranilic acid, N,N-diethylanthranilic acid, and 3,7-diaza-3,7-dimethyl-1,5-diphenyl-9-hydroxybicyclo[3.3.1] nonane, Haslam and Eyring [1023] obtained the order O—H...N > O—H...O ~ N—H...N > N—H...O.

One of the few direct studies of rates of intramolecular hydrogen bonding is the work of Yasunaga et al. [2646] on salicylates and salicylaldehyde. Ultrasonic relaxation absorption methods were used in the frequency range 2.5 to 95 Mc/sec. Rate constants for dissociation reaction, Structure (4), were determined at several temperatures

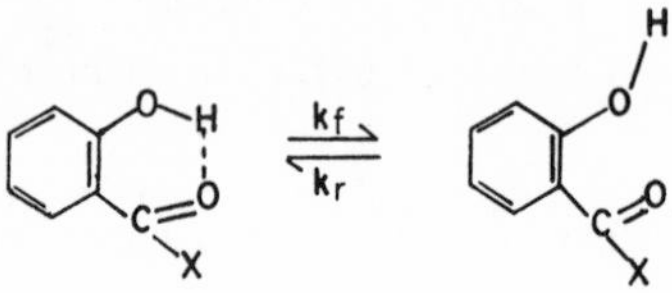

(4)

where X = H, CH$_3$, and C$_2$H$_5$. The results are summarized in Table 3-11.

TABLE 3-11

Rate Data at 30°C

for Dissociation of Intramolecular Hydrogen Bonds[a]

	Salicylaldehyde	Methyl salicylate	Ethyl salicylate
k_F x 10^5 sec^{-1}	4.0	12.	4.7
k_R x 10^7 sec^{-1}	2.6	3.1	2.2
$E_F^{\ddagger}$ kcal/mole	~9	~8	~9
$E_R^{\ddagger}$ kcal/mole	5.6	5.8	6.8

[a]Data are for Structure (4), taken from Yasunaga et al. [2646].

I. FREQUENCY SHIFT CORRELATIONS

In Chapter 1 the sensitivity of the A—H stretching frequency to
hydrogen bond formation was described. The wide use of the frequency
shift as a probe for studying hydrogen bond formation has led to the
publication of frequency shifts for a large number of hydrogen-bonded
systems. However, the reference stretching frequency is affected by
solvents, and it is necessary to be familiar with solvent effects as
well as concentration effects for a given system before correlations
are used.

A. Solvent Effects

Solvent shift theories and experimental studies on solvent effects
have been reviewed by Williams [2603], Hallam [980,981], and Horák
et al. [1084].

The frequency shift, $\Delta\nu$, is generally represented as the difference between the stretching frequency for the monomeric A—H in an
"inert" solvent and the lowered stretching frequency for A—H...B in
the same "inert" solvent. In this section shifts will be represented
as ν^o-ν^s since solvent effect studies use the stretching frequency of
A—H in the vapor phase as the reference point. Although direct com-
parisons of $\Delta\nu$ and ν^o-ν^s cannot be made, tabulations of ν^o-ν^s can be
used to calculate $\Delta\nu$ values by using ν^s for the appropriate solvent as
the reference point.

Early solvent theories were based on dielectric effects. The
first of these were theoretical treatments by Kirkwood [1273] and by
Bauer and Magat [153]. The Kirkwood-Bauer-Magat relationship (KBM) is

generally represented as

$$\frac{\nu^o - \nu^s}{\nu^o} = C(\varepsilon-1)/(2\varepsilon+1) \ , \qquad\qquad (4\text{-}1)$$

where ε is the macroscopic dielectric constant and C is a constant that depends upon the dimensions and electrical properties of the vibrating solute dipole.

Equation (4-1) has been tested by a number of workers, see Refs. [64,981,1205,1207,1995,2603]. Equation (4-1) and its modified form [157,1207]

$$\frac{\nu^o - \nu^s}{\nu^o} = C(n^2-1)/(2n^2+1) \ , \qquad\qquad (4\text{-}2)$$

where n is the refractive index, are not satisfactory for predicting the magnitude of solvent shifts. This is illustrated in Fig. 4-1 for the frequency shift of ν_{NH} pyrrole in various solvents [983].

Equation (4-3), derived by Pullin [1995],

$$\frac{\nu^o - \nu^s}{\nu^o} = C_o + A\{[(n^2-1)/(2n^2+1)R] + B[(\varepsilon-1)/(2\varepsilon+1)R]\} \ , \qquad (4\text{-}3)$$

and Eqs. (4-4) and (4-5), derived by Buckingham [355], give better agree-

$$\frac{\nu^o - \nu^s}{\nu^o} = C + C_\varepsilon(\varepsilon-1)/(2+1) + C_n(n^2-1)/(2n^2+1) \text{ for polar solvents,}$$
$$(4\text{-}4)$$

$$\frac{\nu^o - \nu^s}{\nu^o} = C + 1/2\left(C_\varepsilon + Cn\right)\left(\frac{\varepsilon-1}{2\varepsilon+1}\right) \text{ for nonpolar solvents,} \qquad (4\text{-}5)$$

ment than Eq. (4-1) for solvent shifts of X=O stretching frequencies. However, they are not satisfactory for explaining ν_{AH} frequency shifts since they do not consider specific association or hydrogen bonding.

Although no theoretical treatment is completely satisfactory to account for solvent shifts of ν_{AH} stretching frequencies, there is general agreement that hydrogen bonding has a major influence on ν_{AH}

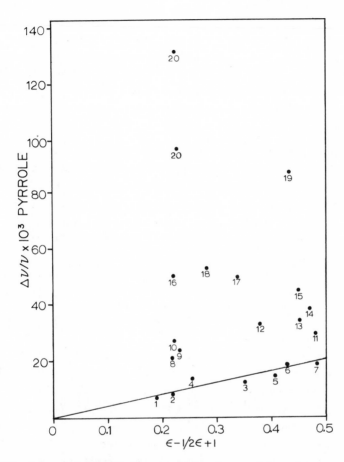

FIG. 4-1.   Relative frequency shift of ν_{NH} of pyrrole as a function of dielectric constant in (1) C_6H_{14}, (2) CCl_4, (3) $CHCl_3$, (4) CS_2,
(5) $CHCl_2CHCl_2$, (6) CH_2ClCH_2Cl, (7) CH_3NO_2, (8) C_6H_6, (9) toluene, (10)
mesitylene, (11) CH_3CN, (12) $CH_3COOC_2H_5$, (13) $C_6H_5COCH_3$, (14) $(CH_3)_2CO$,
(15) cyclohexanone, (16) dioxane, (17) $(C_2H_5)_2O$, (18) (n-$C_4H_9)_2O$, (19)
C_5H_5N, (20) $(C_2H_5)_3N$.   Hallam and Ray [983].

frequency shifts with bulk properties of the solvent such as dielectric
constant becoming important only for nonpolar solvents.   Several workers have investigated solvent effects on ν_{AH}.   These will be described
for a number of A—H systems in order to illustrate the magnitude of
solvent shifts and the empirical linear correlations which have been
proposed for ν_{AH} solvent shifts.

Gordy and Stanford [890-892] were among the first to study solvent effects on ν_{AH} frequencies. They measured ν_{OD} for methanol-d in over seventy different solvents and found a linear relationship between pKa of the solvents, as measured in water, and the ν_{OD} solvent shifts. In 1954 Josien and Fuson [1207] examined solvent shifts of ν_{NH} of pyrrole in connection with their investigation of the validity of the KBM relationship.

In 1958 Bellamy, Hallam, and Williams [181] began an extensive study of relative frequency shifts of different solutes in the same series of solvents. Although the work is concerned with solvent effects on all types of chromophoric groups, we will discuss only A—H groups. Some typical results are shown in Fig. 4-2 where $\Delta\nu/\nu$ for the N—H band of pyrrole is plotted versus the $\Delta\nu/\nu$ values for A—H stretching frequencies of aniline, water, methanol-d, and phenol. The linear plots and the variation in slopes are evidence for the importance of solute-solvent interactions.

The lack of importance of the dielectric constant of the solvent can be demonstrated by a comparison of frequency shifts for a sterically hindered phenol, 2,6,-t-butyl-4-methyl phenol, with those for phenol in the same solvents. The frequency shifts observed for the sterically hindered phenol were much smaller than those observed for phenol. If the dielectric constant were an important factor, the frequency shifts would be similar to those obtained for phenol. Josien and Fuson [1207] found that the frequency of the N—H band of pyrrole is a function of the dielectric constant for nonpolar solvents, but found large deviations for polar solvents. The results of these studies indicate that the effect of the dielectric constant on A—H frequency shifts has to be considered only for nonpolar solvents. In addition, the smooth progression of the lines in Fig. 4-2 suggests that specific association effects are still important in nonpolar solvents. The complexes with nonpolar solvents have been referred to as collision complexes to emphasize that, although the interaction is very weak, it is sufficient to influence ν_{AH}.

Horák, Moravec, and Pliva [1083] have argued that the observation of a closely spaced doublet for ν_{OH} of phenol in mixtures of different nonpolar solvents, e.g., phenol in perfluorooctane and hexane or per-

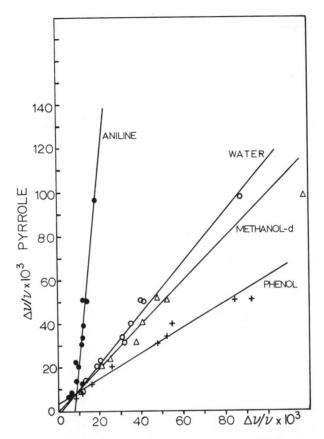

FIG. 4-2.   Correlation of relative frequency shift of ν_{AH} of aniline, water, methanol-d, and phenol with $\Delta\nu_{NH}/\nu_{NH}$ of pyrrole.   Solvents in order of increasing shift are C_6H_{14}, CCl_4, $CHCl_3$, CS_2, C_6H_6, $C_6H_5NO_2$, CH_3CN, $CH_3COOC_2H_5$, $(CH_3)_2CO$, dioxane, $(C_2H_5)_2O$, $(C_2H_5)_3N$. Bellamy, Hallam, and Williams [181].

fluorooctane and carbon tetrachloride [40] is additional evidence for the presence of collision complexes. They interpret the doublet as being due to the collision complex of phenol with both nonpolar solvents. The collision complexes are proposed to be solvated by a mixture of the molecules of the two solvents. The bands of the individual collision complexes shift with change in composition of the secondary solvation shell according to the Kirkwood-Bauer-Magat relation.

Mixed solvent studies are in support of specific solute-solvent complexes. Bellamy and Hallam [180] found that ν_{NH} of pyrrole in solvent mixtures appeared at the frequencies characteristic of each solvent in the pure state. Figure 4-3 illustrates this for solvent

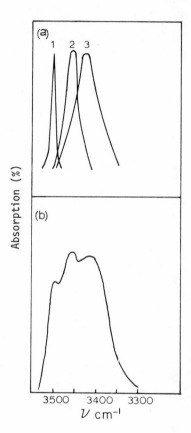

FIG. 4-3. ν_{NH} stretching frequency of pyrrole in mixed solvents. (a) ν_{NH} in (1) CCl_4, (2) C_6H_6, (3) CH_3CN. (b) 0.01 M pyrrole in 60% CCl_4 + 0.01 M CH_3CN in 40% C_6H_6. Bellamy and Hallam [180].

mixtures of carbon tetrachloride, acetonitrile, and benzene. The interaction is competitive, so quantitative intensity measurements of such systems can lead to a direct determination of the proton-accepting power of solvents.

Allerhand and Schleyer [40] were the first to examine solvent effects on frequency shifts for intramolecular, intermolecular, and self-associated hydrogen-bonded systems. They obtained evidence that stretching frequencies of OH groups already engaged in hydrogen bonding are solvent sensitive. This is an important observation, since many correlations involving frequency shifts have been made without consideration of this effect. Data are given in Table 4-1 to illustrate the magnitude of change. Although all of the solvents listed are much weaker bases than the base B in the hydrogen bond being examined, ROH...B, the frequency shift values show considerable variation.

The solvent dependence of ν_{OH} in phenol adducts and ν_{NH} in pyrrole adducts has been interpreted as evidence for the existence of solvent molecules hydrogen-bonded to the adduct [1110]. Bellamy and Pace [186] observed a larger proportionate change in ν_{CO} than in ν_{OH} when excess ROH was added to 1:1 adducts of the type ROH...OCR_1R_2. This was attributed to the formation of 2:1 adducts with both alcohol molecules attached to the carbonyl group.

B. Concentration Effects

Allerhand and Schleyer [40] also examined the effect of base concentration on ir spectra of hydrogen-bonded systems. The "free" peaks of the proton donor are essentially constant, but the "bonded" peaks vary with concentration in several solvents. Therefore, frequency shifts will often be dependent upon the concentration of base used, even if values are measured in the same solvent. For this reason Allerhand and Schleyer recommend that frequency shift measurements be made in an inert solvent with as low a concentration of base as possible. The most reliable procedure is to obtain frequency shifts at several different concentrations and then extrapolate to infinite dilution.

Nitriles are examples of bases which show a large concentration dependence. Figure 4-4 illustrates the effect of nitrile concentration on phenol frequency shifts in carbon tetrachloride solution. At lower nitrile concentrations the frequency shifts of the three nitriles agree with the anticipated order of basicity, acetonitrile < t-butyl cyanide < cyclohexyl cyanide. At higher nitrile concentrations t-butyl cyanide

TABLE 4-1

Solvent Effects on Intramolecular, Intermolecular, and Dimer Hydrogen Bonding[a]

Solvent	Intramolecular $CH_3O-CH-CH_2CH_2OH$ (CH$_3$) $\Delta\nu$, cm^{-1}	Intermolecular $CH_3OH\cdots O(C_2H_5)_2$ $\Delta\nu$, cm^{-1}	Dimer $(CH_3OH)_2$ $\Delta\nu$, cm^{-1}
Gas phase	—	124	90
Hexane	91	133	103
Carbon tetrachloride	104	140	121
Benzene	81	121	107
Dichloromethane	115	143	133
Chloroform	128	183	152

[a]Allerhand and Schleyer [40].

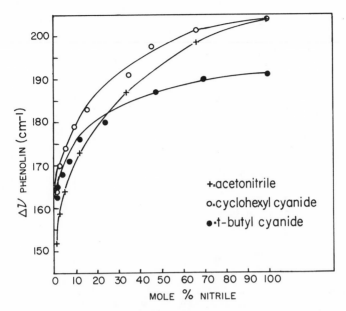

FIG. 4-4.   Effect of nitrile concentration on Δν of phenol in CCl₄.   Reprinted from Allerhand and Schleyer [40], p. 374, by courtesy of the American Chemical Society.

appears to be the weakest base. Frequency shifts for ether adducts of methanol and phenol are not affected much by concentration.

C. Δν/ν Correlations

Cutmore and Hallam [552] have related the slopes of the Δν/ν lines of Fig. 4-2 to the acidity of the solutes relative to pyrrole. When the slopes of the Δν/ν solvent lines are plotted against pKa, the dissociation constant for the conjugate acid, a straight line is obtained for each structural family of bases. Cutmore and Hallam suggest that the equations of these lines can be used to predict pKa values.

David and Hallam [571] obtained Δν/ν plots for various substituted thiophenols relative to Δν/ν for pyrrole. Again, the plots are linear (Fig. 4-5) and the slopes of the lines correlate with Taft parameters. In support of their claim that the slope, S, of Δν/ν lines is a measure of the intrinsic proton-accepting power of the base, they also found a correlation between S and pKa values measured in methanol.

$$pKa = 1.52\, S + 4.65 \text{ (in methanol).} \qquad\qquad (4\text{-}6)$$

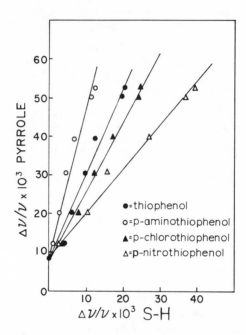

FIG. 4-5. Correlation of relative frequency shifts of ν_{SH} for thiophenol derivatives with relative frequency shift of ν_{NH} of pyrrole. Solvents in order of increasing shift are CCl_4, $CHCl_3$, C_6H_6, CH_3CN, $(CH_3)_2CO$, dioxane, $(n\text{-}C_4H_9)_2O$. David and Hallam [571].

Cole, Little, and Michell [485] have obtained $\Delta\nu/\nu$ plots for t-butanol, phenol, thiophenol, and t-butylmercaptan relative to $\Delta\nu/\nu$ for methanol. These are given in Fig. 4-6. Since a $\Delta\nu/\nu$ plot for pyrrole versus methanol has a slope of almost one, this series can be related to those pictured in Figs. 4-2 and 4-5. The points for thiols show considerable scatter from a straight line. Intensities and band widths are reported to correlate with $\Sigma(\sigma_I+\sigma_R)$, the sum of the inductive and resonance factors for the substituent groups of the solvents. Kagiya, Sumida, and Tachi [1218] have extended the use of $\Delta\nu/\nu$ plots to methanol or methanol-d [1217] as the reference acid. Benzene was the reference solvent. The proton-donating powers were evaluated by means of the ratios of the O—H shifts of the acids to those of methanol. The steeper the slope in Fig. 4-7, the stronger the acid. The observed order is t-butanol < methanol < trimethylsilanol < phenol < p-nitrophenol.

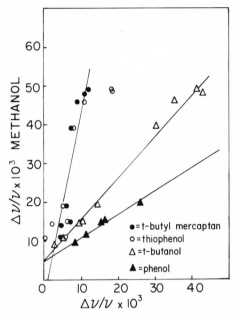

FIG. 4-6. Correlation of relative frequency shift of ν_{OH} of methanol with relative frequency shifts of t-butyl mercaptan, thiophenol, t-butanol, and phenol. Solvents in order of increasing shift are n-C_7H_{16}, CCl_4, $CHCl_3$, CS_2, C_6H_6, CH_3CN, $(CH_3)_2CO$, $(C_2H_5)_2O$, and dioxane. Cole, Little, and Michell [485].

Allerhand and Schleyer [40] propose an empirical linear free energy relationship

$$\nu^o - \nu^s / \nu^o \;=\; \alpha G \tag{4-7}$$

for correlation of solvent sensitive ir vibrations. If A—H...B vibrations are being examined, ν^o refers to the vapor phase value for A—H...B. The value for α is a measure of the solvent sensitivity of a particular ir vibration and is related to the "S" constants derived by David and Hallam [571]. G is a function of the solvent only. Since solvent shifts of ν_{CO} and ν_{SO} are proportional to solvent shifts of $\nu_{A—H...B}$, G values were calculated from solvent shifts for carbonyl bands of dimethylformamide [193] and benzophenone [193] and the sulfonyl band of dimethylsulfoxide [178]. An arbitrary value of 100 was assigned to dichloromethane to fix the scale. Values of G are given in Table 4-2 for a number of solvents.

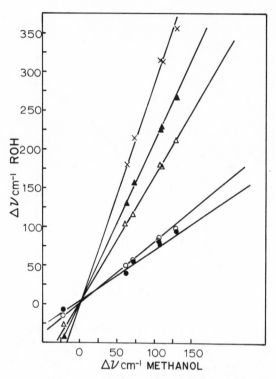

FIG. 4-7. Correlation of $\Delta\nu_{OH}$ for methanol adducts with $\Delta\nu_{OH}$ for adducts of p-nitrophenol (X), phenol (▲), trimethylsilanol (Δ), t-butanol (O), CH_3OD (●). Kagiya, Sumida, and Tachi [1218].

Allerhand and Schleyer [40] propose that Eq. (4-7) gives more accurate results than Eqs. (4-3), (4-4), or (4-5) and is easier to use. Figure 4-8 illustrates the correlation between $\nu_{O-H...O}$ and G values for methanol-ether. It is important to emphasize that Eq. (4-7) is of value only for $\nu_{A-H...B}$ bands and not for ν_{AH} solvent sensitivity. We have already shown how ν_{AH} is highly sensitive to the hydrogen bonding capabilities of solvents, and since Eq. (4-7) is simply a modified version of the KBM relationship, it would not be expected to give good correlations for A—H bands and G values.

These studies emphasize the importance of examining solvent and concentration effects in frequency shift studies. The G values proposed by Allerhand and Schleyer can aid in these studies since they have observed that $\nu_{A-H...B}$ will be independent of concentration if

TABLE 4-2

Solvent Shift Parameters[a]

Solvent	G	Solvent	G
Vacuum	0	Benzene	80
Perfluorooctane	36	Dioxane	86
Hexane	44	Methyl iodide	89
Cyclohexane	49	Acetonitrile	93
Dibutyl ether	61	Pyridine	94
Triethyl amine	62	1,2-Dichloroethane	95
Diethyl ether	64	Nitromethane	99
Tetrachloroethylene	64	Dichloromethane	100
Carbon tetrachloride	69	Chloroform	106
Carbon disulfide	74	Dibromomethane	108
Toluene	74	Bromoform	108

[a]Allerhand and Schleyer [40].

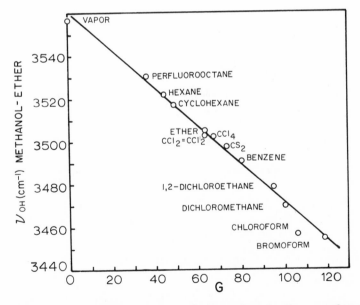

FIG. 4-8. Correlation of ν_{OH} of the methanol-ether complex with G values. Reprinted from Allerhand and Schleyer, [40], p. 377, by courtesy of the American Chemical Society.

the inert solvent and the base have similar G values. Carbon tetrachloride and ether have similar G values and no variation of $\Delta\nu$ with
ether concentration is observed.

D. $\Delta\nu$-ΔH Correlation

In 1937 Badger and Bauer [107] proposed that a linear relationship
exists between the enthalpy of the hydrogen bond and the frequency
shift of the H—A stretching vibration. During the past ten years,
this correlation has been challenged by several research groups, see
Refs. [70,915,1190,1981,2579], and supported by others, see Refs. [650,
1182,1778,1779,2024,2026,2533]. Much of the difficulty in testing the
validity of the $\Delta\nu$-ΔH correlation has now been removed with the availability of accurate enthalpy data and information on the concentration
dependence of $\Delta\nu_{OH}$. In the present section the extensive literature
on $\Delta\nu$-ΔH correlations will be reviewed with emphasis on both the use-
fulness and the limitations of $\Delta\nu$-ΔH relationships.

First, what does the frequency shift measure and how is it related
to the enthalpy of hydrogen bond formation? Bernstein [215] has considered this question by using an energy cycle for dissociation energies of A—H and A—HB in the gas phase:

$$D(A—H) = \Delta H_F(A) + \Delta H_F(H) - \Delta H_F(AH) \; , \tag{4-8}$$

where D is bond dissociation energy and ΔH_F is the heat of formation in
the gas phase. For the complex

$$D(A—HB) = \Delta H_F(A) + \Delta H_F(HB) - \Delta H_F(AHB) \; , \tag{4-9}$$

and the enthalpy of hydrogen bond formation may be represented as

$$-\Delta H = D(AH—B) = \Delta H_F(B) + \Delta H_F(AH) - \Delta H_F(AHB) \; . \tag{4-10}$$

The cycle is completed by using

$$D(H—B) = \Delta H_F(H) + \Delta H_F(B) - \Delta H_F(HB) \; . \tag{4-11}$$

By combining Eqs. (4-8) - (4-11) one obtains

$$D(H—B) = D(A—H) - D(A—HB) - \Delta H \; . \tag{4-12}$$

Bernstein [215] points out that an approximation of D(A—H) and D(A—HB) may be obtained by treating A—H and A—HB as diatoms and using the values for ν_{AH} and $\nu_{A—H...B}$ in the Morse potential. If the approximation is valid, Eq. (4-12) gives a relationship between $\Delta\nu$ and ΔH. However, it does imply that the relationship will hold only for a given donor site, B.

Purcell and Drago [2024] have examined the enthalpy-frequency shift correlation in terms of δE_{OH}, the change in O—H bond energy when O—H...B forms, and E_{HB}, the energy of the hydrogen bond H...B. The enthalpy change is represented as

$$\Delta H = \delta E_{OH} + E_{HB} .\qquad\qquad(4\text{-}13)$$

The equations

$$\Delta H = -(hcN/4\chi_e)\,\delta\omega_{OH} + E_{HB}\qquad\qquad(4\text{-}14)$$

and

$$\delta E_{OH} = -(hcN/4\chi_e)\,\delta\omega_{OH}\qquad\qquad(4\text{-}15)$$

were derived where h is Planck's constant, c is the speed of light, N is Avogadro's number, χ_e is the anharmonicity constant, and $\delta\omega_{OH}$ is the anharmonic energy in cm^{-1}. Values for both ν_{OH} and $\nu_{H...B}$ are needed to calculate $\delta\omega_{OH}$. However, $\nu_{H...B}$ is known for only a limited number of adducts so $\Delta\nu_{OH}$ has been used as an approximation of $\delta\omega_{OH}$. Purcell and Drago [2024] estimate a cumulative error of approximately 5% when $\Delta\nu_{OH}$ is used in place of $\Delta\omega_{OH}$. The basic problem here is that $\Delta\nu_{OH}$ differs from $\delta\omega_{OH}$ because of vibrational coupling of the OH and H...B coordinates.

In order for the frequency shift-enthalpy correlation to be substantiated there must be a linear relationship between E_{HB} and δE_{OH}. Equations (4-14) and (4-15) were used to obtain values for E_{HB} and δE_{OH} by using $\Delta\nu_{OH}$ in place of $\delta\omega_{OH}$ for phenol adducts. A linear relationship between E_{HB} and δE_{OH} was observed as shown in Fig. 4-9.

Purcell and Drago [2024] also carried out extended Hückel molecular orbital calculations in an attempt to determine why E_{HB} and δE_{OH} are linearly related. The system $(FHF)^-$ was used as a model and the

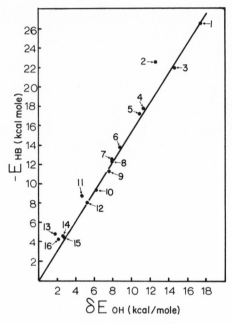

FIG. 4-9. Energy of hydrogen bond (E_{HB}) as a function of change in energy (δE_{OH}) of O—H bond for phenol adducts of (1) $(C_2H_5)_3N$, (2) NH_3, (3) $(CH_3)_3PO$, (4) $(CH_3)_2SO$, (5) $CH_3C(O)N(CH_3)_2$, (6) $(C_2H_5)_2O$, (7) $(C_2H_5)_2S$, (8) $(n\text{-}C_4H_9)_2S$, (9) $(n\text{-}C_4H_9)_2Se$, (10) $(CH_3)_2CO$, (11) CH_3CN, (12) $CH_3C(O)OC_2H_5$, (13) $C_6H_{11}F$, (14) $C_6H_{11}Br$, (15) $C_6H_{11}I$, (16) $C_6H_{11}Cl$. Purcell and Drago [2024].

effect that movement of fluorine atoms had on the H—F bonds was examined. The Lippincott-Schroeder [1481] model for O—H...O systems was used to allow for contraction of the H—F bond as the second fluorine was removed. A plot of the overlap energy of the original hydrogen bond as a function of the overlap energy of the new hydrogen bond is given in Fig. 4-10. This same type of plot is observed for phenol adducts and comparisons with $\Delta\nu$-ΔH plots based upon experimental data will be made later.

The $\Delta\nu$-ΔH relationship has been examined for both adducts of reference acids and reference bases. Most of the work concerns the adducts of reference acids with a series of organic bases. Examples include the reference acids phenol [719,1182,1778], substituted phenols [650], 2,2,2-trifluoroethanol [2255], 1,1,1,3,3,3-hexafluoro-2-propanol [2026], t-butanol [652], perfluoro-t-butanol [2252], and pyrrole [1811].

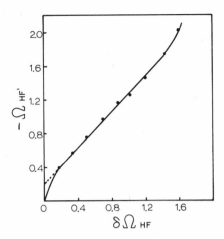

FIG. 4-10. The change in overlap energy of the original hydrogen bond as a function of the overlap energy of the new hydrogen bond. Reprinted from Purcell and Drago, [2024], p. 2878, by courtesy of the American Chemical Society.

The data for phenol and substituted phenols fall on the same line within experimental error. All of the OH reference acids give lines with essentially the same slope but a different intercept. The equations for the least-squares lines are given in Table 4-3.

The accuracy of using these equations for new bases depends upon (1) the strength of the hydrogen bond, (2) the donor site, (3) the reliability of frequency shift data, and (4) the solvent used. Hydrogen bond energies to within 0.3 kcal/mole can be predicted if the hydrogen bond energy is between -2 and -10 kcal/mole, the donor site is nitrogen or oxygen, the concentration effects of $\Delta\nu_{OH}$ have been checked, and the solvent is "inert".

Purcell *et al.* [2026,2255] have interpreted the nonzero intercepts for the linear correlations in Table 4-3 in terms of the Lippincott-Schroeder model for hydrogen bonding. Lippincott and Schroeder [1481] predicted a linear relationship between $\Delta\nu$ and ΔH for the enthalpy range of -2 to -14 kcal mole^{-1}. However, a slight curvature was predicted for very weak hydrogen bonds (less than 2 kcal mole^{-1}). This prediction has received further support from the theoretical work of Purcell and Drago (see Fig. 4-10).

TABLE 4-3

Reference-Acid Equations for ΔH-$\Delta \nu$ Correlations

Reference acid	Equation[a]	Reference
Phenol	$-\Delta H(\pm 0.2) = 0.0105 \,\Delta \nu + 3.0$	[650]
Substituted phenols	$-\Delta H(\pm 0.2) = 0.0103 \,\Delta \nu + 3.1$	[650]
2,2,2-Trifluoroethanol	$-\Delta H(\pm 0.2) = 0.0121 \,\Delta \nu + 2.7$	[2255]
1,1,1,3,3,3-Hexafluoro-2-propanol	$-\Delta H(\pm 0.3) = 0.0115 \,\Delta \nu + 3.6$	[2026]
t-Butanol	$-\Delta H(\pm 0.1) = 0.0106 \,\Delta \nu + 1.65$	[652]
Perfluoro-t-butanol	$-\Delta H(\pm 0.2) = 0.0106 \,\Delta \nu + 3.9$	[2252]
Pyrrole	$-\Delta H(\pm 0.1) = 0.0123 \,\Delta \nu + 1.8$	[1811]

[a]ΔH in kcal/mole; $\Delta \nu$ in cm^{-1}.

The nonzero intercept was predicted by Lippincott and Schroeder by comparing the different response of $\Delta\nu$ and ΔH to changes in the O...O distances for O—H...O systems. The frequency shift increases more rapidly than ΔH with a decrease in the O...O distance. Purcell, Stikeleather, and Brunk [2026] extended this approach by constructing ΔH versus R_{OB} and $\Delta\nu$ versus R_{OB} curves for phenol and 1,1,1,3,3,3-hexafluoro-2-propanol. Again, $\Delta\nu$ increases more rapidly than ΔH with a decrease in the O...O distance. These differences were explained by considering the factors which influence ΔH and $\Delta\nu$. Equation (4-13) was expanded to include V_{OB}, the van der Waals repulsion between O and B:

$$\Delta H = E_{HB} + (\delta E_{OH} + V_{OB}) \ . \tag{4-16}$$

The equation

$$\Delta\nu = \delta K_{OH} + K_{HB} \tag{4-17}$$

was used to represent the frequency shift, where δK_{OH} is the decrease in OH frequency as a result of weakening the O—H bond and K_{HB} is the increase in the OH frequency caused by forming the HB bond and mixing of OH and HB motions.

The curves in Fig. 4-11 suggest that K_{HB} has less effect on δK_{OH} than $(\delta E_{OH} + V_{OB})$ have on E_{HB}. This is proposed as an explanation for

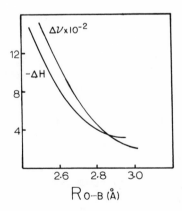

FIG. 4-11. Lippincott-Schroeder relations of $-\Delta H$ and $\Delta\nu$ versus oxygen donor distance. $\Delta\nu$ in hundreds of cm^{-1}, ΔH in kcal/mole. Purcell, Stikeleather, and Brunk [2026].

the change in sensitivity of ΔH and Δν for weak hydrogen bonds and the
nonzero intercepts for Δν-ΔH correlations.

In a recent review Pimentel and McClellan [1937] have presented a
plot for phenol adducts which illustrates very nicely the different
response of Δν and ΔH to changes in the O...O distance. Literature
values of ΔH [1751] within ±0.5 kcal/mole of an integer value were averaged and the resulting points were plotted versus Δν. The tendency to
curve over at higher ΔH (smaller O...O distances) is shown in Fig. 4-12.

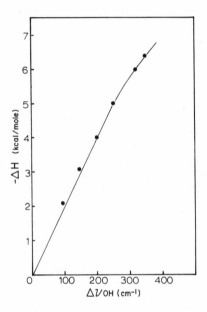

FIG. 4-12. ΔH versus Δν for phenol adducts of oxygen donors.
Values within ±0.5 kcal/mole of an integer value of ΔH were averaged
together. Pimentel and McClellan [1937].

The least-squares equation of the curve is

$$-\Delta H = 0.0221 \ \Delta\nu - 1.0 \times 10^{-5} \ \Delta\nu^2 \ . \tag{4-18}$$

This represents the first statistical treatment of Δν-ΔH data in terms
of a second-order polynomial.

The general form of the equations listed in Table 4-3 is $-\Delta H =$
$m\Delta\nu_{OH} + b$. Drago, O'Bryan, and Vogel [652] have proposed that b is a

property of the acid and is related to the maximum amount of electron density that the hydrogen atom can accumulate in forming a hydrogen bond. The stronger the acid, the larger b will be. This gives the order of acidity for the acids listed in Table 4-3 as t-butanol < pyrrole < 2,2,2-trifluoroethanol < phenol < 1,1,1,3,3,3-hexafluoro-2-propanol < perfluoro-t-butanol. In addition, bases which are "soft" or polarizable will not fit the same relationship as "hard" bases, since covalent contributions to the bonding are more important in the former case. As a result, the proton becomes saturated at a lower enthalpy with a "soft" base such as diethyl sulfide. The model proposes that the electron density which is transferred into the acid is distributed between the proton and the functional group attached to the proton. The former effect gives rise to the intercept and the latter to the change in Δv_{OH}.

The slopes of the equations in Table 4-3 do not correlate with acidity of the acid. Sherry and Purcell [2252,2255] have attributed this lack of correlation to the moderating influence of $\delta E_{OH} + V_{OB}$ [Eq. (4-16)] on the strength of the hydrogen bond. The similarity in ΔH versus ΔH plots for 2,2,2-trifluoroethanol and phenol when 1,1,1,3,3,3-hexafluoro-2-propanol is used as a reference is cited as evidence for the similar effect which $(\delta E_{OH} + V_{OB})$ terms have on E_{HB} [Eq. (4-16)]. The least-squares equations are

$$\Delta H = 0.86 \ \Delta H_{ref} - 0.9 \tag{4-19}$$

for 2,2,2-trifluoroethanol and

$$\Delta H = 0.85 \ \Delta H_{ref} - 0.6 \tag{4-20}$$

for phenol. However, the frequency correlations do have a different slope with equations

$$\Delta v = 0.79 \ \Delta v_{ref} - 36 \tag{4-21}$$

for 2,2,2-trifluoroethanol and

$$\Delta v = 0.89 \ \Delta v_{ref} - 41 \tag{4-22}$$

for phenol. The more rapid change in Δv for phenol causes the slope of the Δv-ΔH plot for phenol to be smaller than that of 2,2,2-trifluoroethanol (0.0121 versus 0.0105). Thus, the ΔH-$\Delta \nu$ slopes do not general-ly reflect the order of acidities, but the ΔH-ΔH and $\Delta \nu$-$\Delta \nu$ plots do.

Neerinck and Lamberts [1779] were the first to investigate $\Delta \nu$-ΔH relationships for reference bases. Enthalpies for pyridine adducts of substituted alcohols and phenols correlated well with $\Delta \nu$. Their data have been analyzed by least-squares and the resulting equation is

$$-\Delta H \text{ kcal/mole } (\pm 0.2) = 0.0126\Delta \nu_{OH} + 0.43 \ . \qquad (4\text{-}23)$$

Since data for twenty-one alcohols and phenols are included, Eq. (4-23) is reliable for predicting enthalpies for pyridine adducts of OH acids.

Drago, O'Bryan, and Vogel [652] investigated the $\Delta \nu$-ΔH relation-ship for the reference bases pyridine, N,N-dimethylacetamide, and ethyl acetate. The $\Delta \nu$-ΔH plots are illustrated in Fig. 4-13.

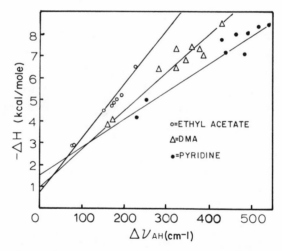

FIG. 4-13. Enthalpy versus frequency shift for the reference bases ethyl acetate, N,N-dimethylacetamide, and pyridine. Data points are included for pyrrole, t-butyl alcohol, p-t-butylphenol, 2,2,2-trifluoroethanol, phenol, p-chlorophenol, p-fluorophenol, 1,1,1,3,3,3-hexafluoro-2-propanol, m-fluorophenol, and m-trifluoromethylphenol. Nozari and Drago [1811].

The equations are

pyridine: $-\Delta H \text{ kcal/mole } (\pm 0.7) = 0.013 \ \Delta \nu_{AH} + 1.4 \ , \qquad (4\text{-}24)$

FREQUENCY SHIFT CORRELATIONS 217

DMA: $-\Delta H$ kcal/mole $(\pm 0.4) = 0.017\ \Delta\nu_{AH} + 1.0$,   (4-25)

ethyl acetate: $-\Delta H$ kcal/mole $(\pm 0.2) = 0.025\ \Delta\nu_{AH} + 0.6$. (4-26)

The acids included pyrrole and substituted alcohols and phenols. The accuracy of Eq. (4-25) has been questioned by Stymne *et al.* [2361], who derived slope and intercept values of 0.023 and -1.92 from $\Delta\nu$-ΔH data for seventeen monosubstituted phenols.

Relationships between reference-acid lines and reference-base lines have been derived that can be used to calculate a reference-acid line for a new hydrogen-bonding acid [1811]. If the hydrogen bonding acid has $\Delta\nu$-ΔH values which fall on the reference base lines for OH acids (Fig. 4-13) and gives a straight-line plot of $\Delta\nu_{AH}$ versus $\Delta\nu_{OH}$ for phenol with a zero intercept, the acid is predicted to give a linear $\Delta\nu$-ΔH relationship. Pyrrole [1811] satisfies these requirements; the intercept for the constant-acid line calculated from the slope of the $\Delta\nu_{AH}$ versus $\Delta\nu_{OH}$ line (Fig. 4-14) is 1.6 compared to the experi-

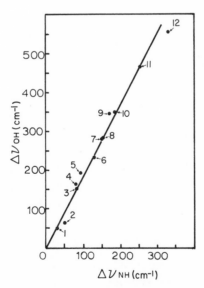

FIG. 4-14. $\Delta\nu_{OH}$ for phenol adducts versus $\Delta\nu_{NH}$ for pyrrole adducts. Bases are (1) C_6H_6, (2) xylene, (3) CH_3CN, (4) $CH_3C(O)OC_2H_5$, (5) $(CH_3)_2CO$, (6) dioxane, (7) $(C_2H_5)_2O$, (8) $(CH_2)_4O$, (9) $CH_3C(O)N(CH_3)_2$, (10) $(CH_3)_2SO$, (11) C_5H_5N, (12) $(C_2H_5)_3N$. Nozari and Drago [1811].

mental value of 1.8 (Table 4-3). Chloroform does not follow these relationships and has been shown not to give a linear $\Delta\nu$-ΔH correlation [2291].

The hydrogen bond energies in the Appendix can be used to derive equations for other reference acids and reference bases. Least-squares analysis of the data in terms of $y = mx + b$ was carried out for acids in the Appendix which have at least six enthalpies of adduct formation with corresponding frequency shifts. The results of the analysis are given in Table 4-4.

Seven of the equations in Table 4-4 have standard deviations of 0.5 kcal/mole or less; twelve more have standard deviations between 0.5 and 1.0 kcal/mole, and three standard deviations are between 1.1 and 1.3 kcal/mole. Since the data used to derive these equations were obtained by a variety of methods in several different laboratories, the standard deviations are not unreasonable. However, only those with standard deviations of 0.5 kcal/mole or less are useful for predicting enthalpies from frequency shifts. The lower standard deviations of the equations in Table 4-3 can be attributed to the following factors: (1) the calorimetric method was used to measure enthalpies; (2) the collection of data for a given acid was carried out in one laboratory; (3) the effect of solvent influence on $\Delta\nu$ was considered; and (4) the range of enthalpies is not as great as those in Table 4-4.

Two equations are given for phenol in Table 4-4. The first equation includes one data point for each base (marked with either a single or a double asterisk in the Appendix). Values measured with carbon tetrachloride as the solvent were used for all bases except amines. The bases which show the largest deviation from this equation (marked with a double asterisk in the Appendix) are pyridines, esters, acid fluorides, sulfides, thioamides, and bases with multiple donor sites. When $\Delta\nu$-ΔH data for eighty-three of these bases are removed, the second equation is obtained with a standard deviation of 0.5 kcal/mole.

In 1967 Murthy and Rao [1751] analyzed $\Delta\nu$-ΔH data for 111 adducts of phenol and reported the equation

$$-\Delta H \text{ (kcal/mole)} = 0.01 \, \Delta\nu_{OH} + 1.75. \tag{4-27}$$

TABLE 4-4

ΔH-$\Delta \nu$ Least-Squares Lines for Reference Acids

$-\Delta H$ (kcal/mole) = $m\Delta\nu_{AH}$ (cm^{-1}) + b

Acid	Number of data points	m	b	Standard deviation, σ
CH_3OH	58[a]	0.0065	2.58	1.0
C_2H_5OH	15	0.0051	1.92	0.93
CF_3CH_2OH	19	0.0138	1.59	0.80
n-C_3H_7OH	11	0.0072	1.82	1.0
$(CH_3)_2CHOH$	15	0.0142	0.02	0.16
$(CF_3)_2CHOH$	17[a]	0.0112	2.42	1.25
$(CH_3)_3COH$	17[a]	0.0162	0.83	0.44
$(CF_3)_3COH$	11[a]	0.0073	4.44	0.54
C_6H_5OH	298[a]	0.0140	1.09	0.93
	215[b]	0.0148	0.87	0.50
C_6H_5OD	8	-0.0027	1.92	0.79
p-F-C_6H_4OH	39	0.0137	1.21	1.0
p-Cl-C_6H_4OH	16	0.013	1.48	1.1
p-CH_3O-C_6H_4OH	6	0.0118	2.22	1.3
m-F-C_6H_4OH	6	0.0105	2.96	0.29
C_6Cl_5OH	52	0.0041	3.46	0.55
α-Naphthol	20	0.0061	3.76	0.48
HNCS	35	0.0102	1.59	0.61
HNCO	7	0.0055	1.89	0.47
HN_3	6	0.0040	2.11	0.31
N-methylaniline	10	0.0394	2.08	0.69
Indole	7	0.0116	0.82	0.76

[a]Enthalpy values used for this analysis are marked with a single or double asterisk in the Appendix. [b]Only enthalpy values marked with a single asterisk in the Appendix were used.

for predicting ΔH. The standard deviation was 0.65. Equation (4-27)
has the same slope as the equations listed for phenol in Table 4-4,
but the intercept is larger. Different correlations were proposed for
weak, medium, and strong hydrogen bonds in phenol adducts [1751]. The
equations are

$$0 - 3 \text{ kcal: } -\Delta H \text{ (kcal/mole)} = 0.014 \; \Delta\nu + 0.62 \; (\pm 0.28) \qquad (4-28)$$

$$3 - 6 \text{ kcal: } -\Delta H \text{ (kcal/mole)} = 0.012 \; \Delta\nu + 1.87 \; (\pm 0.1) \qquad (4-29)$$

$$6 - 10 \text{ kcal: } -\Delta H \text{ (kcal/mole)} = 0.010 \; \Delta\nu + 2.37 \; (\pm 0.55) \; . \qquad (4-30)$$

As the strength of the hydrogen bond increases, the intercept increases
but the slope remains constant. These results suggest that $\Delta\nu$-ΔH correlations for specific base types might be useful. The data for phenol
adducts in the Appendix have been analyzed on this basis, and the results are tabulated in Table 4-5.

E. Limitations on the Use of $\Delta\nu$-ΔH Correlations

    The variation in slope and intercept for phenol adducts of dif-
ferent base types (Table 4-5) illustrates the lack of a general $\Delta\nu$-ΔH
correlation. The largest deviations are found for nitriles [915], sulfides [2533], thiocyanates, and thioamides [915].

    The lack of a general relationship does not preclude the use of
frequency shifts to predict hydrogen bond enthalpies, particularly if
the standard deviations of equations in Tables 4-3, 4-4, and 4-5 provide the desired accuracy. If frequency shifts are measured in an
inert solvent with low concentration of base [40], the use of the appropriate equation in Tables 4-3, 4-4, and 4-5 will provide a reasonable
estimate of ΔH.

    Gramstad and Mundheim [914a] have reported a $\Delta\nu$-ΔH correlation for
chloroform adducts of phosphoryl compounds. However, the enthalpy
range used as a basis for the correlation is very narrow. When a wider
range of bases and enthalpy values are used (Appendix), $\Delta\nu_{CD}$ for
deuterochloroform adducts does not correlate with ΔH for the corresponding chloroform adducts [2291]. Only small shifts in $\Delta\nu_{CD}$ are observed
(<80 cm^{-1}) which are not sensitive to changes in enthalpy.

TABLE 4-5

ΔH-$\Delta\nu$ Equations for Phenol Adducts of Different Base Types

$-\Delta H$ (kcal/mole) $= m\Delta\nu_{AH}$ (cm^{-1}) $+ b$

Base type	Number	m	b	Standard deviation, σ
Aldehydes	12	0.0113	1.92	0.88
Amides	30	0.0147	1.08	0.60
Amines	27	0.0124	0.82	1.1
Esters	15	0.0136	1.75	0.62
Ethers	19	0.0172	0.66	0.63
Ketones	15	0.0127	1.81	0.67
Nitriles	9	0.0276	-0.31	0.63
Phosphoryls	36	0.0125	1.49	1.4
π-Bases	50	0.0115	0.86	0.30
Organic halides	10	0.0127	0.87	0.42
Sulfoxides	15	0.0136	1.45	0.80
Sulfides	9	0.0062	2.04	0.20
Thioamides	6	0.0072	1.57	0.65
Thiocyanates, Isothiocyanates	13	0.0379	-2.22	0.49

F. Double-Scale Enthalpy Equation

Another empirical relationship which is useful for correlating and predicting enthalpies for hydrogen bonds is the Drago-Wayland [647, 654] double-scale enthalpy equation:

$$-\Delta H = E_A E_B + C_A C_B . \tag{4-31}$$

This equation permits comparisons of a wide variety of Lewis acid-base systems. Two empirically determined parameters, E_A and C_A, are assigned to each acid and two, E_B and C_B, are assigned to each base. Substitution into Eq. (4-31) gives enthalpy of adduct formation. The importance of electrostatic (E) and covalent (C) contributions to the bonding are indicated by the magnitude of these constants. Equation (4-31) has been found to correlate the enthalpies of interaction where reversals in donor strength occur. For example, phenol interacts more strongly with oxygen donors than with sulfur donors, but iodine interacts more strongly with sulfur donors than with oxygen donors [1794].

Enthalpy data for adducts of iodine and phenol were used to derive the original set of E and C parameters [654]. Recently, Drago, Vogel, and Needham [653] published a tabulation of E and C parameters for about thirty acids and forty bases. A least-squares analysis was used to find the E and C parameters which gave the best fit between experimental enthalpies and those calculated from Eq. (4-31). Tables 4-6 and 4-7 list the E and C parameters for hydrogen bonding acids and Lewis bases. With these E and C parameters, Eq. (4-31) can be used to reproduce most of the known enthalpies determined in the gas phase or in "inert" solvents such as carbon tetrachloride. In addition,

TABLE 4-6

E and C Parameters for Hydrogen-Bonding Acids[a]

Acid	C_A	E_A
Thiophenol	0.198	0.987
p-tert-Butylphenol	0.387	4.06
p-Methylphenol	0.404	4.18
Phenol	0.442	4.33
p-Fluorophenol	0.446	4.17
p-Chlorophenol	0.478	4.34
m-Fluorophenol	0.506	4.42
m-Trifluoromethylphenol	0.530	4.48
t-Butanol	0.300	2.04
Trifluoroethanol	0.434	4.00
Hexafluoroisopropanol	0.623	5.93
Pyrrole	0.295	2.54
Chloroform[b]	0.159	3.02
Isocyanic acid (HNCO)	0.258	3.22
Isothiocyanic acid (HNCS)	0.227	5.30

[a]Drago, Vogel, and Needham [653].   [b]Slejko, Drago, and Brown [2291].

TABLE 4-7

E and C Parameters for Bases[a]

Base	C_B	E_B
Pyridine	6.40	1.17
Ammonia	3.46	1.36
Methylamine	5.88	1.30
Dimethylamine	8.73	1.09
Trimethylamine	11.54	0.808
Ethylamine	6.02	1.37
Diethylamine	8.83	0.866
Triethylamine	11.09	0.991
Acetonitrile	1.34	0.886
Chloroacetonitrile	0.530	0.940
Dimethycyanamide	1.81	1.10
Dimethylformamide	2.48	1.23
Dimethylacetamide	2.58	1.32
Ethyl acetate	1.74	0.975
Methyl acetate	1.61	0.903
Acetone	2.33	0.987
Diethyl ether	3.25	0.963
Isopropyl ether	3.19	1.11
n-Butyl ether	3.30	1.06
p-Dioxane	2.38	1.09
Tetrahydrofuran	4.27	0.978
Tetrahydropyran	3.91	0.949
Dimethylsulfoxide	2.85	1.34
Tetramethylene sulfoxide	3.16	1.38
Dimethyl sulfide	7.46	0.343
Diethyl sulfide	7.40	0.339
Trimethylene sulfide	6.84	0.352
Tetramethylene sulfide	7.90	0.341
Pentamethylene sulfide	7.40	0.375
Pyridine-N-oxide	4.52	1.34
4-Methylpyridine-N-oxide	4.99	1.36
4-Methoxypyridine-N-oxide	5.77	1.37
Tetramethylurea	3.10	1.20

TABLE 4-7, continued

Base	C_B	E_B
Trimethylphosphine	6.55	0.838
Benzene[b]	0.681	0.525
p-Xylene	1.78	0.416
Mesitylene	2.19	0.574
Dimethyl selenide	8.33	0.217
Hexamethylphosphoramide[b]	3.55	1.52
$CH_3CH_2C(CH_2O)_3P$[b]	6.41	0.55

[a]Drago, Vogel, and Needham [653]. [b]Slejko, Drago, and Brown [2291].

Eq. (4-31) can be used together with available enthalpy data to establish whether a $\Delta\nu$-ΔH correlation exists for a given series of hydrogen-bonded systems.

A recent example of the use of E and C parameters to predict enthalpies is the study of pyrrole adducts [1811]. A calorimetric method was used to determine the enthalpy of hydrogen bonding for adducts of pyrrole with dimethyl sulfoxide, pyridine, and triethylamine. Since these three bases have known E_B and C_B parameters, E_A and C_A values for pyrrole can be calculated using Eq. (4-31). The calculated E_A and C_A values for pyrrole were used with Eq. (4-31) to estimate ΔH for adducts of pyrrole with other bases. When the calculated and measured ΔH values were plotted against corresponding $\Delta\nu$ values, a straight line was obtained. The equation of the least-squares line is given in Table 4-3. This application of Eq. (4-31) is important since it reduces the amount of time for thermodynamic measurements and also increases the accuracy of prediction for weaker hydrogen bonds, which are difficult to measure experimentally.

The E and C parameters can be used to predict whether acids will have a linear $\Delta\nu$-ΔH correlation. Acids with both large E and C parameters and similar C_A/E_A ratios have been found to give linear $\Delta\nu$-ΔH relationships, while those with low C_A/E_A ratios do not. For example, the $\Delta\nu$-ΔH correlation is not valid for chloroform, which has a very small C_A/E_A ratio [2291].

Equation (4-31) can also be used for systems where competing equilibria such as intramolecular hydrogen bonding or self-association are
present if accurate thermodynamic data can be obtained for a series of
bases with known E_B and C_B parameters [949a]. The procedure involves
treating the competing equilibrium as a systematic error and calculating what constant change in $-\Delta H_{obsd}$ gives the best fit for Eq. (4-31).
For example, the intramolecular hydrogen bonding energy in 1,1,1,3,3,3-
hexafluoro-2-propanol (HFIP) was determined to be -1.1 kcal/mole by this
procedure. Experimental enthalpy values for HFIP-base adducts should
be more negative by 1.1 kcal/mole to account for the intramolecular
interaction.

G. Single-Scale Enthalpy Equation

A single-scale enthalpy relationship has been proposed by Sherry
and Purcell [2252,2255] that is useful for detecting substituent inductive effects as well as resonance and steric effects. The relationship
is based on the constant ratios of enthalpy values for two different
reference acids. The equation

$$-\Delta H = \alpha\beta_B \tag{4-32}$$

is used to predict enthalpies in cases where 2,2,2-trifluoroethanol is
the reference acid; α equals $\Delta H/\Delta H_{ref}$; and β_B is the slope of a plot of
$-\Delta H$ versus α for a series of acids interacting with a given base.
Figure 4-15 illustrates ΔH versus α plots for seven alcohols with six
different bases. Table 4-8 compares measured enthalpies with enthalpies calculated from Eq. (4-32). A correlation between α and Taft σ^*
values was also observed [2255].

Terent'ev [2427a] has used literature data to derive constants,
ΔH_A and ΔH_B, for calculating the enthalpy of a reaction

$$\Delta H = \Delta H_A \Delta H_B \ . \tag{4-32a}$$

The phenol-pyridine adduct was used as the reference. The same approach
was used to obtain relationships for $\Delta\nu$ and K.

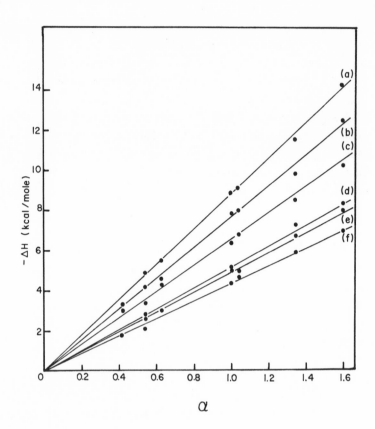

FIG. 4-15. ΔH versus α for (a) $(C_2H_5)_3N$, (b) C_5H_5N, (c) $CH_3C(O)N(CH_3)_2$, (d) $(C_2H_5)_2O$, (e) $(CH_3)_2CO$, (f) CH_3CN. Reprinted from Sherry and Purcell, [2252], p. 1857, by courtesy of the American Chemical Society.

H. Δν-ΔG Correlations

Gramstad *et al.* [907,908,914] have reported a linear correlation between $\Delta\nu_{OH}$ and ΔG for adducts of phenols with Lewis bases for a given base type. We have used the data in the Appendix to test this correlation for a variety of acid-base systems. The least-squares equations along with the deviations are listed in Table 4-9. The results indicate that adducts of reference acids with a given structural type of base do give a good Δν-ΔG correlation. Exceptions were found for phenol-organic halide adducts and phenol-π-base adducts which had standard deviations of 0.35 and 0.53 kcal/mole, respectively.

TABLE 4-8

Application of Eq. (4-32)[a]

Base	$(CF_3)_2CHOH$ ($\alpha = 1.35$) Calc.	Exp.	C_6H_5OH ($\alpha = 1.04$) Calc.	Exp.	CF_3CH_2OH ($\alpha = 1.00$) Calc.	Exp.	$(CH_3)_3COH$ ($\alpha = 0.54$) Calc.	Exp.
CH_3CN	5.9	5.9	4.5	4.7	4.3	4.4	2.3	2.1
$(CH_3)_2CO$	6.6	6.7	5.1	4.9	4.9	5.1	2.6	2.6
$(C_2H_5)_2O$	6.9	7.2	5.4	—	5.1	5.1	2.8	2.8
$CH_3C(O)N(CH_3)_2$	8.8	8.5	6.7	6.8	6.5	6.4	3.5	3.4
C_5H_5N	10.3	9.8	7.9	8.0	7.6	7.8	4.1	4.2
$(C_2H_5)_3N$	11.8	11.5	9.1	9.1	8.7	8.8	4.7	4.9

[a]Sherry and Purcell [2255].

TABLE 4-9

Free Energy - Frequency Shift Correlation

$-\Delta G$ (kcal/mole) $= m\Delta v_{AH} + b$

System	Data points	m	b	Standard deviation, σ
CH_3OH-Amides	32	0.0039	-0.61	0.25
CH_3OH-Amine N-oxides	9	0.0110	-1.38	0.21
CH_3OH-Phosphoryls	15	0.0074	-0.12	0.07
C_6H_5OH-Aldehydes	15	0.0090	-0.70	0.06
C_6H_5OH-Amides	37	0.0079	+0.22	0.16
C_6H_5OH-Amines	27	0.0044	+0.24	0.22
C_6H_5OH-Esters	15	0.0101	-0.48	0.16
C_6H_5OH-Ethers	19	0.0089	-1.14	0.18
C_6H_5OH-Ketones	26	0.0106	-0.88	0.17
C_6H_5OH-Nitriles	31	0.0120	-0.96	0.18
C_6H_5OH-Phosphoryls	57	0.0090	+0.11	0.19
C_6H_5OH-Sulfides	10	0.0035	-0.67	0.08
C_6H_5OH-Sulfates	11	0.0129	-0.72	0.17
C_6H_5OH-Sulfoxides	16	0.0093	-0.29	0.06
C_6H_5OH-Thioamides	6	0.0048	-0.16	0.22
C_6H_5OH-Thiocyanates	6	0.0100	-0.77	0.03
C_6H_5OH-Isothiocyanates	7	-0.0013	-0.04	0.14
p-Cresol-π bases	9	0.0065	-0.86	0.21
m-Cresol-π bases	9	0.0071	-0.88	0.22
o-Cresol-π bases	9	0.0059	-0.84	0.21
C_6Cl_5OH-Amides	28	0.0079	-0.60	0.07
C_6Cl_5OH-Phosphoryls	17	0.0072	-0.48	0.07
α-Naphthol-phosphoryls	16	0.0082	-0.39	0.07
α-Naphthol-π bases	9	0.0045	-0.68	0.17
β-Naphthol-π bases	9	0.0051	-0.73	0.18

[a]Only systems with standard deviations of less than 0.25 kcal/mole are included.

Pullin and Werner [1996] noted that ΔG versus $\Delta \nu$ plots for adducts of indole, pyrrole, N-methylacetamide, and diphenylamine show an asymptotic approach to the ΔG axis as $\Delta \nu$ decreases (Fig. 4-16).

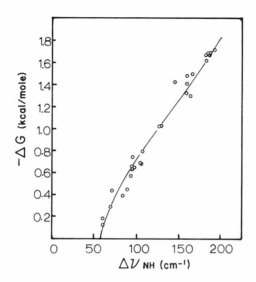

FIG. 4-16.   Correlation of $-\Delta G$ with $\Delta \nu_{NH}$ for indole adducts of carbonyl compounds in CCl_4 at 32°C.   Reprinted from Pullin and Werner [1996], p. 1265, by courtesy of Pergamon Press.

The equation

$$\Delta G = k \log \Delta \nu + C \qquad\qquad\qquad (4\text{-}33)$$

was derived and values of k and C for the acids are given in Table 4-10. By representing $\Delta \nu_{NH}$ in terms of a two-parameter equation,

$$\Delta \nu = D \cdot A \ cm^{-1} , \qquad\qquad\qquad (4\text{-}34)$$

where D and A are constants for the proton donor and acceptor, Eq. (4-33) can be written

$$\Delta G = k \ \ln D + k \ \ln A + C . \qquad\qquad\qquad (4\text{-}35)$$

Values of D and A were obtained by assigning a value of $D = 10.00$ to indole and calculating values of D and A from frequency shift data

TABLE 4-10

Values of k and C^a

Acid	k (kcal/mole)	C (kcal/mole)
Indole	-3.3	+6.0
Pyrrole	-3.2	+5.9
N-Methylacetamide	-3.2	+5.4
Diphenylamine	-2.7	+4.7
N-Methylaniline	-2.5	+4.2

^{a}Pullin and Werner [1996].

(Table 4-11). If D is fixed by using a reference proton donor, then,

$$\Delta G = k \ln A + C_D , \qquad (4-36)$$

where C_D is a constant for a particular acid.

I. Substituent Constant Correlations

1. Hammett aromatic substituent constants (σ)

Correlations of $\Delta \nu$ with σ have been observed for adducts of phenols and substituted phenols, see Refs. [1421,1779,2533,2564]. As a result, one would expect a correlation between σ and ΔH. Such a correlation has been reported for adducts of *meta-* and *para-* substituted phenols [650,1779]. Neerinck and Lamberts [1779] derived the equation

$$-\Delta H = 1.87 \, \sigma + 6.50 \qquad (4-37)$$

from enthalpies for pyridine adducts of *meta-* and *para-* substituted phenols.

Vogel and Drago [2533] used a σ-ΔH correlation and Eq. (4-31) to calculate enthalpies for adducts of diethyl sulfide with substituted phenols. The results are given in Table 4-12. The agreement of enthalpy values calculated by the two methods is quite good.

Assignment of the ν_{OH} peaks for bases which form both a π-complex and a σ-complex with phenol was aided by the correlation between $\Delta \nu_{OH}$ and the Hammett substituent constant [2564]. For example, ν_{OH} bands at

TABLE 4-11

Comparison of Calculated and Observed Solvent Shifts[a]

(observed in parentheses)

	D/A Constant	Indole 10.00	Pyrrole 8.9	N-Methylacetamide 6.2	Diphenylamine 5.4	N-Methylaniline 3.35
N,N-Diethylacetamide	19.2	192 (187)	171 (167)	119 (122)	104 (106)	64 (67)
N,N-Diethylpropionamide	18.7	187 (183)	167 (166)	116 (112)	101 (98)	63 (66)
N,N-Dimethylformamide	16.8	168 (158)	150 (145)	104 (105)	91 (93)	56 (58)
N,N-Diphenylacetamide	14.4	144 (145)	128 (128)	89 (90)	78 (77)	48 (49)
Mesityl oxide	10.8	108 (109)	96 (94)	67 (68)	58 (65)	36 (33)
Methyl acetate	8.2	82 (85)	73 (76)	51 (50)	44 (41)	
Methyl formate	6.3	63 (61)	56 (61)	39 (41)	34 (34)	
Pyridine	27.4	274 (269)	244 (239)	170 (161)	148 (147)	92 (105)

[a]Pullin and Werner [1996].

TABLE 4-12

Comparison of Calculated Enthalpies for Adducts of Diethyl Sulfide[a]

Acid	-ΔH from σ (kcal/mole)	-ΔH from Eq. (4-31) (kcal/mole)
p-$CH_3C_6H_4OH$	4.3	4.4
C_6H_5OH	4.6	4.7
p-FC_6H_4OH	4.7	4.8
p-ClC_6H_4OH	5.0	5.0
p-BrC_6H_4OH	5.0	—
m-FC_6H_4OH	5.2	5.2
m-$CF_3C_6H_4OH$	5.4	5.4

[a]Vogel and Drago [2533].

3552 and 3447 cm^{-1} for the phenol-anisole adduct were assigned to phenol bonded to π-electrons and oxygen electrons, respectively. These assignments were based on the predicted frequencies for π-orbitals (Hammett substituent correlation with $\Delta\nu$) and n-orbitals (Taft σ^* substituent correlation with $\Delta\nu$).

Correlation of σ with log K have been reported for phenol adducts of substituted pyridines [2125,2406], methanol adducts of substituted pyridine N-oxides [847a], and adducts of N,N-dimethylacetamide with substituted phenols [643].

2. Taft alkyl substituent constants (σ^*)

Sara *et al.* [2169b] observed a linear correlation between -ΔH and the sum of Taft's σ^* constants for adducts of phenol with butyl sulfides and butyl sulfoxides. Least-squares equations are

$$\text{sulfides:} \quad -\Delta H \text{ (kcal/mole)} = -2.19 \; \Sigma \; \sigma^* + 2.73 \qquad (4\text{-}38)$$

$$\text{sulfoxides:} \quad -\Delta H \text{ (kcal/mole)} = -2.38 \; \Sigma \; \sigma^* + 5.52 \; . \qquad (4\text{-}38a)$$

Correlations of this type have also been derived for chloroform adducts
in carbon tetrachloride [917a].

sulfoxides, amides: $-\Delta H$ (kcal/mole) = $-0.34 \, \Sigma \, \sigma^* + 1.7$ (4-39)

phosphoryls: $-\Delta H$ (kcal/mole) = $-0.25 \, \Sigma \, \sigma^* + 3.0$ (4-39a)

J. $\Delta\nu$-r Correlation

Correlations between $\Delta\nu_{AH}$ and the A...B distance in A—H...B are
well known. Ratajczak and Orville-Thomas [2062] have reviewed this
relationship for O—H...O systems. Plots of $\Delta\nu$ versus A...B distance
(r) published by Nakamoto *et al.* [1768] are representative of the early
work in this area. Pimentel and Sederholm [1939] derived the equations

$$\Delta\nu_{OH} = 4.43 \times 10^3 \, (2.84 - r)$$ (4-40)

and

$$\Delta\nu_{NH} = 0.548 \times 10^3 \, (3.21 - r)$$ (4-41)

for O—H...O and N—H...O systems. Theoretical treatment of the different response of $\Delta\nu$ to shorter A...B distances was given in I.D.

Recently Bellamy and Pace [187] have examined differences in the
slope and intercept of $\Delta\nu$ versus r plots. In the general equation

$$\Delta\nu = K(R - r)$$ (4-42)

K is a constant characteristic of A and B, r is the final A...B distance, and R is the closest distance AH can approach B without sufficient hydrogen bond formation to affect ν_{AH}. Values of R derived from
$\Delta\nu$ versus r plots are quite close to the sum of van der Waals radii of
A and B atoms (2.7 Å for F—H...F and 2.8 Å for O—H...O). This is
used as evidence for the lack of importance of the effective radius of
the hydrogen atom in calculations of A—H...B distances.

Bellamy and Owen [183] have derived a new empirical relationship
between $\Delta\nu_{AH}$ and the A—H...B distance which is based on the importance

of van der Waals repulsions between the nonbonding orbitals of A and B
atoms. The equation is

$$\Delta\nu \ (cm^{-1}) = 50[(d/R)^{12} - (d/R)^6] \ , \tag{4-43}$$

where values of d are 3.2, 3.35, 3.4, 3.6, 3.85, and 3.9 for the
F—H...F, O—H...O, N—H...F, N—H...O, N—H...N, O—H...Cl, and
N—H...Cl systems, respectively. These values are close to the sum of
the collision radii of the A and B atoms, which provide evidence for
the lack of importance of a repulsion term originating in the hydrogen
atom.

K. $\Delta\nu$-$\nu_{\frac{1}{2}}$ and $\Delta\nu$-B Correlations

Pimentel and McClellan [1938] review the early research in this
area. Correlations of frequency shift with band width ($\nu_{\frac{1}{2}}$) and band
intensity (B) have been noted by several workers, but very little use
has been made of these correlations. Becker [167] compared $\Delta\nu$-$\nu_{\frac{1}{2}}$ and
$\Delta\nu$-ΔB correlations for methanol adducts with those reported by Huggins
and Pimentel [1101a] for a variety of A—H...B systems.

1. $\Delta\nu$-$\nu_{\frac{1}{2}}$

Korobkov [1317] found that band width data for intramolecular
hydrogen bonds fits a different curve than that for intermolecular
hydrogen bonds. Detoni *et al.* [615] have reported a $\Delta\nu$-$\nu_{\frac{1}{2}}$ correlation
for adducts of isothiocyanic acid. Least-squares treatment of the data
gives

$$\nu_{\frac{1}{2}} = 36 + 1.1 \ \Delta\nu - 0.00023 \ \Delta\nu^2 \ . \tag{4-44}$$

2. $\Delta\nu$-B

The correlation between $\Delta\nu$ and the relative integrated intensity
(B/B') for isothiocyanic acid adducts is similar to that observed by
Huggins and Pimentel [1101a] for other A—H...B systems. The equation
is

$$\Delta\nu = 71 \ B/B' - 70 \ . \tag{4-45}$$

II. LINEAR FREE ENERGY RELATIONSHIPS (LFER)

Taft *et al.* [2407] have derived linear free energy relationships
(LFER) for hydrogen-bonded complexes of various reference acids. A
series of straight lines are obtained when log K values for adducts of
p-fluorophenol are plotted against corresponding values for other OH
reference acids (Fig. 4-17). The general equation of the lines is

$$\log K = m \; (pK_{HB}) + c \tag{4-46}$$

where pK_{HB} is defined as the logarithm of the association constant for
the *p*-fluorophenol adduct in carbon tetrachloride at 25°C and m and c
are constants characteristic of the reference acid, the solvent, and
the temperature. Values of m and c have been derived for a number of
OH reference acids (Table 4-13). In view of the good fit of data to
Eq. (4-46) the LFER equations can be used to predict log K values for
adducts of a given base with OH reference acids if log K for the base
with one reference acid is known. Similar correlations relating log K
values for phenol adducts with those for chloroform adducts have been
reported by Gramstad and Vikane [917a].

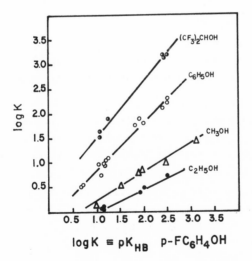

$$\log K \equiv pK_{HB} \quad p\text{-}FC_6H_4OH$$

FIG. 4-17. Log $K_{25°C}$ for alcohol adducts versus log $K_{25°C}$ for
p-fluorophenol adducts. Solvent, CCl$_4$. Reprinted from Taft *et al.*
[2407], p. 4802, by courtesy of the American Chemical Society.

TABLE 4-13

m and c Values for Various Solvents[a]

Solvent	m	c
C_6H_{12}	1.10	0.04
CCl_4	(1.00)	(0.00)
C_6H_5Cl	0.91	-0.16
o-$C_6H_4Cl_2$	0.90	-0.20
CH_2Cl_2	0.82	-0.54
$Cl(CH_2)_2Cl$	0.87	-0.62

[a]Joris, Mitsky, Taft [1676]. Applicable only for weak bases given in Ref. [1676].

Mitsky, Joris, and Taft [1676] have recently reported the extension of Eq. (4-46) to correlations of pK_{HB} with a number of parameters including ΔH for p-fluorophenol adducts, ΔH for phenol adducts, E_B parameters, log K for HNCO adducts, and ^{19}F NMR shifts for 5-fluoroindole. However, the pK_{HB}-log K correlation for 5-fluoroindole is limited by base type.

Another application of pK_{HB} has been reported by Kamlet *et al.* [1227a]. They observed a correlation of pK_{HB} and Δ values with the bathochromic shifts of 4-nitro-aniline and 4-nitrophenol.

The applicability of Eq. (4-46) to solvents other than carbon tetrachloride has also been investigated [1676] by measuring equilibrium constants for adducts of p-fluorophenol in several solvents. Equation (4-46) holds for weaker bases (*e.g.*, dioxane, dimethyl sulfoxide, and N,N-dimethylformamide) and the values of m and c decrease with increasing polarity of the solvent (Table 4-14). This trend can be explained by a more favorable solvation of the free acid and base relative to the hydrogen-bonded complex ($\Delta H_2 + \Delta H_3 > \Delta H_5$ in Scheme 1, p. 163).

In polar solvents strong bases such as pyridine, triethylamine, and quinuclidine give log K values larger than those predicted by

TABLE 4-14

m and c Values at 25°C in $CCl_4{}^a$

Acid	m^b	c^b	Standard deviation
CH_3OH [6]	0.51	−0.39	0.07
C_2H_5OH [6]	0.51	−0.53	0.03
CF_3CH_2OH [6]	0.92	−0.17	0.04
C_6H_5OH [14]	0.97	−0.13	0.09
$(CF_3)_2CHOH$ [6]	1.16	+0.32	0.09

[a]Taft et al. [2407].   [b]The values of m and c for a given acid show a slight dependence on the combination of bases which are used. Values here are for the lines in Fig. 4-17.

Eq. (4-46). This is attributed to the increased amount of proton transfer which causes ΔH_5 to be larger than $\Delta H_2 + \Delta H_3$ (Scheme 1, p. 163).

pK_{HB} values for bases with a common functional group correlate with pK_a values. A plot of pK_a versus pK_{HB} is given in Fig. 4-18 for primary amines, 3- and 4- substituted pyridines, and carbonyl compounds [2407]. These results illustrate the lack of generality of pK_a - pK_{HB} relationships which had been noted previously by Gordon [889]. The values of pK_a show a large variation with family type. For equal values of pK_{HB}, the pK_a value of the primary amine is 10^5 times that of the substituted pyridine and 10^{13} times that of the carbonyl base.

These differences have generally been attributed to solvation effects. However, a recent study by Arnett and Mitchell [71] provides independent evidence that different base types will not respond the same way to hydrogen bonding and proton transfer. The reference acid for hydrogen bonding was p-fluorophenol, and the protonation enthalpy was determined by measuring the heat of transfer of a base from carbon tetrachloride to high dilution in fluorosulfuric acid at 25°C. The results are illustrated in Fig. 4-19. If bases showed the same response, only one line would be obtained. The difference is believed to be too large to attribute to solvation differences in fluorosulfuric acid.

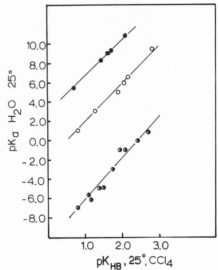

FIG. 4-18. Family linear free energy relationships for adducts of
p-FC$_6$H$_4$OH with primary amines (●), 3- and 4- substituted pyridines (○),
and carbonyl compounds (◑). Taft *et al*. [2407].

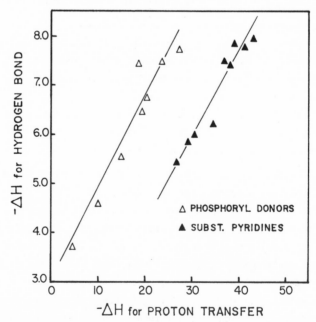

FIG. 4-19. Correlation of enthalpies of hydrogen bonds with enthalpies of proton transfer. Units are kcal/mole. Solvents are carbon
tetrachloride for hydrogen bonding and fluorosulfuric acid for proton
transfer. Arnett and Mitchell [71].

In a related study Hadži, Klofutar, and Oblak [972] have compared
A—H frequency shifts for A—H...B adducts with equilibrium constants
for B + $HClO_4$ $\xrightleftharpoons[\text{acid}]{\text{acetic}}$ BH^+ + ClO_4^- for a series of oxygen bases. The
results are given in Table 4-15. Values for K_{BH^+} increase in the
order sulfoxides < phosphine oxides < pyridine N-oxide < selenoxides
< arsine oxide. Examination of the frequency shift data indicate that
within a group of similar bases, the sequence observed for K_{BH^+} is
followed. However, the relationship is not general, as shown by the
similar shifts of pyridine N-oxide and trialkylphosphine oxides. The
data do allow comparison of protonation with hydrogen bonding. If
protonation is considered as a full charge transfer from the base to
the acid, and the hydrogen bond as a partial charge transfer, then $\Delta\nu$
can be considered a measure of charge transfer. For weaker acids $\Delta\nu$
is affected by other factors such as ion-dipole and dipole-dipole
interactions. For partial charge transfer the oxo-bases are comparable
with the amines. However, in full charge transfer (protonation) they
are weaker.

Firestone [764] attributes the lack of generality of pK_a - log K
(hydrogen bond) correlations to the difference in L-strain for different functional groups. In the context of Linnett's double-quartet
theory [1474] L-strained bonds occur when the two spin sets about one
or more atoms do not coincide. The strength of a hydrogen bond is
attributed to the combined effects of basicity and L-strain. As L-
strain increases (F > O > N > C), hydrogen bond strength decreases for
systems where basicity effects cancel (A—H...A). One example cited
by Firestone as support for L-strain is the fact that pK_{HB} values
[2407] for amines are much smaller than would be predicted from pK_a
values (Fig. 4-18). This is attributed to the high L-strain of tri-
valent nitrogen.

III. NMR CHEMICAL SHIFT CORRELATIONS

The "hydrogen bond shift" or the difference in chemical shifts for
unassociated and associated HA can be used as an indication of hydrogen
bond strength. In Chap. 1, Section IX, the method for calculating hydrogen bond shifts was described.

TABLE 4-15

K_{BH^+} and $\Delta\nu_{AH}$ for Oxygen Bases[a,b]

Base	K_{BH^+}	HCNS $\Delta\nu$ (cm^{-1})	PhOD $\Delta\nu$ (cm^{-1})	MeOH $\Delta\nu$ (cm^{-1})	Indole $\Delta\nu$ (cm^{-1})	PhNHCH$_3$ $\Delta\nu$ (cm^{-1})	PhC≡CH $\Delta\nu$ (cm^{-1})
$(C_6H_5)_2SO$	9.34	570	197	169	172	58	80
$(C_6H_5CH_2)_2SO$	2.1×10^2	700	237	198	200	70	90
$(CH_3)_2SeO$	$>10^8$	—	397	338	315	130	145[c]
$(C_6H_5)_3PO$	2.72×10^2	860	277	238	252	100	120
$C_5H_3Br_2NO$	3.34×10^2	—	—	158	170	—	—
$(CH_3)_2SO$	3.50×10^2	820	262	206	212	87	106
$(CH_3)_3PO$	3.44×10^3	980	332	273	275	108	125
$(C_8H_{17})_3PO$	6.54×10^3	1020	332	285	285	120	140
C_5H_5NO	2.82×10^5	—	324	278	280	105	108
$(C_6H_5)_2SeO$	4.18×10^5	—	312	268	270	100	130[c]
$C_5H_3(CH_3)_2NO$	2.73×10^6	—	357	288	235	100	125[c]
$(C_6H_5)_3AsO$	4.66×10^6	—	432	353	340	153	165[c]

[a]Hadži, Klofutar, Oblak [972].   [b]$\Delta\nu_{AH}$ measured for solutions in CCl$_4$.   [c]Binary mixtures.

Hydrogen bond shifts and observed chemical shifts for hydrogen-
bonded systems have been correlated with pK_a [2308-2310], ir frequency
shifts, see Refs. [1639,1975,2073,2252,2255], bond lengths [918],
integrated ir intensities [498], and hydrogen bond enthalpies, see
Refs. [731,1220,2026,2255,2290].

A. pK_a-Chemical Shift Correlations

Figure 4-20 illustrates the correlation between the chemical
shift of the hydroxy resonance and pK_a for fifty-five aromatic hydroxy-
substituted compounds [2308]. The least-squares line is

$$pK_a = -0.560 \ \delta \ (CCl_4) + 12.40 \qquad\qquad (4-47)$$

for carbon tetrachloride as the solvent and

$$pK_a = -1.492 \ \delta \ (DMSO) + 23.64 \qquad\qquad (4-48)$$

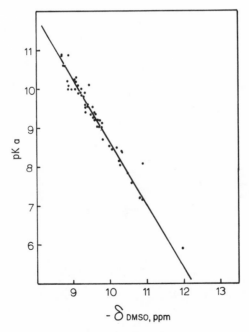

FIG. 4-20. Correlation of pK_a with $-\delta_{DMSO}$ for aromatic hydroxy-
substituted compounds. Socrates [2308].

for dimethyl sulfoxide as the solvent. The correlation is much better
for dimethyl sulfoxide solutions. This is attributed to the greater
similarity of dimethyl sulfoxide to water, the reference solvent for
pK_a values.

A linear correlation between the chemical shift of the hydrogen-
bonded proton in phenol-base adducts and pK_a of the protonated base
has also been observed [2310]. The least-squares line for the plot
in Fig. 4-21 is

$$pK_a = -4.59 \ \tau \ (CCl_4) + 7.76 \ . \tag{4-49}$$

A similar correlation was reported for cresols [2309].

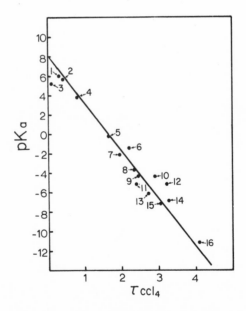

FIG. 4-21. pK_a versus chemical shift (τ) for adducts of phenol
with (1) $4\text{-}CH_3\text{-}C_5H_4N$, (2) $3\text{-}CH_3\text{-}C_5H_4N$, (3) C_5H_5N, (4) $4\text{-}Cl\text{-}C_5H_4N$,
(5) $CH_3C(O)N(CH_3)_2$, (6) $CH_3C(O)NH_2$, (7) $(CH_2)_4O$, (8) $(C_2H_5)_2O$, (9)
dioxane, (10) CH_3CN, (11) C_6H_5CHO, (12) $CH_3C(O)OC_2H_5$, (13) $C_6H_5C(O)CH_3$,
(14) cyclohexanone, (15) $(CH_3)_2CO$, (16) $C_6H_5NO_2$. Socrates [2310].

B. ΔH-Chemical Shift Correlations

Eyman and Drago [731] have reported a correlation between $\delta_{A—H...B}$ and ΔH for phenol adducts. The equation of the correlation illustrated in Fig. 4-22 is

$$\delta'_{obsd} = 0.748 \ (\Delta H) - 4.68 \ , \qquad\qquad (4\text{-}50)$$

where δ'_{obsd} is the corrected chemical shift for the completely associated proton and ΔH is calculated from the Δν-ΔH correlation. An anistropy correction was applied to the chemical shifts of carbonyl bases and pyridine.

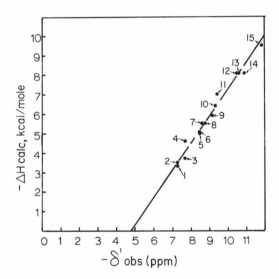

FIG. 4-22. Enthalpy versus corrected chemical shifts (δ') for phenol adducts of (1) $CH_3C(O)OC_2H_5$, (2) CH_3CN, (3) $(CH_3)_2CO$, (4) $(C_2H_5)_2S$, (5) $CH_3C(O)NH_2$, (6) $(C_2H_5)_2O$, (7) $(CH_2)_4O$, (8) $(C_2H_5O)_2POH$, (9) $(C_2H_5O)_3PO$, (10) $(CH_3)_2SO$, (11) $(CH_2)_4SO$, (12) $(CH_3)_3PO$, (13) $[(CH_3)_2N]_3PO$, (14) C_5H_5N, (15) $(C_2H_5)_3N$. Eyman and Drago [731].

Since then, correlations of enthalpy change with $\Delta(\delta_{A—H...B} - \delta_{HA})$ have been reported for several reference acid and reference base systems. Purcell, Stikeleather, and Brunk [2026] observed a linear enthalpy-chemical shift correlation for adducts of 1,1,1,3,3,3-

hexafluoro-2-propanol with seven bases. The least-squares equation is

$$-\Delta H \ (kcal/mole) = 0.89 \ \Delta + 3.6 \ (\pm 0.3) \ . \qquad (4-51)$$

A correlation was also found for 2,2,2-trifluoroethanol adducts [2255] which can be represented by

$$-\Delta H \ (kcal/mole) = 0.98 \ \Delta + 2.3 \ (\pm 0.3) \ . \qquad (4-52)$$

For chloroform adducts enthalpy versus chemical shift is more linear than enthalpy versus frequency shift [2291], and equations have been derived for different base types [917a,2624b].

Reference bases which give linear enthalpy-chemical shift correlations include quinuclidine [2290], tetrahydrothiophene [2290a], and 1-phospha-2,6,7-trioxa-4-ethylbicyclo[2.2.2]octane [2290a]. However, acetonitrile, tetrahydrofuran, diethyl ether, pyridine, and acetone do not give linear correlations. The lack of a general correlation between ΔH and Δ has been accounted for in terms of the magnitude of the electric field of the base and the importance of neighbor magnetic anistropy effects [2290a].

C. Log K-Chemical Shift Correlations

Gurka and Taft [955] observed a linear correlation of the ^{19}F NMR limiting chemical shift (Δ_F) with log K for adducts of p-fluorophenol with forty-eight bases (Fig. 4-23). The correlation is limited to bases with no steric effects. The general equation is

$$\log K = m'\Delta_F + b \ . \qquad (4-53)$$

Small changes in m' are observed when the solvent is changed (Table 4-16).

When 5-fluoroindole is the reference acid, the Δ_F versus log K relationship is limited by base type [1676]. However, general correlations are observed for Δ_F (5-fluoroindole) versus pK_{HB} and Δ_F (5-fluoroindole) versus Δ_F (p-fluorophenol). Mitsky, Joris, and Taft [1676] attribute the family dependence of the Δ_F-log K and log K - pK_{HB}

relationships for 5-fluoroindole to a steric entropy effect associated
with the change in bond lengths for different N—H...B systems.

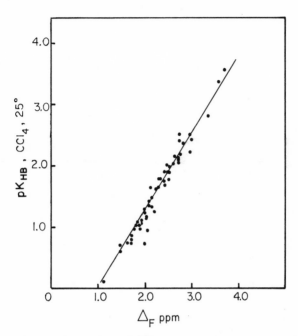

FIG. 4-23. Correlation of ^{19}F NMR limiting chemical shift (Δ_F)
with log K for adducts of p-fluorophenol in carbon tetrachloride at
25°C. Gurka and Taft [955].

TABLE 4-16

Correlation of ^{19}F Chemical Shift with Log K[a]

Solvent	m'	b
C_6H_{12}	1.21	-1.21
CCl_4	1.25	-1.28
C_6H_5Cl	1.17	-0.75
o-$C_6H_4Cl_2$	1.33	-1.16
$Cl(CH_2)_2Cl$	1.43	-1.63
CH_2Cl_2	1.25	-1.50

[a]Joris, Mitsky, and Taft [1676].

D. Δν-Chemical Shift Correlations

The correlation between frequency shift and proton chemical shift
is illustrated in Fig. 4-24 for phenol adducts [918]. A correlation
between proton chemical shift and the N—H stretching frequency for a
number of β-amino, α,β-unsaturated ketones and esters has also been
observed [663].

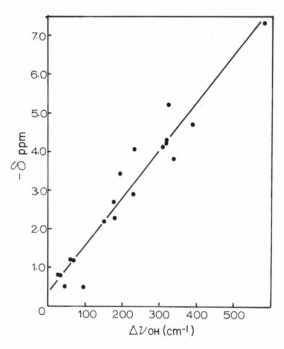

FIG. 4-24. Correlation of $\Delta\nu_{OH}$ with proton chemical shift for
phenol adducts. Gränacher [918].

IV. CORRELATION OF ΔH WITH INTEGRATED BAND INTENSITY

Becker [167] observed a correlation between ΔH and B for adducts
of methanol and phenol with a variety of bases. Least-squares treatment of data for twenty-seven adducts gave

$$-\Delta H = 4.33 \times 10^{-4} B + 1.83 \text{ kcal/mole} \qquad (4\text{-}54)$$

with a standard deviation of 0.4 kcal. The Δv-ΔH correlation for the same adducts was not linear.

Iogansen [1136,1139,2055] has presented evidence in support of a linear correlation between ΔH and $\Delta \Gamma^{\frac{1}{2}}$, the difference between the square roots of integrated intensities of free v_{OH} and bonded v_{OH}. For phenol

$$-\Delta H = 5.3 \ \Delta \Gamma^{\frac{1}{2}} \ . \tag{4-55}$$

The relationship between Γ and B, the integrated band intensity, is

$$\Gamma = 2.3 \ B/v \ (cm^2/mole) \ . \tag{4-56}$$

Equation (4-55) is proposed as a better correlation than Δv-ΔH [1139]. However, additional data on a wide variety of adducts are needed to test the generality of the relationship since Eq. (4-55) is based on data for adducts of phenol with oxygen donors and the range of hydrogen bond energies is narrow (2-7 kcal/mole). In addition, the correlation is not as convenient to use as Δv-ΔH correlations because of the diffi-culty encountered in measuring integrated intensities of the bonded O—H band.

V. ΔH-ΔS CORRELATIONS

Pimentel and McClellan [1937] have recently reviewed the evidence for a monotonic relationship between ΔH and ΔS for phenol adducts of specific base types. Data for adducts of phenol with ethers, amines, amides, aldehydes, and ketones gave straight line plots. The equa-tions and standard deviations are given in Table 4-17.

The entropy effect causes equilibrium constants to depend upon the base type when the hydrogen bond enthalpy is the same. For example, when -ΔH is 5 kcal/mole, the calculated equilibrium constants vary from 3.9 ℓ/mole for aldehydes to 62 ℓ/mole for amides. Although ΔH-ΔS correlations have been observed for acid adducts of specific base types, see Refs. [25,139,907,908,1182,1496], exceptions have been reported for phenol [70,847] and p-fluorophenol adducts [1674]. The data in the

TABLE 4-17

Enthalpy-Entropy Relationship for Phenol Adducts in CCl_4[a]

ΔS (cal/deg mole) = $m\Delta H$ (kcal/mole) + b

Base	m	b	σ
Ethers	1.86	-3.81	0.93
Aldehydes	3.10	+1.43	0.69
Ketones, esters	2.18	-0.82	0.74
Amines	2.25	+0.59	0.88
Amides	2.36	+3.21	0.89

[a]Pimentel and McClellan [1937].

Appendix have been used to test the ΔH-ΔS correlation for different
acids. In Table 4-18 the least-squares equations for phenol adducts of
specific base types are given. Equations for the same base types in
Tables 4-17 and 4-18 are within the standard deviation. Table 4-19
lists the least-squares equations for a number of acids in the Appendix.
The standard deviation is generally larger since all base types are
included in one equation.

   Since there is general agreement that enthalpies measured by the
calorimetric method are the most reliable, the equations in Tables
4-18 and 4-19 might prove useful for calculating ΔS from experimental
ΔH values. Then ΔG could be calculated by using $\Delta G = \Delta H - T\Delta S$. As a
test of the reliability of such calculations, calorimetric enthalpy
data from the Appendix have been used to obtain ΔS and ΔG for systems
where ΔG was measured by a spectroscopic method. A comparison of experimental and calculated values for selected systems is given in
Table 4-20.

   The predicted free energies are in fair agreement with the experimental free energies. In most instances the agreement is as good as
that observed for experimental free energies obtained by different
spectroscopic methods.

   In conclusion, the standard deviations of the least-squares equations listed in Tables 4-17, 4-18, and 4-19 indicate that ΔH-ΔS corre-

lations for a variety of acids are generally valid. The best correlations are obtained for acid adducts of specific base types. The equations can best be used for qualitative estimates of free energy (Table 4-20).

TABLE 4-18

Enthalpy-Entropy Least-Squares Lines for Phenol Adducts

ΔS (cal/deg mole) = $m \Delta H$ (kcal/mole) + b

Base type[a]	m	b	σ
Aldehydes (15)	2.81	+0.7	1.1
Amides (42)	2.33	+3.0	0.9
Amines (42)	2.56	+2.1	1.2
Esters (14)	2.38	+0.04	0.6
Ethers (33)	2.06	-3.1	1.2
Ketones (30)	2.45	+0.5	1.3
Nitriles (19)	1.89	-2.9	1.5
Phosphoryl donors (58)	2.50	+5.4	1.6
Organic halides (14)	0.47	-6.8	1.3
π-Bases (49)	1.49	-5.4	2.2
Sulfides (13)	1.91	-4.4	0.7
Sulfoxides (18)	1.69	-0.9	1.4
Thioamides (6)	1.91	-1.4	0.8
Thiocyanates (6)	3.32	+2.3	0.2

[a]Number of bases in parentheses.

TABLE 4-19

Enthalpy-Entropy Least-Squares Lines for Acids

ΔS (cal/deg mole) = $m\Delta H$ (kcal/mole) + b

Acid[a]	m	b	σ
HOH (10)	3.63	+4.4	0.5
CH_3OH (68)	2.15	-1.7	2.5
C_2H_5OH (16)	2.66	-1.1	1.0
CF_3CH_2OH (14)	1.86	-2.7	1.6
CF_2HCF_2OH (8)	2.29	-3.8	1.4
n-C_3H_7OH (11)	3.02	-1.0	2.4
$(CH_3)_2CHOH$ (24)	2.66	-2.2	1.2
n-C_4H_9OH (8)	1.93	-4.6	1.1
$(CH_3)_3COH$ (17)	1.98	-3.5	0.8
C_6H_5OH (353)	1.39	-4.1	2.5
C_6Cl_5OH (54)	1.28	-2.5	1.9
C_6H_5OD (7)	2.06	+4.4	1.9
p-F-C_6H_4OH (30)	1.34	-4.7	2.1
p-Cl-C_6H_4OH (21)	2.08	-0.6	3.1
m-CF_3-C_6H_4OH (7)	0.68	-10.2	2.2
2,4-NO_2-C_6H_3OH (40)	2.22	+5.3	3.2
α-Naphthol (23)	1.27	-2.2	2.0
β-Naphthol (11)	0.92	-7.6	0.9
C_6H_5SH (7)	0.77	-7.7	2.3
$C_6H_5NH_2$ (14)	1.74	-3.5	2.1
HNCS (32)	1.57	-4.2	0.8
$C_6H_5NH(CH_3)$ (17)	1.64	-5.7	1.6
Pyrrole (8)	2.82	1.5	2.0
Indole (7)	1.75	-1.3	0.4
HN_3 (6)	2.16	-2.0	1.0
HNCO (7)	1.19	-4.7	0.5
$CHCl_3$ (19)	2.56	-2.8	2.3

[a]Number of acids in parentheses.

TABLE 4-20

Use of ΔH-ΔS Correlations to Predict Free Energies

Adduct	Calc. ΔS, (cal/deg mole)	Calc. ΔG, (kcal/mole)	Exp. ΔG, (kcal/mole)
$C_2H_5OH-(C_2H_5)_3N$	-11.2	-0.5	-0.5
$C_2H_5OH-(CH_3)_2CO$	-10.1	-0.4	-0.3
$n-C_4H_9OH-(C_2H_5)_3N$	-15.3	-0.9	-0.7
$n-C_4H_9OH-C_5H_5N$	-13.1	-0.5	-0.4
$(CH_3)_3COH-C_5H_5N$	-10.8	-0.5	-0.2
$C_6H_5OH-CH_3C(O)N(CH_3)_2$	-14.2	-3.2	-2.8
$C_6H_5OH-(C_2H_5)_3N$	-20.6	-2.7	-2.5
$C_6H_5OH-C_5H_5N$	-14.5	-2.2	-2.3
$C_6H_5OH-(C_2H_5)_2O$	-14.3	-1.2	-1.3
$C_6H_5OH-(CH_2)_4O$	-14.9	-1.4	-1.4
$C_6H_5OH-p-dioxane$	-13.6	-1.1	-1.2
$C_6H_5OH-CH_3C(O)C_2H_5$	-12.2	-1.6	-1.3
$C_6H_5OH-C_6H_{11}Cl$	-7.7	+0.3	+0.5
$C_6H_5OH-cumene$	-8.4	+1.0	+0.6
$C_6H_5OH-(CH_3)_2SO$	-12.9	-3.4	-3.1
$p-F-C_6H_4OH-C_5H_5N$	-13.9	-2.8	-2.6

CHAPTER 5

INTRAMOLECULAR AND
HOMO-INTERMOLECULAR
HYDROGEN BONDS

I. INTRAMOLECULAR HYDROGEN BONDING

Both intra- and intermolecular hydrogen bonds may be represented
by A—H...B with the limitation that B and A—H be in a favorable
spatial configuration in the same molecule for intramolecular hydrogen
bonding. This representation helps to emphasize the fact that the same
groups which enhance intermolecular hydrogen bonds will form intramolecular hydrogen bonds if the geometric arrangement is favorable.
Since intramolecular hydrogen bonds minimize the amount of molecular
association, the behavior of substances containing intramolecular hydrogen bonds is much closer to that of normal non-hydrogen-bonded
substances. Another difference is the tendency for intramolecular
hydrogen bonds to be bent rather than linear.

Although thermodynamic data for intermolecular hydrogen bonds are
quite plentiful, this is not the case for intramolecular hydrogen
bonds. Because of this lack of thermodynamic data, research on intramolecular hydrogen bonding relies very heavily on comparisons of ir
frequency shifts ($\Delta\nu$) and NMR chemical shifts attributed to hydrogen
bonding (Δ). Such comparisons can be misleading since $\Delta\nu$ and Δ are
affected by geometric effects as well as by the acidity of the proton
in HA and the basicity of B.

The importance of intramolecular hydrogen bonding to conformational analysis has been reviewed by Tichy [2451]. Infrared spectroscopy
has been used extensively for this purpose; Tichy has tabulated $\Delta\nu_{OH}$
values for over 1600 compounds which exhibit intramolecular hydrogen
bonding.

Specific examples will be presented here for the various A—H...B classes of intramolecular hydrogen bonds in order to illustrate the variety of intramolecular hydrogen bonds and the methods which have been used to estimate the relative importance of geometric effects and hydrogen bond strength.

A. O—H...B

1. O—H...Halogen

One of the first examples of intramolecular hydrogen bonding to be studied was *ortho*-chlorophenol [2632]. The doubling of the O—H stretching frequency was interpreted by Pauling [1884] in terms of *cis* and *trans* isomers as shown in Structure (1). The less intense band at

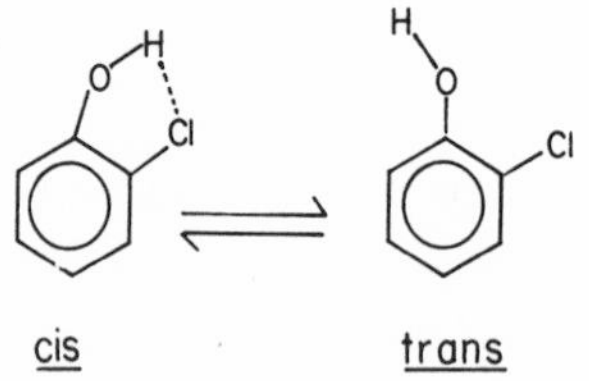

cis   trans

(1)

higher frequency was assigned to the *trans* isomer and the other band was assigned to the *cis* isomer. Pauling proposed that the *cis* isomer was more stable than the *trans* isomer by 1400 cal/mole.

The shifts in $\Delta\nu_{OH}$ for *ortho*-halophenols increase with the size of the halogen atom, with values of 18, 61, 78, and 105 cm^{-1} for F, Cl, Br, and I, respectively [115]. If the Badger-Bauer rule applies, the energy of the hydrogen bond would be expected to follow the same order. However, thermodynamic data measured in the gas phase and in solution give the order Cl > Br > I (Table 5-1). (See Ref. [2103] for a comprehensive survey of thermodynamic data for the intramolecular O—H...X bond.)

The relative strength of the intramolecular hydrogen bond in *ortho*-fluorophenol was uncertain until the recent work of Carlson *et al.* [402a]. They assigned O—H torsional frequencies for the *cis* and *trans* isomers of *ortho*-halophenols and then used these frequencies to calculate the energy difference between the *cis* and *trans* isomers.

TABLE 5-1

Enthalpy Values for *ortho*-Halophenols

Solvent	Method	$-\Delta H$, kcal/mole				Reference
		F	Cl	Br	I	
Gas phase	ir (ν_s)	—	3.41	3.13	2.75	[1467]
	Far ir (ν_t)	1.63	1.63	1.53	1.32	[402a]
CCl_4	NMR	—	2.36	2.14	1.65	[32]
	ir	—	1.44	1.21	1.08	[118]
C_2Cl_4	ir	—	1.27	1.86	0.99	[1189]
C_2H_{12}	Far ir	1.44	1.62	1.57	1.45	[402a]

The enthalpy values obtained by Carlson *et al.* are probably the
most reliable since they do not depend on intensity measurements. The
relative order of intramolecular hydrogen bond strength is F = Cl > Br
> I in the vapor phase and Cl > Br > I = F in solution. The apparent
stabilization of the *trans* isomer of *ortho*-fluorophenol in cyclohexane
is attributed to the importance of dimerization. The lack of change
of the order Cl > Br > I in going from gas phase to solution indicates
a negligible stabilization of the *trans* isomer of these phenols by
dimerization.

Torsional frequencies have also been used to determine enthalpy
differences between *cis* and *trans* isomers of *ortho*-phenyl-, *ortho*-
ethoxy-, *ortho*-methoxy-, and *ortho*-cyano-phenol [402b].

Baker and Shulgin [117] obtained enthalpy values for the
trans ⇌ *cis* conversion of *ortho*-iodophenol in a variety of solvents.
As expected, the *trans* form becomes more stable with increased hydrogen
bonding ability of the solvent (Table 5-2). Of the solvents studied,
methylene bromide caused the greatest stabilization of the *trans* form.

Doddrell, Wenkert, and Demarco [632] observed a large upfield
chemical shift for the OH proton in *ortho*-CF_3-C_6H_4OH when it was diluted with CCl_4. This shift is attributed to Structure (2a). The
lower resonance position for the *ortho* versus the *meta* and *para* isomers

TABLE 5-2

Enthalpy Data for *trans-cis* Conversion of *ortho*-Iodophenol[a]

Solvent	$-\Delta H$, kcal/mole	$\Delta\nu$, cm^{-1}
CH_2Br_2	0.18	57
$CDCl_3$	0.30	84
C_6F_6	0.31	109
C_7F_8	0.38	109
CCl_4	0.99	95
C_4Cl_6	0.93	93
C_2Cl_4	0.99	94
$CBrF_2CBrF_2$	1.03	105
$CFCl_2CF_2Cl$	1.17	109
iso-C_8H_{18}	1.40	103
C_7F_{14}	2.24	120
$C_{10}F_{18}$	2.28	120

[a]Baker and Shulgin [117].

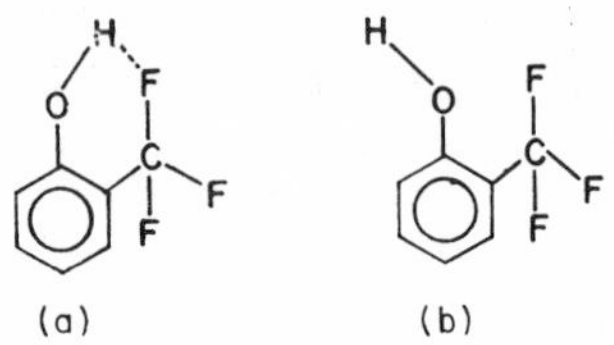

(a)   (b)

(2)

also supports Structure (2a) for the *ortho* isomer (*ortho*, -5.54 ppm;
meta, -5.04 ppm; *para*, -4.98 ppm). Infrared work by Baker and Shulgin
[119] also supports Structure (2a). The ν_{OH} band is double with a
major peak at 3624.6 cm^{-1} and a shoulder at 3605 cm^{-1}. The higher peak
is assigned to the *cis* isomer (2a) and has a band area several times
that of the *trans* isomer at 3605 cm^{-1}.

Robinson *et al.* [2103] have applied the Schroeder-Lippincott potential function for bent hydrogen bonds [2206] to the O—H...halogen
intramolecular hydrogen bonds. Their results are in agreement with

the lack of correlation between frequency shift and enthalpy for intramolecular O—H...halogen bonds. The lack of correlation is attributed
to geometric effects.

2. *O—H...O*

   a. *Alcohols.* Kuhn and Wires [1363] obtained thermodynamic data
for intramolecular hydrogen bonding in 2-methoxyethanol, 3-methoxy-
propanol-1, and 4-methoxybutanol-1 in carbon tetrachloride from measurements of the ir intensities of the free and bonded OH bands. The
thermodynamic and frequency shift data are given in Table 5-3.

TABLE 5-3

Intramolecular Hydrogen Bonding[a]

Compound	$-\Delta H$ kcal/mole	$-\Delta G$ kcal/mole	$-\Delta S$ e.u.	$\Delta\nu_{OH}$ cm^{-1}
$HO(CH_2)_4OCH_3$	2.7 ± 0.3	-.23	10.8	180
$HO(CH_2)_3OCH_3$	2.1 ± 0.3	+.22	6.9	86
$HO(CH_2)_2OCH_3$	2.2 ± 0.5	1.36	3.0	30

[a]Kuhn and Wires [1363].

Although $\Delta\nu_{OH}$ changes by a factor of six, ΔH hardly changes at all.
This provides another example of the failure of the Badger-Bauer rule
for intramolecular hydrogen bonds. The lack of a $\Delta\nu$-ΔH correlation is
attributed to the influence which conformation effects have on $\Delta\nu$.

   Kuhn and Wires [1363] use the difference between the predicted ΔH
from the Joesten-Drago equation [1182] and the experimental ΔH as a
rough measure of the repulsive interactions which are lost when an
intramolecular hydrogen bond forms in 2-methoxyethanol [Structure (3)].
The interactions between nonbonded atoms on adjacent atoms consist of
six interactions for each conformation. In the *gauche* there are
(OH—OCH$_3$), (H$_3$CO—H), and 3(H—H), while the *anti* has 2(H$_3$CO—H),
2(H—H), and 2(H—OH). The interactions accompanying hydrogen bond
formation are (OH—OCH$_3$) - (H$_3$CO—H) + (H—H) - (H—OH). The first
term is exothermic, while the other three terms are endothermic.

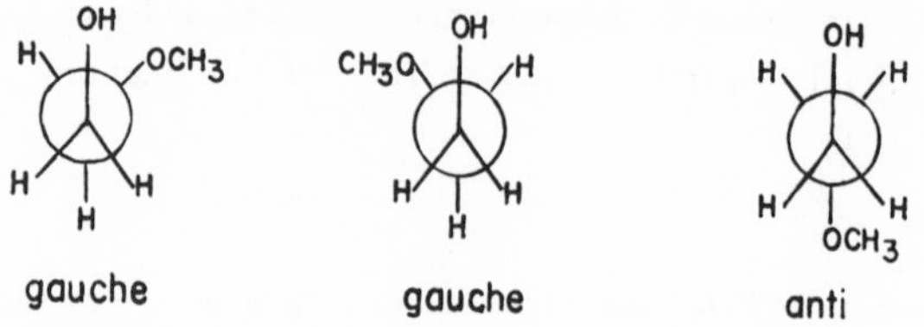

gauche   gauche   anti

(3)

Wittstruck and Cronan [2611] derived a number of mathematical expressions to represent equilibria involving free monomer, intramolecularly bonded monomer, and cyclic dimer for 3-methyl-3-hydroxy-2-butanone [Structure (4a)], 4-methyl-4-hydroxy-2-pentanone [Structure (4b)], and 3-methyl-4-hydroxy-2-butanone [Structure (4c)]. Values for K_1 (intramolecular) and K_2 (cyclic dimer) obtained at 28°C are 1.38 ℓ/mole and 0.61 ℓ^2/mole^2 for (4a); 1.33 ℓ/mole and 0.73 ℓ^2/mole^2 for (4b); and 0.73 ℓ/mole and 3.02 ℓ^2/mole^2 for (4c).

In 1,2-diols intramolecular hydrogen bonding can occur although the O...O distances are larger than those for favorable intramolecular bonding. Kuhn [1358] was the first to propose a correlation between Δv_{OH} and the O...O distance for 1,2-diols. His results emphasize the importance of the azimuthal angle Φ between the two C—O bonds on

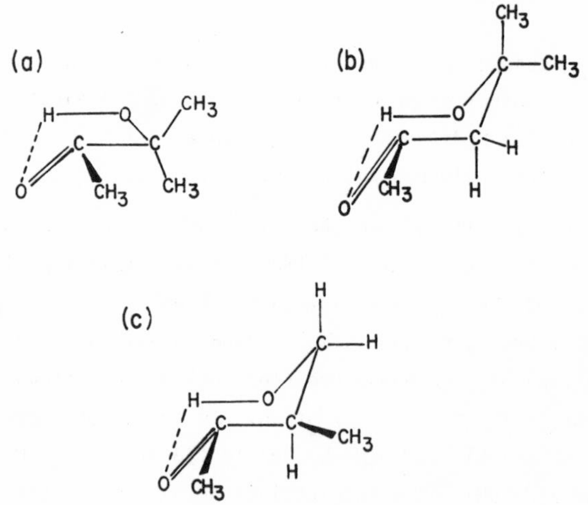

(a)   (b)

(c)

(4)

adjacent carbon atoms. In cyclopentane-1,2-diol, where Φ is between 90° and 120°, the oxygen atoms are too far apart for intramolecular hydrogen bonding, while cis-exo-norbornane-2,3-diol (Φ = 0) has a $\Delta\nu_{OH}$ value of 103 cm^{-1}. Most 1,2-diols have azimuthal angles near 60° and $\Delta\nu_{OH}$ values between 30 and 50 cm^{-1}.

The effectiveness of substituents in promoting intramolecular hydrogen bonding in 1,4-diols has been studied by several groups [1362, 2354]. Infrared spectroscopic studies were used to demonstrate that the population of intramolecularly hydrogen-bonded nonchair conformations was negligible for cis-1,4-cyclohexanediol but very large for cis, cis, cis-2,5-di-t-alkyl-1,4-cyclohexanediols. Stolow, McDonagh, and Bonaventura [2354] examined the intermediate cases by varying the alkyl substituents in cis, cis, cis-2,5-dialkyl-1,4-cyclohexanediols [Structure (5)]. The results are summarized in Table 5-4.

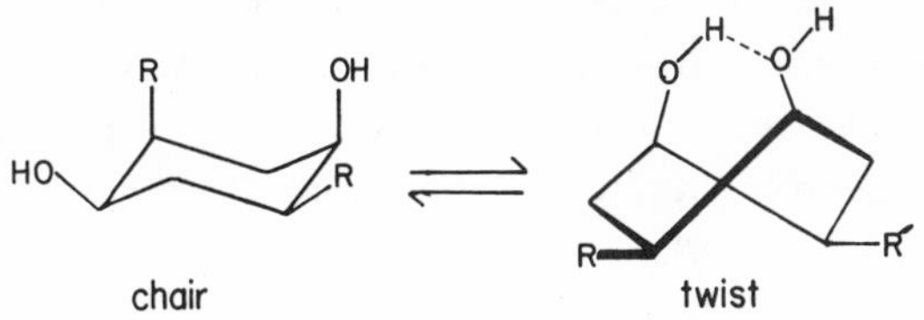

chair   twist

(5)

TABLE 5-4

Estimated Percentage of Twist Conformer[a]

1,4-Diol	Twist form, %[b]
R = H	1
R = t-Bu	100
R = t-pentyl	100
R = cyclohexyl	80
R = i-propyl	80
R = methyl	5
R = methyl, i-propyl	13
R = methyl, t-butyl	14

[a] Stolow, McDonagh, and Bonaventura [2354].
[b] Structure (5).

Extensive studies on intramolecular hydrogen bonding in fifty-seven butane-1,4-diols substituted in the 2- and 3- positions were carried out by Kuhn et $al.$ [1362]. Differences between intramolecular hydrogen bonding ability in 1,2- and 1,4-diols were discussed. The size of the hydrogen-bonded ring is much larger in 1,4-diols, and the addition of two more carbon atoms to the ring gives more conformational flexibility. Although $\Delta\nu_{OH}$ increases with a decrease in azimuthal angle for 1,2-diols, 1,4-diols show a maximum $\Delta\nu_{OH}$ at $\Phi = 90°$ with a decrease in $\Delta\nu_{OH}$ on either side. The $\Delta\nu_{OH}$ values are much larger for 1,4-diols (100-160 cm^{-1}) because of the more favorable O...O distance and O—H...O angle (closer to 180°).

Fishman and Chen [768] obtained ir spectra of 1,2-, 1,3-, 2,3-, and 1,4-butanediols in the vapor phase, neat, and in dilute solutions of carbon tetrachloride and carbon disulfide. For dilute solutions in these solvents three bands were observed which were assigned to the free O—H, intramolecularly bound O—H, and intermolecularly bound O—H stretching frequencies. An ir method was used to determine intramolecular hydrogen bond enthalpies. A listing of enthalpies and frequency shifts is given in Table 5-5. The strength of intramolecular bonds is 1,4-diol > 1,3-diol > 1,2- and 2,3-diols.

Wright and Marchessault [2628] applied ir spectroscopy to a conformational study of cyclopentane- and cyclohexane-1,2-diol monoacetates. These compounds can serve as models for an intramolecular hydrogen bonding situation that is common to polysaccharides, steroids, and other natural products. Two hydrogen bond acceptor sites are available at alkoxy and carbonyl oxygen atoms. Bruice and Fife [345] had interpreted the spectra of cyclopentane-1,2-diol monoacetates in terms of a hydrogen bond to the alkoxy oxygen atom in the cis-isomer and to the carbonyl oxygen atom in the $trans$-isomer. Wright and Marchessault [2628] disagree with the assignment for the cis-isomer and interpret the single O—H band at 3610 cm^{-1} in terms of an equilibrium between two hydrogen bonded conformers of cis-cyclopentane-1,2-diol monoacetate as shown in Structure (6).

   $b.$ $Phenols$ and $naphthols$——PMR. A number of workers have correlated hydrogen bond PMR chemical shifts (Δ) with $\Delta\nu_{OH}$ for intramolecular hydrogen bonds. The hydrogen bond PMR chemical shift in this case is

TABLE 5-5

ΔH and $\Delta \nu_{OH}$ for Intramolecular Bonds in Butanediols[a]

Compound	$-\Delta H$, kcal/mole	$\Delta \nu$, cm^{-1}	Physical state
1,4-diol	3.00 ± 0.20	110	vapor
	2.00 ± 0.10	159	CCl$_4$
	2.49 ± 0.10	152	CS$_2$
1,4-diol-d$_2$	2.00 ± 0.20	80	vapor
	1.68 ± 0.13	115	CCl$_4$
	2.02 ± 0.10	107	CS$_2$
1,3-diol	1.35 ± 0.10	61	vapor
	0.75	81	CCl$_4$
	1.06 ± 0.10	70	CS$_2$
1,2-diol	0.76 ± 0.40	42	vapor
	-0.23 ± 0.40	42	CCl$_4$
	0.26 ± 0.40	40	CS$_2$
2,3-diol	0.75 ± 0.40	43	vapor
	-0.20 ± 0.40	44	CCl$_4$
	0.22 ± 0.40	40	CS$_2$

[a]Fishman and Chen [768].

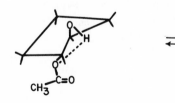 $\rightleftarrows$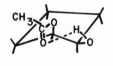

(equatorial, axial)   (axial, equatorial)

(6)

represented as the difference between the infinite dilution shift of
the parent compound in an inert solvent and the chemical shift for the
ortho-substituent in 1-2 mole % in the same solvent.

Porte, Gutowsky, and Hunsberger [1975] used this technique to obtain hydrogen bond chemical shifts for *ortho*-substituted derivatives
of phenol, β-naphthol, and 9-phenanthrol. The chemical shifts caused
by intramolecular hydrogen bonding are quite large, with hydrogen bond
shifts (Δ) of 5.5 to 9.7 ppm for the chelated *ortho*-substituted derivatives. A linear correlation between Δ and $\Delta\nu_{CO}$ was observed [1975].
Reeves, Allan, and Strommer [2073] found a linear correlation between
Δ and $\Delta\nu_{OH}$ for a series of *ortho*-substituted phenols and naphthols.
The hydrogen bond chemical shifts, Δ, were corrected for diamagnetic
anisotropies.

    c. Phenols and naphthols—electronic. A detailed study of the
effect of intramolecular hydrogen bonding on the electronic spectra of
ortho-substituted phenols and anilines has been carried out by Dearden
and Forbes [591]. In the *ortho*-substituted phenols and anilines
studied, a bathochromic displacement of between 5 and 12.5 nm (690-1780
cm^{-1}) was observed in the *B*-band compared to the *B*-band absorption for
the corresponding *meta* isomer. All except *ortho*-nitroaniline (690 cm^{-1})
and *ortho*-nitrophenol (1780 cm^{-1}) fall within the limits 1175 ± 215
cm^{-1}. Bathochromic shifts are also observed in the *C*-band along with
an increase in intensity. However, some of this change must be due to
steric effects, since an increase in intensity is also observed for
compounds which cannot form intramolecular hydrogen bonds.

Changes in intensity of ultraviolet absorption bands for 3-
substituted, 2-nitrophenols were used by Dearden [590] to calculate
O...O distances in the intramolecular hydrogen bond. Braude and Sondheimer [317] had previously derived a relationship for the angle of
twist of a chromophoric group subjected to steric hindrance on a benzene ring. The equation is

$$\cos^2 \Theta = \varepsilon/\varepsilon_0 \qquad (5\text{-}1)$$

where Θ is the angle of twist, ε is the molar absorptivity of the
primary absorption band and ε_0 is the molar absorptivity of the same
band for the reference compound. The results for the substituted
2-nitrophenols are given in Table 5-6. These data have been used

TABLE 5-6

Calculated O...O Distances for Some Substituted 2-Nitrophenols[a]

Substituent	ε, uv Band in cyclohexane	Angle of twist of C—N bond (degrees)	O...O Distance (Å)	ν_{OH} 0.01 M in CCl$_4$ (cm^{-1})
None	7400	0	2.50	3241
3-Methoxy-	6040	26	2.61	3259
3,4-Dimethyl-	5000	35	2.69	3377
3-Trifluoromethyl-	3950	44	2.78	3340
3,6-Dichloro-	3810	45	2.79	3327
3-Methyl-4,6-di-t-butyl-	3350	48	2.83	3391
3,4,6-Trichloro-	3020	51	2.86	3370
3,5-Di-t-butyl-	3000	51	2.86	3485

[a]Dearden [590].

together with some previous data of Nakamoto *et al.* [1768] to test the correlation between v_{AH} and A...B distances for intramolecular and intermolecular hydrogen bonds (Fig. 5-1). At O...O distances greater

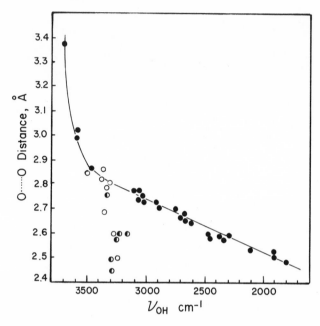

FIG. 5-1. Correlation of O...O distance with v_{OH}. ⦿, intermolecular hydrogen bonds, [1768]; ○, intramolecular hydrogen bonds, [590]; ◐, intramolecular hydrogen bonds [1768].

than 2.8 Å the data for both intramolecular and intermolecular hydrogen bonds fall on the same line. This is used as evidence for the linearity of long intramolecular hydrogen bonds (> 2.8 Å).

   d. Phenols and naphthols—solvent effects. Dabrowska and Urbanski [555] propose that solvent sensitivity is not a reliable criterion for determining the strength of intramolecular hydrogen bonds on the basis of observations of v_{OH} solvent shifts for a series of *ortho*-nitrophenols. Intramolecular hydrogen bonds of similar strength give different v_{OH} shifts in different solvents. In addition, methylsalicylate, which has a weaker hydrogen bond than *ortho*-nitrophenol, did not show any change in v_{OH} in the various solvents (see Table 5-7).

TABLE 5-7

ν_{OH} of Substituted Nitrophenols in Various Solvents[a]

Compound/Solvent	CCl$_4$	C$_6$H$_5$NO$_2$	Ethyl acetate	CH$_3$CN	Acetone	Dioxane
2-Nitrophenol	3237	3245	3305 3248(sh)	3300 3240(sh)	3309	3283
2,4-Dinitrophenol	3211	3260 3230(sh)	3290	3283	3226	3244
2,5-Dinitrophenol	3247	3283	3309	3307	3302	3270
2,6-Dinitrophenol	3164	3200	3250	3230	3244	3213
Picric acid	3138(sh)	3200	3226	3205	3210	3164
2,4-Dinitroresorcinol	3100 3022	3283	3206	3178	3200	3135
2,4,6-Trinitroresorcinol	3123	3218	3224	3218	3212	3172
2,4,6-Trinitrofluoroglucinol	3100	3155	3190	3182	3178	3147
Methyl salicylate	3197	3188	3188	3191	3189	3191
4-Nitrophenol	3595	3433	3331	3313	3240	3214 3240(sh)

[a]Dabrowska and Urbanski [555].

e. *Phenols and naphthols—ir band separation techniques.* Baker
and Yeaman [122] have analyzed the possible errors which may occur
when different band separation techniques are used to obtain intramolecular hydrogen bond energies. Lorentzian and Gaussian band profiles
are used as models. In the *half-band* technique the outer halves of
each band are used to obtain values for area (bound) and area (free)
at different temperatures. In the *minimum* technique a vertical line
is dropped from the minimum between the two bands and the resulting
areas are used. Enthalpy values for intramolecular hydrogen bonding
in a number of ferrocenyl alcohols were measured by this procedure.

f. *β-Diketones and β-ketoesters.* Kol'tsov and Kheifets have
recently reviewed the applications of NMR to investigations of both
keto-enol tautomerism [1306a] and other types of tautomerism [1306b].
Separate resonance signals are commonly observed for the protons of
the various groups in the keto and enol isomers [Structure (7)].

keto   enol

(7)

Burdett and Rogers [367] found for a series of acyclic β-diketones and
β-ketoesters [Structure (7)] that the resonance position for the acetyl-
methyl protons and the alkoxy protons does not vary. However, the enol
OH protons give shifts to lower field as expected for intramolecular
hydrogen bonds. A comparative study of proton chemical shifts for β-
diketones and β-ketoesters also illustrates this [1034]. The β-
ketoesters exist primarily in the keto form while the β-diketones favor
the enol tautomer. The equilibrium constants for the keto ⇌ enol
equilibrium were determined by integration of proton signals (Table
5-8) [367].

Rogers and Burdette [2106] also studied the effect of polar solvents on the degree of tautomerization of acetylacetone and ethyl
acetoacetate. As the polarity or hydrogen bonding ability of the solvent increases, the percentage of enol tautomer decreases (Table 5-9).

TABLE 5-8

Keto-Enol Data for Some β-Diketones and β-Ketoesters[a]

	% Enol	K^b, 33°C
A. β-*Diketones*		
Acetylacetone	81	4.3
α-Bromoacetylacetone	46	0.85
Hexafluoroacetylacetone	100	—
α-Methylacetylacetone	30	0.43
B. β-*Ketoesters*		
Butyl acetoacetate	15	0.18
t-Butyl acetoacetate	17	0.21
Ethyl acetoacetate	20	0.25

[a]Burdette and Rogers [367], spectra are of pure compounds.
[b]K = [enol]/[keto].

TABLE 5-9

Effect of Solvents on the Tautomerization of Acetylacetone
and Ethyl Acetoacetate

Solvent	Acetylacetone		Ethyl acetoacetate	
	% enol	K^b, 33°C	% enol	K^b
Hexane	95	19	39	0.64
Diethyl ether	95	19	22	0.29
Benzene	89	8.1	16	0.19
Dioxane	82	4.6	11	0.12
Ethanol	82	4.6	10	0.11
Pure solute	81	4.3	7.5	0.081
Methanol	74	2.8	5.8	0.062
Acetic acid	67	2.0	1.9	0.019
Dimethyl sulfoxide	62	1.6	2.2	0.023

[a]Rogers and Burdette [2106]. [b]K = [enol]/[keto].

Also of interest is the fact that inert solvents favor the enol tautomer to a greater extent than the pure compound does.

Free energies, enthalpies, and entropies of tautomerization have
been obtained by investigating the temperature dependence of the proton
chemical shifts for the pure liquids of several substituted β-diketones
and β-ketoesters [366]. These data are given in Table 5-10.

TABLE 5-10

Thermodynamic Data for Tautomerization of β-Diketones
and β-Ketoesters[a]

Compound	$-\Delta G$ (33°) kcal/mole	$-\Delta H$ kcal/mole	$-\Delta S$ (33°) cal/deg mole
Acetylacetone	0.80 ± 0.07	2.84 ± 0.20	6.66 ± 0.88
α-Chloroacetylacetone	1.61 ± 0.11	5.92 ± 0.20	14.1 ± 1.0
Ethyl trifluoroacetoacetate	1.09 ± 0.18	3.91 ± 0.20	9.21 ± 1.24
Ethyl α-chloroacetoacetate	-0.26 ± 0.10	0.88 ± 0.10	3.69 ± 0.57
α-Methylacetylacetone	-0.40 ± 0.10	1.33 ± 0.10	5.65 ± 0.45

[a]Burdett and Rogers [366].

g. *Carboxylic acids*. Schellenberger, Beer, and Oehme [2189] obtained ir spectra of vapors of pyruvic acid, methyl-, dimethyl-, and
trimethyl-pyruvic acid at temperatures between 90 and 180°. Enthalpies
of formation for the stability of the proton chelate form [Structure
(8)] are given in Table 5-11. Carboxy-deuterated trimethyl-pyruvic
acid was used to provide a comparison of the deuteron and proton
chelate.

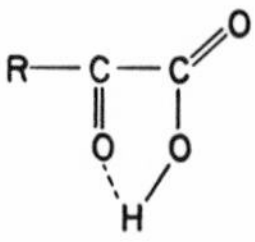

(8)

TABLE 5-11

Enthalpies, Proton Chelate Form of Pyruvic Acids[a]

Acid	-ΔH, kcal/mole
Pyruvic	2.34 ± 0.32
Methylpyruvic	3.23 ± 0.42
Dimethylpyruvic	2.86 ± 0.13
Trimethylpyruvic	3.21 ± 0.13
Trimethylpyruvic-d	3.02 ± 0.23

[a]Schellenberger, Beer, and Oehme [2189].

The use of ^{13}C chemical shifts as an indicator of intramolecular
hydrogen bonding was first reported by Lauterbur [1422] for a series
of *ortho*-substituted benzoates. For methyl salicylate two peaks were
observed: one about 4 ppm downfield and the other about 4 ppm upfield
from the resonance peak of the carbonyl carbon atom in methyl benzoate.
Lauterbur [1422] assigned the lower one to the carbonyl carbon and the
upper one to the hydroxyl group carbon. Maciel and Savitsky [1577]
verified this assignment by investigating a series of compounds where
a carbonyl group attached to a benzene ring forms a strong intramolecular hydrogen bond with a hydroxyl group *ortho* to it. The intramolecular hydrogen bond causes downfield shifts of 4.4 to 7.3 ppm in the ^{13}CO
resonance.

Ōki and co-workers have made an extensive study of intramolecular
hydrogen bonding in alkoxy- and aryloxy-aliphatic acids [1830,1832,
1834-1836] as well as *o*-aryloxy- and *o*-alkoxy-benzoic acids [1830,1831,
1833,1837]. Although the self-association of carboxylic acids is quite
strong, evidence for intramolecular bonding can be found by examining
the ν_{OH} and ν_{CO} regions of α-substituted aliphatic acids or *o*-substi-
tuted aromatic acids. For α-substituted aliphatic acids of the type
shown in Structure (9) two maxima are observed in both the ν_{OH} region
and the $\nu_{C=O}$ region. The higher ν_{OH} frequency (3525 cm^{-1}) is assigned
to the *cis*-isomer on the assumption that the hydrogen bond in this
isomer is much weaker than the hydrogen bond to an alkoxy or aryloxy
group in the *trans*-isomer. The value of ν_{OH} for the *trans*-isomer is
in the range of 3480-3400 cm^{-1}. The higher $\nu_{C=O}$ frequency (1790 cm^{-1})

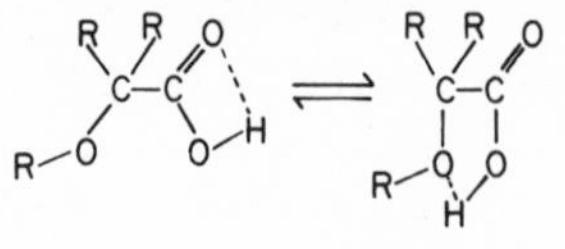

cis   trans

(9)

is assigned to the *trans*-isomer since the carbonyl group in this isomer
is not as close to the carboxylic hydroxyl group. The $\nu_{C=O}$ frequency
for the *cis*-isomer appears at 1760 cm^{-1}.

The temperature dependence of the intensity ratio A_{trans}/A_{cis} of
the two ν_{OH} bonds was used to determine the enthalpy difference of the
cis and *trans* forms [1832]. The *ortho*-substituent has a greater influence than the α-substituent on the value of ΔH although both have about
the same influence on Δν (Table 5-12).

TABLE 5-12

Cis ⇌ *Trans* Equilibrium for α-Substituted Carboxylic Acids[a]

Acid	Δν, cm^{-1}	-ΔH, kcal/mole
CH_3OCH_2COOH	77.0	0.76
$CH_3CH_2OCH_2COOH$	84.8	1.11
$(CH_3)_2CHOCH_2COOH$	97.8	1.53
$(CH_3)_3COCH_2COOH$	103.8	—
$CH_3OCH(CH_3)COOH$	97.2	1.15
$CH_3OC(CH_3)_2COOH$	119.3	1.28
$C_6H_5OCH_2COOH$	39.4	0.98
$C_6H_5OCH(CH_3)COOH$	51.2	0.94
$C_6H_5OC(CH_3)_2COOH$	93.9	1.05
$p\text{-X-}C_6H_4OC(CH_3)_2COOH$		
X = CH_3	99.2	1.35
H	93.9	1.05
Br	79.5	0.88

[a]Ōki and Hirota [1832].

Six-membered chelate rings have been shown to exist in o-methoxy-benzoic acid [577] and in o-aryloxybenzoic acids [1831,1833]. Values of ΔH have been reported for R = CH$_3$ (-3.3 kcal/mole [577]) and R = C$_6$H$_5$ (-1.89 kcal/mole [1836]). Although these ΔH values are larger than the corresponding values for α-substituted aliphatic acids, the accuracy of these values is open to question, particularly in the case of o-methoxybenzoic acid. No evidence has been found for intramolecular hydrogen bonding in β-substituted propionic acid where six-membered chelate rings could form [1836].

Infrared evidence indicates that the five-membered ring is more stable for the aliphatic acids even though the six-membered ring is stable in o-benzoic acid derivatives. Ōki and Hirota [1832] have attributed these differences to stabilization of o-alkoxy benzoic acids by conjugation and to unfavorable steric effects in β-alkoxypropionic acids.

Possible intramolecular hydrogen bond isomers for salicylic and mandelic acids are shown in Structure (10). Mori et al. [1689] have observed two ν_{OH} bands at 3530 and 3200 cm^{-1} and a $\nu_{C=O}$ band at 1690 cm^{-1}. The position of these bands supports Structure (10a) since

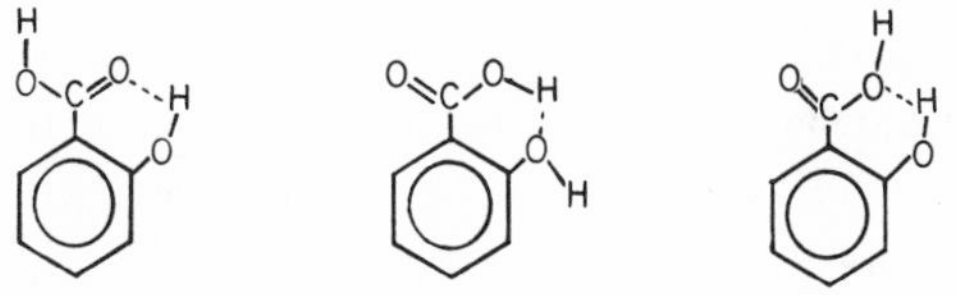

(a)   (b)   (c)

Salicylic Acid

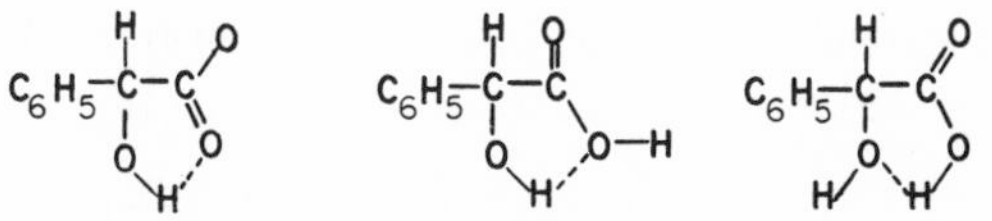

(d)   (e)   (f)

Mandelic Acid

(10)

Structures (10b) and (10c) would be expected to give a ν_{CO} value of
1750 cm^{-1} and ν_{OH} values of 3365 and 3610 cm^{-1} for (b) and 3530 and
3470 cm^{-1} for (c). The same approach resulted in assignment of Structure (10d) for mandelic acid.

Mori, Asano, and Tsuzuki [1690] observed a correlation between ν_{OH}
and Hammett σ-constants for intramolecular hydrogen-bonded methyl salicylates. The best linear plots are obtained for ν_{OH} versus σ(OH) - 0.38
σ(COOMe), and ν_{CO} versus σ(COOMe) - 0.49 ν_{OH}, where σ(OH) and σ(COOMe)
are σ-constants of a substituent relative to the OH and COOMe groups,
respectively.

Evidence for strong intramolecular hydrogen bonding in monoanions
of dicarboxylic acids has been obtained from PMR. Eberson and Forsen
[692] observed a separate peak for the carboxyl hydrogen of the monoanions of maleic, phthalic, and several substituted acids dissolved in
dimethylsulfoxide. A chemical shift of -15 ppm from water was observed.
Silver *et al.* [2269] extended this study to proton chemical shifts of
noncarboxylic hydrogen atoms. For phthalic, pyromellitic, and maleic
acids, they observed a downfield shift upon half-neutralization of the
acid.

3. *O—H...N*

Amino alcohols have been investigated by a number of groups, see
Refs. [800,1065,1346,1364,1545,1692]. Again, an effort was made to
estimate the effects which geometry and the donor site have on $\Delta\nu_{OH}$.
Values of $\Delta\nu_{OH}$ for intermolecular hydrogen bonds involving O—H...N are
used as a guide to the importance of steric effects.

Kuhn *et al.* [1364] have calculated the O—H...N angle and O...N
distance for three hydroxyalkylpyridines from Dreiding models in an
attempt to assess the relative importance of steric effects to intramolecular O—H...N bonding as shown in Structure (11). A large difference in Δν is obtained for linear bonds versus bent bonds, but Δν is
not sensitive to small changes in the O—H...N angle.

A series of 1,2-, 1,3-, and 1,4-aminoalcohols were examined to see
whether influences of basicity, bond angle, and bond distance could be
detected. The data in Table 5-13 indicate that the 1,2- and 1,3-
aminoalcohols experience cyclic bonding which is weaker than the intermolecular bond (*e.g.*, methanol-pyridine, 275 cm^{-1}; methanol-diethylamine,

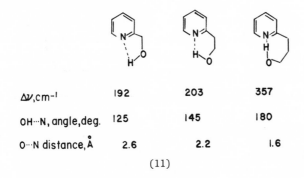

$\Delta\nu$,cm^{-1}	192	203	357
OH···N, angle, deg.	125	145	180
O···N distance, Å	2.6	2.2	1.6

(11)

TABLE 5-13

Intramolecular Frequency Shifts of ν_{OH} in Aminoalcohols[a]

Aminoalcohol	Compound	$\Delta\nu$, cm^{-1}
1,4	4-Diethylaminobutanol-1	485
3,4	3-(2-Piperidyl)propanol-1	450
1,4	3-(2-Pyridyl)propanol-1	357
1,3	3-Aminopropanol-1	235
1,3	3-Diethylaminopropanol-1	360
1,3	2-(2-Piperidyl)ethanol-1	307
1,2	2-(2-Pyridyl)ethanol-1	203
1,2	Aminoethanol	114
1,2	Diethylaminoethanol	170
1,2	Methylaminoethanol	126
1,2	3-Hydroxypiperidine	100
1,2	2-Piperidylmethanol	130
1,2	2-Pyridylmethanol	192

[a]Kuhn *et al.* [1364].

430 cm^{-1}), but that the intramolecular hydrogen bond in 1,4-amino-alcohols is stronger than the corresponding intermolecular hydrogen bond. This situation is somewhat different from that for the 1,2-, 1,3-, and 1,4-diols. Although the 1,2- and 1,3-diols have smaller $\Delta\nu_{OH}$ values than those for intermolecular hydrogen bonding of alcohols, the 1,4-diols have $\Delta\nu_{OH}$ values in the same range as intermolecular hydrogen bonding values.

Within a given series the order of increasing $\Delta\nu_{OH}$ is as expected
for the increase in basicity in going from primary to tertiary amines.
Kuhn *et al.* [1364] attribute the large value of $\Delta\nu_{OH}$ for 2-pyridylmeth-
anol to the shorter O—H...N distance in this compound. The distance
is largely determined by the dihedral angle formed by the C—O and C—N
bonds. Proposed dihedral angles are 0° for 2-piperidylcarbinols and
60° for 2-pyridylcarbinols.

4. *O—H...π*

Ōki and Iwamura [1838] have studied the effect of steric hindrance
on the O—H...π interaction in 2-hydroxylbiphenyl. A peak at 3607 cm^{-1}
has been assigned to free ν_{OH} and a peak at 3566 cm^{-1} has been assigned
to ν_{OH} for the OH group association with π electrons on the phenyl
group in the *ortho* position.

Iwamura [1152] derived an equation based on a correlation between
ν_{OH} and the geometric relationship of the OH group to the aromatic π-
orbitals in β-phenyl ethanols:

$$\text{bonded } \nu_{OH} \text{ (cm}^{-1}) = 3625 - 50 \cos 2(\Theta-30°) . \qquad (5-2)$$

The angle Θ is defined as the dihedral angle between the plane of the
benzene ring and the plane through C_α and C_β of the side chain, Structure (12).

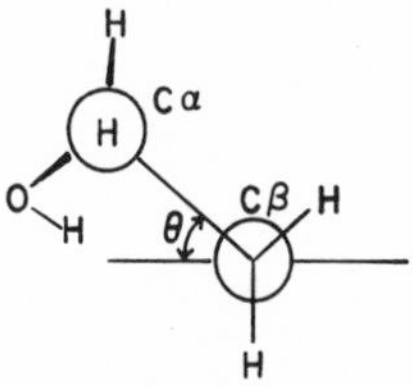

(12)

Allerhand and Schleyer [39] have reported ir evidence for intramolecular hydrogen bonding in α- and β-hydroxynitriles between the OH
group and the π-electrons of the C≡N triple bond.

The cyclopropane ring is a weak basic site for intramolecular hydrogen bonding, as indicated by the small frequency shifts of 11-17 cm^{-1}
for cyclopropylcarbinols [1202]. The preferred site for proton donor

interaction appears to be the "edge" of the cyclopropane ring. Interpretation of spectra is complicated by the effect of different conformers.

Murty and Curl [1755] investigated the microwave spectrum of allyl
alcohol. The rotational constants of the deuterated sample and ^{18}O-
substituted sample, Structure (13), indicate the assigned conformation

(13)

is the *gauche* form, which allows a hydrogen bond between the π electrons and the OH group. Although the presence of two concentration-
independent O—H stretching bands in the ir region is used as evidence
for intramolecular hydrogen bonding, Joris, Schleyer, and Osawa [1203]
have pointed out that conformational heterogeneity may be a contributing factor since several saturated alcohols incapable of hydrogen
bonding also have doublet character. Compounds with weak intramolecular hydrogen bonds such as allyl alcohols, benzyl alcohols, and cyclopropylcarbinols give spectra shifts comparable to those displayed by
some saturated alcohols. In order to determine whether hydrogen bonding is present it is necessary to compare the alcohol in question with
a model saturated alcohol. The difference in the spectrum of the
intramolecularly hydrogen-bonded alcohol is the greater displacement
of the low frequency band and the greater ratio of the intensity of the
low frequency band to that of the high frequency band. An extensive
analysis of conformational heterogeneity in saturated alcohols which
may serve as suitable models was carried out by Joris *et al.* [1203].

The complex band structure of ν_{OH} for Structure (14) has been
assigned to (1) free O—H, (2) O—H...π, and (3) O—H...Fe, where d-
orbital electrons are utilized [114].

B. N—H...B

The first overtone N—H symmetric bands of several *ortho*-substi-
tuted anilines are split into two components [1402]. Since doublets

$$\begin{array}{c} \text{OH} \\ | \\ \boxed{\bigcirc}\!-\!\overset{|}{\underset{|}{C}}\!-\!R_1 \\ \text{M} \quad R_2 \\ \boxed{\bigcirc} \end{array}$$

(14)

are also observed for *meta-* and *para-*substituted compounds, the doublet cannot be attributed to a double-minimum potential or to intramolecular hydrogen bonding [1342,1343]. However, the separation of the doublet does change with the strength of intramolecular bonding in *ortho-* anilines. Lady and Whetsel [1402] assign the low frequency component to the first overtone of the symmetric N—H stretching vibration and the high frequency component to the combination of asymmetric and symmetric N—H stretching vibrations.

Krueger [1341] has discussed the changes in ν_{NH} that occur when intramolecular hydrogen bonding is present. These include (1) narrowing of both the fundamental symmetric and asymmetric N—H stretching vibrations, (2) higher intensity for the asymmetric band, and (3) a positive deviation of ν_{asym} from the linear correlation of ν_{asym} with ν_{sym}: $[\nu_{asym} - \nu_{sym} = 0.42\ \nu_{sym} - 1348\ cm^{-1}]$. Krueger [1340] also investigated intramolecular hydrogen bonding in aliphatic diamines such as N,N'-dimethyl-1,3-propanediamine and N,N'-dimethyl- and N,N,N'-trimethyl-ethylenediamine. When the intervening methylene chain exceeds three, intramolecular hydrogen bonding disappears.

Solvent dependence of $^1J(^{15}NH)$ in *ortho-*substituted anilines was examined by Axenrod and Wieder [96]. As the electron-withdrawing ability of the *ortho* substituent increases, the difference between the coupling constants observed in $CDCl_3$ and $(CH_3)_2SO[\Delta^1J(^{15}NH)]$ decreases. Previous work [174,2037] had demonstrated that *ortho-*nitroanilines are intramolecularly hydrogen-bonded in $CDCl_3$ and intermolecularly hydrogen-bonded in dimethylsulfoxide. In addition, intra- and intermolecular hydrogen bonding in ring-substituted anilines had been shown to enhance the electron-donating ability of the amino group. On this basis, Axenrod and Wieder [96] have proposed that $\Delta^1J(^{15}NH)$ is a measure of the strength of the intramolecular hydrogen bond in $CDCl_3$. If $\Delta^1J(^{15}NH)$ is small, a strong intramolecular hydrogen bond forms when the substituted aniline is dissolved in $CDCl_3$. The order of increasing strength

of the intramolecular hydrogen bond according to this correlation is
NO_2 > C ≈ O > CF_3 ≈ Br ≈ OCH_3 > F.

Andrews, Rae, and Reichert [58] obtained PMR evidence for intramolecular hydrogen bonding between the amide proton and the *ortho*-
substituent for *ortho*-substituted acetanilides [Structure (15)]. The

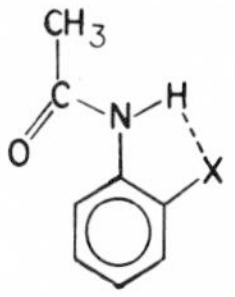

(15)

aromatic proton adjacent to the amide group is strongly deshielded.
The extent to which an acylation shift exceeds the shift observed for
unsubstituted acetanilides is used as a measure of the strength of the
intramolecular hydrogen bond.

Andrews, Rae, and Reichert [58] also studied PMR spectra of amides
derived from 2-substituted-1,3-phenylenediamines in which certain hydrogen bonding groups can simultaneously hydrogen bond to two amide
protons [Structure (16)]. The acylation shifts of H-4' and H-6' were
approximately equal to the sum of one *ortho* and one *para* acylation
shift.

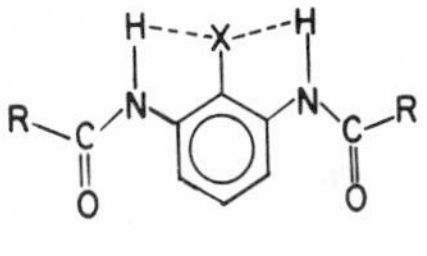

(16)

Hydrogen bonding studies involving both NH and OH groups as proton
donors and both N and O atoms as proton acceptors are of interest since
such multiplicity is prevalent in many biological systems. In ethanol-
amine and N-methylethanolamine there is competition between O—H...N
and N—H...O bonds. The enthalpy difference between *trans* and O—H...N
hydrogen-bonded *gauche* forms of N,N-dimethylethanolamine (ΔH = -2.8
kcal/mole) is much greater than the corresponding difference for ethan-
olamine and for N-methylethanolamine (ΔH = -0.7, -0.7 kcal/mole) [1346].

The increase in the relative stability of the *gauche* conformer has been attributed to both the change in basicity of the nitrogen atom and the Thorpe-Ingold effect. The latter effect involves a considera- tion of the increase in the C—N—C angle caused by crowding of two methyl groups, which leads to a decrease in the angle that the nitrogen lone pair orbitals make with the C—N bond.

C. S—H...B

Infrared spectral data indicate that benzyl mercaptan, ω- ethoxyalkyl mercaptans, and ethyl α- and β-mercaptoalkanoates form no intramolecular hydrogen bond [1691]. Some evidence exists for intra- molecular hydrogen bonds in thiosalicylic acid between the S—H group and the carbonyl group [869].

Nyquist [1814] observed two ir bands for ν_{SH} of $(RO)_2P(S)(SH)$ and $(C_6H_5O)_2P(S)(SH)$ in the 2500-2600 cm^{-1} region. Bands at 2550 and 2588 cm^{-1} for $(CH_3O)_2P(S)SH$ in 1% carbon tetrachloride are assigned to rota- tional isomers, with the band at 2550 cm^{-1} assigned to the intramolecu- lar bond in Structure (17a).

(17)

David and Hallam [570] observed two peaks for solutions of thio- phenol in carbon tetrachloride at about 1M concentration. The peak at 2571 cm^{-1} is assigned to the intramolecular hydrogen bond, while the peak at 2590 cm^{-1} is assigned to the free S—H group.

Muller and Hyne [1724] have examined the proton chemical shifts for sulfanes (H_2S_x). The various terminal protons in a sequence of low-molecular-weight sulfanes $(H_2S \longrightarrow H_2S_6)$ are characterized by dis- crete and identifiable PMR signals. When the chain length is plotted versus τ, H_2S_3 falls on the low field side. This has been attributed to intramolecular hydrogen bonding in H_2S_3 [2197]. Evidence for this is the difference in upfield shift for H_2S_3 and H_2S_4 when solutions are diluted with CCl_4. The H_2S_3 signal shifts upfield by 8.8 ± 0.4 cps while the H_2S_4 signal shifts upfield 10.0 ± 0.4 cps.

Two bands of equal intensity are observed at 2613 and 2547 cm^{-1} for ortho-aminothiophenol. The peak at 2547 cm^{-1} is sensitive to concentration changes and is assigned to the intramolecular S—H...N bond. In ortho-hydroxylthiophenol two bands are also observed at lower concentrations, and these are assigned in the same way as above. At higher concentrations the cyclic dimer is proposed as the stable species [570].

II. SELF-ASSOCIATION

A. Determination of Thermodynamic Data

Uncertainty about the degree of polymerization and the importance of linear and cyclic structures has resulted in confusion and controversy in the interpretation of experimental data for self-associated systems. As a result, thermodynamic data for a particular self-associated species will often cover a wide range of enthalpy values, since the number of hydrogen bonds formed must be known in order to obtain the energy/bond. Two recent reviews include tabulations of thermodynamic data for self-association [579,1751]. Much of the solution data were obtained by NMR methods.

1. Saunders-Hyne method

One procedure is to assume that only one polymeric species makes a significant contribution to the self-association. The Saunders-Hyne method [2182] and variations of it [1597] are based on this assumption. Marcus and Miller [1597] give a detailed description of the method in their research on the self-association of thiols.

Experimental data for the self-association of 1-butanethiol are plotted in Fig. 5-2 to illustrate the good fit for the dimer. The studies of alcohols by this method and a modified version by Feeney and Walker [745] support the predominance of the methanol tetramer, ethanol trimer or tetramer, t-butanol trimer, and methyl cellosolve dimer.

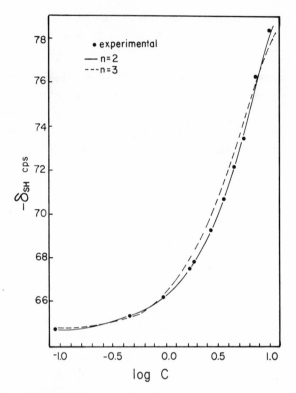

FIG. 5-2. ^{1}H NMR chemical shift versus log C for SH in n-C₄H₉SH.
TMS reference.   Marcus and Miller [1597].

2.   LaPlanche-Thompson-Rogers method [1414a]

This method assumes that the monomer-dimer equilibrium constant
is distinct from the equilibrium constants for higher polymers, but
that all the equilibrium constants for

$$n\text{-mer} + \text{monomer} \rightleftharpoons (n + 1)\text{-mer} , \tag{5-3}$$

where $n > 1$ are equal.   This model was applied to the self-association
of N-substituted amides in an inert solvent and in hydrogen bonding
solvents.

Graham and Chang [900,901] extended this method to the determination of equilibrium constants at several temperatures for the self-
association of N-methylacetamide, N-isopropylacetamide, and N-*tert*-

butylacetamide in carbon tetrachloride and in dioxane. Values of the equilibrium constant for dimer formation, K dimer, and the constant $\bar{K}$ for Eq. (5-3) are given in Table 5-14 along with enthalpies and entropies calculated from the temperature dependence of K and $\bar{K}$. The results indicate that the formation of higher aggregates [Eq. (5-3)] is favored. The enthalpy values are smaller in dioxane because of the competition between dioxane and amide molecules for N—H.

Tucker and Becker [2486a] have expanded this approach in a critical analysis of the self-association of t-butanol in n-hexadecane. Results of vapor pressure, 220 MHz PMR, and ir investigations are consistent with an association model based on a monomer-trimer equilibrium and the equilibrium given in Eq. (5-3) with $n = 3$.

3. Goldman-Emerson method [882a]

Goldman and Emerson [882a] have proposed a quantitative model for hydrogen-bonded species of acetic acid in carbon tetrachloride which includes three competing equilibria involving monomers, cyclic dimers, linear dimers, and linear polymers. Values for K_c, cyclic dimer; K_ℓ, linear dimer; and $K_{\ell p}$, linear polymer (Eq. 5-3) were obtained by finding the set of equilibria constants and chemical shifts which give the best fit for the plot of chemical shift versus concentration of acetic acid. Values of 3550, 1130, and 59.0 $(\text{mf})^{-1}$ were obtained at 16.5°C for K_c, K_ℓ, and $K_{\ell p}$, respectively. Although the cyclic dimer is the most important species, linear polymers are present in significant amounts. In deriving the quantitative model Goldman and Emerson assumed that $K_{\ell p}$ (Eq. 5-3) is independent of chain length. Therefore, $K_{\ell p}$ is a weighted average of individual polymerization constants.

This method offers several advantages over other methods since it not only provides quantitative information about several species, but also is applicable to a wide concentration range.

B. Types of Self-Association

In Fig. 5-3 the various types of self-association are represented with the range of energies which have been reported [1751]. Examples of each class are discussed in the following sections.

TABLE 5-14

Thermodynamic Parameters[a]

Amide	Solvent	K dimer, mf^{-1} 20°C	$-\Delta H$ dimer, kcal/mole	$\bar{K}$, mf^{-1} 20°C	$-\Delta\bar{H}$ kcal/mole
N-Methylacetamide	CCl_4	17 ± 0.6	3.7 ± 0.3	270 ± 9	4.5 ± 0.3
	dioxane	4.3	1.07 ± 0.07	11.8	1.14 ± 0.09
N-Isopropylacetamide	CCl_4	11 ± 0.4	3.0 ± 0.3	180 ± 13	4.3 ± 0.5
	dioxane	3.7	0.80 ± 0.5	8.4	0.91 ± 0.03
N-tert-Butylacetamide	CCl_4	7.6 ± 0.3	2.4 ± 0.2	67 ± 1.3	3.3 ± 0.1
	dioxane	3.2	0.66 ± 0.07	4.8	0.79 ± 0.05

[a]Graham and Chang [900,901].

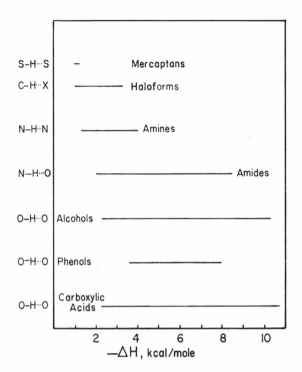

FIG. 5-3.   Ranges of hydrogen bond energies of self-association.
Murthy and Rao [1751].

1. O—H...O

a. *Phenols and alcohols*. Puttnam [2029] evaluated the hindering
effect of various alkyl groups on the self-association of *ortho*-
substituted phenols by comparing O—H stretching frequency shifts.
Two bands are observed for self-associated phenols and alcohols that
are at lower frequencies than the monomer O—H band.   Puttnam assigned
the band at about 3500 cm^{-1} to the dimer and the other band to higher
polymers.   He observed that the frequency shift for the dimer is less
than one half that for the polymer.   If we assume a correlation of bond
strength with frequency shift, the hydrogen bonds in the polymer are
stronger than those in the dimer.

Substituents in the *meta* or *para* position have no effect on the
strength of the hydrogen bond, but substituents in the *ortho* positions
reduce the strength of the hydrogen bond.   Puttnam attributes the

reduction in hydrogen bond strength for *ortho*-substituted phenols either to the screening effect of the substituent or to the steric hindrance of the substituent toward coplanarity of the OH group with the aromatic ring. When two *ortho*-alkyl groups are present, the values of $\Delta\nu$ dimer and $\Delta\nu$ polymer are still smaller, and in the case of 2,6-di-*t*-butylphenol only one sharp absorption band is observed.

Singh and Rao [2279] used ir and NMR methods to obtain thermodynamic data for the self-association of a series of alcohols. Equilibrium constants were calculated on the basis of closed dimers. Values of K dimer, $-\Delta H$, and $\Delta\nu_{OH}$ are given in Table 5-15. No correlation between ΔH and $\Delta\nu_{OH}$ was observed.

TABLE 5-15

Thermodynamic Parameters for Dimerization of Alcohols and Phenols[a]

ROH	Solvent	$K_{25°C}$, ℓ/mole	$-\Delta H$, kcal/mole	$\Delta\nu_{OH}$, cm^{-1}
C_2H_5OH	C_6H_6	0.39	5.0	125
CF_3CH_2OH	C_6H_6	0.19	3.8	230
n-C_3H_7OH	CCl_4	1.25	2.6	125
$CF_2HCF_2CH_2OH$	C_6H_6	0.22	1.2	195
C_6H_5OH[b]	CCl_4	0.74	5.1	120
4-ClC_6H_4OH	CCl_4	2.20	3.2	170
4-$CH_3OC_6H_4OH$	CCl_4	3.00	3.8	150
$2,6$-di-i-$C_3H_7C_6H_3OH$	CCl_4	0.53	8.4	260
$(i$-$C_3H_7)_2$-C-C_2H_5OH	CCl_4	1.00	10.0	127

[a]Singh and Rao [2279].   [b]Maguire and West [1586].

The effect of intramolecular bonding versus self-association was examined for a series of halogen-substituted phenols by Griffiths and Socrates [937] who measured the concentration dependence of the hydroxyl PMR signal in carbon tetrachloride. They analyzed the data by the Saunders-Hyne curve-fitting method. The best agreement of theoretical curves with experimental data was obtained for the trimer model in all cases except *o*-fluorophenol, *o*-iodophenol, and *o*-hydroxy-acetophenone.

The equilibrium constants are given in Table 5-16. The small values
of K for the *ortho*-substituted phenols can be attributed to the strong
intramolecular bond formed by the *ortho* substituent.

TABLE 5-16

Self-Association of Phenols[a]

Phenol	Trimer K_3, $\ell^3/mole^3$	Phenol	Dimer K_2, $\ell^2/mole^2$
o-Cresol	1.15	*o*-Fluorophenol	0.06
m-Cresol	12.6	*o*-Iodophenol	0.02
p-Cresol	12.6	*o*-Hydroxyacetophenone	0.05
m-Fluorophenol	3.98		
p-Fluorophenol	2.00		
o-Chlorophenol	0.07		
m-Chlorophenol	2.51		
p-Chlorophenol	0.30		
o-Bromophenol	0.02		
m-Bromophenol	1.00		
p-Bromophenol	2.00		

[a]Griffiths and Socrates [937].

Bellamy and Pace [188] have examined the frequency shifts of
dimers versus polymers in an attempt to determine the cause of the
large difference in frequency shifts. Previous workers [1459,2514] had
explained the difference by assuming a cyclic structure for the dimer.
Such a structure has bent hydrogen bonds, which are weaker than linear
hydrogen bonds. The polymers, whether open chain or cyclic, would have
stronger hydrogen bonds because the O—H...O atoms are more nearly co-
linear. Bellamy and Pace [188] propose that the results can also be
explained by assuming that the dimers are open-chain systems with normal hydrogen bonds, and that the strong bonds in the polymers are due
to a cyclic structure. They suggest that an oxygen atom of an OH group
whose proton is engaged in hydrogen bonding is more basic than the
original monomer. Conversely, the hydrogen atom of an OH group whose

oxygen is acting as a proton acceptor should be more acidic than in the monomer. This would favor a cyclic structure for the polymer. Difference spectra were interpreted on the basis of the greater acidity of the free OH group in a dimer as compared to the OH group in a monomer [188]. The band at 3393 cm^{-1} for methanol-phenol mixtures in carbon tetrachloride was assigned to the open-chain dimer.

Fletcher and Heller [772] have discussed the danger in assuming that the ir band at 3500 cm^{-1} is always the dimer band. They find that the peak at 3500 cm^{-1} for 1-octanol in n-decane is not due to the dimer. However, later work by Hammaker et al. [991] supports the assignment of the 3500 cm^{-1} band to the dimer in the case of methanol, t-butanol, and di-t-butyl carbinol. Fletcher and Heller [771] concur with these assignments and provide additional evidence that the bands near 6536 cm^{-1} are due to a combination of v_{OH} and v_{CH} stretching modes of the monomer. The problems of assuming a specific self-association model and then fitting the experimental data to this model are discussed [772]. The tetramer is assumed to be the predominant polymer in the self-association of 1-octanol in n-decane.

The work of Fletcher and Heller points out the main reason for the confusion about the structure and stoichiometry of the predominant polymers in solution. Since most mathematical analyses are based on the Saunders-Hyne curve-fitting procedure, an assumption of a particular self-association model is always necessary for this approach. Saunders and Hyne [2182] emphasized that the curve-fitting approach relies upon the postulation of a predominant polymer throughout the total concentration range.

Tucker and Becker [2486a] argue that more consistent association models are possible if data from several types of experiments are used for testing the models. In alcohol association studies they used vapor pressure measurements as a check on the monomer concentration obtained by ir measurements. This check is necessary since ir bands of OH end groups may overlap the monomer band. Results of measurements verified the trimer as the major associated species of t-butanol in n-hexadecane. Data from measurements on association of t-butanol were consistent with a monomer-trimer-higher polymer model (see II.A.2). Association data for methanol were examined and found to fit a monomer-trimer-higher

polymer model rather than the monomer-tetramer model proposed previously [629].

   b. *Carboxylic acids*. Eberson [691] has reviewed the various
types of hydrogen bonding which are important in the chemistry of carboxyl groups. Carboxylic acids are generally assumed to form cyclic
dimers in the gas phase or in nonpolar solvents, although some evidence
does exist for the open-chain structure (see II.A.3). Improved band
resolution techniques which include factor analysis of digitized ir
spectra have provided evidence for the presence of a chain structure in
addition to the monomer and dimer [364a].

   Bellamy, Lake, and Pace [182] suggest that carboxylic acids may
exist as an equilibrium mixture of open-chain and cyclic forms, and
consider both the appreciable dipole moments and the anomalies in the
structure of the ν_{OH} band of carboxylic acids as evidence for the importance of the open-chain form. If the ir spectra of the cyclic and
linear forms of anhydrous oxalic acid are superimposed, the resulting
spectrum has a pattern similar to those of monocarboxylic acids.

   Muller and Rose [1727,1728] have made a number of dilution studies
of carboxylic acids by using NMR spectroscopy. Solvents which have
donor sites capable of forming hydrogen bonds were used. One problem
encountered in studies of this kind is the effect which small traces
of water have on the dilution shifts. According to previous reports
[1102,2072], the resonance for the carboxyl proton in the acetic acid-
acetone adduct appeared to be at least 10 ppm upfield from the dimer
signal. However, when carefully dried acetone was used, the resonance
for the adduct was only 2 ppm upfield from the dimer signal [1727].
The dimerization constants and solvent association constants for acetic
acid in acetic anhydride, acetone, and dioxane are given in Table 5-17.

   Allen, Watkinson, and Webb [34] studied the ir spectra of benzoic
acid in the vapor phase and in cyclohexane, carbon tetrachloride, and
benzene. They assigned bands at 1770 and 1670 cm^{-1} to the free and
hydrogen-bonded carbonyl groups, respectively. Values of K_2 are 5830
ℓ/mole in cyclohexane at 35°C, 3660 ℓ/mole in carbon tetrachloride at
30°C, and 462 ℓ/mole in benzene at 30°C. Values for the enthalpy of
dimerization of benzoic acid are 8.1 kcal/mole in the vapor phase, 6.4
kcal/mole in cyclohexane, and 5.5 kcal/mole in benzene.

TABLE 5-17

Dimerization and Solvent Association Constants of Acetic Acid[a]

Solvent	K_2, $(mf)^{-1}$	K_{assn}, $(mf)^{-1}$
Acetic anhydride	800	14
Acetone	800	34
Dioxane	4000	100

[a]Muller and Rose [1727].

Affsprung, Christian, and Melnick [11] examined the hetero-association of acetic acid and trichloroacetic acid in carbon tetra-chloride by measuring the peak absorbances of ν_{OH} at 2.822 and 2.847 μ, respectively. Thermodynamic values obtained for the self-association of acetic acid are $K_2 = 3200$ molal^{-1} at 25°C and $\Delta H = -10.7$ kcal/mole. For trichloroacetic acid the values are $K_2 = 430$ molal^{-1} at 25°C and $\Delta H = -7.9$ kcal/mole. The values for the 1:1 adduct of the two acids are $K_{25°} = 7200$ molal^{-1} and $\Delta H = -7.7$ kcal/mole.

A recent example of gas-phase studies is the use of ir techniques to obtain thermodynamic data [460] for the dimerization of carboxylic acids (Table 5-18).

TABLE 5-18

Thermodynamic Data for Dimerization of Carboxylic Acids in Gas Phase[a]

Acid	$-\Delta H$, kcal/mole	$-\Delta S$, e.u.
Formic	14.8 ± 0.5	36.2 ± 2.0
Acetic	14.6 ± 0.5	34.8 ± 2.0
Propionic	14.9 ± 0.6	37.0 ± 2.5
n-Butyric	15.7 ± 0.7	38.9 ± 2.5
Isobutyric	15.2 ± 0.7	36.6 ± 2.5
Trimethylacetic	14.6 ± 0.5	36.5 ± 2.0

[a]Clague and Bernstein [460].

2. *N—H...N*

   Springer and Meek [2328] determined the equilibrium constant for
the self-association of diethylamine in cyclohexane by using the curve-
fitting procedure of Saunders and Hyne [2182]. The monomer-tetramer
model gave the best fit with the experimental data; the value found
for K_4 was 0.00175 M^{-3} at 40°C. Murphy and Davis [1738] propose that
the dimer of diethyl amine becomes more important at low concentrations.
The constant chemical shift reported for the NH proton at concentrations
less than 0.11 mole-fraction of amine [2328] was attributed to the presence of a ^{13}C satellite which obscures changes in the NH proton resonance. For piperidine the ^{13}C satellite and NH proton resonance could
be distinguished and a linear variation of the NH chemical shift with
concentration was observed [1738].

   Crook and Schug [543] used the method of Saunders and Hyne [2182]
to study self-association of hydrazines in cyclohexane solution. For
the hydrazines $(CH_3)_2NNH_2$, $CH_3NHNHCH_3$, and $(CH_3)_2NNHCH_3$ the experimental data fit a monomer-dimer model. The absence of appreciable amounts
of trimers and higher polymers is considered support for the cyclic
dimer since the linear dimer would polymerize further. Values of K_2
$(mf)^{-1}$ at 28°C are 1.8, 1.0, 0.8 for $(CH_3)_2NNH_2$, $CH_3NHNHCH_3$, and
$(CH_3)_2NNHCH_3$, respectively.

   Durig *et al.* [675] studied the self-association of N_2H_4 and N_2D_4.
They found that the Raman spectral bands at 2348 cm^{-1} and 3189 cm^{-1} for
N_2D_4 and N_2H_4, respectively, decreased in intensity with dilution. The
Raman intensities of the two lines 3189 and 3260 cm^{-1} were measured at
several different temperatures between 30°C and 70°C. From the slope
of a plot of log I_{3189}/I_{3260} versus 1/T a ΔH of -1.45 kcal/mole was
calculated.

   Lady and Whetsel [1403] used the first overtone N—H symmetric
stretching band of aniline to study the self-association process. A
model based on monomer, dimer, and tetramer is consistent with the
experimental data. Hydrogen bond energies of -1.64 and -7.7 kcal/mole
were calculated for the dimer and the tetramer, respectively. A cyclic
tetramer based upon an average ΔH/bond energy of -2 kcal/mole is proposed.

Anderson, Duncan, and Rossotti [54] have studied the self-association of imidazole and pyrazole. The self-association of pyrazole is primarily cyclic dimers and trimers, while imidazole gives a mixture of polymers up to the dodecamer. Equilibrium constants determined for the dimerization and trimerization of pyrazole are 47.5 ℓ/mole and 7540 ℓ^2/mole^2. A value of 234 ℓ/mole was determined for the dimerization of imidazole. The cyclic trimer of pyrazole would be favored over the cyclic dimer because of the greater stability of linear hydrogen bonds. Substitution at the 3,5 positions of pyrazole increase the stability of the dimer; in the case of 3,5-diphenyl pyrazole a dimer and linear polymer are formed with no evidence for the presence of a trimer.

3. N—H...O

Itoh and Shimanouchi [1147] measured the far-infrared spectra of mono-substituted amides in the liquid state with particular attention to N-methylacetamide since it is the simplest molecule with a peptide bond. Two bands at 120 and 201 cm^{-1} are assigned to the CO...H—N bond and the C—N torsional vibration, respectively. The 201 cm^{-1} band shifts to 195 cm^{-1} for $CD_3CONHCH_3$ and to 185 cm^{-1} for $CD_3CONHCD_3$.

4. P—O...H

Ferraro and Peppard [750] studied the self-association of organophosphorus acids using PMR techniques. The hydroxyl proton resonance was measured as a function of concentration in carbon tetrachloride. Only one PMR peak was observed, which indicates a rapid exchange between the protons of the various polymeric species. In addition, the hydroxyl proton does not exhibit spin-spin coupling due to the presence of ^{31}P. These results were interpreted in terms of a dimer with an eight-membered ring structure.

The ratio of the coupling constants J_{PH}/J_{PD} for a series of dialkyl phosphonates varies with the hydrogen bonding ability of the solvent. Since dialkyl phosphonates form dimers and higher aggregates, the variation in J_{PH}/J_{PD} has been ascribed to the difference in the strength of P—O...H versus P—O...D hydrogen bonds [2341].

THERMODYNAMIC DATA FOR HYDROGEN BOND ADDUCTS

The appendix is a noncritical compilation of data reported during the 1960-1973 period for intermolecular adducts. Chapter 3 describes the reliability of data measured by various spectroscopic and calorimetric methods. Selected data for self-associated systems and intramolecular hydrogen bonding are given in Chapter 5.

In columns 1 and 2 a blank space means that the entry above is to be carried down. The phase or solvent is given in column 3. If a binary mixture was used, "base" is entered as the solvent. The abbreviation TMP stands for 2,2,4-trimethylpentane.

Column 4 gives the method. Abbreviations are IR, infrared; CAL, calorimetric; NIR, near-infrared; GLC, gas liquid chromatography; NMR, nuclear magnetic resonance; UV, ultraviolet; DEC, dielectric constant; VP, vapor pressure; MP, molar polarization.

Column 5 gives the value of K in ℓ/mole at 25°C. The temperature is given in parentheses if data were obtained at a centigrade temperature other than 25°. The units are given in parentheses when they are not ℓ/mole.

Column 6 lists $-\Delta G$ at 25°C in kcal/mole. Values of ΔG were calculated using $\Delta G = -RT\ln K$. In those cases where K values are at temperatures other than 25°C, the log K versus 1/T relationship was assumed to be linear and log K at 25°C was calculated. In cases where ΔG was reported at a temperature other than 25°C, the temperature is given in parentheses.

Columns 7 and 8 list $-\Delta H$ in kcal/mole and $-\Delta S$ at 25°C in cal deg^{-1} mole^{-1}. A number of literature reports list $K_{20°}$ and $K_{50°}$ (e.g., 908-

910, 914). In these cases ΔH was calculated from

$$-\Delta H \text{ (kcal/mole)} = 4.58 \times 10^{-3} \left[\frac{\log K_1 - \log K_2}{T_1^{-1} - T_2^{-1}} \right].$$

Enthalpy values marked with asterisks were used in Δν-ΔH correlations (see Table 4-5).

The frequency shift for the A—H stretching frequency is given in column 9 and the reference number in the bibliography is given in column 10.

Grouping of acids is OH, SH, CH, NH, and XH. Bases are listed in the following order: amides, amines, esters, ethers, organic halides, ketones, nitriles, phosphoryl donors, π-bases, and miscellaneous.

Error limits are given when values cited in the literature included them.

THERMODYNAMIC DATA AND A—H FREQUENCY SHIFTS FOR HYDROGEN BONDS
BETWEEN REFERENCE ACIDS AND LEWIS BASES

Acid	Base	Phase or Solvent	Method	$K_{25°C}$, ℓ/mole	$-\Delta G_{25°C}$, kcal/mole	$-\Delta H$, kcal/mole	$-\Delta S_{25°C}$, cal deg^{-1} mole^{-1}	$\Delta\nu_{AH}$ cm^{-1}	Ref.
HOH	HC(O)N(CH$_3$)$_2$	CCl$_4$	IR	4.4 ±1.7	0.9 ±0.2	3.2 ±0.7	8 ±3		1046a
	CH$_3$C(O)N(CH$_3$)$_2$	CCl$_4$	IR	8.7 ±2.4	1.3 ±0.2	3.2 ±0.4	6.3 ±1.9		1046a
	CH$_3$C(O)N(C$_4$H$_9$)(CH$_3$)	CCl$_4$	IR	6.5 ±1.5	0.8 ±0.1	4.4 ±1.0	12 ±4		1046a
	CH$_3$C(O)N(C$_4$H$_9$)(C$_6$H$_4$CH$_3$)	CCl$_4$	IR	4.9 ±1.3	0.9 ±0.2	5.7 ±1.2	16 ±5		1046a
	CH$_3$C(O)N(CH$_3$)(C$_6$H$_5$)	CCl$_4$	IR	5.2 ±0.8	1.0 ±0.1	3.8 ±0.5	9.4 ±2.0		1046a
	C$_2$H$_5$C(O)N(CH$_3$)$_2$	CCl$_4$	IR	5.1 ±2.3	1.0 ±0.3	3.7 ±0.6	9 ±3		1046a
	n-C$_3$H$_7$C(O)N(CH$_3$)$_2$	CCl$_4$	IR	5.7 ±1.0	1.0 ±0.1	4.0 ±0.6	10 ±2		1046a
	(CH$_3$)$_2$CHC(O)N(CH$_3$)$_2$	CCl$_4$	IR	5.9 ±2.3	1.1 ±0.1	4.3 ±0.8	11 ±3		1046a
	i-C$_4$H$_9$C(O)N(CH$_3$)$_2$	CCl$_4$	IR	5.5 ±1.6	1.0 ±0.2	4.7 ±0.8	13 ±3		1046a
	C$_6$H$_5$C(O)N(CH$_3$)$_2$	CCl$_4$	IR	3.0 ±1.0	0.7 ±0.2	7.6 ±0.5	23 ±2		1046a
	AMIDES								
CH$_3$OH	HC(O)N(CH$_3$)$_2$	CCl$_4$	IR	5.5	1.0	3.7	9.1	160	167
		CCl$_4$	CAL			3.6*			1779a
		CCl$_4$	IR	4.2(28°)				116	1350
	AMINES								
	(CH$_3$)$_3$N	vapor	IR			7.1 ±0.2		302	759
		vapor	VP			7.5 ±0.3			759
		vapor	NMR			5.8 ±0.7			461
	(C$_2$H$_5$)$_3$N	CCl$_4$	IR	6.8(31°)	1.3	6.0	15.8	430	1058
		CCl$_4$	CAL			4.2*			1779
		C$_6$H$_5$Cl	IR	6.4(20°)	1.1	3.8	9.1	427	909
		C$_6$H$_5$Cl	IR	6.0(24°)	1.1	4.3	10.7		
		CH$_2$Cl$_2$	IR	6.8(24°)	1.1	3.0	6.4	328	1058

APPENDIX (continued)

Acid	Base	Phase or Solvent	Method	$K_{25°C}$, ℓ/mole	$-\Delta G_{25°C}$, kcal/mole	$-\Delta H$, kcal/mole	$-\Delta S_{25°C}$, cal deg^{-1} mole^{-1}	$\Delta\nu_{AH}$ cm^{-1}	Ref.
CH$_3$OH	(C$_2$H$_5$)$_3$N	i-C$_8$H$_{18}$	UV	3.4	0.7	7.6 ±0.5			2347
		vapor	IR						1058
	(n-C$_3$H$_7$)$_3$N	CCl$_4$	NIR	3.4(20°)	0.7	2.7*	6.7	370	909
	(n-C$_4$H$_9$)$_3$N	CCl$_4$	NIR	4.1(20°)	0.8	3.5*	9.0	413	909
	(C$_6$H$_5$CH$_2$)$_3$N	CCl$_4$	NIR	3.4(20°)	0.7	1.5	2.7	413	909
	C$_6$H$_5$N(C$_2$H$_5$)$_2$	CCl$_4$	NIR	1.7(20°)	0.31	1.5	4.0		909
	Pyridines								
	C$_5$H$_5$N	CCl$_4$	IR	3.0	0.65	3.9	10.8	286	167
		CCl$_4$	IR	2.9 ±0.1	0.64 ±0.02	3.84 ±0.12*	10.8 ±0.6	284	1906
		CCl$_4$	NIR	6.0(20°)	1.0	3.2	7.4	300	909
	2-CH$_3$-	CCl$_4$	IR	2.3	0.5	4.4	13.1	287	1277
		CCl$_4$	IR	3.05	0.66	4.0*	11.2	317	1277
	2-C$_2$H$_5$-	CCl$_4$	IR	3.13	0.68	4.1*	11.5	312	1906
	2-(CH$_3$)$_2$CH-	CCl$_4$	IR	2.63	0.57	4.3*	11.5	315	1277
	2-(CH$_3$)$_3$C-	CCl$_4$	IR	1.73	0.32	4.2*	12.3	311	1277
	2-Br-	CCl$_4$	IR	0.22	0.22	4.0*	12.7	295	1277
	2-Cl-	CCl$_4$	IR	1.15 ±0.1	0.08 ±0.02	2.8 ±0.1*	9.2	198	1906
	2-F-	CCl$_4$	IR	1.1 ±0.1	0.07 ±0.02	2.5 ±0.1*	8.0	198	1906
	3-CH$_3$-	CCl$_4$	IR	1.1 ±0.1	0.06 ±0.02	2.2 ±0.1*	7.0	172	1906
		CCl$_4$	IR	3.0 ±0.1	0.65 ±0.02	4.1 ±0.1*	11.6	301	1906
	3-Br-	CCl$_4$	IR	1.8 ±0.1	0.35 ±0.02	3.2 ±0.1*	9.5	243	1906
	3-Cl-	CCl$_4$	IR	1.35 ±0.1	0.18 ±0.02	3.0 ±0.1*	9.7	241	1906
	4-CH$_3$-	CCl$_4$	IR	3.1 ±0.1	0.67 ±0.02	4.2 ±0.1*	12.0	305	1906
	2,4-CH$_3$-	CCl$_4$	IR	3.25 ±0.1	0.70 ±0.02	5.0 ±0.1*	14.4	332	1906
	2,5-CH$_3$-	CCl$_4$	IR	3.2 ±0.1	0.69 ±0.02	4.9 ±0.1*	14.4	332	1906
	2,6-CH$_3$-	CCl$_4$	IR	4.42	0.88	4.0*	10.5	338	1277
		CCl$_4$	IR	3.20	0.69	5.0*	14.4	333	1906

Compound	Solvent	Method						Ref.
2,6-C₂H₅-	CCl₄	IR	2.96	0.64	3.9*	11.3	339	1277
2,6-(CH₃)₂CH-	CCl₄	IR	0.76	-0.17	3.8*	14.0	329	1277
2-C₂H₅-6-CH₃-	CCl₄	IR	4.14	0.84	4.0*	10.6	339	1277
3,4-CH₃-	CCl₄	IR	3.2 ±0.1	0.69 ±0.02	4.9 ±0.1*	14.3	315	1906
3,5-CH₃-	CCl₄	IR	3.2 ±0.1	0.68 ±0.02	4.9 ±0.1*	14.2	318	1906
2,4,6-CH₃-	CCl₄	NIR	7.6(20°)	1.1	3.6*	8.4	356	909
	CCl₄	IR	3.32 ±0.1	0.71 ±0.02	5.37 ±0.12*	15.8 ±0.6	346	1906
ESTERS								
CH₃C(O)OC₂H₅	CCl₄	IR	1.4	0.20	2.5*	7.8	84	167
	CCl₄	CAL			2.5			1779a
ETHERS								
(C₂H₅)₂O	CCl₄	IR	1.27(21.7°)	0.14	3.73 ±0.27*	12.2	150	1717
	vapor	IR			4.7 ±0.7		124	1128
[(CH₃)₂CH]₂O	CCl₄	IR	1.16(28°)		4.31 ±0.14*		140	1350
	CCl₄	IR	1.62(21.7°)	0.2	2.8*	13.7	170	1717
p-dioxane	CCl₄	IR	1.5	0.24		8.6	126	167
	CCl₄	IR	1.1(28°)				109	1350
(n-octyl)₂O	base	GLC	1.9(22.5°)	0.34	4.69 ±0.50	14.6 ±1.7		1608
HALIDES								
n-C₄H₉Cl	CCl₄	IR	1.1(28°)				31	1350
n-C₄H₉Br	CCl₄	IR	0.39(28°)				33	1350
n-C₄H₉I	CCl₄	IR	0.28(28°)				45	1350
(CH₃)₃CCl	CCl₄	IR	0.27(28°)				35	1350
(CH₃)₃CBr	CCl₄	IR	0.28(28°)				44	1350
CH₃(CH₂)₄F	CCl₄	IR	1.5(28°)				17	1350
CH₃(CH₂)₄Br	CCl₄	IR	0.38(28°)				38	1350
(C₇H₁₅)₄N⁺I⁻	CCl₄	IR	42	2.2	9.0	22.8	255	2282
C₆H₅CH₂Cl	CCl₄	IR	0.49(28°)				21	1350
C₆H₅CH₂Br	CCl₄	IR	0.59(28°)				32	1350

APPENDIX (continued)

Acid	Base	Phase or Solvent	Method	$K_{25°C}$, ℓ/mole	$-\Delta G_{25°C}$, kcal/mole	$-\Delta H$, kcal/mole	$-\Delta S_{25°C}$, cal deg^{-1} mole^{-1}	$\Delta\nu_{AH}$ cm^{-1}	Ref.
	KETONES								
CH$_3$OH	(CH$_3$)$_2$CO	CCl$_4$	IR	1.8	0.35	2.5*	7.3	112	167
		CCl$_4$	IR	1.6(28°)					1350
		CCl$_4$	CAL			3.6			1408
	(n-octyl)$_2$CO	base	GLC	2.84(50°)				112	1608
	(C$_6$H$_5$)$_2$CO	CCl$_4$	IR	1.5	0.24	2.2*	6.5	88	167
		CCl$_4$	CAL			1.8			1779a
		base	GLC			2.6			2521
	diphenylcyclopropanone	CCl$_4$	IR			8.1			1855
	benzil	base	GLC			1.6			2521
	NITRILES								
	CH$_3$CN	CCl$_4$	NMR	2.6(20°)	0.55	0.90*	1.2		1511
		CCl$_4$	IR			2.2*		81	1675
		CCl$_4$	IR	1.2(28°)				82	1350
	CH$_3$NC	CCl$_4$	NMR	3.0(30°)	0.7	2.0	4.4		1511
	PHOSPHORYL DONORS								
	HPO(OCH$_3$)$_2$	CCl$_4$	NIR	9.8(20°)	1.3	3.3*	6.7	169	910
	HPO(OC$_2$H$_5$)$_2$	CCl$_4$	NIR	11(20°)	1.4	3.5*	7.0	170	910
	HPO[OCH(CH$_3$)$_2$]	CCl$_4$	NIR	15(20°)	1.5	4.8*	11.1	180	910
	CH$_3$PO(OC$_2$H$_5$)$_2$	CCl$_4$	NIR	18(20°)	1.6	4.7*	10.4	211	910
	CH$_3$PO(OCH$_2$CH$_2$CHCl)$_2$	CCl$_4$	NIR	14(20°)	1.5	4.8*	11.1	182	910
	ClCH$_2$PO(OC$_2$H$_5$)$_2$	CCl$_4$	NIR	11(20°)	1.4	4.2*	9.4	165	910
	Cl$_2$CHPO(OC$_2$H$_5$)$_2$	CCl$_4$	NIR	7.5(20°)	1.2	2.5*	4.7	148	910
	Cl$_3$CPO(OC$_2$H$_5$)$_2$	CCl$_4$	NIR	6.8(20°)	1.1	3.6*	8.7	.130	910
	(CH$_3$)$_3$PO	CCl$_4$	NIR	40(20°)	2.1	6.0*	13.1	260	910

Compound	Solvent	Method						
(CH₃O)₃PO	CCl₄	NIR	11(20°)	1.4	3.4*	6.7	175	910
	CCl₄	NMR			3.9 ±0.6	6.5 ±1.3		2235
C₂H₅PO(OC₂H₅)₂	CCl₄	NIR	18(20°)	1.6	4.6*	10.1	211	910
(C₂H₅)₃PO	CCl₄	NIR	46(20°)	2.2	6.3*	13.7	270	910
(C₂H₅O)₃PO	CCl₄	NIR	16(20°)	1.6	4.4*	9.4	189	910
(C₂H₅)₂NPO(OC₂H₅)₂	CCl₄	NIR	18(20°)	1.6	3.7*	7.0	217	910
(C₆H₅)₃PO	CCl₄	NIR	24(20°)	1.8	5.2*	11.4	231	910
(i-C₃H₇O)₃PO	CCl₄	NMR			7.4 ±1.1	16 ±3		2235
(n-C₄H₉O)₃PO	CCl₄	NMR			4.0 ±0.6	5.7 ±1.2		2235
tri-p-tolylphosphate	CCl₄	NMR			8.6 ±1.3	20 ±4		2235
π-BASES								
benzene	CCl₄	IR	0.80(28°)				23	1350
toluene	CCl₄	IR	0.44(28°)				34	1350
m-xylene	CCl₄	IR	0.21(28°)				41	1350
mesitylene	CCl₄	IR	0.24(28°)				46	1350
durene	CCl₄	IR	0.45(28°)				53	1350
hexamethylbenzene	CCl₄	IR	0.50(28°)				69	1350
fluorobenzene	CCl₄	IR	0.11(28°)				20	1350
chlorobenzene	CCl₄	IR	0.16(28°)				19	1350
bromobenzene	CCl₄	IR	0.15(28°)				22	1350
iodobenzene	CCl₄	IR	0.23(28°)				24	1350
MISCELLANEOUS								
(CH₃)₂SO	CCl₄	IR	7.6(28°)	1.4 ±0.2	5.4 ±0.3*	13.4 ±0.5	27	1350
CH₃NO₂	CCl₄	IR	0.2(28°)	1.3 ±0.2	4.6 ±0.3*	11.3 ±0.5	277	1350
C₅H₅NO	C₂Cl₄	IR	9.2(30°)	1.1	3.3 ±0.2*	7.4	225	847a
2-Cl-C₅H₄NO	C₂Cl₄	IR	7.6(30°)	1.5 ±0.2	5.7 ±0.3*	13.9 ±0.5		847a
2-CN-C₅H₄NO	TMP	UV	6.6					2469
2-C₆H₅CH₂-C₅H₄NO	C₂Cl₄	IR	10.5(30°)	1.8 ±0.2	5.5 ±0.3*	12.3 ±0.5	285	847a
3-CH₃-C₅H₄NO	C₂Cl₄	IR	18.3(30°)	1.3 ±0.2	4.9 ±0.3*	12.0 ±0.5	279	847a
3-Cl-C₅H₄NO	C₂Cl₄	IR	7.9(30°)	1.3	4.8 ±0.1	11.7	250	847a
	TMP	UV	8.4					2469
3-I-C₅H₄NO	TMP	UV	16.1	1.6	4.4 ±0.2	9.4		2469
4-CH₃-C₅H₄NO	C₂Cl₄	IR	45.9(30°)	2.4 ±0.2	6.6 ±0.3*	14.1 ±0.5	315	847a

APPENDIX (continued)

Acid	Base	Phase or Solvent	Method	$K_{25°C}$, ℓ/mole	$-\Delta G_{25°C}$, kcal/mole	$-\Delta H$, kcal/mole	$-\Delta S_{25°C}$, cal deg^{-1} mole^{-1}	$\Delta\nu_{AH}$ cm^{-1}	Ref.
CH_3OH	$4\text{-}C_2H_5\text{-}C_5H_4NO$	C_2Cl_4	IR	22.5(30°)	2.0 ±0.2	6.7 ±0.2*	16.0 ±0.5	307	847a
	$4\text{-}CH(CH_3)_2\text{-}C_5H_4NO$	C_2Cl_4	IR	23.6(30°)	2.1 ±0.2	6.8 ±0.3*	16.3 ±0.5	305	847a
	$2,4\text{-}CH_3\text{-}C_5H_3NO$	C_2Cl_4	IR	25.4(30°)	2.0 ±0.2	6.3 ±0.3*	14.4 ±0.5	317	847a
	$3,4\text{-}CH_3\text{-}C_5H_3NO$	TMP	UV	28.5	2.0	5.2 ±0.2	10.7		2469
	$3,5\text{-}CH_3\text{-}C_5H_3NO$	C_2Cl_4	IR	23.0(30°)	1.8 ±0.2	5.7 ±0.3	13.0 ±0.5		847a
	$HC(O)N(CH_3)_2$	CCl_4	IR	3.5	0.74	3.9	10.5	155	167
		CCl_4	CAL			4.0			1779a
C_2H_5OH	$CH_3C(O)N(CH_3)_2$	CCl_4	IR	5.0(20°)	0.9	3.4	8.4	185	915
	$CH_3C(S)N(CH_3)_2$	CCl_4	IR	1.7(20°)	0.3	2.2	6.4	183	915
	$C_6H_5C(S)N(CH_3)_2$	CCl_4	IR	1.5(20°)	0.2	1.6	4.7	158	915
	$CH_3OC(S)N(CH_3)_2$	CCl_4	IR	0.97(20°)	-0.04	1.1	3.6	148	915
	$CH_3SC(S)N(CH_3)_2$	CCl_4	IR	0.93(20°)	-0.04	1.2	3.9	142	915
	$(CH_3)_2NC(S)N(CH_3)_2$	CCl_4	IR	1.9(20°)	0.3	2.4	7.0	187	915
	$NCC(S)N(CH_3)_2$	CCl_4	IR	0.78(20°)	-0.2	2.3	7.0	54	915
	$(C_2H_5)_3N$	$i\text{-}C_8H_{18}$	UV	2.6	0.57				2347
	C_5H_5N	CCl_4	IR	2.4	0.52	3.7	10.5	276	167
		CCl_4	CAL			3.8		287	1779
	$CH_3C(O)OC_2H_5$	CCl_4	IR	1.0	0.02	2.3	7.8	80	167
		CCl_4	CAL			2.4			1779a
	$(CH_3CO)_2O$	CCl_4	IR	1.7	0.31			100	943
	$(C_2H_5)_2O$	CCl_4	IR	2.0	0.41			56	943
	$[(CH_3)_2CH]_2O$	CCl_4	IR	0.79(21.7°)	-0.17	2.9 ±0.2	10.4	144	1717
	$(n\text{-}octyl)_2O$	CCl_4	IR	1.1(21.7°)	0.01	4.2 ±0.4	14.0	163	1717
	$p\text{-dioxane}$	base	GLC	1.20(22.5°)	0.08	3.79 ±0.16	12.5 ±0.5		1608
		CCl_4	IR	1.0	0.03	3.1	10.3	123	167
	$(CH_3)_2CO$	CCl_4	IR	1.2	0.12	3.5	11.2	109	167
		C_6H_6	UV	1.6	0.28				2278
		CCl_4	CAL			3.4			1408

298

Proton donor	Base	Solvent	Method						Ref.
	(n-octyl)₂CO	base	GLC	2.16(50°)		3.2			1608
	(C₆H₅)₂CO	CCl₄	IR	1.2	0.11	1.6	10.5	84	167
		CCl₄	CAL			2.3			1779a
	cyclopentanone	base	GLC					123	2521
		CCl₄	IR	8.0	1.2	1.3			1751
		C₆H₆	UV	9.3 ±1.4	1.2	7.0	18.8	240	1751
	benzil	base	GLC	11	1.4			155	2521
	(C₇H₁₅)₄N⁺I⁻	CCl₄	IR	24 ±2.8	1.9				2282
CClH₂CH₂OH	cyclopentanone	C₆H₁₂	UV	42 ±4.5	2.2			262	1751
CCl₂HCH₂OH	cyclopentanone	C₆H₁₂	UV	4.5	0.9			283	1751
CCl₃CH₂OH	cyclopentanone	CCl₄	IR	60 ±5	2.4			513	1751
CF₃CH₂OH	cyclopentanone	C₆H₁₂	IR	41	2.2			428	1751
CF₃CH₂OH	HC(O)N(CH₃)₂	CCl₄	CAL	70	2.5	6.1 ±0.2	13.1	503	2255
	CH₃C(O)N(CH₃)₂	CCl₄	CAL	71	2.5	6.4 ±0.2	13.1	150	2255
	(C₂H₅)₃N	C₆H₁₄	CAL			8.8 ±0.2	21.1	232	2255
	C₅H₅N	C₆H₁₄	CAL	59	2.4	7.8 ±0.2	18.1	214	2255
	collidine	CCl₄	CAL	105	2.8	6.7 ±0.02	20.1	253	2254
	CH₃C(O)OC₂H₅	C₆H₁₄	CAL	6.9	1.1	8.8 ±0.2	11.4	201	2255
	(C₂H₅)₂O	CCl₄	CAL	5.2	1.0	7.5 ±0.01	13.7	133	2254
	p-dioxane	CCl₄	CAL	6.1	1.1	4.5 ±0.2	10.8	130	2255
	(CH₂)₄O	CCl₄	IR	10.1	1.4	5.1 ±0.2	12.6		2255
	(CH₃)₂CO	CCl₄	IR	7.3	1.2	4.3	13.1		1279
	(C₆H₅)₂CO	C₆H₆	CAL	7.6	1.2	5.1	11.2		1279
	CH₃CN	CCl₄	IR	5.5	1.0	5.1 ±0.2	14.4		2255
	[(CH₃)₂N]₃PO	CCl₄	CAL	312	3.4			405	2278
	(CH₃)₂SO	CCl₄	CAL	141	2.9	4.3 ±0.2	11.4	303	2255
	(CH₃)₂NC(S)N(CH₃)₂	CCl₄	CAL			7.7 ±0.2		290	2255
	(n-C₄H₉)₃PS	CCl₄	CAL			6.3 ±0.2		273	2255
	CH₃C(S)N(CH₃)₂	CCl₄	CAL			4.6 ±0.1		280	2253
	(C₈H₁₇)₃PS	CCl₄	CAL			4.4 ±0.1		272	2253
	(CH₂)₄S	CCl₄	CAL			4.3 ±0.1		227	2253
	(C₂H₅)₂S	CCl₄	CAL			4.3 ±0.1		225	2253
	C₉H₁₈NO	CCl₄	CAL			3.9 ±0.3			2253
	(C₂H₅)₃N	CCl₄	CAL			3.6 ±0.3			2253
	C₉H₁₈NO	C₆H₁₂	CAL			6.2			1463
n-C₃H₇OH	(C₂H₅)₃N	CCl₄	IR	2.0	0.41	4.5	13.7	465	2278

299

APPENDIX (continued)

Acid	Base	Phase or Solvent	Method	$K_{25°C}$, ℓ/mole	$-\Delta G_{25°C}$, kcal/mole	$-\Delta H$, kcal/mole	$-\Delta S_{25°C}$, cal deg^{-1} mole^{-1}	$\Delta\nu_{AH}$ cm^{-1}	Ref.
n-C$_3$H$_7$OH	C$_5$H$_5$N	CCl$_4$	IR	1.8	0.35	4.3 ±0.9	11.9	286	763
		CCl$_4$	CAL			3.8		279	1779
	(C$_2$H$_5$)$_2$O	CCl$_4$	IR	0.91(21.7°)	-0.06	3.58 ±0.16	12.3	147	1717
	[(CH$_3$)$_2$CH]$_2$O	CCl$_4$	IR	1.0(21.7°)	0.01	4.67 ±0.43	18.9	163	1717
	(n-octyl)$_2$O	base	GLC	1.12(22.5°)	0.04	3.52 ±0.10	11.7 ±0.3		1608
	(CH$_3$)$_2$CO	CCl$_4$	IR	0.50	-0.41	2.4	9.4	110	2278
	[(CH$_3$)$_2$CH]$_2$CO	CCl$_4$	IR	1.0	0.0	1.4	4.7	85	2278
	(n-octyl)$_2$CO	base	GLC	1.98(50°)	0.7	3.57 ±0.15	9.7 ±0.5		1608
	(C$_6$H$_5$)$_2$CO	CCl$_4$	IR	0.95	-0.03	3.5	11.8	85	2278
	C$_6$H$_5$CH$_2$Cl	CCl$_4$	IR	0.30	-0.71	2.4	10.4	30	2278
	C$_7$H$_{15}$Br	CCl$_4$	IR	0.22	-0.90	0.7	3.3	45	2278
	(C$_7$H$_{15}$)$_4$N$^+$I$^-$	CCl$_4$	IR	13	1.5	3.0	5.0	250	2282
	thiocamphor	CCl$_4$	IR	0.51	-0.40	1.8	7.4	120	2278
(CH$_3$)$_2$CHOH	CH$_3$C(O)NH(CH$_3$)	CCl$_4$	NMR	5.0(24°)	0.95	4.1	10.6		2412
		CCl$_4$	CAL			4.5			1779a
	CH$_3$C(O)N(CH$_3$)$_2$	CCl$_4$	NMR	2.9(21°)	0.6	2.4	6.0		2412
		CCl$_4$	CAL			3.9			1779a
	(C$_2$H$_5$)$_3$N	i-C$_8$H$_{18}$	UV	1.4	0.20				2347
	Pyridines								
	C$_5$H$_5$N	C$_6$H$_6$	DEC	2.0	0.4	4.5	13.7	269	2171
		CCl$_4$	IR	1.8(30.5°)	0.5	6.1 ±0.4	18.9 ±1.3	275	763
		CCl$_4$	CAL			4.1			1779
		CCl$_4$	IR	1.20	0.11	3.7	11.9	265	1806
	2-CH$_3$-	CCl$_4$	IR	1.2 ±0.1	0.12 ±0.02	3.9 ±0.1	11.7	286	1906
	3-CH$_3$-	CCl$_4$	IR	1.2 ±0.1	0.11 ±0.02	3.8 ±0.1	11.3	279	1906
	4-CH$_3$-	CCl$_4$	IR	1.2 ±0.1	0.11 ±0.02	4.0 ±0.1	12.0	282	1906
	2-Br-	CCl$_4$	IR	0.6 ±0.1	-0.27 ±0.02	2.8 ±0.1	10.1	175	1906

2-Cl-	CCl₄	IR	0.4 ±0.1	-0.57 ±0.02	2.4 ±0.1	9.9	173	1906
2-F-	CCl₄	IR	0.4 ±0.1	-0.60 ±0.02	2.2 ±0.1	9.3	153	1906
3-Br-	CCl₄	IR	0.6 ±0.1	-0.33 ±0.02	3.1 ±0.1	11.4	227	1906
3-Cl-	CCl₄	IR	0.5 ±0.1	-0.44 ±0.02	3.0 ±0.1	11.6	223	1906
2,4-CH₃-	CCl₄	IR	1.3 ±0.1	0.15 ±0.02	4.2 ±0.1	13.7	302	1906
2,5-CH₃-	CCl₄	IR	1.3 ±0.1	0.15 ±0.02	4.2 ±0.1	13.7	300	1906
2,6-CH₃-	CCl₄	IR	1.2 ±0.1	0.13 ±0.02	4.3 ±0.1	13.7	300	1906
3,4-CH₃-	CCl₄	IR	1.2 ±0.1	0.12 ±0.02	4.3 ±0.1	14.0	294	1906
3,5-CH₃-	CCl₄	IR	1.3 ±0.1	0.14 ±0.02	4.3 ±0.1	14.1	291	1906
2,4,6-CH₃-	CCl₄	IR	1.4 ±0.1	0.22 ±0.02	4.7 ±0.1	15.1	314	1906
(C₂H₅)₂O	CCl₄	IR	0.8 (21.7°)	-0.14	3.14 ±0.15	10.1	139	1717
[(CH₃)₂CH]₂O	CCl₄	IR	1.2 (21.7°)	0.10	4.4 ±0.5	14.4	151	1717
(n-octyl)₂O	base	GLC	2.4	0.52	2.1 ±0.4	6.0 ±2.3		257
poly(propylene oxide)	CCl₄	IR	0.89 (22.5°)	-0.10	3.77 ±0.10	13.0 ±0.3		1608
poly(n-butyl methacrylate)	base	IR	0.6	-0.26	3.8 ±0.8	13.7 ±3.0		257
(CF₃)₂CHOH								
(C₇H₁₅)₄N⁺I⁻	CCl₄	IR	0.4	-0.61	2.4 ±0.5	10.0 ±1.7	230	257
(n-octyl)₂CO	CCl₄	IR	10	1.4	1.10	-1.3		2282
CH₃C(O)N(CH₃)₂	base	GLC	1.70 (50°)	0.59	3.54 ±0.27*	9.9 ±0.8	428	1608
(C₂H₅)₃N	CCl₄	CAL	825	4.0	8.5 ±0.1*	14.1		2026
	C₆H₁₄	CAL			8.2 ±0.1			949a
	CCl₄	CAL			10.0 ±0.1			2026
C₅H₅N	C₆H₁₄	CAL			11.5 ±0.1		1038	2026
	CCl₄	CAL	731	3.9	8.4 ±0.1			2026
2,-collidine	C₆H₁₄	CAL			9.8 ±0.1*	19.8	542	2026
	CCl₄	CAL			9.8 ±0.1*			2254
quinuclidine	C₆H₁₄	CAL			9.67 ±0.03			2254
	C₆H₁₄	CAL			11.12 ±0.03			949a
	CCl₄	CAL	322	3.4	11.3 ±0.1*	26.5	224	2026
CH₃C(O)OC₂H₅	C₆H₁₄	CAL			6.5 ±0.1*			949a
(C₂H₅)₂O	CCl₄	CAL			5.9 ±0.1*			2026
	CCl₄	CAL	37	2.1	7.2 ±0.1*	12.8		
p-dioxane	CCl₄	IR	35.4	2.1	5.3*	10.6	357	1279
(CH₂)₄O	CCl₄	IR	77.7	2.6	6.3*	12.4	342	1280
7-oxabicyclo[2.2.1]-heptane	CCl₄	CAL	86	2.6	6.6	13.4	389	949a

APPENDIX (continued)

Acid	Base	Phase or Solvent	Method	$K_{25°C}$, ℓ/mole	$-\Delta G_{25°C}$, kcal/mole	$-\Delta H$, kcal/mole	$-\Delta S_{25°C}$, cal deg^{-1} mole^{-1}	$\Delta\nu_{AH}$ cm^{-1}	Ref.
$(CF_3)_2CHOH$	$(CH_3)_2CO$	CCl_4	CAL			6.7 ±0.1*		280	2026
	CH_3CN	CCl_4	CAL			5.9 ±0.1*		208	2026
	$(CH_3)_2SO$	CCl_4	CAL			8.7 ±0.1*		449	2026
	$(C_2H_5)_2S$	CCl_4	CAL			5.1 ±0.1*		254	2026
		CCl_4	CAL			4.9 ±0.1*		348	2253
	$(CH_2)_4S$	C_6H_{14}	CAL	6.4 ±1(22°)	1.0	5.8 ±0.1*	16.1		2533
		CCl_4	CAL			5.2 ±0.1*		361	2253
		C_6H_{12}	NMR	14.7 ±1.9	1.6				2290a
	$[(CH_3)_2N]_3PO$	CCl_4	CAL			9.9 ±0.1*		540	2026
	$(CH_3)_2NC(S)N(CH_3)_2$	CCl_4	CAL			6.5 ±0.1*		445	2253
	$(n\text{-}C_4H_9)_3PS$	CCl_4	CAL			6.3 ±0.1*		423	2253
	$CH_3C(S)N(CH_3)_2$	CCl_4	CAL			6.0 ±0.1*		430	2253
	$(C_8H_{17})_3PS$	CCl_4	CAL			5.9 ±0.1*		414	2253
	$(C_6H_5)_3PS$	CCl_4	CAL			4.6 ±0.1*		312	2253
	$C_9H_{18}NO$	C_6H_{12}	CAL			7.9			1463
	$CH_3CH_2C(CH_2O)_3P$	C_6H_{12}	NMR	16.2 ±1.7	1.6				2290a
$3\text{-}C_6H_5\text{-}1\text{-}$ propanol	cyclohexene	CCl_4	IR	0.08	-1.5	1.2	9.2	48	2542
	isopropylbenzene	CCl_4	IR	0.14	-1.2	1.0	7.3	30	2542
	$(CH_3)_2CHCN$	CCl_4	IR	0.79	-0.14	3.2	11.6	82	2542
CF_2HCF_2OH	$(C_2H_5)_3N$	CCl_4	IR	20	1.8	9.1	24.5	540	2278
	$(CH_3)_2CO$	CCl_4	IR	7.0	1.2	7.0	19.5	165	2278
	$[(CH_3)_2CH]_2CO$	CCl_4	IR	3.8	0.79	7.4	22.2	135	2278
	$(C_6H_5)_2CO$	CCl_4	IR	6.6	1.1	6.5	18.1	140	2278
	thiocamphor	CCl_4	IR	2.8	0.61	3.6	10.0	170	2278
	$C_6H_5CH_2Cl$	CCl_4	IR	0.44	-0.49	2.6	10.4	35	2278
	$C_7H_{15}Br$	CCl_4	IR	0.23	-0.87	2.6	11.6	70	2278
	$(C_7H_{15})_4N^+I^-$	CCl_4	IR	2.1	0.44	3.8	11.3	350	2282

n-C₄H₉OH

Base	Solvent	Method						
HC(O)NH(CH₃)	base	CAL			3.99 ±0.12			70
HC(O)N(CH₃)₂	base	CAL			4.14 ±0.11			70
CH₃C(O)N(CH₃)₂	base	CAL			4.43 ±0.13			70
(C₂H₅)₃N	base	CAL			5.53 ±0.11			70
	CCl₄	IR	3.2	0.69	4.9	14.1	406	1113
	CCl₄	IR	3.2	0.70			423	2686
C₅H₅N	base	CAL			4.4 ±0.12			70
	CCl₄	CAL			3.9	15.4	281	1779
	CCl₄	IR	2.3(21°)	0.4 ±0.1	5.0 ±1.0		274	763
	CCl₄	IR	2.6	0.55	3.5	11.9	278	2686
	CCl₄	IR	2.5(30°)				254	306
	C₆H₆	DEC	2.0	0.4	4.5	13.7		2171
quinoline	base	CAL			4.57 ±0.12			70
4-picoline	base	CAL			4.62 ±0.13			70
C₆H₅N(CH₃)₂	base	CAL			2.02 ±0.12			70
(C₂H₅)₂O	base	CAL			2.96 ±0.13			70
	CCl₄	IR	0.73(21.7°)	0.2	3.8 ±0.1	12.0	147	1717
[(CH₃)₂CH]₂O	CCl₄	IR	0.80(21.7°)	0.2	4.3 ±0.2	13.8	161	1717
(n-octyl)₂O	base	GLC	1.11(50°)	0.1	4.21 ±0.32	14.0 ±1.0		1608
(C₆H₅)₂O	base	CAL			1.10 ±0.14			70
anisole	base	CAL			1.69 ±0.14			70
p-dioxane	base	CAL			3.11 ±0.11			70
(CH₂)₄O	base	CAL			3.06 ±0.11			70
CH₃C(O)OC₂H₅	base	CAL			2.43 ±0.13			70
(C₂H₅)₂CO	CCl₄	CAL			3.4			1779a
CH₃C(O)C₆H₅	CCl₄	CAL			3.8			1779a
2-butanone	base	CAL			3.05 ±0.13			70
(n-octyl)₂CO	CCl₄	GLC	1.91(50°)	0.7	3.98 ±0.10	11.0 ±0.3		1608
cyclohexanone	base	CAL			4.66 ±0.11			70
	CCl₄	CAL			4.1			1779a
benzene	base	CAL			0.50 ±0.14			70
toluene	base	CAL			0.50 ±0.16			70
mesitylene	base	CAL			0.54 ±0.12			70
cumene	base	CAL			0.55 ±0.13			70
[n-C₇H₁₅]₄N⁺I⁻	CCl₄	IR	13	1.5	1.5	0	245	2282
N-methylmorpholine	CCl₄	IR	2.3	0.50	3.7	11.1		2686

APPENDIX (continued)

Acid	Base	Phase or Solvent	Method	$K_{25°C}$, ℓ/mole	$-\Delta G_{25°C}$, kcal/mole	$-\Delta H$, kcal/mole	$-\Delta S_{25°C}$, cal deg^{-1} mole^{-1}	$\Delta\nu_{AH}$ cm^{-1}	Ref.
n-C_4H_9OH	diazabicyclooctane	CCl_4	IR	8.1	1.25	4.7	11.6	405	2686
2-butanol	$(C_2H_5)_3N$	CCl_4	IR	1.7	0.31				1751
	C_5H_5N	CCl_4	IR	1.3(31°)	0.2	4.1 ±0.4	13.1	275	763
		CCl_4	IR	2.1(30°)				245	306
		CCl_4	CAL			4.0		262	1779
	$(n\text{-octyl})_2O$	base	GLC	1.13(22.5°)	0.04	4.08 ±0.22	13.6 ±0.7		1608
	$(n\text{-octyl})_2CO$	base	GLC	1.91(50°)	0.7	4.06 ±0.10	11.3 ±0.3		1608
2-methyl-1-propanol	$(C_2H_5)_3N$	CCl_4	IR	0.79	-0.14		10.7	376	1751
	C_5H_5N	CCl_4	IR	0.57	-0.33		11.9	244	1751
	$(C_2H_5)_2O$	CCl_4	IR	0.89(21.7°)	-0.1	3.1 ±0.2		150	1717
	$[(CH_3)_2CH]_2O$	CCl_4	IR	0.93(21.7°)	-0.1	3.5 ±0.5		163	1717
	$(n\text{-octyl})_2O$	base	GLC	0.77(22.5°)	-0.2	3.96 ±0.27	13.9 ±0.9		1608
	$(n\text{-octyl})_2CO$	base	GLC	1.42(50°)	0.5	3.45 ±0.05	10.0 ±0.2		1608
$(CH_3)_3COH$	**AMIDES**								
	$HC(O)N(CH_3)_2$	CCl_4	IR	2.9	0.63	3.9*	11.0	143	167
		CCl_4	CAL			3.8*			1779a
		base	NMR			4.1*			2516
	$CH_3C(O)N(CH_3)_2$	CCl_4	CAL			3.8*		160	652
		base	NMR			4.3			2516
	$CH_3NC(O)CH_2CH_2CH_2$	base	NMR			4.2			2516
	AMINES								
	$(C_2H_5)_3N$	i-C_8H_{18}	UV	1.0	0.0	4.0	12.6	252	2347
	C_5H_5N	CCl_4	IR	1.4	0.2			244	167
		CCl_4	CAL			3.7			1779

304

	Solvent	Method						
C_5H_5N	CCl_4	IR	$1.7(30°)$				230	306
	CCl_4	IR	$24(mf^{-1})$				231	2117
	C_6H_{12}	CAL			4.2*			652
ESTERS								
$CH_3C(O)OC_2H_5$	CCl_4	IR	0.84	-0.10	2.9*	10.1	75	167
	CCl_4	CAL			2.5*			1779a
ETHERS								
$(C_2H_5)_2O$	CCl_4	IR	$0.74(21.7°)$	-0.2	2.6 ± 0.1*	9.5	127	1717
	CCl_4	IR	$3(mf^{-1})$					2117
$[(CH_3)_2CH]_2O$	CCl_4	IR	$1.1(21.7°)$	0.03	2.7 ± 0.3*	9.0	132	1717
$(n\text{-octyl})_2O$	base	GLC	$0.61(22.5°)$	-0.3	3.93 ± 0.15	14.3 ± 0.5		1608
p-dioxane	CCl_4	IR	1.1	0.7	2.9*	9.6	118	167
furan	CCl_4	IR	$0.4(mf^{-1})$					2117
$CH_3OC_2H_4OCH_3$	base	NMR			3.1			2516
$CH_3OC_2H_4OC_2H_4OCH_3$	base	NMR			3.1			2516
KETONES								
$(CH_3)_2CO$	CCl_4	IR	$5(mf^{-1})$	0.02	2.9*	9.8	101	2117
	CCl_4	IR	1.0		2.85 ± 0.15		100	167
	CCl_4	IR			3.0			2429
	base	NMR						2516
$(C_6H_5)_2CO$	CCl_4	IR	1.1	0.01	2.7*	8.9	77	167
	CCl_4	CAL			1.2*			1779a
$(n\text{-octyl})_2CO$	base	GLC	$1.31(50°)$	0.43	3.47 ± 0.16	10.2 ± 0.5		1608
cyclohexanone	C_7H_{16}	UV	0.75	-0.17				422
NITRILES								
CH_3CN	C_2Cl_4	IR	0.81	-0.13	2.0*	7.0	63	1496
	CCl_4	IR	$19(mf^{-1})$					2117
$ClCH_2CN$	C_2Cl_4	IR	0.72	-0.19	1.6*	6.0	47	1496
Cl_3CCN	C_2Cl_4	IR	0.15	-1.14	0.7*	6.3	21	1496
$CH_2{=}CHCN$	C_2Cl_4	IR	0.70	-0.21	2.3	8.3	55	1496

APPENDIX (continued)

Acid	Base	Phase or Solvent	Method	$K_{25°C}$, ℓ/mole	$-\Delta G_{25°C}$, kcal/mole	$-\Delta H$, kcal/mole	$-\Delta S_{25°C}$, cal deg^{-1} mole^{-1}	$\Delta\nu_{AH}$ cm^{-1}	Ref.
$(CH_3)_3COH$	$(CH_3)_3CCN$	C_2Cl_4	IR	1.05	0.03	1.9*	6.2	67	1496
	$(CH_3)_2NCN$	C_2Cl_4	IR	1.64	0.29	2.3*	6.9	101	1496
	C_6H_5CN	C_2Cl_4	IR	0.83	-0.11	1.6*	5.8	61	1496
	PHOSPHORYL DONORS								
	$(CH_3OCH_2)_2P(O)CH_3$	base	NMR			5.1			2516
	$(C_2H_5OCH_2)_2P(O)CH_3$	base	NMR			4.8			2516
	$(CH_3OC_2H_4OCH_2)_2P(O)CH_3$	base	NMR			5.1			2516
	$[(CH_3)_2N]_2P(O)CH_3$	base	NMR			5.5			2516
	$[(CH_3)_2N]_3PO$	base	NMR			5.7			2516
	π-BASES								
	toluene	CCl_4	IR	0.19(26°)				35	147
	o-xylene	CCl_4	IR	0.15(26°)				38	147
	m-xylene	CCl_4	IR	0.24(26°)				40	147
	p-xylene	CCl_4	IR	0.20(26°)				39	147
	mesitylene	CCl_4	IR	0.26(26°)				49	147
	durene	CCl_4	IR	0.19(26°)				50	147
	isodurene	CCl_4	IR	0.19(26°)				52	147
	pentamethylbenzene	CCl_4	IR	0.23(26°)				58	147
	hexamethylbenzene	CCl_4	IR	0.37(26°)				69	147
	MISCELLANEOUS								
	$(CH_3)_2SO$	C_2Cl_4	IR	5.63	1.02	4.5*	11.6	181	1496
		CCl_4	CAL			3.6*		183	652
		base	NMR			4.5			2516
	$(C_6H_5)_2SO$	C_2Cl_4	IR	2.40	0.52	3.5	9.9	139	1496

Alcohol	Base	Solvent	Method						
(CF₃)₃COH	(C₂H₅)₂NNO	C₇H₁₆	UV	1.6	0.28				421
	[(CH₃)₂N]₂CO	base	NMR	7.2	1.2				2516
	(n-C₇H₁₅)₄N⁺I⁻	CCl₄	IR						2282
	HC(O)N(CH₃)₂	CCl₄	CAL			3.9		220	2252
	CH₃C(O)N(CH₃)₂	CCl₄	CAL			1.0		510	2252
	(C₂H₅)₃N	C₆H₁₄	CAL			8.6 ±0.2*			2252
	(C₂H₅)₃N	CCl₄	CAL			8.9 ±0.2*		540	2252
	C₅H₅N	C₆H₁₄	CAL			14.3 ±0.2*			2252
	C₅H₅N	CCl₄	CAL			10.3 ±0.2*		980	2252
	collidine	C₆H₁₄	CAL			12.5 ±0.2			2252
	collidine	CCl₄	CAL			11.0 ±0.2*		835	2252
	CH₃C(O)OC₂H₅	CCl₄	CAL			14.1 ±0.2		968	2252
	(C₂H₅)₂O	C₆H₁₄	CAL			6.7 ±0.2*			2252
	(C₂H₅)₂O	CCl₄	CAL			7.1 ±0.2*		320	2252
	(CH₂)₄O	CCl₄	CAL			8.3 ±0.2*		480	2252
	(CH₃)₂CO	C₆H₁₄	CAL			7.7 ±0.2*			2252
	(CH₃)₂CO	CCl₄	CAL			7.2 ±0.2*		506	2252
	CH₃CN	CCl₄	CAL			8.0 ±0.2		380	2252
	(C₂H₅)₂S	C₆H₁₄	CAL			6.4 ±0.2*			2252
	(C₂H₅)₂S	CCl₄	CAL			7.0 ±0.2*		306	2252
	(CH₃)₂SO	CCl₄	CAL			9.5 ±0.2*		578	2252
	[(CH₃)₂N]₃PO	CCl₄	CAL			10.5 ±0.2*		770	2252
1-pentanol	C₅H₅N	CCl₄	IR	2.7(20°)				257	306
	C₆H₅CH₂CN	CCl₄	IR	5.8(mf⁻¹)(26°)					753
	C₆H₅CH₂NC	CCl₄	IR	6.0(mf⁻¹)(26°)					753
1-nonanol	C₅H₅N	CCl₄	IR	2.7(30°)				260	306
1-octanol	hexyl hexanoate	CCl₄	IR	1.09(21.7°)	0.03	2.2			2295
	C₅H₅N	CCl₄	CAL			4.0	7.3	287	1779
trans-dihydrocryptol	p-dioxane	C₂Cl₄	IR	0.89	-0.07	3.2 ±0.3	11.0		486
neopentyl alcohol	(C₂H₅)₂O	CCl₄	IR	1.0(21.7°)	-0.04	3.8 ±0.1	12.6	150	1717
	[(CH₃)₂CH]₂O	CCl₄	IR	1.0(21.7°)	-0.05	4.6 ±0.2	15.4	160	1717
[(CH₃)₃C]₂CHOH	CH₃C(O)N(CH₃)₂	CCl₄	NIR	2.07(-10°)					2095b
	C₅H₅N	CCl₄	NIR	3.52(-10°)					2095b
	CH₃C(O)OCH₃	CCl₄	NIR	1.33(-10°)					2095b
	(C₂H₅)₂O	CCl₄	NIR	0.61(-10°)					2095b
	(CH₂)₄O	CCl₄	NIR	1.54(-10°)					2095b

APPENDIX (continued)

Acid	Base	Phase or Solvent	Method	$K_{25°C}$ ℓ/mole	$-\Delta G_{25°C}$ kcal/mole	$-\Delta H$, kcal/mole	$-\Delta S_{25°C}$, cal deg^{-1} mole^{-1}	$\Delta\nu_{AH}$ cm^{-1}	Ref.
[(CH$_3$)$_3$C]$_2$CHOH	(CH$_2$)$_4$O	CCl$_4$	IR	0.7	-0.21	2.6	9.4	145	2281
	p-dioxane	CCl$_4$	NIR	1.70(-10°)	-0.13			110	2095b
	p-dioxane	CCl$_4$	IR	0.8	-0.13				2281
	(CH$_3$)$_2$CO	CCl$_4$	NIR	1.76(-10°)					2095b
	C$_6$H$_5$C(O)CH$_3$	CCl$_4$	IR	1.8	0.35			85	2281
	(C$_6$H$_5$)$_2$CO	CCl$_4$	NIR	1.66(-10°)					2095b
		CCl$_4$	IR	1.8	0.35			80	2281
	cyclopentanone	CCl$_4$	NIR	2.07(-10°)					2095b
	C$_6$H$_5$CN	CCl$_4$	NIR	1.39(-10°)					2095b
	(CH$_3$O)$_3$P	CCl$_4$	NIR	2.73(-10°)					2095b
	(CH$_3$)$_2$SO	CCl$_4$	NIR	3.84(26°)					2095b
	(C$_6$H$_5$)$_2$SO	CCl$_4$	NIR	3.47(0°)					2095b
	(CH$_2$)$_4$SO	CCl$_4$	NIR	9.92(-10°)					2095b
	(CH$_2$)$_4$SO$_2$	CCl$_4$	NIR	1.27(26°)					2095b
	(CH$_3$O)$_2$SO	CCl$_4$	NIR	2.01(-10°)					2095b
	(CH$_3$O)$_2$CO	CCl$_4$	NIR	1.30(-10°)					2095b
	azobenzene	CCl$_4$	IR	0.6	-0.30			130	2281
	CH$_3$NO$_2$	CCl$_4$	IR	0.8	-0.54			45	2281
	ethylene trithiocarbonate	CCl$_4$	IR	0.8	-0.13			102	2281
2,4-dimethyl-3-ethyl-3-pentanol	p-dioxane	CCl$_4$	IR	0.7	-0.21	2.2	8.8	102	2281
	(CH$_2$)$_4$O	CCl$_4$	IR	0.5	-0.41			123	2281
	C$_6$H$_5$C(O)CH$_3$	CCl$_4$	IR	0.7	-0.21			60	2281
	(C$_6$H$_5$)$_2$CO	CCl$_4$	IR	0.6	-0.30			62	2281
	azobenzene	CCl$_4$	IR	1.0	0.0			15	2281
	CH$_3$NO$_2$	CCl$_4$	IR	0.5	-0.41			10	2281

ethylene trithiocarbonate	CCl₄	IR	1.0	0.0			10	2281
√ 2,2,4-tetra-methyl-3-isopropyl-3-pentanol								
p-dioxane	CCl₄	IR	0.5	-0.41	2.8		85	2281
	C₆H₁₂	NMR			1.8	8.4		2281
(CH₂)₄O	CCl₄	IR	0.3	-0.71			45	2281
C₆H₅C(O)CH₃	CCl₄	IR	0.2	-0.95			50	2281
(C₆H₅)₂CO	CCl₄	IR	0.4	-0.54			21	2281
azobenzene	CCl₄	IR	0.6	-0.30			9	2281
CH₃NO₂	CCl₄	IR	0.2	-0.95				2281
ethylene trithiocarbonate	CCl₄	IR	0.7	-0.21			10	2281
cholesterol								
triacetin	CCl₄	IR	2.9 (23°)	0.6	4.3	12.4	80	1876
tributyrin	CCl₄	IR	2.4 (23°)	0.5	3.5	10.1	75	1876
trilaurin	CCl₄	IR	3.7 (23°)	0.7	5.4	15.8	75	1876
(CH₃)₃SiOH								
C₅H₅N	CCl₄	IR	70 (mf⁻¹)					2117
quinoline	CCl₄	IR	68 (mf⁻¹)					2117
CH₃C(O)C₂H₅	base	IR	1.50 (28°)					1219
(C₂H₅)₂O	CCl₄	IR	8 (mf⁻¹)					2117
furan	CCl₄	IR	5 (mf⁻¹)					2117
(CH₂)₄O	base	IR	3.46 (28°)					1219
dioxane	base	IR	2.97 (28°)					1219
(CH₃)₂CO	CCl₄	IR	16 (mf⁻¹)					2117
CH₃CN	base	IR	1.76 (28°)					1219
	CCl₄	IR	33 (mf⁻¹)					2117
benzene	base	IR	0.22 (28°)					1219
chlorobenzene	base	IR	0.19 (28°)					1219
toluene	base	IR	0.23 (28°)					1219
p-xylene	base	IR	0.25 (28°)					1219
C₆H₅OH — ALDEHYDES								
CH₃CH₂CHO	CCl₄	NIR	4.5 (20°)	0.8	4.3*	11.7	170	908
CH₃CH₂CH₂CHO	CCl₄	NIR	5.1 (20°)	0.9	4.9	13.4	170	908

APPENDIX (continued)

Acid	Base	Phase or Solvent	Method	$K_{25°C}$, ℓ/mole	$-\Delta G_{25°C}$, kcal/mole	$-\Delta H$, kcal/mole	$-\Delta S_{25°C}$, cal deg^{-1} mole^{-1}	$\Delta\nu_{AH}$ cm^{-1}	Ref.
C_6H_5OH	$CH_3CH_2CH_2CHO$	CCl_4	NIR	4.9	0.94	4.67 ±0.07*	12.5	180	1981
	$CH_3(CH_2)_6CHO$	CCl_4	NIR	4.4(20°)	0.8	3.3*	8.4	170	908
	$CH_3(CH_2)_8CHO$	CCl_4	NIR	5.1(20°)	0.9	4.5*	12.1	170	908
	C_6H_5CHO	CCl_4	NIR	4.8(20°)	0.9	4.3*	11.4	168	908
	$o-ClC_6H_4CHO$	CCl_4	NIR	3.6(20°)	0.7	6.3	18.8	166	908
	$o-NO_2C_6H_4CHO$	CCl_4	NIR	2.2(20°)	0.4	2.9	8.4	125	908
	$m-NO_2C_6H_4CHO$	CCl_4	NIR	2.1(20°)	0.4	2.6*	7.4	124	908
	$p-NO_2C_6H_4CHO$	CCl_4	NIR	2.1(20°)	0.4	3.5*	7.4	117	908
	$p-ClC_6H_4CHO$	CCl_4	NIR	5.2(20°)	0.9	3.1*	7.4	183	908
	$p-(CH_3)_2NC_6H_4CHO$	CCl_4	NIR	21(20°)	1.7	4.8*	10.4	260	908
	$C_6H_5CH{=}CHCHO$	CCl_4	NIR	13(20°)	1.4	4.5*	10.4	236	908
	cumaldehyde	CCl_4	NIR	5.6(20°)	1.0	3.9*	9.7	188	908
	2-naphthaldehyde	CCl_4	NIR	4.9(20°)	0.9	3.7	9.4	178	908
	$HC(O)NH(CH_3)$	base	CAL			6.38 ±0.07			70
	$HC(O)N(CH_3)_2$	CCl_4	UV	64 ±1	2.5	6.1 ±0.4	12.1 ±0.7	294	1182
		CCl_4	NIR	79(20°)	2.5	5.4	9.7	292	914
		CCl_4	NIR	68	2.6	6.2 ±0.4	12.5 ±1.4	285	807
		CCl_4	CAL			6.3*			1779a
		CCl_4	IR	75.5 ±2.0	2.6	6.86 ±0.08	13.6 ±0.3	297	70
		base	CAL			5.9	10.7	277	70
	$CH_3C(O)NH(CH_3)$	CS_2	UV	116.7(20°)	2.7	4.0	4.0		914
	$CH_3C(O)N(CH_3)_2$	CCl_4	UV	134(20°)	2.8				2411
		CCl_4	UV	134 ±3	2.9	6.4 ±0.3	11.7 ±0.6	342	1182
		CCl_4	CAL	107 ±15	2.8	6.8 ±0.1	13.4		719
		CCl_4	IR	134 ±3	2.9				70
		CCl_4	IR	152(20°)	2.9	6.2	11.1	336	915
		CCl_4	IR		2.9	5.8		341	2361
		CCl_4	NMR	145					1772

Compound	Solvent	Method						Ref.
$CH_3C(O)N(C_2H_5)_2$	CCl_4	CAL	157(20°)	2.9	6.5*	7.0	340	1779a
	base	CAL			7.36 ±0.08			70
$CH_3C(O)N[CH(C_2H_5)CH_3]$	CCl_4	NIR	259	3.2	5.0*	12.1	331	914
$CH_3C(O)N(C_6H_{11})_2$	CCl_4	CAL	155 ±17	3.0	6.6*	13.1	342	1779a
	CS_2	NIR	167(20°)	2.9	6.8	10.7		914
$CH_3C(O)N(C_6H_5)_2$	CCl_4	UV	257(20°)	3.2	6.9 ±3.1	12.7	394	1170
$CH_3C(O)N(CH_2)_5$	CCl_4	NIR	59(20°)	2.3	6.1*	9.4	270	914
$ClCH_2C(O)N(CH_3)_2$	CS_2	NIR	135(20°)	2.8	7.0*	7.4	330	914
$ClCH_2C(O)N(C_2H_5)_2$	CCl_4	NIR	38 ±2	2.2	5.1*	8.5 ±1.6	272	914
$ClCH_2C(O)N(C_6H_5)_2$	CCl_4	UV	46(20°)	2.2	5.0*	8.0	263	914
	CCl_4	CAL			4.7 ±0.5*			1182
$ClCH_2C(O)N(C_6H_{11})_2$	CCl_4	NIR	21(20°)	1.7	5.9	6.0	217	1779a
$ClCH_2C(O)N(CH_2)_5$	CCl_4	NIR	42(20°)	2.1	4.6*	10.1	250	914
$Cl_3C(O)N(CH_3)_2$	CCl_4	NIR	48(20°)	2.2	3.5*	8.7	252	914
$C_2H_5C(O)N(CH_3)_2$	CCl_4	NIR	32 ±3	2.1	5.1*	5.5 ±1.8	171	914
$C_2H_5C(O)N(C_6H_{11})_2$	CCl_4	UV	107 ±2	2.8	4.8*	12.1 ±0.6	325	914
	CCl_4	UV			3.8 ±0.5*			1182
	CCl_4	CAL	105(20°)	2.7	6.4 ±0.2	8.4		1182
$C_2H_5C(O)N(C_2H_5)_2$	CCl_4	NIR	134(20°)	2.8	6.3*	9.1		1779a
$C_2H_5C(O)N(C_6H_5)_2$	CCl_4	NIR	44(20°)	2.2	5.2*	8.7	307	914
$C_2H_5C(O)N(CH_2)_5$	CCl_4	NIR	122(20°)	2.8	5.5*	8.0	329	914
$CH_3CH_2CH_2C(O)N(C_2H_5)_2$	CCl_4	NIR	124(20°)	2.8	4.8*	8.7	250	914
$CH_3C(O)N(C_3H_7)_2$	CCl_4	NIR	179 ±8	3.1	5.2*	15.4	320	914
$CH_3C(O)N[CH(CH_3)_2]_2$	CCl_4	UV	168 ±11	3.0	5.4*	11.7	325	914
	CCl_4	UV			7.7 ±0.9			1170
	CCl_4	NIR			6.5 ±3.6			1170
$(n\text{-}C_3H_7)C(O)N(CH_2)_5$	CCl_4	NIR	129(20°)	2.8	5.8*	10.1	328	914
$(n\text{-}C_3H_7)C(O)N(C_6H_{11})_2$	CCl_4	NIR	112(20°)	2.7	5.7*	10.1	313	914
$(n\text{-}C_3H_7)C(O)N(C_6H_5)_2$	CCl_4	NIR	46(20°)	2.2	4.4*	7.4	250	914
$C_6H_5C(O)N(C_2H_5)_2$	CCl_4	NIR	97(20°)	2.6	5.5*	9.7	298	914
$C_6H_5C(O)N(C_6H_5)_2$	CCl_4	NIR	35(20°)	2.0	3.9*	6.4	240	914
$C_6H_5C(O)N(C_6H_{11})_2$	CCl_4	NIR	106(20°)	2.7	5.8*	10.4	311	914
$C_6H_5C(O)N(CH_2)_5$	CCl_4	NIR	103(20°)	2.7	5.7*	10.1	306	914
$p\text{-}NO_2C_6H_4C(O)N(C_6H_{11})_2$	CCl_4	NIR	41(20°)	2.1	4.2**	7.0	272	914
$p\text{-}NO_2C_6H_4C(O)N(C_2H_5)_2$	CCl_4	NIR	47(20°)	2.2	5.9**	12.4	267	914
$p\text{-}NO_2C_6H_4C(O)N(CH_2)_5$	CCl_4	NIR	45(20°)	2.2	5.4	10.7	260	914

APPENDIX (continued)

Acid	Base	Phase or Solvent	Method	$K_{25°C}$, ℓ/mole	$-\Delta G_{25°C}$, kcal/mole	$-\Delta H$, kcal/mole	$-\Delta S_{25°C}$, cal deg^{-1} mole^{-1}	$\Delta\nu_{AH}$ cm^{-1}	Ref.
C_6H_5OH	p-$NO_2C_6H_4C(O)N(C_6H_5)_2$	CCl_4	NIR	15(20°)	1.5	4.4*	9.7	190	914
	N-methyl-2-pyridone	CCl_4	NIR	202(20°)	3.0	6.0*	10.1	345	914
	N-methyl-γ-butyrolactam	CCl_4	NIR	163(20°)	2.9	5.9	10.1		913
		CCl_4	UV	150 ±3	2.95 ±0.15	5.95 ±0.15	10.0 ±0.5		2200
		CCl_4	CAL			6.6			1779a
	δ-valerolactam	CCl_4	NIR	146(20°)	2.9	5.3	8.1		913
		CCl_4	CAL			6.0			1779a
	N-methyl-δ-valerolactam	CCl_4	UV	186 ±3	3.11 ±0.12	5.62 ±0.12	8.4 ±0.4		2200
	ε-caprolactam	CCl_4	UV	154 ±5	2.97 ±0.12	5.67 ±0.12	9.1 ±0.5		2200
		CCl_4	NIR	152(20°)	2.9	5.2	7.7		913
		CCl_4	CAL			5.8			1779a
	N-methyl-ε-caprolactam	CCl_4	UV	149 ±3	2.98 ±0.11	6.41 ±0.11	11.5 ±0.4		2200
	AMINES								
	$(C_2H_5)_3N$	n-C_7H_{16}	UV	89 ±4	2.7	9.2 ±0.1	21.7 ±0.1	553	1182
		C_6H_{12}	UV	81.3	2.6				674a
		C_6H_{12}	CAL	90 ±10	2.7	9.08 ±0.05*	21.4		719
		CCl_4	NIR	91(20°)	2.5	7.8	17.8	380	909
		CCl_4	IR	65.6(27°)					481a
		CCl_4	IR	69	2.5	8.35 ±0.50	20.0 ±1.7		807
		CCl_4	NIR	58	2.4	9.2	22	450	2280
		CCl_4	IR	76	2.65	7.3	15.6		2686
		base	CAL			8.85 ±0.08			70
	$(n$-$C_3H_7)_3N$	CCl_4	NIR	22(20°)	1.7	5.9*	14.1	330	909
		CCl_4	IR	32.4(27°)					481a
	$(n$-$C_4H_9)_3N$	CCl_4	NIR	29(20°)	1.9	6.9*	16.8	350	909
		CCl_4	IR	29.2(27°)					481a
		CS_2	NIR	34(20°)	2.0	7.9	19.8		909

$(n\text{-}C_5H_{11})_3N$	CCl_4	IR	32.7 (27°)					481a
$(i\text{-}C_5H_{11})_3N$	CCl_4	IR	38.1 (27°)					481a
$(n\text{-}C_8H_{17})_3N$	CCl_4	IR	38.1 (27°)					481a
$(CH_2{=}CHCH_2)_3N$	CCl_4	IR	19.3 (27°)					481a
$(C_6H_5)_3N$	CCl_4	IR	1.5	0.2				235
$(C_6H_5CH_2)_3N$	CCl_4	NIR	2.7 (20°)	0.6	1.6*	3.4	90	909
	CCl_4	IR	13.3 (27°)					481a
$(C_6H_5CH_2)(CH_3)_2N$	CCl_4	IR	35.4 (27°)					481a
$(C_6H_{11})(CH_3)_2N$	CCl_4	IR	91.0 (27°)					481a

Pyridines

C_5H_5N	CCl_4	NIR	60(20°)	2.3	7.0	15.8	492	909
	CCl_4	IR	35.2(30°)	2.0	6.5 ±0.4	15.0	465	671
	CCl_4	NIR	59(20°)	2.3	6.5	14.1	473	2124
	CCl_4	IR	50	2.3	5.6	11.1		2686
	CCl_4	CAL	49	2.3	6.5 ±0.4	14.4 ±1.3	471	807
	CCl_4	IR	49.7 ±1	2.3	6.5			1779
	CCl_4	NIR	41	2.2	9.0	23	475	70
	CCl_4	DEC	45 ±5	2.3	7 ±2	17 ±3		2280
	CS_2	NIR	72(20°)	2.4	7.5	17.1		251
	CS_2	NIR	73	2.5				909
	C_6H_6	CAL	79 ±10	2.6	8.0 ±0.1*		465	2124
	base	CAL			7.3 ±0.10*			719
	$CHCl_3$	NIR	17	1.7				70
	CH_2Cl_2	NIR	17	1.7				2124
	CH_2Cl_2	NIR	19.4 (24°)	1.8	5.9 ±0.2	13.8		2124
	C_6H_6	DEC	36 ±2	2.1	9 ±2	24 ±3		1366a
	$n\text{-}C_7H_{16}$	DEC	80 ±5	2.6	6 ±2	11 ±3		251
	$n\text{-}C_7H_{16}$	UV	88 (22°)	2.6	6	11.4		251
2-CH₃-	CCl_4	NIR	75(20°)	2.4	6.9**	15.1	520	420
	CCl_4	CAL			6.9**		496	909
3-CH₃-	CCl_4	NIR	73(20°)	2.4	6.5**	13.7	491	1778
	CCl_4	CAL			6.6**		481	909
	CCl_4	NIR	65(20°)	2.4	6.7	14.4		1778
								278

313

Acid	Base	Phase or Solvent	Method	$K_{25°C}$ ℓ/mole	$-\Delta G_{25°C}$ kcal/mole	$-\Delta H$, kcal/mole	$-\Delta S_{25°C}$, cal deg⁻¹ mole⁻¹	$\Delta \nu_{AH}$ cm⁻¹	Ref.
C_6H_5OH	**Pyridines**								
	4-CH_3-	CCl_4	NIR	80(20°)	2.5	6.6	13.7	500	909
		CCl_4	NIR	82(20°)	2.5	6.4	13.1		2124
		base	CAL			7.66 ±0.08*		485	70
		CCl_4	CAL			6.7		501	1778
	3-CN-	CCl_4	NIR	14(20°)	1.5	4.9	11.4		2124
		CCl_4	CAL			4.7**		375	1778
	4-CN-	CCl_4	NIR	12(20°)	1.4	3.2	6.0		2124
		CCl_4	CAL			4.9**		385	1778
	4-C_2H_5-	CCl_4	NIR	77(20°)	2.5	5.9	11.4	485	2124
		CCl_4	CAL			6.8**		510	1778
	4-i-C_3H_7-	CCl_4	IR	60	2.4			490	234
	2-t-C_4H_9-	CCl_4	CAL			6.1**		440	1778
	4-t-C_4H_9-	CCl_4	NIR	84(20°)	2.5	7.1**	15.4		2124
		CCl_4		64				490	234
	4-$C_6H_5CH_2$-	CCl_4	NIR	78(20°)	2.5	6.2	12.4	535	909
	3-Br-	CCl_4	NIR	15(20°)	1.5	4.5	10.1		2124
		CCl_4	CAL			5.4			1778
	2,6-CH_3-	CCl_4	NIR	95(20°)	2.6	6.9**	14.4		909
		CCl_4	CAL			7.0			1778
	2,4-CH_3-	CCl_4	NIR	104(20°)	2.6	6.8	14.1	516	909
		CCl_4	CAL			7.2**			1778
	2,4,6-CH_3-	CCl_4	NIR	137(20°)	2.8	7.5**	15.8	531	909
		CCl_4	CAL			7.5**			1778
	2-NH_2-	CCl_4	IR	62	2.4				234
	2,6-t-C_4H_9-	CCl_4	CAL			3.3			1778
	2-Cl-	CCl_4	CAL			4.8**		346	1778

Compound								
2-Br-	CCl₄	CAL			4.8**		351	1778
3-Cl-	CCl₄	CAL			5.5**		416	1778
3,5-CH₃-	CCl₄	CAL			7.0**		511	1778
quinuclidine	C₆H₁₂	CAL			9.03 ±0.08			2290
quinolizidine	C₂Cl₄	IR	409	3.6	8.6	16.9		1084a
	C₂Cl₄	IR	78	2.6	7.4	16.3		1084a
quinoline	CCl₄	NIR	57(20°)	2.3	7.2**	16.4	498	909
	base	CAL			6.7			70
	CCl₄	CAL			7.44 ±0.08			1778
isoquinoline	CCl₄	NIR	61(20°)	2.3	6.9	15.4	529	909
	CCl₄	CAL			6.7**			1778
2-methylquinoline	CCl₄	NIR	80(20°)	2.5	7.1	15.4	532	909
	CCl₄	CAL			7.1**			1778
acridine	CCl₄	NIR	67(20°)	2.4	6.6	14.1	520	909
	CCl₄	CAL			7.1**			1778
(C₆H₅)N(C₂H₅)₂	CCl₄	NIR	2.6(20°)	0.5	1.3*	2.7	67	909
(C₆H₅)N(CH₃)₂	base	CAL			3.89 ±0.08			70
pyrimidine	CCl₄	IR	9.0	1.3	5.4 ±0.5**	13.8 ±1.7	385	807
pyridazine	CCl₄	IR	25	1.9	5.6 ±0.7**	12.6 ±2.4	390	807
(C₂H₅)₂NH	CCl₄	IR	96.1(27°)					481a
(i-C₃H₇)₂NH	CCl₄	IR	56.6(27°)					481a
(n-C₃H₇)₂NH	CCl₄	IR	87.0(27°)					481a
(n-C₄H₉)₂NH	CCl₄	IR	87.0(27°)					481a
	C₂Cl₄	UV	128	2.9				278
(i-C₄H₉)₂NH	CCl₄	IR	41.6(27°)					481a
(n-C₅H₁₁)₂NH	CCl₄	IR	84.6(27°)					481a
(C₆H₅CH₂)₂NH	CCl₄	IR	22.6(27°)					481a
n-C₃H₇NH₂	CCl₄	IR	81.6(27°)					481a
i-C₃H₇NH₂	CCl₄	IR	101.5(27°)					481a
n-C₄H₉NH₂	CCl₄	IR	91.6(27°)					481a
	C₂Cl₄	UV	114	2.8				278
i-C₄H₉NH₂	CCl₄	IR	74.9(27°)					481a
sec-C₄H₉NH₂	CCl₄	IR	84.8(27°)					481a
t-C₄H₉NH₂	CCl₄	IR	90.7(27°)					481a
n-C₅H₁₁NH₂	CCl₄	IR	79.2(27°)					481a
i-C₅H₁₁NH₂	CCl₄	IR	76.5(27°)					481a

Acid	Base	Phase or Solvent	Method	$K_{25°C}$, ℓ/mole	$-\Delta G_{25°C}$, kcal/mole	$-\Delta H$, kcal/mole	$-\Delta S_{25°C}$, cal deg^{-1} mole^{-1}	$\Delta\nu_{AH}$ cm^{-1}	Ref.
C_6H_5OH	$C_6H_{13}NH_2$	CCl_4	IR	73.8(27°)					481a
	$C_7H_{15}NH_2$	CCl_4	IR	71.0(27°)					481a
	$C_8H_{17}NH_2$	CCl_4	IR	84.3(27°)					481a
	$C_9H_{19}NH_2$	CCl_4	IR	80.2(27°)					481a
	aniline	CCl_4	IR	4.0(27°)				341	625
	p-F-aniline	CCl_4	IR	3.8(27°)				346	625
	p-Cl-aniline	CCl_4	IR	3.2(27°)				326	625
	p-Br-aniline	CCl_4	IR	3.2(27°)				331	625
	m-F-aniline	CCl_4	IR	2.4(27°)				317	625
	o-Cl-aniline	CCl_4	IR	2.1(27°)				300	625
	o-F-aniline	CCl_4	IR	1.9(27°)				306	625
	o-Br-aniline	CCl_4	IR	2.1(27°)				297	625
	o-toluidine	CCl_4	IR	3.6(27°)				346	625
	m-toluidine	CCl_4	IR	4.6(27°)				361	625
	p-toluidine	CCl_4	IR	5.6(27°)				366	625
	p-anisidine	CCl_4	IR	7.4(27°)				381	625
	AMINE OXIDES, NITROSOAMINES								
	$(CH_3)_2NNO$	CCl_4	NIR	14.7(20°)	1.5	4.2*	9.1	240	907
	$(C_6H_5)_2NNO$	CCl_4	NIR	3.2(20°)	0.6	2.6*	6.7	152	907
	$(CH_3)_3NO$	CH_2Cl_2	UV	3680(20.5°)	4.8	7.9	10.4		1353
	C_5H_5NO	CCl_4	NIR	371(20°)	3.4	7.1*	12.4	441	907
		CH_2Cl_2	NIR	63.3(24°)	2.45 ±0.02	5.8 ±0.1	11.2		1366a
	$C_9H_{18}NO$	C_6H_{12}	CAL			6.9			1463
	pyrazine N-oxide	CH_2Cl_2	NIR	8.0(24°)	1.23 ±0.01	4.2 ±0.3	10.0		1366a
	pyridazine N-oxide	CH_2Cl_2	NIR	10.0(24°)	1.36 ±0.01	5.0 ±0.3	12.2		1366a
	pyrimidine N-oxide	CH_2Cl_2	NIR	13.0(24°)	1.52 ±0.01	4.9 ±0.3	11.3		1366a

Compound	Solvent	Method						Ref
ESTERS								
$CH_3C(O)OCH_3$	CCl_4	NIR	8.4(20°)	1.2	4.5**	11.1	170	908
$CH_3C(O)OC_2H_5$	CCl_4	NIR	9.8(20°)	1.3	4.2	9.7	182	908
	CCl_4	NIR	9.3 ±0.3	1.3	3.2 ±0.5	6.3 ±1.6	164	1182
	CCl_4	IR		1.5	4.82 ±0.10	11.2	168	2656
	CCl_4	CAL			4.6			1779a
	base	CAL			4.75 ±0.08			70
	CCl_4	IR	9.0 ±0.4	1.3	4.8 ±0.1**	12.1	138	70
	CCl_4	CAL	7.1	1.2	2.5**	5.7	176	334
$ClCH_2C(O)OC_2H_5$	CCl_4	NIR	4.3(20°)	0.8	4.95 ±0.04**	11.4	174	908
$C_2H_5C(O)OCH_3$	CCl_4	IR		1.55	4.3**	10.4	166	2656
$C_2H_5C(O)OC_2H_5$	CCl_4	NIR	8.8(20°)	1.2	3.7**	9.1	142	908
$C_6H_5C(O)OC_2H_5$	CCl_4	NIR	6.4(20°)	1.0	3.3**	8.4	150	908
$C_6H_5C(O)OC_6H_5$	CCl_4	NIR	4.6(20°)	0.8	2.7**	6.0	100	908
$p\text{-}NO_2C_6H_4C(O)OC_2H_5$	CCl_4	NIR	5.2(20°)	0.9	2.5**	6.7	143	908
$p\text{-}NO_2C_6H_4OC(O)CH_3$	CCl_4	NIR	2.5(20°)	0.5		10.0	163	908
$CH_3C(O)OCH{=}CH_2$	CCl_4	IR		1.04	4.00 ±0.09**	10.0	123	2656
$CH_2{=}CHC(O)OCH_3$	CCl_4	IR		1.23	4.55 ±0.06**	12.1		2656
β-propionlactone	CCl_4	NIR	5.1(20°)	0.9	4.5*			913
	CCl_4	CAL			4.8			1779a
γ-butyrolactone	CCl_4	NIR	17(20°)	1.6	4.5**	9.7	190	913.
	CCl_4	CAL			4.9			1779a
δ-valerolactone	CCl_4	NIR	16(20°)	1.6	4.5**	9.7	185	913
	CCl_4	CAL			5.1			1779a
β-CH_3-δ-valerolactone	CCl_4	NIR	21(20°)	1.7	4.0**	7.7	235	913
ACID FLUORIDES								
$C_2H_5C(O)F$	CCl_4	NIR	0.7(20°)	−0.2	2.8**	10.1	50	908
$C_3H_7C(O)F$	CCl_4	NIR	0.7(20°)	−0.2	2.9**	10.4	44	908
$C_6H_5C(O)F$	CCl_4	NIR	0.5(20°)	−0.4	2.6**	10.1	30	908
ETHERS								
$(C_2H_5)_2O$	CCl_4	NIR	8.9	1.29	5.41 ±0.06*	13.8	271	2578
	CCl_4	NIR	9.6(20°)	1.3	4.8	11.7		908

APPENDIX (continued)

Acid	Base	Phase or Solvent	Method	$K_{25°C}$, ℓ/mole	$-\Delta G_{25°C}$, kcal/mole	$-\Delta H$, kcal/mole	$-\Delta S_{25°C}$, cal deg^{-1} mole^{-1}	$\Delta\nu_{AH}$ cm^{-1}	Ref.
C_6H_5OH	$(C_2H_5)_2O$	base	CAL			5.45 ±0.08			70
		CCl_4	IR	6.0(29°)	1.09			288	179
	$(n\text{-}C_4H_9)_2O$	CCl_4	NIR	6.3		5.71 ±0.09*	15.5		2578
		CCl_4	NIR	9.4(20°)		5.3	13.7	286	908
		CCl_4	IR	4.9(29°)	1.2			292	179
	$C_6H_5OC_2H_5$	C_6H_{12}	CAL	8.5(23°)	1.3	6.0 ±0.1	15.8		2533
		CCl_4	NIR	1.2	0.12	3.18 ±0.13*	10.3		2578
		CCl_4	NIR	1.1(20°)	0.07	2.0	6.6	145	908
	$(C_6H_5CH_2)_2O$	CCl_4	NIR	3.7	0.77	4.28 ±0.11*	11.8		2578
		CCl_4	NIR	4.6(20°)	0.84	3.9	10.3	233	908
	$(C_6H_5)_2O$	base	CAL			4.43 ±0.10			70
		CCl_4	NIR	0.79	-0.14	2.06 ±0.17*	7.4		2578
		CCl_4	NIR	1.2(20°)	0.03	4.2	14.0	130	908
	$CH_3OC_6H_5$	base	CAL			1.81 ±0.12			70
		CCl_4	NIR	1.0(20°)	-0.05	3.2*	10.6	137	908
	$(CH_2)_3O$	base	CAL			3.08 ±0.13			70
		CCl_4	NIR	16	1.63	5.16 ±0.06	11.8		2578
	$(CH_2)_4O$	CCl_4	NIR	16	1.65	5.29 ±0.08	12.2		2578
		CCl_4	NIR	13(20°)	1.4	5.7	14.4	283	908
		CCl_4	NIR	16	1.65	5.50 ±0.35*	13.1 ±1.2	285	807
		base	CAL			5.75 ±0.12			70
	$(CH_2)_5O$	CCl_4	IR	13.3 ±0.4	1.5			301	179
		CCl_4	IR	10.4(29°)					179
		CCl_4	NIR	7.1	1.16	5.19 ±0.08*	13.5		2578
		CCl_4	NIR	14(20°)	1.5	5.9	14.8	290	908
	$(t\text{-}C_4H_9)_2O$	CCl_4	NIR	3.5	0.75	7.31 ±0.11**	22.0		2578
		CCl_4	IR	3.9(29°)				325	179
	$C_2H_5O(t\text{-}C_4H_9)$	CCl_4	NIR	10	1.40	6.52 ±0.06	17.2		2578

318

Compound	Solvent	Method						Ref.
n-C₄H₉OCH=CH₂	CCl₄	IR		1.11	4.68 ±0.04*	12.0	260	2656
C₂H₅O(n-C₄H₉)	CCl₄	IR		1.30	5.17 ±0.05*	13.0	217	2656
furan	CCl₄	NIR	0.7(20°)	-0.2	2.0*	7.4	103	908
2,5-dimethylfuran	CCl₄	NIR	0.85(20°)	-0.15	3.3*	11.6	113	908
dihydropyran	CCl₄	NIR	2.5(20°)	0.47	4.2*	12.5	180	908
nerolin	CCl₄	NIR	1.3(20°)	0.1	3.9*	12.7	143	908
benzofuran	CCl₄	NIR	0.6(20°)	-0.3	1.81*	7.1	103	908
(CH₃CH₂CH₂)₂O	CCl₄	NIR	7.0(20°)	1.1	4.2*	10.4	282	908
[(CH₃)₂CH]₂O	CCl₄	NIR	9.5(20°)	1.2	5.4*	14.1	287	908
	CCl₄	IR	7.2(29°)				308	179
p-dioxane	CCl₄	NIR	4.8	0.93	5.00 ±0.20*	13.9 ±0.7	235	807
	base	CAL			5.11 ±0.08	13.6		70
	CCl₄	NIR	7.8	1.21	5.26 ±0.13			2578
	iso-octane	UV	16.4(20°)	1.55 ±0.01	5.5 ±0.1	13.2		102

KETONES

Compound	Solvent	Method						Ref.
(CH₃)₂CO	CCl₄	UV	13.5 ±1.0	1.5	3.3 ±0.5	6.2 ±1.2	193	1182
	CCl₄	NIR	12.3(20°)	1.4	4.5	10.4	213	908
	CCl₄	UV	10.1(30°)	1.5	4.7 ±0.3	10.7	230	671
	CCl₄	CAL			4.8*			1779a
	CCl₄	NIR	14	1.6	4.5	10	210	2280
	CCl₄	NMR	10.7	1.4				1772
CH₃C(O)C₂H₅	CCl₄	NIR	10.3 (20°)	1.3	5.2	13.1	200	908
	CCl₄	NIR			5.34 ±0.09*		198	1981
	base	CAL			5.23 ±0.09*			70
	CCl₄	IR			4.9	12.8		70
(C₂H₅)₂CO	CCl₄	CAL	10.6 ±0.4	1.4	5.81 ±0.10	14.6		1779a
	CCl₄	IR		1.46	5.0			2656
CH₃C(O)CH₂CH₂CH₃	CCl₄	NIR	12.1(20°)	1.4	4.6*	10.7	210	1779a
	CCl₄	CAL			4.9*			908
CH₃C(O)CH(CH₃)₂	CCl₄	NIR	9.3(20°)	1.3	4.0*	9.1	200	1779a, 908
CH₃C(O)C(CH₃)₃	CCl₄	NIR	9.1(20°)	1.2	4.1**	9.7	200	908
CH₃C(O)CH=CH₂	CCl₄	IR		1.39	5.50 ±0.11**	13.8	210	2656

APPENDIX (continued)

Acid	Base	Phase or Solvent	Method	$K_{25°C}$, ℓ/mole	$-\Delta G_{25\ C'}$, kcal/mole	$-\Delta H$, kcal/mole	$-\Delta S_{25\ C'}$, cal deg^{-1} mole^{-1}	$\Delta\nu_{AH}$ cm^{-1}	Ref.
C_6H_5OH	$(CH_3)_3CC(O)C(CH_3)_3$	CCl_4	NIR	9.2(20°)	1.2	3.9*	9.1	195	908
	$CH_3C(O)C_6H_5$	CCl_4	NIR	8.3(20°)	1.2	2.6	4.7	205	908
		CCl_4	NIR	7.8(30°)	1.3	5.05 ±0.35*	12.4	225	672
		CCl_4	CAL			4.7			1779a
		CCl_4	IR	8.0	1.2			185	2281
	$ClCH_2C(O)CH_2Cl$	CCl_4	NIR	2.7(20°)	0.5	2.9*	8.0	125	908
	$C_6H_5C(O)C_6H_5$	CCl_4	NIR	7.8(20°)	1.2	3.3*	7.0	189	908
	$(C_6H_5)_2CO$	CCl_4	NIR	5.6(30°)	1.1	4.37 ±0.20*	11.3	195	672
		CCl_4	CAL			4.5			1779a
		CCl_4	IR	7.7	1.2	4.40	10.7	175	2278
	Benzophenone Derivatives								
	3,3'-Cl-	CCl_4	IR	2.0	0.4			140	2278
	4-Cl-	CCl_4	IR	4.7	0.9			160	2278
	4-CH₃-4'-Cl-	CCl_4	IR	4.0	0.8			170	2278
	4-CH₃-	CCl_4	IR	5.8	1.0			180	2278
	4-CH₃O-	CCl_4	IR	9.0	1.3			190	2278
	4,4'-CH₃-	CCl_4	IR	8.6	1.3			195	2278
	p-benzoquinone	CCl_4	IR	2.2(30°)	0.6	5.05 ±0.55**	14.9	160	672
		CCl_4	CAL			4.9**			1779a
	fluorenone	CCl_4	IR	7.2(30°)	1.2	4.65 ±0.40*	11.6	215	671
	cyclopentanone	CCl_4	NIR	18(20°)	1.6	4.5*	9.7	243	908
	cyclohexanone	CCl_4	NIR	17(20°)	1.6	5.0*	11.4	238	908
		CCl_4	CAL			4.7*			1779a
	3-CH₃-cyclohexanone	CCl_4	NMR	15.3	1.6				1772
	$CH_3C(O)C(O)CH_3$	CCl_4	NIR	3.9(20°)	0.7	3.8**	10.4	121	908
	$CH_3C(O)CH_2C(O)CH_3$	CCl_4	NIR	5.4(20°)	0.9	4.1*	10.7	214	908
	camphor	CCl_4	NIR	9.5(20°)	1.3	3.8*	8.4	208	908

320

Compound	Solvent	Method						
2,6-dimethyl-4-pyrone	CCl₄	NIR	196(20°)	3.0	6.2*	10.7	350	905
phorone	CCl₄	NMR	14.7	1.6				1772
isophorone	CCl₄	NMR	29.7	2.0				1772
mesityl oxide	CCl₄	NMR	15.6	1.6				1772
C₆H₅C(O)CH=CHC₆H₅	CCl₄	NIR	11(20°)	1.4	3.8	8.0	209	908

NITRILES

Compound	Solvent	Method						
CH₃CN	C₂Cl₄	IR	5.69	1.03	5.22	14.0	150	1496
	CCl₄	CAL	4.8 ±0.2	0.93	4.65 ±0.06*	12.5	150	719
	CCl₄	UV	5.0 ±0.2	0.95	3.2 ±0.5	7.6	178	1182
	CCl₄	IR	4.8(30°)	1.00	4.3 ±0.35	11.0 ±1.2	160	671
	CCl₄	IR	4.9	0.94				2590
	CCl₄	IR		1.12	4.74 ±0.08	12.2	154	2654
	CCl₄	NIR			3.92			1675
C₂H₅CN	CCl₄	IR	5	0.95	2.3	8	160	2280
	CCl₄	IR	5.5(20°)	0.9	4.5	12.1	160	915
	CCl₄	IR	4.8	0.93			167	2590
	C₂Cl₄	IR	7.25	1.29	3.80 ±0.07*	8.4	153	2656
	CCl₄	IR	5.2	1.17	5.62	11.6	153	1496
n-C₃H₇CN	CCl₄	IR	5.77(30°)	0.98	3.99 ±0.14*	9.7	169	2590
n-C₄H₉CN	CCl₄	IR	2.0	1.1			161	8
ClCH₂CN	C₂Cl₄	IR	2.32	0.41	4.31	12.8	122	2590
	CCl₄	IR	0.9	0.50			107	1496
Cl₂CHCN	CCl₄	IR	0.5	-0.06			93	2590
Cl₃CCN	C₂Cl₄	IR	0.52	-0.41	3.54	13.2	63	2590
C₆H₅CN	CCl₄	IR	3.4	-0.38			60	1496
	C₂Cl₄	IR	4.70	0.72	4.62	12.4	156	2590
	CCl₄	IR	3.7(20°)	0.92	3.3*	8.7	144	1496
C₆H₅CH₂CN	CCl₄	IR	4.4	0.72			154	915
	CCl₄	IR	4.8(20°)	0.88	3.9*	10.4	160	2590
(C₆H₅)₂CHCN	CCl₄	IR	3.7	0.86			156	915
	CCl₄	IR		0.77	3.45 ±0.08**	8.5	152	2590
CH₂=CHCN	CCl₄	IR	3.2	0.91			135	2656
	CCl₄	IR		0.69			146	2590
	C₂Cl₄	IR	4.63	0.91	4.54	12.2	136	1496

APPENDIX (continued)

Acid	Base	Phase or Solvent	Method	$K_{25°C}$, ℓ/mole	$-\Delta G_{25°C}$, kcal/mole	$-\Delta H$, kcal/mole	$-\Delta S_{25°C}$, cal deg^{-1} mole^{-1}	$\Delta \nu_{AH}$ cm^{-1}	Ref.
C_6H_5OH	CH_2=CHCN	CCl_4	IR		0.84	3.60		145	1675
	CH_2=CHCH$_2$CN	CCl_4	IR	4.6(20°)	0.92	3.6*	9.4	155	915
	$(CH_3)_3$CCN	CCl_4	IR	4.7	1.2			162	2590
		C_2Cl_4	IR	7.65		4.28	10.3	161	1496
	BrCN	CCl_4	IR			2.60*		102	1675
	$(CH_2$=CH$)_2$NCN	CCl_4	IR	10.9	1.4			218	2590
	$(CH_3)_2$NCN	CCl_4	IR	12.8	1.5			219	2590
		C_2Cl_4	IR	19.0	1.75	5.72**	13.4	216	1496
	PHOSPHORYL DONORS								
	$(CH_3)_3$PO	CCl_4	UV	1480 ±20	4.3	7.4 ±0.5*	10.2 ±2.0	464	1182
		CCl_4	NIR	1836(20°)	4.3	6.6*	7.7	470	25
	$(C_2H_5)_3$PO	CCl_4	NIR	2522(20°)	4.5	6.8*	7.7	510	25
	$(C_6H_5)_3$PO	CCl_4	NIR	1055(20°)	4.0	6.7*	9.1	430	25
		CCl_4	IR	627	3.8	9.6	19.4	410	22
	$(C_2H_5O)_3$PO	CCl_4	NIR	351(20°)	3.4	6.7*	11.1	345	25
		CCl_4	IR	218	3.2	8.1	16.5	325	22
		CCl_4	NIR	268(20°)	3.2	5.9*	9.1	330	263
	$(CH_3O)_3$PO	CCl_4	NIR	183(20°)	3.0	5.3*	7.7	315	25
	$[(C_2H_5)_2N]P(O)(OC_2H_5)_2$	CCl_4	NIR	518(20°)	3.6	5.6**	6.7	395	25
		CCl_4	NIR	491(20°)	3.6	6.5	9.7	374	263
	$CH_3P(O)(OC_2H_5)_2$	CCl_4	NIR	381(20°)	3.4	6.3*	9.7	360	25
	$C_2H_5P(O)(OCH_3)_2$	CCl_4	NIR	330(20°)	3.3	6.0*	9.1	355	25
	$CH_3P(O)(OC_3H_6Cl)_2$	CCl_4	NIR	247(20°)	3.2	6.0*	9.4	335	25
	$(C_6H_5O)_3$PO	CCl_4	NIR	41(20°)	2.1	3.9*	6.0	230	263
	(p-$tert$-butylphenyl-phosphate	CCl_4	GLC			5.7 ±0.6			2530
	$ClCH_2P(O)(OC_2H_5)_2$	CCl_4	NIR	239(20°)	3.1	6.5*	11.4	325	25

Compound	Solvent	Method						Ref.
Cl₂CHP(O)(OC₂H₅)₂	CCl₄	NIR	133(20°)	2.80	4.9*	9.7	275	25
Cl₃CP(O)(OC₂H₅)₂	CCl₄	NIR	73(20°)	2.5	3.1*	2.0	260	25
HP(O)(OCH₃)₂	CCl₄	NIR	123(20°)	2.8	5.2*	8.0	300	25
HP(O)(OC₂H₅)₂	CCl₄	NIR	162(20°)	2.9	4.7*	6.0	310	25
HP(O)(OC₃H₇)₂	CCl₄	NIR	200(20°)	3.1	5.2*	7.0	320	25
HP(O)[OCH(CH₃)₂]₂	CCl₄	NIR	237(20°)	3.1	5.4*	7.7	330	25
FP(O)(C₂H₅)₂	CCl₄	NIR	129(20°)	2.8	4.7**	6.4	305	25
FP(O)(OCH₃)(C₂H₅)	CCl₄	NIR	83(20°)	2.5	4.9*	8.0	270	25
(C₂H₅S)₃PO	CCl₄	NIR	23(20°)	1.8	4.7	9.7	288	907
	CCl₄	NIR	75(20°)	2.5	4.8	7.7		263
(EtO)₂P(O)OCH₂CH₂NMe₂	CCl₄	NIR	348(20°)	3.3	7.7**	14.4	345	906
(C₂H₅)₂P(O)OC₂H₅	CCl₄	IR	513	3.7	8.2**	15.2	410	22
(C₂H₅)P(O)(OC₂H₅)₂	CCl₄	IR	325	3.4	7.9**	15.1	360	22
(5-membered ring: O=P–OC₂H₅)	CCl₄	IR	169	3.0	7.8**	15.9	320	22
(5-membered ring: O=P–OC₂H₅)	CCl₄	IR	96	2.7	7.6**	16.3	270	22
(6-membered ring: O=P–OC₂H₅)	CCl₄	IR	307	3.4	7.6**	14.2	341	22
(6-membered ring: O=P(O)(OC₂H₅))	CCl₄	IR	181	3.1	7.6**	15.0	280	22
(ring with CH₃ groups: P(O)(OC₂H₅))	CCl₄	IR	331	3.4	8.3**	16.4	370	22

APPENDIX (continued)

Acid	Base	Phase or Solvent	Method	$K_{25°C}$, ℓ/mole	$-\Delta G_{25°C}$, kcal/mole	$-\Delta H$, kcal/mole	$-\Delta S_{25°C}$, cal deg^{-1} mole^{-1}	$\Delta\nu_{AH}$ cm^{-1}	Ref.
C_6H_5OH	$(C_6H_5)_2(CH_3)P(O)$	CCl_4	IR	682	3.9	9.5**	19.2	420	22
	$(C_6H_5)(CH_3)_2P(O)$	CCl_4	IR	910	4.0	9.6**	18.7	430	22
	$(n\text{-}C_4H_9)_3P(O)$	CCl_4	IR	1000	4.1	9.6**	18.5	430	22
	(cyclohexyl)$P(O)(C_6H_5)$	CCl_4	IR	1758	4.4	10.5*	20.5	450	22
	(cyclopentyl P=O)$(n\text{-}C_3H_7)$	CCl_4	IR	2148	4.5	12.8**	27.7	450	22
	(bicyclic CH$_3$ structure) $P(O)(n\text{-}C_3H_7)$	CCl_4	IR	191	3.1	8.9**	19.4	310	22
	$i\text{-}C_3H_7P(O)(OC_2H_5)_2$	CCl_4	NIR	457(20°)	3.5	6.0*	8.5	370	264
	$(C_3H_7)_2P(O)OC_2H_5$	CCl_4	NIR	824(20°)	3.9	6.6*	9.1	400	264
	$i\text{-}C_3H_7P(O)(OC_6H_5)_2$	CCl_4	NIR	100(20°)	2.6	4.8*	7.4	280	264
	$(C_3H_7)_2P(O)OC_6H_5$	CCl_4	NIR	308(20°)	3.3	5.9*	8.7	355	264
	$i\text{-}C_3H_7P(O)(SC_2H_5)_2$	CCl_4	NIR	165(20°)	2.9	5.7*	9.4	335	264
	$(n\text{-}C_3H_7)_2P(O)SC_2H_5$	CCl_4	NIR	386(20°)	3.4	6.6*	10.7	390	264
	$i\text{-}C_3H_7P(O)(SC_6H_5)_2$	CCl_4	NIR	152(20°)	2.9	5.6*	9.1	305	264
	$(n\text{-}C_3H_7)_2P(O)SC_6H_5$	CCl_4	NIR	360(20°)	3.4	6.6*	10.7	370	264
	$(n\text{-}C_3H_7)_3PO$	CCl_4	NIR	2341(20°)	4.5	8.0*	11.7	500	264
	$[(CH_3)_2N]_3PO$	CCl_4	NIR	1874(20°)	4.3	7.0*	9.1	460	263
	$(C_6H_5O)_2P(O)N(C_2H_5)_2$	CCl_4	NIR	109(20°)	2.7	5.4*	9.1	298	263

	Solvent	Method							Ref.
$C_6H_5OP(O)[N(C_2H_5)_2]_2$	CCl_4	NIR	360(20°)		6.3*	3.4	9.7	395	263
$(C_6H_5S)_3PO$	CCl_4	NIR	69(20°)		4.4*	2.4	6.7	260	263
$(C_6H_5S)_2P(O)N(C_2H_5)_2$	CCl_4	NIR	114(20°)		5.8*	2.7	10.4	318	263
$C_6H_5SP(O)[N(C_2H_5)_2]_2$	CCl_4	NIR	445(20°)		6.3*	3.5	9.4	382, 412	263
$C_2H_5OP(O)[N(C_2H_5)_2]_2$	CCl_4	NIR	1248(20°)		6.9*	4.1	9.4	444	263
$(C_2H_5S)_2P(O)N(C_2H_5)_2$	CCl_4	NIR	148(20°)		6.0*	2.9	10.4	343	263
$C_2H_5SP(O)[N(C_2H_5)_2]_2$	CCl_4	NIR	631(20°)		7.0*	3.7	11.1	404, 424	263

ORGANIC HALIDES

	Solvent	Method							Ref.
$C_6H_{11}F$	CCl_4	NIR	0.89	−.071	2.84 ±0.05*		9.8	53	1431a
$C_6H_{11}Cl$	CCl_4	NIR	0.43	−.495	1.95 ±0.09		8.2	66	1431a
	base	CAL			2.03 ±0.13*				70
$C_6H_{11}Br$	CCl_4	NIR	0.41	−.529	1.78 ±0.07		7.8	82	1431a
	base	CAL			1.81 ±0.09*				70
$C_6H_{11}I$	CCl_4	NIR	0.39	−.555	1.46 ±0.14		6.8	86	1431a
	base	CAL			1.35 ±0.13*				70
$n\text{-}C_7H_{15}F$	C_2Cl_4	IR			2.13*			40	1190
$n\text{-}C_7H_{15}Cl$	C_2Cl_4	IR			1.65*			59	1190
$n\text{-}C_7H_{15}Br$	C_2Cl_4	IR			1.57*			69	1190
$n\text{-}C_7H_{15}I$	C_2Cl_4	IR			1.25*			71	1190
$(n\text{-}C_7H_{15})_4N^+\,I^-$	CCl_4	IR	382		5.9*	3.5	8.0	380	2280
$C_6H_5CH_2Cl$	CCl_4	IR	0.51		0.8*	−0.4	4.0	75	2278
$n\text{-}C_4H_9Cl$	base	CAL			1.71 ±0.10				70
	CCl_4	NIR	0.44	−0.491	2.22 ±0.12		9.0		1431a
$n\text{-}C_4H_9Br$	base	CAL			1.71 ±0.08				70
	CCl_4	NIR	0.34	−0.635	1.59 ±0.13		7.5		1431a
$n\text{-}C_4H_9I$	base	CAL			1.3 ±0.2				70
	CCl_4	NIR	0.37	−0.585	1.69 ±0.09		7.6		1981a
$CH_3CH_2CH(CH_3)Br$	CCl_4	NIR	0.37	−0.583	1.66 ±0.06		7.5		1981a
$CH_3CH_2CH(CH_3)Cl$	CCl_4	NIR		−0.677	1.83 ±0.18		8.4		1981a
$CH_3CH(CH_3)CH_2Cl$	CCl_4	NIR		−0.521	2.72 ±0.20		10.9		1981a
$CH_3CH(CH_3)CH_2Br$	CCl_4	NIR	0.48	−0.439	2.28 ±0.14		9.1		1431a
$(CH_3)_3CCl$	CCl_4	NIR	0.44	−0.480	1.80 ±0.11		7.7		1431a
$(CH_3)_3CBr$	CCl_4	NIR		−0.372	2.64 ±0.05		10.1		1431a
$CH_3(CH_2)_5F$	CCl_4	NIR	0.32	−0.673	1.69 ±0.11		7.9		1431a
$CH_3(CH_2)_5Cl$									

APPENDIX (continued)

Acid	Base	Phase or Solvent	Method	$K_{25°C}$, ℓ/mole	$-\Delta G_{25°C}$, kcal/mole	$-\Delta H$, kcal/mole	$-\Delta S_{25°C}$, cal deg^{-1} mole^{-1}	$\Delta\nu_{AH}$ cm^{-1}	Ref.
C_6H_5OH	$CH_3(CH_2)_5Br$	CCl_4	NIR	0.35	-0.626	1.61 ±0.10	7.5		1431a
	$CH_3(CH_2)_5I$	CCl_4	NIR		-0.691	1.76 ±0.21	8.2		1431a
	π-BASES								
	benzene	CCl_4	IR	0.33(29°)	-0.66(29°)	1.18 ±0.07*		49	2660
		C_6H_6	CAL						70
		CCl_4	IR	0.39(40°)		1.56 ±0.11	7.8	48	2398
		CCl_4	NIR	0.28				47	1981
		CCl_4	IR	0.39(29°)	-0.76			58	2660
		CCl_4	IR	0.49(40°)	-0.56(29°)			65	2398
	toluene	base	CAL			1.33 ±0.08*			70
		CCl_4	NIR	0.34	-0.63	1.65 ±0.14	7.7	59	1431a
	ethylbenzene	CCl_4	IR	0.35(29°)	-0.62(29°)			56	2660
		CCl_4	IR		-0.61	1.52 ±0.12*	7.1	55	2656
	cumene	CCl_4	IR	0.39(29°)	-0.56(29°)			57	2660
		base	CAL			1.55 ±0.11*	7.1		70
		CCl_4	IR	0.32	-0.67	2.6		57	2542
		CCl_4	IR		-0.61	1.52 ±0.10			2656
	t-butylbenzene	CCl_4	IR	0.35(29°)	-0.62(29°)			60	2660
	laurylbenzene	CCl_4	IR	0.43(29°)	-0.50(29°)			62	2660
	o-xylene	CCl_4	IR	0.35(29°)	-0.62(29°)			68	2660
		CCl_4	IR		-0.47(26°)			67	2655
	m-xylene	CCl_4	IR	0.41(29°)	-0.53(29°)			69	2660
		CCl_4	NIR	0.40	-0.54	2.09 ±0.17*	8.8		1431a
	p-xylene	CCl_4	IR		-0.51(26°)				2655
		CCl_4	IR	0.39(29°)	-0.56(29°)			69	2660
		CCl_4	IR	0.43(40°)				72	2398
		CCl_4	IR		-0.45(26°)			69	2655

	CCl_4	NIR		-0.55	2.16 ±0.10*	9.1		1981a
mesitylene	CCl_4	IR	0.57(29°)	-0.33(29°)			78	2660
	CCl_4	IR	0.53(40°)				84	2398
	base	CAL		1.57 ±0.15*				70
	CCl_4	NIR		-0.425	2.19 ±0.12	8.8		1431a
	CCl_4	IR		-0.38(26°)	1.92 ±0.16	7.8	73	2654
durene	CCl_4	IR	0.60(29°)	-0.30(29°)			85	2660
	CCl_4	IR	0.60(40°)				95	2398
	CCl_4	NIR	0.61	-0.297	2.25 ±0.29*	8.5		1431a
hexamethylbenzene	CCl_4	IR	0.63(29°)	-0.27(29°)			106	2660
	CCl_4	IR	0.95	-0.03	1.65 ±0.5*	5.5 ±1.7	102	807
naphthalene	CCl_4	IR	0.58(29°)	-0.32(29°)			48	2660
	CCl_4	NIR	0.43	-0.50	1.46 ±0.27*	6.6		1431a
α-methylnaphthalene	CCl_4	IR	0.69(29°)	-0.22(29°)			55	2660
	CCl_4	IR	0.48(40°)				48	2398
	CCl_4	NIR	0.51	-0.40	1.53 ±0.15*	6.5		1431a
β-methylnaphthalene	CCl_4	IR	0.63(29°)	-0.27(29°)			54	2660
2,3-dimethylnaphthalene	CCl_4	IR	0.71(29°)	-0.20(29°)			58	2660
2,6-dimethylnaphthalene	CCl_4	IR	0.69(29°)	-0.22(29°)			59	2660
acenaphthene	CCl_4	IR	0.71(29°)	-0.20(29°)			58	2660
1,2,3,6,7,8-hexahydropyrene	CCl_4	IR	0.96(29°)	-0.02(29°)			69	2660
fluoranthene	CCl_4	IR	0.69(29°)	-0.22(29°)			49	2660
biphenyl	CCl_4	IR	0.57(29°)	-0.33(29°)			52	2660
	CCl_4	NIR	-0.51	-0.41	1.74 ±0.15*	7.3		1431a
fluorene	CCl_4	IR	0.42(29°)	-0.51(29°)			57	2660
4,5,9,10-tetrahydropyrene	CCl_4	IR	0.78(29°)	-0.15(29°)			69	2660
o-terphenyl	CCl_4	IR	1.0(29°)	0.0(29°)			52	2660
m-terphenyl	CCl_4	IR	0.98(29°)	-0.01(29°)			56	2660
	CCl_4	NIR	0.69	-.22	1.13 ±0.14*	4.5		1431a
phenanthrene	CCl_4	IR	0.58(29°)	-0.32(29°)			48	2660
	CCl_4	NIR	0.52	-0.39	1.01 ±0.45*	4.7		1431a
	CCl_4	IR		-0.23(26°)			47	2655
pyrene	CCl_4	IR	0.75(29°)	-0.17(29°)				2660
ferrocene	CCl_4	IR	0.59(29°)	-0.31(29°)			35	2660

APPENDIX (continued)

Acid	Base	Phase or Solvent	Method	$K_{25°C}$, ℓ/mole	$-\Delta G_{25°C}$, kcal/mole	$-\Delta H$, kcal/mole	$-\Delta S_{25°C}$, cal deg^{-1} mole^{-1}	$\Delta \nu_{AH}$ cm^{-1}	Ref.
C_6H_5OH	1-methyl-1,3-butadiene	CCl₄	IR	0.30(29°)	-0.71(29°)			56	2660
	2-methyl-1,3-butadiene	CCl₄	IR	0.27(29°)	-0.78(29°)			56	2660
	2,3-dimethyl-1,3-butadiene	CCl₄	IR	0.47(29°)	-0.45(29°)			92	2660
	1,3-cyclooctadiene	CCl₄	IR	0.29(29°)	-0.73(29°)			74	2660
	1,4-hexadiene	CCl₄	IR	0.34(29°)	-0.64(29°)			68	2660
	trans,trans,cis-1,5,9-cyclododecatriene	CCl₄	IR	0.47(29°)	-0.45(29°)			103	2660
	cyclohexene	CCl₄	IR	0.21(29°)	-0.92(29°)			99	2660
		CCl₄	NIR	0.15	-1.14	2.02 ±0.25*	10.6		1431a
		CCl₄	IR	0.23	-0.87	2.83			2542
	furan	CCl₄	IR	0.34(29°)	-0.64(29°)			48	2660
	2-methylfuran	CCl₄	IR	0.36(29°)	-0.61(29°)			66	2660
	2,5-dimethylfuran	CCl₄	IR	0.76(29°)	-0.16(29°)			83	2660
	dibenzofuran	CCl₄	IR	0.79(29°)	-0.14(29°)			50	2660
	thiophene	CCl₄	IR	0.38(29°)	-0.57(29°)			52	2660
	2-methylthiophene	CCl₄	IR	0.31(29°)	-0.69(29°)			64	2660
	2,5-dimethylthiophene	CCl₄	IR	0.40(29°)	-0.54(29°)			75	2660
	dibenzothiophene	CCl₄	IR	0.49(29°)	-0.42(29°)			42	2660
	N-methylpyrrole	CCl₄	IR	1.40(29°)	0.20(29°)			114	2660
	N-ethylcarbazole	CCl₄	IR	1.16(29°)	0.09(29°)			62	2660
	6,6-dimethylfulvene	CCl₄	IR	0.74(29°)	-0.18(29°)			86	2660
	6,6-methylethylfulvene	CCl₄	IR	0.70(29°)	-0.21(29°)			88	2660
	6,6-methyl, n-propylfulvene	CCl₄	IR	0.88(29°)	-0.08(29°)			87	2660
	6,6-methyl, isopropylfulvene	CCl₄	IR	0.96(29°)	-0.02(29°)			87	2660
	$6,6\text{-}C_5H_4{=}C(CH_3)(CH{=}C(CH_3)_2$	CCl₄	IR	1.1(29°)	0.56(29°)			104	2660

Compound	Solvent	Method					
$C_5H_4{=}C(CH_2)_5$	CCl$_4$	IR	0.76(29°)	-0.16(29°)		90	2660
6,6-methylphenylfulvene[a]	CCl$_4$	IR	0.30(29°)	-0.71(29°)		57	2660
6,6-methylphenylfulvene[b]	CCl$_4$	IR	1.0(29°)	0.0(29°)		87	2660
6,6-diphenylfulvene[a]	CCl$_4$	IR	0.25(29°)	-0.82(29°)		51	2660
6,6-diphenylfulvene[b]	CCl$_4$	IR	0.79(29°)	-0.14(29°)		80	2660
6,6-$(CH_3)[C_6H_4CH(p)]$-fulvene[a]	CCl$_4$	IR	0.21(29°)	-0.92(29°)		63	2660
6,6-$(CH_3)[C_6H_4CH(p)]$-fulvene[b]	CCl$_4$	IR	1.0(29°)	0.0(29°)		83	2660
6,6-$(CH_3)[C_6H_4Cl(p)]$-fulvene[a]	CCl$_4$	IR	0.1(29°)	-1.4(29°)		22	2660
6,6-$(CH_3)[C_6H_4Cl(p)]$-fulvene[b]	CCl$_4$	IR	0.79(29°)	-0.14(29°)		80	2660
6,6-$(CH_3)[C_6H_4NO(m)]$-fulvene	CCl$_4$	IR	3.7(29°)	0.77(29°)		71	2660
6,6-$(CH_3)[\alpha{-}C_{10}H_7]$-fulvene[a]	CCl$_4$	IR	0.45(29°)	-0.47(29°)		50	2660
6,6-$(CH_3)[\alpha{-}C_{10}H_7]$-fulvene[b]	CCl$_4$	IR	0.86(29°)	-0.09(29°)		94	2660
6-α-furylfulvene	CCl$_4$	IR	0.52(29°)	-0.39(29°)		78	2660
azulene[a]	CCl$_4$	IR	0.19(29°)	-0.98(29°)		45	2660
azulene[b]	CCl$_4$	IR	0.53(29°)	-0.38(29°)		84	2660
4,6,8-trimethylazulene[a]	CCl$_4$	IR	0.31(29°)	-0.69(29°)		79	2660
4,6,8-trimethylazulene[b]	CCl$_4$	IR	1.1(29°)	0.06(29°)		116	2660
4,8-dimethyl-6-t-butylazulene[a]	CCl$_4$	IR	0.46(29°)	-0.46(29°)		72	2660
4,8-dimethyl-6-t-butylazulene[b]	CCl$_4$	IR	1.4(29°)	0.20(29°)		115	2660
guaiazulene[a]	CCl$_4$	IR	0.25(29°)	-0.82(29°)		62	2660
guaiazulene[b]	CCl$_4$	IR	0.87(29°)	-0.08(29°)		105	2660
nitrobenzene	CCl$_4$	IR	3.5(29°)	0.74(29°)		66	2660
styrene	CCl$_4$	IR		-0.42	1.34 ±0.13*	5.9	46	2656
α-methylstyrene	CCl$_4$	IR		-0.59	1.47 ±0.13*	5.9	49	2656

[a]Hydrogen bond to α-ring. [b]Hydrogen bond to β-ring.

Acid	Base	Phase or Solvent	Method	$K_{25°C}$, ℓ/mole	$-\Delta G_{25°C}$, kcal/mole	$-\Delta H$, kcal/mole	$-\Delta S_{25°C}$, cal deg^{-1} mole^{-1}	$\Delta\nu_{AH}$ cm^{-1}	Ref.
C$_6$H$_5$OH	2-pentene	CCl$_4$	IR		-1.01(29°)			82	2653
	1-hexene	CCl$_4$	IR		-1.46(29°)			60	2653
	2-methyl-1-pentene	CCl$_4$	IR		-1.04(29°)			87	2653
	4-methyl-2-pentene	CCl$_4$	IR		-1.03(29°)			81	2653
	1-heptene	CCl$_4$	IR		-1.37(29°)			60	2653
	1-octene	CCl$_4$	IR		-1.47(29°)			61	2653
	2-ethyl-1-hexene	CCl$_4$	IR		-0.79(29°)			100	2653
	cyclopentene	CCl$_4$	IR		-1.40(29°)			78	2653
	cyclohexene	CCl$_4$	IR		-0.93(29°)			92	2653
	4-methyl-cyclohexene	CCl$_4$	IR		-0.93(29°)			93	2653
	cycloheptene	CCl$_4$	IR		-1.05(29°)			88	2653
	cis-cyclooctene	CCl$_4$	IR		-1.19(29°)			88	2653
	n-propylcyclopropane	CCl$_4$	IR		-2.09 ±0.10	1.30 ±0.12*	11.4 ±0.7	40	2657
	n-pentylcyclopropane	CCl$_4$	IR		-1.92	1.34*	11.3	40	2657
	isopropylcyclopropane	CCl$_4$	IR		-2.04	1.39*	11.5	41	2657
	trans-1-methyl-2-isopropylcyclopropane	CCl$_4$	IR		-2.14	1.47*	12.1	42	2657
	cis-1-methyl-2-isopropylcyclopropane	CCl$_4$	IR		-2.24	1.50*	12.6	41	2657
	bicyclo[3.1.0]hexane	CCl$_4$	IR		-1.88	1.78*	12.3	43	2657
	bicyclo[4.1.0]heptane	CCl$_4$	IR		-1.86	1.63*	11.7	42	2657
	bicyclo[6.1.0]nonane	CCl$_4$	IR		-1.83	1.76*	12.0	44	2657
	phenylcyclopropane	CCl$_4$	IR		-0.57	1.45*	6.8	53	2657
	isopropylbenzene	CCl$_4$	IR		-0.61	1.52*	7.1	57	2657
	1-pentene	CCl$_4$	IR		-1.16(29°)			58	2653
	1,3-pentadiene	CCl$_4$	IR		-1.10	1.71 ±0.14*	9.4	65	2656
	1-hexyne	CCl$_4$	IR		-0.60	1.60 ±0.11*	7.4	58	2656
		CCl$_4$	IR		-0.48	1.52*	6.7	78.5	2658

	Solvent	Method						Ref.
1-heptyne	CCl_4	IR		−0.32	1.56*	5.3	81	2658
1-octyne	CCl_4	IR		−0.15	1.62*	5.9	83	2658
2-octyne	CCl_4	IR		0.12	2.15*	6.8	115	2658
3-hexyne	CCl_4	IR		0.13	2.16*	6.8	116.5	2658
3-octyne	CCl_4	IR		0.38	2.56*	7.3	120	2658
4-octyne	CCl_4	IR		0.24	2.42*	7.1	118.5	2658
2-methyl-1-buten-3-yne	CCl_4	IR		−0.32	1.42*	5.9	62	2658
phenylacetylene α	CCl_4	IR		−0.63	0.90*	5.1	35	2658
β				−0.92	1.64*	8.6	67	2658
diphenylacetylene α	CCl_4	IR		−0.51	0.93*	4.8	33.5	2658
β				−0.93	1.66*	8.7	71.5	2658
1,2-hexadiene	CCl_4	IR		−0.89	1.49*	8.0	55.5	2658
3-methyl-1,2-pentadiene	CCl_4	IR		−0.76	1.51*	7.6	77.5	2658
2,3-hexadiene	CCl_4	IR		−0.73	1.59*	7.8	81	2658
2-methyl-1,2,3-pentadiene	CCl_4	IR		−0.72	1.61*	7.8	83.5	2658
2,4-dimethyl-1,2,3-pentadiene	CCl_4	IR		−0.63	1.78*	8.1	100.5	2658
1,2-cyclononadiene	CCl_4	IR		−0.55	1.57*	7.1	79	2658
phenylallene	CCl_4	IR		−0.39	1.54*	5.8	46	2658
1-phenyl-1-methylallene	CCl_4	IR		−0.30	1.50*	6.0	57	2658
1,1-diphenylallene	CCl_4	IR		−0.34	1.42*	5.9	52	2658
SULFIDES								
$(C_2H_5)_2S$	CCl_4	CAL	1.1 ±0.1	0.06	3.6 ±0.1*	11.9	256	2533
	C_6H_{12}	CAL	2.0 ±0.1	0.4	4.6 ±0.1*	14.1	274	2533
$(CH_2)_4S$	CCl_4	CAL	1.4 ±0.1	0.2	3.7 ±0.1*	11.7		2533
	C_6H_{12}	CAL	8.5 (23°)	1.3	6.0 ±0.1	15.8		2533
$(t\text{-}C_4H_9)_2S$	CCl_4	NIR	2.0	0.42	4.87 ±0.2	12.2	254	2578
$(n\text{-}C_4H_9)_2S$	CCl_4	NIR	1.3	0.17	4.19 ±0.3**	13.5	252	2578
$(sec\text{-}C_4H_9)_2S$	CCl_4	IR	1.48	0.23	3.4 ±0.1**	10.6	250	2169b
$(i\text{-}C_4H_9)S(n\text{-}C_4H_9)$	CCl_4	IR	1.60	0.28	3.5 ±0.1**	10.9	245	2169b
$(i\text{-}C_4H_9)S(t\text{-}C_4H_9)$	CCl_4	IR	1.35	0.17	3.2 ±0.1**	9.9	270	2169b
	CCl_4	IR	1.71	0.32	3.5 ±0.1**	10.6		2169b

331

APPENDIX (continued)

Acid	Base	Phase or Solvent	Method	$K_{25°C}$, ℓ/mole	$-\Delta G_{25°C}$, kcal/mole	$-\Delta H$, kcal/mole	$-\Delta S_{25°C}$, cal deg^{-1} mole^{-1}	$\Delta \nu_{AH}$ cm^{-1}	Ref.
C_6H_5OH	$(n\text{-}C_4H_9)S(sec\text{-}C_4H_9)$	CCl_4	IR	1.46	0.22	3.6 ±0.1**	11.3	260	2169b
	$(n\text{-}C_4H_9)S(t\text{-}C_4H_9)$	CCl_4	IR	1.73	0.32	3.6 ±0.1**	11.0	270	2169b
	$(sec\text{-}C_4H_9)S(t\text{-}C_4H_9)$	CCl_4	IR	1.60	0.27	4.0 ±0.1**	12.5	263	2169b
	SULFATES, SULFONATES, SULFONES								
	$(CH_3O)_2SO$	CCl_4	IR	1.3	0.16			75	246
	ethylbenzenesulfonate	CCl_4	IR	2.2	0.47			100	246
	phenylbenzenesulfonate	CCl_4	IR	1.9	0.38			92	246
	p-chlorobenzenesulfonate	CCl_4	IR	1.7	0.31			85	246
	$(C_2H_5)_2SO_2$	CCl_4	IR	13.9	1.6			157	246
	$(CH_3)(n\text{-}C_3H_7)SO_2$	CCl_4	IR	11.7	1.5			164	246
	$(n\text{-}C_4H_9)_2SO_2$	CCl_4	IR	15.0	1.6			197	246
	$(t\text{-}C_4H_9)_2SO_2$	CCl_4	IR	15.8	1.6			197	246
	$(C_6H_5)(CH_3)SO_2$	CCl_4	IR	7.7	1.2			141	246
	$(C_6H_5)(C_6H_5CH_2)SO_2$	CCl_4	IR	8.1	1.2			142	246
	$(C_6H_5)_2SO_2$	CCl_4	IR	6.0	1.1			135	246
	$(CH_2)_4SO_2$	CCl_4	UV	17.1 ±0.7	1.7	4.9 ±0.3	10.7	135	655
	SULFOXIDES								
	$(CH_3)_2SO$	CCl_4	NIR	230(20°)	3.1	8.0 ±0.2	16.4	350	907
		CCl_4	UV	182 ±1	3.1	6.5 ±0.2	11.4		655
		base	CAL	202 ±2.0		7.21 ±0.08*	16.5 ±0.3	257	70
		CCl_4	IR	188	3.1		13.8		70
	$[(CH_3)_2CH]_2SO$	C_2Cl_4	IR	210(20°)	3.1	6.93	12.8	353	1496
	$(C_6H_5)_2SO$	CCl_4	NIR	70(20°)	2.4	5.8*	9.1	360	907
	$(CH_3C_6H_4)_2SO$	CCl_4	NIR	105(20°)	2.7	5.8*	11.4	294	907
		CCl_4	NIR			6.2*	11.7	320	907

Compound	Solvent	Method						Ref.
$(C_6H_5CH_2)_2SO$	CCl_4	NIR	131(20°)	2.8	5.8*	10.1	330	907
$(CH_3CH_2CH_2CH_2)_2SO$	CCl_4	NIR	264(20°)	3.2	6.3*	10.4	373	907
	CCl_4	IR	218	3.2	6.2*	10.2	380	2169b
$(CH_3CH_2O)_2SO$	CCl_4	NIR	7.9(20°)	1.2	4.0*	9.4	154	907
$(CH_3CH_2CH_2O)_2SO$	CCl_4	NIR	6.9(20°)	1.1	3.4*	7.7	154	907
$(CH_3CH_2CH_2CH_2O)_2SO$	CCl_4	NIR	8.2(20°)	1.2	3.0*	6.0	162	907
$(CH_2)_4SO$	CCl_4	NIR	233(20°)	3.1		12.7	370	907
$(n\text{-}C_4H_9)S(O)(sec\text{-}C_4H_9)$	CCl_4	IR	250	3.3	6.9*	10.2	380	2169b
$(i\text{-}C_4H_9)S(O)(t\text{-}C_4H_9)$	CCl_4	IR	234	3.2	6.3*	10.3	385	2169b
$(n\text{-}C_4H_9)S(O)(t\text{-}C_4H_9)$	CCl_4	IR	240	3.2	6.5*	10.8	380	2169b
$(sec\text{-}C_4H_9)S(O)(t\text{-}C_4H_9)$	CCl_4	IR	245	3.3	6.9*	12.3	383	2169b
THIOAMIDES, THIOUREAS								
$CH_3C(S)N(CH_3)_2$	CCl_4	IR	12.0(20°)	1.4	4.0**	8.7	315	915
$(CH_3)_2NC(S)N(CH_3)_2$	CCl_4	IR	16.0(20°)	1.6	4.6**	10.1	330	915
$C_6H_5C(S)N(CH_3)_2$	CCl_4	IR	8.1(20°)	1.2	3.0**	6.0	275	915
$CH_3OC(S)N(CH_3)_2$	CCl_4	IR	5.1(20°)	0.9	3.5**	8.7	256	915
$CH_3SC(S)N(CH_3)_2$	CCl_4	IR	3.7(20°)	0.7	2.5**	6.0	247	915
$NCC(S)N(CH_3)_2$	CCl_4	IR	3.0(20°)	0.6	2.9**	7.7	125	915
THIOCYANATES, ISOTHIOCYANATES								
CH_3SCN	CCl_4	IR	3.11	0.7	3.79 ±0.25*	10.50 ±0.6	146	1119
C_2H_5SCN	CCl_4	IR	3.38	0.7	3.81 ±0.25*	10.48 ±0.6	152	1119
$n\text{-}C_3H_7SCN$	CCl_4	IR	3.61	0.8	4.14 ±0.25*	11.35 ±0.6	156	1119
$i\text{-}C_3H_7SCN$	CCl_4	IR	3.85	0.8	3.43 ±0.25*	8.84 ±0.6	157	1119
$n\text{-}C_4H_9SCN$	CCl_4	IR	3.69	0.8	4.06 ±0.25*	11.05 ±0.6	154	1119
$sec\text{-}C_4H_9SCN$	CCl_4	IR	3.62	0.8	2.85 ±0.25*	7.03 ±0.6	158	1119
CH_3NCS	CCl_4	IR	0.72	-0.19	1.7 ±0.3*	6.4 ±2.5	107	2337a
C_2H_5NCS	CCl_4	IR	0.93	-0.04	1.9 ±0.2*	6.4 ±1.9	113	2337a
$i\text{-}C_3H_7NCS$	CCl_4	IR	0.75	-0.17	2.8 ±0.3*	10.0 ±2.6	120	2337a
$n\text{-}C_4H_9NCS$	CCl_4	IR	0.46	-0.45	1.9 ±0.5*	7.8 ±4.2	116	2337a
$sec\text{-}C_4H_9NCS$	CCl_4	IR	0.65	-0.26	2.2 ±0.3*	8.2 ±3.2	126	2337a
$(CH_3)_2CHCH_2NCS$	CCl_4	IR	0.85	-0.10	2.7 ±0.3*	9.5 ±2.7	117	2337a
$(CH_3)_3CNCS$	CCl_4	IR	0.76	-0.16	2.3 ±0.3	8.2 ±3.1	130	2337a

APPENDIX (continued)

Acid	Base	Phase or Solvent	Method	$K_{25°C}$, ℓ/mole	$-\Delta G_{25°C}$, kcal/mole	$-\Delta H$, kcal/mole	$-\Delta S_{25°C}$, cal deg^{-1} mole^{-1}	$\Delta\nu_{AH}$ cm^{-1}	Ref.
C_6H_5OH	MISCELLANEOUS DONORS								
	$(C_6H_5)_3P$	CCl_4	NIR	9.8	1.4			430	235
		CCl_4	NIR	6	1.1	1.3	0.8	360	2280
	$(C_6H_5)_3As$	CCl_4	NIR	1.5	0.24			460	235
	$(CH_3)_3SiOC(CH_3)_3$	CCl_4	NIR	1.3	.15	6.83 ±0.24	22.6		1431a
	$(n\text{-}C_4H_9)_2Se$	CCl_4	NIR	1.1	0.04	3.69 ±0.33*	12.2	240	2578
	$(CH_3)_3SiOC_2H_5$	CCl_4	NIR	4.4	0.88	5.76 ±0.04	16.4		1431a
	$(CH_3)_3SiSC(CH_3)_3$	CCl_4	NIR	4.0	0.82	1.93 ±0.22	3.7		1431a
	$sym\text{-}(CH_2)_4$ disiloxane	CCl_4	NIR	3.7	0.78	5.47 ±0.07	15.7		1431a
	$[(CH_3)_3SiO]_4$	CCl_4	NIR			3.2*		144	2574
	$(CH_3)_3SiOSi(CH_3)_3$	CCl_4	NIR			2.9*		169	2574
		CCl_4	IR	0.4	-0.54			168	179
	$[(CH_3)_2N]_2CS$	CCl_4	NIR	14	1.6	3.1**	5.0	310	2280
	ethylenetrithiocarbonate	CCl_4	UV	0.59	-0.3	2.7 ±0.6	10.1		1723
	$(CH_3)_2NC(O)N(CH_3)_2$	CCl_4	IR	164(20°)	2.9	6.5*	12.1	340	915
	$(CH_3)_3SiOH$	CCl_4	IR	109(mf^{-1})					2117
	$(CH_3)_3COH$	CCl_4	IR	460(mf^{-1})					2117
	diazabicyclooctane	CCl_4	IR	300	3.35	6.6	10.9		2686
	$(C_6H_5)_3P\text{=}CHC(O)CH_3$	CS_2	IR	1180(20°)	4.1	7.9	12.8	440	703a
	$(C_6H_5)_3P\text{=}CHC(O)OCH_3$	CS_2	IR	457(20°)	3.5	6.9	11.4	370	703a
	$(C_6H_5)_3P\text{=}CHC(O)OC_2H_5$	CS_2	NIR	462(20°)	3.5	6.9	11.4	400	903
	$(C_6H_5)_3P\text{=}CHC(O)OC_6H_5$	CS_2	IR	201(20°)	3.0	6.6	12.1	305	703a
	$(C_6H_5)_3P\text{=}CHC(O)C_6H_5$	CS_2	IR	1408(20°)	4.2	8.1	13.1		703a
	$(C_6H_5)_3P\text{=}CHP(O)(OC_6H_5)_2$	CS_2	IR	1632(20°)	4.2	8.4	14.1	405	703a
	$(C_6H_5)_3P\text{=}CHC(O)N(C_6H_5)_2$	CS_2	IR	743(20°)	3.8	7.8	13.4	400	703a
	N-methyl-morpholine	CCl_4	IR	43	2.2	5.5	11.1		2686

Compound	Solvent	Method						
1:1								
Cl⁻	CH₂Cl₂	IR	400(40°)					2423
	CH₂Cl₂	IR	100(40°)					2423
Br⁻	CCl₄	IR	1300(40°)					2423
	CH₃NO₂	IR	30(40°)					2423
I⁻	CH₂Cl₂	IR	30(40°)					2423
picrate⁻	CH₂Cl₂	IR	20(40°)					2423
2 Phenol:1X⁻								
Cl⁻	CH₂Cl₂	IR	25(40°)					2423
Br⁻	CH₂Cl₂	IR	20(40°)					2423
I⁻	CCl₄	IR	150(40°)					2423
picrate⁻	CH₂Cl₂	IR	10(40°)					2423
	CH₂Cl₂	IR	5(40°)					2423
3 Phenol:1X⁻								
Cl⁻	CH₂Cl₂	IR	25(40°)					2423
Br⁻	CH₂Cl₂	IR	7(40°)					2423
	CCl₄	IR	30(40°)					2423
C₆H₅OD								
(C₂H₅)₃N	CCl₄	IR	32	2.1	2.7	2.0	180	2280
C₅H₅N	CCl₄	IR	33	2.1	1.1	-3.3	310	2280
(CH₃)₂CO	CCl₄	IR	27	2.0	0.8	-3.7	150	2280
CH₃CN	CCl₄	IR	7.2	1.2	0.6	-2.0	100	2280
(C₆H₅)₃P	CCl₄	IR	5.6	1.0	1.1	0.3	230	2280
(CH₃)₂NC(S)N(CH₃)₂	CCl₄	IR	16	1.6	1.1	-2.0	220	2280
(n-C₇H₁₅)₄N⁺I⁻	CCl₄	IR	89	2.7	1.5	-4.0	225	2280
n-C₇H₁₅F	CCl₄	IR			2.54		28	1190
AMIDES								
p-F-C₆H₄OH								
HC(O)N(CH₃)₂	CCl₄	IR,CAL	116 ±3	2.81 ±0.01	6.6 ±0.1	12.7 ±0.3	308	70
	CCl₄	NMR	115 ±2	2.81				955
	base	CAL			6.97 ±0.11			70
	C₆H₁₂	NMR	200	3.14				1201

APPENDIX (continued)

Acid	Base	Phase or Solvent	Method	$K_{25°C}$, ℓ/mole	$-\Delta G_{25°C}$, kcal/mole	$-\Delta H$, kcal/mole	$-\Delta S_{25°C}$, cal deg⁻¹ mole⁻¹	$\Delta\nu_{AH}$ cm⁻¹	Ref.
p-F-C₆H₄OH	HC(O)N(CH₃)₂	C₆H₅Cl	NMR	55	2.37				1201
		o-C₆H₄Cl₂	NMR	50	2.32				1201
		ClCH₂CH₂Cl	NMR	18.6	1.73				1201
		CH₂Cl₂	NMR	15.2	1.61				1201
	CH₃C(O)N(CH₃)₂	CCl₄	IR	260 ±12	3.29 ±0.03			356	70
		CCl₄	NMR	242 ±6	3.25				955
		base				7.44 ±0.13	13.9 ±0.3		70
		CCl₄	IR		3.3	6.8	11.7	354	2361
	CClCH₂C(O)N(CH₃)₂	CCl₄	NMR	48 ±5	2.3				955
		CCl₄	CAL			6.9 ±0.3	15.4		1674
	HC(O)NH(CH₃)	CCl₄	IR,CAL	90.0 ±5.6	2.67 ±0.03	5.5 ±0.1	9.5	271	70
		base	CAL			6.44 ±0.08	9.5 ±0.3		70
	C₆H₅C(O)N(CH₃)₂	CCl₄	NMR	167 ±16	3.0				955
		CCl₄	CAL			6.9 ±0.2	13.1		1674
	p-NO₂-C₆H₄C(O)N(CH₃)₂	CCl₄	IR,NMR	48 ±6	2.3				955
	CF₃C(O)N(CH₃)₂	CCl₄	IR,NMR	7.7	1.2				1201
		CCl₄	CAL			3.2 ±0.2	6.7		1674
	CF₃C(O)N(C₂H₅)₂	CCl₄	IR,NMR	10.1	1.4				1201
	CH₃C(O)N(C₂H₅)₂	CCl₄	IR,NMR	295	3.4				1201
	t-C₄H₉C(O)(CH₃)₂	CCl₄	IR,NMR	141	2.9				1201
	AMINES								
	(C₂H₅)₃N	CCl₄	IR	85.2 ±1.9	2.63 ±0.02				70
		CCl₄	NMR	85 ±2	2.63				955
		base				8.92 ±0.09	21.0 ±0.3		70
		C₆H₁₂	NMR	98.0	2.7				1201
		C₆H₅Cl	NMR	69.0	2.5				1201
		o-C₆H₄Cl₂	NMR	85.0	2.6				1201

Compound	Solvent	Method						Ref
	ClCH$_2$CH$_2$Cl	NMR	50	2.3				1201
	CH$_2$Cl$_2$	NMR	47	2.3				1201
n-C$_3$H$_7$N(CH$_3$)$_2$	CCl$_4$	NMR	95 ±1	2.7				955
(n-C$_4$H$_9$)$_3$N	CCl$_4$	NMR	37 ±3	2.1				955
C$_6$H$_5$CH$_2$NH$_2$	CCl$_4$	NMR	56 ±2	2.4				955
aniline	CCl$_4$	IR	4.7 (27°)	2.1				625
C$_6$H$_5$CH$_2$N(CH$_3$)$_2$	CCl$_4$	IR,NMR	36	0.5				955
(C$_6$H$_5$CH$_2$)$_3$N	CCl$_4$	IR,NMR	2.3	1.8				1201
(CH$_2$=CHCH$_2$)$_3$N	CCl$_4$	NMR	20	0.95				955
(HC≡CCH$_2$)$_3$N	CCl$_4$	IR,NMR	5	2.0				1201
CH$_3$C≡CNH$_2$	CCl$_4$	NMR	30 ±1	1.8				955
pyrimidine	CCl$_4$	NMR	22.5 ±0.5	3.6				1201
quinuclidine	o-C$_6$H$_4$Cl$_2$	NMR	470	3.6				1201
	CCl$_4$	IR,NMR	427					1201
	CCl$_4$	CAL			9.54 ±0.18	19.8		1674
CF$_3$CH$_2$NH$_2$	CCl$_4$	NMR	3.9 ±0.1	0.81				955
CF$_3$CH$_2$N(C$_2$H$_5$)$_2$	CCl$_4$	IR,NMR	1.7	0.31				1201

Pyridines

Compound	Solvent	Method						Ref
C$_5$H$_5$N	CCl$_4$	IR,CAL	76.2 ±1.1	2.56 ±0.01	7.1 ±0.1	15.2 ±0.3	485	70
	CCl$_4$	CAL	76	2.57	6.9		489	1779
	CCl$_4$	NMR						955
	base	CAL	69 (27°)		7.4 ±0.09		489	70
	CCl$_4$	IR	107					625
	C$_6$H$_{12}$	NMR	40	2.77				1201
	C$_6$H$_5$Cl	NMR	43	2.19				1201
	o-C$_6$H$_4$Cl$_2$	NMR	19.5	2.23				1201
	ClCH$_2$CH$_2$Cl	NMR	18.0	1.76				1201
	CH$_2$Cl$_2$	NMR		1.72				1201
4-CH$_3$-	CCl$_4$	IR,CAL	109 ±5	2.78 ±0.03	7.3 ±0.1	15.2 ±0.3	495	70
	CCl$_4$	NMR	107 ±2	2.77				955
	base	CAL			7.59 ±0.08			70
4-CH$_3$O-	CCl$_4$	NMR	139 ±2	2.92				955
2-n-C$_4$H$_9$	CCl$_4$	NMR	76 ±2	2.57				955
2-Cl-	base	CAL			5.93 ±0.07			71

APPENDIX (continued)

Acid	Base	Phase or Solvent	Method	$K_{25°C}$, ℓ/mole	$-\Delta G_{25°C}$, kcal/mole	$-\Delta H$, kcal/mole	$-\Delta S_{25°C}$, cal deg^{-1} mole^{-1}	$\Delta\nu_{AH}$ cm^{-1}	Ref.
p-F-C$_6$H$_4$OH	2-Br-	base	CAL			5.83 ±0.09		340	71
		CCl$_4$	NMR	8.8 ±0.5	1.29				955
	3-Br-	CCl$_4$	IR,CAL	30.2 ±0.09	2.02 ±0.03	6.2 ±0.2	14.8 ±0.7	421	70
		CCl$_4$	IR,NMR	20.0	1.78	5.8	13.5		955,
									1201
		C$_6$H$_{12}$	NMR	12.6	1.50				1201
		C$_6$H$_5$Cl	NMR	11.5	1.45				1201
		o-C$_6$H$_4$Cl$_2$	NMR	8.9	1.30				1201
		ClCH$_2$CH$_2$Cl	NMR	5.3	0.99				1201
	4-(CH$_3$)$_2$N-	CCl$_4$	NMR,IR	650 ±90	3.79 ±0.01	7.8 ±0.1	13.4 ±0.3		70
		CCl$_4$	CAL			8.4 ±0.2			1674
		C$_6$H$_{12}$	NMR	1400	4.29				1201
		C$_6$H$_5$Cl	NMR	240	3.24				1201
		o-C$_6$H$_4$Cl$_2$	NMR	410	3.57				1201
		ClCH$_2$CH$_2$Cl	NMR	138	2.92				1201
		CH$_2$Cl$_2$	NMR	126	2.87				1201
	3,5-Cl-	CCl$_4$	CAL			5.4 ±0.3			71
		CCl$_4$	IR,NMR	6.3	1.09				1201
		CCl$_4$	NMR	5.6 ±0.2	1.02				955
	2,6-CH$_3$-	base	CAL			8.44 ±0.11			71
		CCl$_4$	CAL			7.8 ±0.3			71
	2,4,6-CH$_3$-	base	CAL			8.36 ±0.12			71
		CCl$_4$	CAL			7.9 ±0.1			71
	N-methylpyridone	CCl$_4$	IR,NMR	240	3.25				1201
	quinoline	CCl$_4$	IR,CAL	72.3 ±1	2.54 ±0.01	7.35 ±0.1	16.1 ±0.3	498	70
		CCl$_4$	NMR	71 ±3	2.53				955
		base	CAL			7.47 ±0.09			70

Compound	Solvent	Method						Ref.
N-methylimidazole	CCl_4	IR,NMR	316					1201
$C_6H_5N(CH_3)_2$	CCl_4	CAL	3.5 ±0.4	0.74 ±0.07	4.0 ±0.4	10.9 ±1.3	382	70
	CCl_4	NMR	2.7 ±0.2	0.59				955
	base	CAL			4.02 ±0.08			70
cyclopropylamine	CCl_4	IR,CAL	44 ±2	2.24 ±0.01	7.5 ±0.3	17.6 ±1.0	304	70
	CCl_4	NMR	44 ±2	2.24				955
	C_6H_{12}	NMR	15.0	1.60				1201
$C_6H_{11}N(CH_3)_2$	CCl_4	NMR	118 ±2	2.8				955
AMINE OXIDES								
C_5H_5NO	CCl_4	CAL			7.5 ±0.2			1674
$(CH_3)_3NO$	CCl_4	CAL			8.8 ±0.3			1674
ESTERS								
$CH_3C(O)OC_2H_5$	CCl_4	IR	12.3 ±0.4	1.49 ±0.02	4.7	10.8	199	70
	CCl_4	NMR	12.0 ±0.2	1.47		10.9 ±0.3		955
	base	CAL			4.74 ±0.12			70
ETHERS								
$C_6H_5OCH_3$	CCl_4	CAL	2.3 ±0.2	0.49 ±0.05	3.1 ±0.2	8.8 ±0.6	43	70
	base	CAL			3.13 ±0.08			70
$(C_6H_5CH_2)_2O$	CCl_4	CAL	6.3 ±0.6	1.09 ±0.09	4.5 ±0.3	11.4 ±1.0	169	70
	CCl_4	NMR	5.3 ±0.2	0.99		12.4	249	955
	base	CAL			4.7			70
$(C_2H_5)_2O$	CCl_4	CAL	10.3 ±1.0	1.38 ±0.05	4.59 ±0.09	14.2 ±0.3	285	955
	CCl_4	NMR	9.5 ±0.2	1.33	5.6 ±0.1			70
	base	CAL			5.57 ±0.12			955
$[(CH_3)_3C]_2O$	CCl_4	NMR	5.1 ±0.4	0.97				70
$(C_6H_5)_2O$	CCl_4	CAL	1.9 ±0.2	0.38 ±0.06	1.9 ±0.4	5.1 ±1.3	132	70
	base	CAL			1.89 ±0.10			70
p-dioxane	base	CAL			5.10 ±0.11			955, 1201
	CCl_4	IR,NMR	5.4	1.0	4.8	12.7	252	1201
C_6H_{12}	C_6H_{12}	NMR	6.7	1.13				1201

APPENDIX (continued)

Acid	Base	Phase or Solvent	Method	$K_{25°C}$, ℓ/mole	$-\Delta G_{25°C}$, kcal/mole	$-\Delta H$, kcal/mole	$-\Delta S_{25°C}$, cal deg^{-1} mole^{-1}	$\Delta\nu_{AH}$ cm^{-1}	Ref.
p-F-C$_6$H$_4$OH	p-dioxane	C$_6$H$_5$Cl	NMR	3.5	0.74				1201
		o-C$_6$H$_4$Cl$_2$	NMR	2.8	0.61				1201
		ClCH$_2$CH$_2$Cl	NMR	1.23	0.12				1201
		CH$_2$Cl$_2$	NMR	1.38	0.19				1201
	1,2-dimethoxyethane	base	CAL			5.75 ±0.10			1674
	(CH$_2$)$_4$O	CCl$_4$	IR,CAL	17.7 ±0.5	1.70 ±0.02	5.6 ±0.1	13.1 ±0.3	292	70
		CCl$_4$	NMR	18.4 ±0.5	1.73	5.7	13.3		955
		base	CAL			5.75 ±0.08			70
	(CNCH$_2$CH$_2$)$_2$O	CCl$_4$	NMR	11.2 ±0.5	1.4				955
	KETONES								
	(CH$_3$)$_2$C(O)	base	CAL			5.59 ±0.08			1674
	C$_2$H$_5$C(O)CH$_3$	CCl$_4$	IR	15.6 ±0.5	1.63 ±0.02	4.8	10.7	221	70
		CCl$_4$	NMR	15.1 ±0.3	1.61	5.20 ±0.13	12.0 ±0.3		955
	C$_6$H$_5$C(O)CH$_3$	base	CAL			5.49 ±0.09			1674
		CCl$_4$	NMR	13.6 ±0.3	1.5				955
	p-CH$_3$O-C$_6$H$_4$C(O)CH$_3$	CCl$_4$	IR,NMR	21	1.80				1674
	cyclopentanone	base	CAL						70
	cyclohexanone	CCl$_4$	IR,CAL	20.5 ±0.7	1.79 ±0.02	5.8 ±0.2	13.4 ±0.7	229	70
		CCl$_4$	NMR	21.4 ±0.5	1.82				955
		base	CAL			5.68 ±0.09			70
		o-C$_6$H$_4$Cl$_2$	NMR	10.5	1.39				1201
		ClCH$_2$CH$_2$Cl	NMR	3.3	0.71				1201
		CH$_2$Cl$_2$	NMR	3.2	0.69				1201
	anthrone	CCl$_4$	IR,CAL	17.6 ±0.2	1.70 ±0.01	5.6 ±0.4	13.1 ±1.3	218	70
	xanthone	CCl$_4$	IR,NMR	23	1.86				1201
	flavone	CCl$_4$	NMR	98 ±6	2.7				955

Compound	Solvent	Method						Ref.
2,6-dimethyl-γ-pyrone	CCl_4	NMR	318 ±18	3.4	6.86 ±0.16			955
	CCl_4	CAL						1674
NITRILES								
CH_3CN	CCl_4	IR,NMR	8.0	1.2				1201
C_6H_5CN	CCl_4	NMR	6.1 ±0.6	1.1				955
	C_6H_{12}	NMR	10.0	1.4				1201
	C_6H_5Cl	NMR	3.0	0.65				1201
	$o\text{-}C_6H_4Cl_2$	NMR	2.5	0.54				1201
	$ClCH_2CH_2Cl$	NMR	0.71	-0.2				1201
$p\text{-}CH_3O\text{-}C_6H_4CN$	CCl_4	IR,NMR	9.4	1.3				1201
$p\text{-}(CH_3)_2N\text{-}C_6H_4CN$	CCl_4	IR,NMR	17.0	1.7				1201
$p\text{-}Br\text{-}C_6H_4CN$	CCl_4	IR,NMR	3.7	0.78				1201
$m\text{-}Br\text{-}C_6H_4CN$	CCl_4	IR,NMR	3.4	0.73				1201
$(CH_3)_2NCH_2CH_2CN$	CCl_4	IR,CAL	17.1 ±0.9	1.68 ±0.03	5.8 ±0.1	13.8 ±0.3		70
	base	CAL			6.8 ±0.1		164	1674
ORGANIC HALIDES								
$C_6H_{11}Cl$	base	CAL			2.12 ±0.11		71	70
$C_6H_{11}Br$	base	CAL			1.96 ±0.10		90	70
$C_6H_{11}I$	base	CAL			1.46 ±0.11		95	70
$n\text{-}C_4H_9Cl$	base	CAL			1.93 ±0.08		62	70
$n\text{-}C_4H_9Br$	base	CAL			1.82 ±0.07		71	70
$n\text{-}C_4H_9I$	base	CAL			1.55 ±0.11		78	70
PHOSPHORYL DONORS								
$POCl_3$	base	CAL			3.71 ±0.08			71
$C_6H_5P(O)Cl_2$	base	CAL			4.59 ±0.08			71
$(C_2H_5O)_2P(O)Cl$	base	CAL			5.55 ±0.07			71
$(CH_3O)_3PO$	base	CAL			6.44 ±0.09			71
	CCl_4	IR,NMR	282 ±8	3.3				1201
$(C_2H_5O)_3PO$	base	CAL			6.59 ±0.09			71
$(C_6H_5O)_3PO$	CCl_4	CAL			6.76 ±0.19			1674
$(C_2H_5O)_2(C_2H_5)PO$	base	CAL			7.46 ±0.12			71

APPENDIX (continued)

Acid	Base	Phase or Solvent	Method	$K_{25°C}$, ℓ/mole	$-\Delta G_{25°C}$, kcal/mole	$-\Delta H$, kcal/mole	$-\Delta S_{25°C}$, cal deg^{-1} mole^{-1}	$\Delta\nu_{AH}$ cm^{-1}	Ref.
p-F-C$_6$H$_4$OH	(CH$_3$)$_3$PO	CCl$_4$	CAL			7.7 ±0.2			71
	(C$_6$H$_5$)$_3$PO	CCl$_4$	IR,CAL	1456 ±80	4.32± 0.01	7.4 ±0.1	10.3 ±0.3	422	70
		CCl$_4$	IR,NMR	1445	4.31				1201
		o-C$_6$H$_4$Cl$_2$	NMR	450	3.62				1201
		ClCH$_2$CH$_2$Cl	NMR	110	2.79				1201
		CH$_2$Cl$_2$	NMR	80	2.60				1201
	[(CH$_3$)$_2$N]$_3$PO	CCl$_4$	IR,CAL	3600	4.85 ±0.02	8.0 ±0.1	10.6 ±1.3	479	70
		base	CAL			8.73 ±0.11			70
		C$_6$H$_{12}$	NMR	6300	5.18				1201
		C$_6$H$_5$Cl	NMR	1150	4.18				1201
		o-C$_6$H$_4$Cl$_2$	NMR	1150	4.18				1201
		ClCH$_2$CH$_2$Cl	NMR	355	3.48				1201
		CH$_2$Cl$_2$	NMR	234	3.23				1201
SULFOXIDES									
	(CH$_3$)$_2$SO	CCl$_4$	IR,CAL	346 ±8.0	3.46 ±0.02	6.6 ±0.1	10.9 ±0.3	367	70
			NMR	338 ±7	3.45				955
		base	CAL			7.21 ±0.08			70
		C$_6$H$_{12}$	NMR	360	3.49				1201
		C$_6$H$_5$Cl	NMR	160	3.01				1201
		o-C$_6$H$_4$Cl$_2$	NMR	150	2.97				1201
		ClCH$_2$CH$_2$Cl	NMR	44.7	2.25				1201
		CH$_2$Cl$_2$	NMR	27.6	1.97				1201
		CCl$_4$	NMR	141 ±4	2.9				955
	(CH$_3$)(C$_6$H$_5$)SO	CCl$_4$	IR,CAL	105 ±1	2.76 ±0.01	6.3 ±0.3	11.9 ±1.0	311	70
	(C$_6$H$_5$)$_2$SO	CCl$_4$	NMR	106 ±2	2.76				955
		CH$_2$Cl$_2$	NMR	13.2	1.5				1201
	(n-C$_4$H$_9$)$_2$SO	CCl$_4$	CAL			6.89 ±0.10			1674

Compound	Solvent	Method					Ref.
$(CH_2)_4SO$	base	CAL			7.64 ±0.10		1674

MISCELLANEOUS

Compound	Solvent	Method					Ref.
$(C_6H_5)_3AsO$	CCl₄	IR,NMR 4570	5.0	1.2 ±0.2			1201
$p\text{-}NO_2\text{-}C_6H_4S(O)CH_3$	CCl₄	NMR 38 ±5	2.2				955
$SOCl_2$	base	CAL		1.23 ±0.14			1674
benzene	base	CAL		0.50 ±0.11		49	70
o-chlorobenzene	base	CAL		1.27 ±0.11			1674
toluene	base	CAL		1.60 ±0.11		57	70
mesitylene	base	CAL		2.15		76	70
$n\text{-}C_7H_{15}F$	C₂Cl₄	IR				42	1190
$(CH_3)(C_6H_5)S$	base	CAL		1.67 ±0.16			1674
$(C_2H_5)_2S$	CCl₄	NMR 1.3 ±0.2	0.16			263	955
$(C_2H_5)_2S$	base	CAL		3.63 ±0.10		261	1674
$(n\text{-}C_4H_9)_2S$	base	CAL		3.44 ±0.08			1674
$(C_6H_5)_2S$	base	CAL		1.41 ±0.17			1674
$(ClCH_2)(CH_3)S$	base	CAL		1.89 ±0.15			1674
$(CH_2)_4S$	base	CAL		3.71 ±0.13		282	1674
$(CH_3O)_2SO$	base	CAL		2.82 ±0.11		77	1674
$(CH_2)_4SO_2$	base	CAL		4.25 ±0.10			1674
$(C_6H_5O)PCl_2$	base	CAL		4.29 ±0.08			1674
dimethyl sulfite	base	CAL		4.07 ±0.12			1674
diethyl sulfite	base	CAL		4.40 ±0.10			1674
propylene carbonate	base	CAL		4.53 ±0.12			1674
N-methyl-2-pyrrolidinone	base	CAL		7.38 ±0.11			1674
C_6H_5CHO	CCl₄	IR,NMR 6.0	1.06				955
$p\text{-}CH_3O\text{-}C_6H_4CHO$	CCl₄	NMR 12.6 ±0.2	1.5				955
$p\text{-}N(CH_3)_2\text{-}C_6H_4CHO$	CCl₄	IR,NMR 34	2.1				1201
acetyl ferrocene	CCl₄	NMR 44 ±1	2.2				955
$[(CH_3)_2N]_2C(S)$	CCl₄	IR,NMR 261 ±5	3.3				955
$[(CH_3)_2N]_2C(O)$	CCl₄	NMR 200	3.1				1201
$[(CH_3)_2N]_2C(O)$	base	CAL					1674
$C_2H_5OC(O)N(C_2H_5)_2$	CCl₄	IR,NMR 63	2.46				1201
$CH_3SC(O)N(C_2H_5)_2$	CCl₄	IR,NMR 36	2.12				1201
$NCC(O)N(C_2H_5)_2$	CCl₄	IR,NMR 11	1.42				1201
$[(CH_3)_2N]_2C(NH)$	CCl₄	IR,NMR 1380	4.3	7.75 ±0.17	15.6		1201

APPENDIX (continued)

Acid	Base	Phase or Solvent	Method	$K_{25°C}$, ℓ/mole	$-\Delta G_{25°C}$, kcal/mole	$-\Delta H$, kcal/mole	$-\Delta S_{25°C}$, cal deg^{-1} mole^{-1}	$\Delta \nu_{AH}$ cm^{-1}	Ref.
p-Cl-C_6H_4OH	$CH_3C(O)N(CH_3)_2$	CCl_4	CAL	207 ±25	3.2	7.3 ±0.1	13.8	376	650
	$(C_2H_5)_3N$	CCl_4	IR	130	3.4	6.8	11.4	373	2361
		CCl_4	IR	160	2.9	9.8	23.1	555	2278
		CCl_4	IR	226 ±26	3.0	7.4	14.3		2686
		C_6H_{12}	CAL	194	3.2	9.5 ±0.1	21.1	586	650
		C_6H_{12}	UV	1550	4.35				674a
		C_6H_{12}	UV		4.35				1464
	C_5H_5N	CCl_4	IR	113(27°)				510	625
		CCl_4	CAL			7.0			1779
		C_6H_{12}	CAL	211 ±59	3.2	8.1 ±0.1	16.4	491	650
		CCl_4	CAL	116 ±30	2.8	7.0 ±0.2	14.1	491	650
	$(n\text{-}C_3H_7)_3N$	C_6H_{12}	UV	264	3.3				1464
	$(CH_3)_2CHNH_2$	C_6H_{12}	UV	2420	4.6				1464
	$n\text{-}C_4H_9NH_2$	C_6H_{12}	UV	2036	4.4 ±0.1	7.2 ±1.2	9.60	375	1464
	aniline	CCl_4	IR	6.2(27°)					625
	$(n\text{-}C_4H_9)_3N$	C_6H_{12}	UV	245	3.3				1464
	$(n\text{-}C_4H_9)_2NH$	C_6H_{12}	UV	2760	4.7				1464
	$CH_3C(O)OC_2H_5$	CCl_4	CAL	15.8 ±1.4	1.6	5.0 ±0.2	11.4	179	650
	$(C_2H_5)_2O$	CCl_4	IR	12.1(20°)	1.4	5.0	12.1	300	862
		C_6H_{12}	UV	13.9(27°)					238
	$(C_2H_5)O(n\text{-}C_3H_7)$	CCl_4	IR	16.2(20°)	1.6	5.3	12.4	300	862
	$(n\text{-}C_4H_9)_2O$	CCl_4	IR	13.6(20°)	1.5	5.4	13.1	305	862
	$(CH_2)_4O$	CCl_4	IR	34.5(20°)	2.0	5.8	12.7	316	862
		C_6H_{12}	UV	26.0(27°)					238
		C_6H_{12}	UV	15.6(27°)					238
	p-dioxane	C_6H_{12}	UV	56	2.41 ±0.02	3.8 ±0.3	4.5	·	1464
	$(C_6H_5)_2CO$	CCl_4	IR	14.3	1.6	4.9	11.1	190	2278
	$(C_6H_5)_2SO$	C_2Cl_4	IR	190	3.0	7.7	15.9	325	847

Proton donor	Base	Solvent	Method	Value					Ref.
	$(C_2H_5)_2S$	CCl_4	IR	3.0(20°)	0.6	3.5	9.7	274	862
	ethylene trithiocarbonate	CCl_4	UV	1.69	0.3	8.0 ±0.7	25.8		1723
	$(C_2H_5)S(n\text{-}C_3H_7)$	CCl_4	IR	3.4(20°)	0.7	3.7	10.1	276	862
	$(CH_2)_4S$	C_6H_{12}	NMR	10.0 ±1.6	1.4	3.7	10.4	280	2290a
	$(n\text{-}C_4H_9)_2S$	CCl_4	IR	3.2(20°)	0.6	4.2 ±0.16	9.1	177	862
	$n\text{-}C_4H_9CN$	CCl_4	IR	10.87(30°)	1.5	3.9	9.1	282	8
	thiophene	C_2Cl_4	IR	8.3(20°)	1.2	2.3		45	862
	$n\text{-}C_7H_{15}F$	C_6H_{12}	IR						1190
	morpholine	C_6H_{12}	UV	524	3.7				1464
	N-methyl morpholine	CCl_4	IR	72	2.9	5.4	8.4		2686
	N-ethyl morpholine	C_6H_{12}	UV	177	3.1				1464
	diazabicyclooctane	CCl_4	IR	660	4.4	6.6	7.4		2686
	$CH_3CH_2C(CH_2O)_3P$	C_6H_{12}	NMR	17.5 ±1.9	1.7				2290a
	$(CH_3)_3COH$	CCl_4	IR	604(mf^{-1})					2117
	$(CH_3)_3SiOH$	CCl_4	IR	110(mf^{-1})					2117
$p\text{-}Br\text{-}C_6H_4OH$	$CH_3C(O)N(CH_3)_2$	CCl_4	IR	355.2(27°)	3.5	7.0	11.7	369	643, 2361
	$(C_2H_5)_3N$	CCl_4	IR	137(27°)		7.2		511	1530a
	C_5H_5N	CCl_4	IR	110(27°)					625
	aniline	CCl_4	CAL	5.6(27°)		2.3		371	1779
	$n\text{-}C_7H_{15}F$	C_2Cl_4	IR					45	625
	benzene	CCl_4	IR	0.50(40°)				48	1190
	p-xylene	CCl_4	IR	0.55(40°)				101	2398
	mesitylene	CCl_4	IR	0.61(40°)				82	2398
$p\text{-}I\text{-}C_6H_4OH$	$(C_2H_5)_3N$	CCl_4	IR	171(24°)				512	2398
	C_5H_5N	C_2Cl_4	IR	79(27°)				46	1530a
	$n\text{-}C_7H_{15}F$	CCl_4	IR						625
	aniline	CCl_4	IR	4.8(27°)		2.4		376	1190
$p\text{-}CH_3O\text{-}C_6H_4OH$	$CH_3C(O)N(CH_3)_2$	CCl_4	IR	43	2.7	5.7	10.1	328	2361
	$(C_2H_5)_3N$	CCl_4	IR	33(27°)	2.2	8.4	20.8	430	2278
	C_5H_5N	CCl_4	IR			6.1		464	625
	aniline	CCl_4	CAL	2.6(27°)	1.2			319	625
	$(C_6H_5)_2CO$	CCl_4	IR	7.3		4.4	10.7	170	2278

APPENDIX (continued)

Acid	Base	Phase or Solvent	Method	$K_{25°C}$, ℓ/mole	$-\Delta G_{25°C}$, kcal/mole	$-\Delta H$, kcal/mole	$-\Delta S_{25°C}$, cal deg^{-1} mole^{-1}	$\Delta\nu_{AH}$ cm^{-1}	Ref.
$p\text{-}CH_3O\text{-}C_6H_4OH$	$(C_6H_5)_2SO$	C_2Cl_4	IR	51	2.3	7.0	15.6	282	847
	$n\text{-}C_7H_{15}F$	C_2Cl_4	IR			2.0		37	1190
$p\text{-}CN\text{-}C_6H_4OH$	$CH_3C(O)N(CH_3)_2$	CCl_4	IR		4.4	8.3	13.1	432	2361
	$(C_2H_5)_3N$	C_6H_{12}	UV	926	4.0				674a
$p\text{-}CHO\text{-}C_6H_4OH$	$(C_2H_5)_3N$	CCl_4	IR	514(27°)					1530a
		C_6H_{12}	UV	525	3.7				674a
$p\text{-}NO_2\text{-}C_6H_4OH$	$CH_3C(O)N(CH_3)_2$	CCl_4	IR		4.7	8.2	11.7	431	2361
	$(C_2H_5)_3N$	CCl_4	UV	659(27°)					1530a
		C_6H_{12}	UV		4.14 ±0.03	10.3 ±0.4	20.6 ±1.5		1100
		C_6H_{12}	UV		4.2				674a
	$n\text{-}C_4H_9NH_2$	C_6H_{12}	UV	1260	4.44	9.2 ±0.6	16.0 ±2.3		1100
	aniline	CCl_4	IR	14.5(27°)				482	1457a
	4-F-aniline	CCl_4	IR	9.1(27°)				462	1457a
	3-F-aniline	CCl_4	IR	2.5(27°)				402	1457a
	4-Cl-aniline	CCl_4	IR	6.9(27°)				422	1457a
	3-Cl-aniline	CCl_4	IR	4.1(27°)				397	1457a
	4-Br-aniline	CCl_4	IR	6.6(27°)				412	1457a
	3-Br-aniline	CCl_4	IR	2.7(27°)				402	1457a
	4-I-aniline	CCl_4	IR	4.2(27°)				407	1457a
	3-I-aniline	CCl_4	IR	3.1(27°)				402	1457a
	p-anisidine	CCl_4	IR	42.1(27°)				542	1457a
	m-anisidine	CCl_4	IR	17.9(27°)				467	1457a
	p-toluidine	CCl_4	IR	24.4(27°)				507	1457a
	m-toluidine	CCl_4	IR	15.5(27°)				492	1457a
	$(C_2H_5)_2O$	C_6H_{12}	UV	24.6(20°)					2231
	p-dioxane	C_6H_{12}	UV		2.41 ±0.07	7.03 ±0.84	15.5 ±3.1		1100
	$(C_6H_5)_2SO$	C_2Cl_4	IR	1300	4.3	9.0	15.8	416	847
	$n\text{-}C_7H_{15}F$	C_2Cl_4	IR			3.3		59	2278

Phenol	Base	Solvent	Method	K				Δν	Ref.
	benzene	CCl$_4$	IR	0.52(40°)				58	2398
	p-xylene	CCl$_4$	IR	0.67(40°)				83	2398
	mesitylene	CCl$_4$	IR	1.02(40°)				97	2398
	p-NO$_2$-toluene	CCl$_4$	IR	6.1	1.1			136	2361
	nitrobenzene	CCl$_4$	IR	5.2	1.0			126	2361
	p-Cl-nitrobenzene	CCl$_4$	IR	4.1	0.84			115	2361
	m-Cl-nitrobenzene	CCl$_4$	IR	3.1	0.67			111	2361
p-cresol	CH$_3$C(O)N(CH$_3$)$_2$	CCl$_4$	IR		2.6	5.7	10.4	327	2361
	(C$_2$H$_5$)$_3$N	C$_7$H$_{16}$	UV	49.2(27°)					1530a
		C$_6$H$_{12}$	UV	55					278
		CCl$_4$	UV	60.7					674a
	C$_5$H$_5$N	CCl$_4$	IR	34(27°)	2.4			468	625
		CCl$_4$	CAL		2.4	6.0			1779
	(n-C$_4$H$_9$)$_2$NH	C$_7$H$_{16}$	UV	83	2.6				278
	n-C$_4$H$_9$NH$_2$	C$_7$H$_{16}$	UV	68	2.5				278
	aniline	CCl$_4$	IR	2.7(27°)				378	674a
		C$_6$H$_{12}$	UV	9.7(27°)					238
	(C$_2$H$_5$)$_2$O	C$_6$H$_{12}$	UV	18(27°)					238
	(CH$_2$)$_4$O	C$_6$H$_{12}$	UV	11(27°)					238
	p-dioxane	C$_6$H$_{12}$	UV	13(30°)					2585
	(CH$_3$)$_2$CO	C$_6$H$_{12}$	IR	18(30°)					2585
	cyclohexanone	C$_6$H$_{12}$	IR	4.72(30°)	1.0				8
	n-C$_4$H$_9$CN	CCl$_4$	IR	2.1 ±0.1	0.4	3.98 ±0.16	10.1	152	2533
	(CH$_2$)$_4$S	C$_6$H$_{12}$	CAL			4.6 ±0.1	14.1	264	1723
	ethylene trithiocarbonate	CCl$_4$	UV	1.00	0.0	7.0 ±0.5	23.4		2655
p-t-butylphenol	benzene	CCl$_4$	IR		-0.78(26°)			47	2655
	toluene	CCl$_4$	IR		-0.69(26°)			55	2655
	m-xylene	CCl$_4$	IR		-0.54(26°)			64	2655
	o-xylene	CCl$_4$	IR		-0.53(26°)			65	2655
	p-xylene	CCl$_4$	IR		-0.49(26°)			66	2655
	mesitylene	CCl$_4$	IR		-0.39(26°)			75	2655
	hexamethylbenzene	CCl$_4$	IR		-0.04(26°)			102	2655
	naphthalene	CCl$_4$	IR		-0.30(26°)			43	2655
	phenanthrene	CCl$_4$	IR		-0.27(26°)			46	2655
	CH$_3$C(O)N(CH$_3$)$_2$	CCl$_4$	CAL	79 ±3	2.6	6.4 ±0.1	12.8	320	650
		C$_6$H$_{12}$	CAL	274 ±46	3.3	8.1 ±0.2	16.1		650

347

APPENDIX (continued)

Acid	Base	Phase or Solvent	Method	$K_{25°C}$, ℓ/mole	$-\Delta G_{25°C}$, kcal/mole	$-\Delta H$, kcal/mole	$-\Delta S_{25°C}$, cal deg^{-1} mole^{-1}	$\Delta\nu_{AH}$ cm^{-1}	Ref.
$p\text{-}t\text{-}butylphenol$	C_5H_5N	C_6H_{12}	CAL	71.7 ±0.3	2.5	7.2 ±0.1	15.8	441	650
		CCl_4	CAL			6.2		469	1779
	$(CH_2)_4O$	C_6H_{12}	NMR	24.7 ±1.8	1.9				2290a
	$n\text{-}C_4H_9CN$	CCl_4	IR	4.44(30°)	0.95	3.37 ±0.11	8.1	152	8
	$(CH_2)_4S$	C_6H_{12}	NMR	3.36 ±0.22	0.72				2290a
	$CH_3CH_2C(CH_2O)_3P$	C_6H_{12}	NMR	9.89 ±0.37	1.36				2290a
	$(C_2H_5)_3N$	C_6H_{12}	CAL	106 ±5	2.8	8.3 ±0.1	18.5	531	650
	benzene	CCl_4	IR		-0.8(26°)			47	2655
	toluene	CCl_4	IR		-0.73(26°)			53	2655
	m-xylene	CCl_4	IR		-0.60(26°)			63	2655
	o-xylene	CCl_4	IR		-0.59(26°)			64	2655
	p-xylene	CCl_4	IR		-0.54(26°)			64	2655
	mesitylene	CCl_4	IR		-0.45(26°)			74	2655
	hexamethylbenzene	CCl_4	IR		-0.13(26°)			101	2655
	naphthalene	CCl_4	IR		-0.31(26°)			43	2655
	phenanthrene	CCl_4	IR		-0.26(26°)			45	2655
$m\text{-}F\text{-}C_6H_4OH$	$CH_3C(O)N(CH_3)_2$	C_6H_{12}	CAL			7.0		384	653
		CCl_4	IR		3.5	7.1	12.1	378	2361
	$(C_2H_5)_3N$	$o\text{-}C_6H_4Cl_2$	CAL	158 ±25	3.00 ±0.10	9.3 ±0.1	21.1		651
		C_6H_6	CAL	120 ±10	2.84 ±0.05	8.6 ±0.1	19.3		651
		$1,2\text{-}C_2H_4Cl_2$	CAL	82 ±8	2.61 ±0.06	8.8 ±0.1	20.8		1811a
	C_5H_5N	C_6H_{12}	CAL	262 ±11	3.30 ±0.03	8.4 ±0.1	17.1	520	1811a
		CCl_4	CAL	106 ±3	2.76 ±0.02	7.5 ±0.1	15.9		1811a
		$o\text{-}C_6H_4Cl_2$	CAL	116 ±10	2.82 ±0.06	6.9 ±0.1	13.7		1811a
		C_6H_6	CAL	53 ±2	2.35 ±0.02	6.3 ±0.1	13.3		1811a
		$1,2\text{-}C_2H_4Cl_2$	CAL	32 ±3	2.05 ±0.05	6.4 ±0.1	14.6		1811a
	$CH_3C(O)OC_2H_5$	C_6H_{12}	CAL	34 ±3	2.09 ±0.05	6.7 ±0.1	15.5	194	1811a
		CCl_4	CAL	19 ±1	1.74 ±0.03	5.2 ±0.1	11.6		651

348

Base	Solvent	Method						Ref
$(n\text{-}C_4H_9)_2O$	$o\text{-}C_6H_4Cl_2$	CAL	10.3 ±0.6	1.38 ±0.04	4.7 ±0.1	11.1		651
	C_6H_6	CAL	5.0 ±0.1	0.95 ±0.01	4.0 ±0.1	10.2		651
	$1,2\text{-}C_2H_4Cl_2$	CAL	2.5 ±0.1	0.54 ±0.02	3.7 ±0.2	10.6		1811a
	C_6H_{12}	CAL	17.0 ±0.2	1.68 ±0.02	6.5 ±0.1	16.2		1811a
	CCl_4	CAL	11.1 ±0.2	1.43 ±0.01	6.0 ±0.1	15.3		1811a
	$o\text{-}C_6H_4Cl_2$	CAL	7.4 ±0.1	1.19 ±0.01	5.7 ±0.2	15.1		1811a
	$1,2\text{-}C_2H_4Cl_2$	CAL	3.3 ±0.1	0.71 ±0.02	4.5 ±0.2	12.7		653
CH_3CN	CCl_4	CAL			4.9		175	651
$(CH_3)_2SO$	CCl_4	CAL	470 ±71	3.65 ±0.24	7.2 ±0.1	11.9	402	651
	$o\text{-}C_6H_4Cl_2$	CAL	321 ±59	3.42 ±0.12	6.7 ±0.1	11.0		1811a
	C_6H_6	CAL	254 ±15	3.28 ±0.03	6.1 ±0.1	9.5		2533
	$1,2\text{-}C_2H_4Cl_2$	CAL	73 ±8	2.54 ±0.07	5.4 ±0.1	9.6		1811a
$(C_2H_5)_2S$	C_6H_{12}	CAL	2.6 ±0.1	0.57	5.2 ±0.1	15.5	262	1811a
C_5H_5N	C_6H_{12}	CAL			8.4		520	1811a
CH_3CN	C_6H_{12}	CAL			4.9		175	1811a
$CH_3C(O)OC_2H_5$	C_6H_{12}	CAL			5.2		194	1811a
$CH_3C(O)N(CH_3)_2$	C_6H_{12}	CAL			7.0		384	1811a
$(CH_3)_2SO$	C_6H_{12}	CAL			7.3		402	1463
$C_9H_{18}NO$	C_6H_{12}	CAL			7.5			1811a
$(C_4H_9)_2O$	C_6H_{12}	CAL			6.0			2290
$C_2H_5C(CH_2O)_3PO$	C_6H_{12}	CAL			5.6			2290a
$C_9H_{18}NO$	CCl_4	CAL	8.8 ±1.2	0.94	5.5 ±0.3	15.3	385	1463
$CH_3C(O)N(CH_3)_2$	CCl_4	IR		3.6	7.5	10.4	525	2361
C_5H_5N	CCl_4	IR	99(27°)		6.7			625
	CCl_4	CAL			7.2			1779
aniline	CCl_4	IR	5.5(27°)				375	1457a
4-F-aniline	CCl_4	IR	5.6(27°)				395	625
3-F-aniline	CCl_4	IR	5.0(27°)				360	1457a
4-Cl-aniline	CCl_4	IR	2.2(27°)					1457a
3-Cl-aniline	CCl_4	IR	3.5(27°)					1457a
4-Br-aniline	CCl_4	IR	2.3(27°)				380	1457a
3-Br-aniline	CCl_4	IR	3.4(27°)				360	1457a
4-I-aniline	CCl_4	IR	2.2(27°)					1457a
3-I-aniline	CCl_4	IR	4.0(27°)					1457a
	CCl_4	IR	3.0(27°)					1457a

Acid (group label, left margin): $m\text{-}Cl\text{-}C_6H_4OH$

APPENDIX (continued)

Acid	Base	Phase or Solvent	Method	$K_{25°C}$, ℓ/mole	$-\Delta G_{25°C}$, kcal/mole	$-\Delta H$, kcal/mole	$-\Delta S_{25°C}$, cal deg^{-1} mole^{-1}	$\Delta\nu_{AH}$ cm^{-1}	Ref.
m-Cl-C$_6$H$_4$OH	p-anisidine	CCl$_4$	IR	15.8(27°)				450	1457a
	m-anisidine	CCl$_4$	IR	8.1(27°)					1457a
	p-toluidine	CCl$_4$	IR	9.9(27°)				425	1457a
	m-toluidine	CCl$_4$	IR	7.2(27°)					1457a
	(C$_6$H$_5$)$_2$SO	C$_2$Cl$_4$	IR	266	3.3	8.0	15.4	338	847
m-Br-C$_6$H$_4$OH	CH$_3$C(O)N(CH$_3$)$_2$	CCl$_4$	IR		3.7	7.3	12.1	384	2361
	C$_5$H$_5$N	CCl$_4$	IR	124(27°)				517	625
		CCl$_4$	CAL			7.0			1779
m-NO$_2$-C$_6$H$_4$OH	aniline	CCl$_4$	IR	6.3(27°)				382	625
	CH$_3$C(O)N(CH$_3$)$_2$	CCl$_4$	IR		4.4	7.7	11.1	418	2361
	(C$_2$H$_5$)$_3$N	C$_6$H$_{12}$	UV	1040	4.1				674a
		CCl$_4$	IR	618(27°)					1530a
	aniline	C$_6$H$_{12}$	UV	13.4(27°)				438	625
	(C$_2$H$_5$)$_2$O	C$_6$H$_{12}$	UV	113(12°)					2231
	(C$_6$H$_5$)$_2$SO	C$_2$Cl$_4$	IR	900	4.0	8.5	15.2	388	847
	n-C$_7$H$_{15}$F	C$_2$Cl$_4$	UV			2.2		40	2278
	benzene	CCl$_4$	IR	0.64(40°)				61	2398
	p-xylene	CCl$_4$	IR	1.08(40°)				86	2398
	mesitylene	CCl$_4$	IR	1.26(40°)				99	2398
m-cresol	CH$_3$C(O)N(CH$_3$)$_2$	CCl$_4$	IR		2.7	5.9	10.7	330	2361
	n-C$_4$H$_9$NH$_2$	C$_7$H$_{16}$	UV	86	2.6				278
	aniline	CCl$_4$	IR	3.3(27°)				326	625
	(n-C$_4$H$_9$)$_2$NH	C$_7$H$_{16}$	UV	97	2.7				278
	(C$_2$H$_5$)$_3$N	C$_7$H$_{16}$	UV	66	2.5				278
	C$_5$H$_5$N	CCl$_4$	IR	36(27°)				469	625
		CCl$_4$	CAL			6.2			1779
	(C$_2$H$_5$)$_2$O	C$_6$H$_{12}$	UV	10(27°)					238
	p-dioxane	C$_6$H$_{12}$	UV	12(27°)					238

Compound	Solvent	Method						
$(CH_2)_4O$	C_6H_{12}	UV	19 (27°)	0.96	3.05 ±0.12	7.0	152	238
$n\text{-}C_4H_9CN$	CCl_4	IR	4.56 (30°)	2.4	7.2	16.3	292	847
$(C_6H_5)_2SO$	C_2Cl_4	IR	54					8
ethylene trithiocarbonate	CCl_4	UV	0.69	-0.22				1723
benzene	CCl_4	IR	0.49 (40°)	-0.85 (26°)			48	2655
	CCl_4	IR		-0.65 (26°)			39	2398
toluene	CCl_4	IR		-0.51 (26°)			55	2655
m-xylene	CCl_4	IR		-0.50 (26°)			65	2655
o-xylene	CCl_4	IR		-0.47 (26°)			67	2655
p-xylene	CCl_4	IR	0.33 (40°)				67	2655
mesitylene	CCl_4	IR	0.49 (40°)	-0.35 (26°)			60	2398
hexamethylbenzene	CCl_4	IR		-0.01 (26°)			76	2655
naphthalene	CCl_4	IR		-0.28 (26°)			70	2398
phenanthrene	CCl_4	IR		-0.25 (26°)			44	2655
	CCl_4	IR					46	2655
$CH_3C(O)N(CH_3)_2$	C_6H_{12}	CAL		4.0	10.3 ±0.4	11.1	391	650
	CCl_4	IR			7.3 ±0.1		391	2361
	CCl_4	CAL	768 ±166	3.9	7.3 ±0.1	11.4	320	650
$m\text{-}CF_3\text{-}C_6H_4OH$	$o\text{-}C_6H_4Cl_2$	CAL	347 ±114	3.5	6.9 ±0.1	11.4		1879
C_5H_5N	C_6H_{12}	CAL	400 ±142	3.5	8.5 ±0.1	16.8		650
$CH_3C(O)OC_2H_5$	C_6H_{12}	CAL	43 ±7	2.2	6.8 ±0.1	15.4		650
cyclohexanone	C_6H_{12}	CAL	71 ±15	2.5	7.4 ±0.1	16.4	544	650
$(C_2H_5)_2S$	C_6H_{12}	CAL	4.9 (20°)	0.85	5.4 ±0.1	15.3		2533
$(CH_2)_4S$	C_6H_{12}	CAL	6.0 (21°)	0.98	5.7 ±0.1	15.8		2533
	CCl_4	CAL	2.9 (24°)	0.61	4.1 ±0.1	11.7		2533
$C_9H_{18}NO$	C_6H_{12}	CAL			7.5			1463
$m\text{-}CH_3O\text{-}C_6H_4OH$	CCl_4	IR	3.4 (29°)	2.9	6.3	11.4	341	2361
$m\text{-}CN\text{-}C_6H_4OH$	CCl_4	IR	3.2 (29°)	4.3	7.6	11.1	410	2361
$m\text{-}N(CH_3)_2\text{-}C_6H_4OH$	CCl_4	IR	2.6 (29°)	2.6	5.6	10.1	336	2361
o-cresol	CCl_4	IR					283	179
$[(CH_3)_2CH]_2O$	CCl_4	IR	3.4 (29°)				301	179
$[n\text{-}C_4H_9]_2O$	CCl_4	IR	3.2 (29°)				288	179
$[(CH_3)_3C]_2O$	CCl_4	IR	2.6 (29°)				309	179
$[(CH_3)_3Si]_2O$	CCl_4	IR	1.7 (29°)				168	179
	CCl_4	IR	0.4 (29°)					179

APPENDIX (continued)

Acid	Base	Phase or Solvent	Method	$K_{25°C}$, ℓ/mole	$-\Delta G_{25°C}$, kcal/mole	$-\Delta H$, kcal/mole	$-\Delta S_{25°C}$, cal deg^{-1} mole^{-1}	$\Delta\nu_{AH}$ cm^{-1}	Ref.
o-cresol	(CH$_2$)$_4$O	CCl$_4$	IR	5.5(29°)				294	179
	ethylene trithiocarbonate	CCl$_4$	UV	0.43	-0.5	3.0	11.7		1723
	benzene	CCl$_4$	IR		-0.81			46	2655
	toluene	CCl$_4$	IR		-0.71			54	2655
	m-xylene	CCl$_4$	IR		-0.58			63	2655
	o-xylene	CCl$_4$	IR		-0.56			65	2655
	p-xylene	CCl$_4$	IR		-0.53			66	2655
	mesitylene	CCl$_4$	IR		-0.41			74	2655
	hexamethylbenzene	CCl$_4$	IR		-0.10			102	2655
	naphthalene	CCl$_4$	IR		-0.31			43	2655
	phenanthrene	CCl$_4$	IR		-0.26			45	2655
2-isopropyl-phenol	(C$_2$H$_5$)$_2$O	CCl$_4$	IR	3.9(29°)				285	179
	[(CH$_3$)$_2$CH]$_2$O	CCl$_4$	IR	3.9(29°)				300	179
	[n-C$_4$H$_9$]$_2$O	CCl$_4$	IR	3.0(29°)				289	179
	[(CH$_3$)$_3$C]$_2$O	CCl$_4$	IR	2.0(29°)				300	179
	(CH$_2$)$_4$O	CCl$_4$	IR	8.0	1.2	5.0	12.8	275	2281
		CCl$_4$	IR	6.7(29°)				296	179
	p-dioxane	CCl$_4$	NMR	14(mf^{-1})					2320
		CCl$_4$	IR	6.3	1.1			225	2281
	[(CH$_3$)$_3$Si]$_2$O	CCl$_4$	IR	0.06(29°)				163	179
	C$_6$H$_5$C(O)CH$_3$	CCl$_4$	IR	6.6	1.1			176	2281
	(C$_6$H$_5$)$_2$CO	CCl$_4$	IR	4.4	0.9			160	2281
	azobenzene	CCl$_4$	IR	0.5	-0.4			250	2281
	CH$_3$NO$_2$	CCl$_4$	IR	1.2	0.1			65	2281
	ethylene trithiocarbonate	CCl$_4$	IR	0.5	-0.4			175·	2281

Compound	Base	Solvent	Method						
2-t-butylphenol	CH3C(O)N(CH3)2	CCl4	UV	80(20°)	2.5	7.0	15.1		2411
	CH3C(O)N(CH3)2	CCl4	NMR	31.6(19°)	2.0	3.5	5.0		2411
	CH3C(O)NH(CH3)	CCl4	UV	57(20°)	2.3	4.0	5.7		2411
	C5H5N	CCl4	CAL			6.2		461	1779
	(C2H5)2O	CCl4	IR	3.7(29°)				278	179
	[(CH3)2CH]2O	CCl4	IR	4.1(29°)				291	179
	(n-C4H9)2O	CCl4	IR	2.9(29°)				280	179
	[(CH3)3C]2O	CCl4	IR	2.3(29°)				280	179
	p-dioxane	CCl4	NMR	6.7(mF^{-1})					2320
	(CH2)4O	CCl4	IR	3.7	0.8		12.8	225	2281
	(CH2)4O	CCl4	IR	7.0	1.2	5.0		270	2281
	C6H5C(O)CH3	CCl4	IR	5.7	1.0			284	179
	(C6H5)2CO	CCl4	IR	3.0	0.65			175	2281
	(C6H5)2CO	CCl4	IR	2.9	0.63			160	2281
	ethylene trithiocarbonate	CCl4	IR	1.1	0.06			165	
	azobenzene	CCl4	IR	1.0	0.0			235	2281
	CH3NO2	CCl4	IR	1.1	0.06			65	2281
2,4-dinitrophenol	n-C4H9NH2	C6H5Cl	UV		4.38 ±0.01	11.80 ±0.27	24.9 ±0.9		156
	n-C4H9NH2	C6H5F	UV		4.19 ±0.01	13.23 ±0.1	30.3 ±0.4		156
	n-C4H9NH2	C6H5Br	UV		4.06 ±0.01	13.23 ±0.2	30.8 ±0.7		156
	n-C4H9NH2	C6H6	UV		4.04 ±0.01	11.86 ±0.1	26.2 ±0.5		156
	n-C4H9NH2	toluene	UV		3.81 ±0.01	11.81 ±0.2	26.8 ±0.5		156
	n-C4H9NH2	anisole	UV		4.97 ±0.01	12.11 ±0.1	24.0 ±0.4		156
	n-C4H9NH2	(n-C4H9)2O	UV		5.46 ±0.01	15.39 ±0.18	33.3 ±0.6		156
	(n-C4H9)2NH	C6H5Cl	UV		5.87 ±0.01	11.89 ±0.14	20.2 ±0.5		156
	(n-C4H9)2NH	C6H5F	UV		5.96 ±0.01	12.72 ±0.1	22.6 ±0.5		156
	(n-C4H9)2NH	C6H5Br	UV		6.08 ±0.01	12.91 ±0.2	22.9 ±0.7		156
	(n-C4H9)2NH	C6H6	UV		5.59 ±0.01	13.13 ±0.1	25.3 ±0.4		156
	(n-C4H9)2NH	toluene	UV		5.35 ±0.01	13.04 ±0.1	25.8 ±0.4		156
	(n-C4H9)2NH	anisole	UV		6.31 ±0.01	13.75 ±0.1	24.9 ±0.5		156
	(n-C4H9)2NH	(n-C4H9)2O	UV		5.74 ±0.01	15.22 ±0.09	31.8 ±0.3		156
	(n-C4H9)3N	C6H5Cl	UV		6.39 ±0.01	14.16 ±0.23	26.1 ±0.8		156
	(n-C4H9)3N	C6H5F	UV		6.33 ±0.01	13.81 ±0.1	25.1 ±0.4		156
	(n-C4H9)3N	C6H5Br	UV		6.33 ±0.01	13.49 ±0.2	24.0 ±0.7		156

APPENDIX (continued)

Acid	Base	Phase or Solvent	Method	$K_{25°C}$, ℓ/mole	$-\Delta G_{25°C}$, kcal/mole	$-\Delta H$, kcal/mole	$-\Delta S_{25°C}$, cal deg^{-1} mole^{-1}	$\Delta\nu_{AH}$ cm^{-1}	Ref.
2,4-dinitrophenol	$(n\text{-}C_4H_9)_3N$	C_6H_6	UV		5.28 ±0.1	11.97 ±0.06	22.4 ±0.2		156
		toluene	UV		5.00 ±0.01	11.71 ±0.1	22.5 ±0.3		156
		anisole	UV		6.03 ±0.01	12.57 ±0.1	21.9 ±0.2		156
		$(n\text{-}C_4H_9)_2O$	UV		4.57 ±0.01	11.24 ±0.1	22.4 ±0.4		156
	pyrrolidine	C_6H_5Cl	UV		6.66 ±0.01	12.17 ±0.22	18.47 ±0.74		155
	piperidine	C_6H_5Cl	UV		6.37	11.24 ±0.19	16.33 ±0.63		155
	hexamethyleneimine	C_6H_5Cl	UV		6.25	11.34 ±0.08	17.07 ±0.28		155
	cyclohexylamine	C_6H_5Cl	UV		4.21	12.49 ±0.13	27.77 ±0.44		155
	tricyclohexylamine	C_6H_5Cl	UV		4.34	9.57 ±0.15	17.55 ±0.49		155
	$N\text{-}n\text{-alkylpyrrolidines}$								
	CH_3-	C_6H_5Cl	UV		6.80	12.6 ±0.1	19.3 ±0.4		155
	C_2H_5-	C_6H_5Cl	UV		7.09	12.8 ±0.1	19.2 ±0.3		155
	$n\text{-}C_3H_7-$	C_6H_5Cl	UV		6.96	13.1 ±0.1	20.8 ±0.5		155
	$n\text{-}C_4H_9-$	C_6H_5Cl	UV		7.00	13.1 ±0.1	20.4 ±0.3		155
	$C_8H_{17}-$	C_6H_5Cl	UV		7.04	12.9 ±0.2	19.7 ±0.6		155
	$N\text{-}n\text{-alkylpiperidines}$								
	CH_3-	C_6H_5Cl	UV		6.56	12.7 ±0.2	20.5 ±0.6		155
	C_2H_5-	C_6H_5Cl	UV		6.85	13.25 ±0.05	21.5 ±0.2		155
	$n\text{-}C_3H_7-$	C_6H_5Cl	UV		6.62	13.1 ±0.1	21.9 ±0.4		155
	$n\text{-}C_4H_9-$	C_6H_5Cl	UV		6.58	12.8 ±0.1	20.8 ±0.4		155
	$C_6H_{13}-$	C_6H_5Cl	UV		6.70	13.0 ±0.1	21.3 ±0.4		155
	$C_8H_{17}-$	C_6H_5Cl	UV		6.73	13.0 ±0.1	21.0 ±0.5		155
	hexadecyl	C_6H_5Cl	UV		6.67	12.7 ±0.2	20.2 ±0.7		155

Phenol	Base	Solvent	Method						Ref.
3,4-dinitro-phenol	N-cyclopentylpiperidine	C_6H_5Cl	UV		6.82	12.9 ±0.1	20.4 ±0.4		155
	N-cyclohexylpiperidine	C_6H_5Cl	UV		7.22	13.6 ±0.1	21.5 ±0.3		155
	$(C_2H_5)_3N$	C_6H_6	UV		4.77 ±0.11	8.1 ±1.7	11.1 ±5.9		1100a
	$(n\text{-}C_3H_7)_3N$	C_6H_6	UV		3.89 ±0.04	9.2 ±1.3	17.1 ±4.4		1100a
	$(n\text{-}C_4H_9)_3N$	C_6H_6	UV		3.98 ±0.01	10.3 ±0.2	21.2 ±0.6		1100a
	$(n\text{-}C_4H_9)_2NH$	C_6H_{12}	UV		5.41 ±0.04	10.9 ±0.6	18.3 ±2.1		1100a
	$n\text{-}C_4H_9NH_2$	C_6H_6	UV		4.77 ±0.03	9.4 ±0.7	15.6 ±2.4		1100a
	aniline	C_6H_6	UV		4.57 ±0.05	7.5 ±1.3	9.8 ±4.5		1100a
	aniline	CCl_4	IR	89(27°)				552	1457a
	4-F-aniline	CCl_4	IR	48(27°)				537	1457a
	3-F-aniline	CCl_4	IR	16(27°)				482	1457a
	4-Br-aniline	CCl_4	IR	13(27°)				477	1457a
	3-Br-aniline	CCl_4	IR	22(27°)				467	1457a
	3-I-aniline	CCl_4	IR	14(27°)				487	1457a
	p-toluidine	CCl_4	IR	174(27°)				587	1457a
	m-toluidine	CCl_4	IR	98(27°)				577	1457a
3,5-dinitro-phenol	$(C_2H_5)_3N$	C_6H_{12}	UV	15,600	5.7	11.2	22.8		674a
picric acid	$(C_6H_5CH_2)_3N$	C_6H_6	UV	1685	4.4				585
3,4-dimethyl-phenol	$(C_2H_5)_3N$	CCl_4	IR	36(27°)	0.54				1530a
2,6-dimethyl-phenol	$(C_2H_5)_2O$	CCl_4	IR	0.67(29°)				220	179
	$[(CH_3)_2CH]_2O$	CCl_4	IR	0.62(29°)				234	179
	$(n\text{-}C_4H_9)_2O$	CCl_4	IR	0.63(29°)				228	179
	$[(CH_3)_3Cl]_2O$	CCl_4	IR	0.22(29°)				205	179
	$[(CH_3)_3Si]_2O$	CCl_4	IR	0.03(29°)					179
	$(CH_2)_4O$	CCl_4	IR	2.00(29°)				229	179
		CCl_4	IR	2.5					2281
2,6-diisopropyl-phenol	$(C_2H_5)_2O$	CCl_4	IR	0.51(29°)				191	179
	$[(CH_3)_2CH]_2O$	CCl_4	IR	0.48(29°)				215	179
	$(n\text{-}C_4H_9)_2O$	CCl_4	IR	0.23(29°)				201	179
	$[(CH_3)_3Cl_2O$	CCl_4	IR	0.19(29°)				197	179
	$(CH_3)_3SiOSi(CH_3)_3$	CCl_4	IR	0.06(29°)					179

APPENDIX (continued)

Acid	Base	Phase or Solvent	Method	$K_{25°C}$, ℓ/mole	$-\Delta G_{25°C}$, kcal/mole	$-\Delta H$, kcal/mole	$-\Delta S_{25°C}$, cal deg^{-1} mole^{-1}	$\Delta\nu_{AH}$ cm^{-1}	Ref.
2,6-diisopropyl-phenol	p-dioxane	CCl$_4$	NMR	7.1(mf^{-1})	0.16			140	2320
		CCl$_4$	IR	1.3	0.35	5.0	15.6	190	2281
	(CH$_2$)$_4$O	CCl$_4$	IR	1.8				201	2281
		CCl$_4$	IR	1.7(29°)					179
	C$_6$H$_5$C(O)CH$_3$	CCl$_4$	IR	1.1	.06			95	2281
	(C$_6$H$_5$)$_2$CO	CCl$_4$	IR	0.4	-0.54			95	2281
	azobenzene	CCl$_4$	IR	0.4	-0.54			140	2281
	CH$_3$NO$_2$	CCl$_4$	IR	0.5	-0.41			40	2281
	ethylene trithiocar-bonate	CCl$_4$	IR	0.1	-1.4			60	2281
2-methyl-6-t-butylphenol	CH$_3$C(O)N(CH$_3$)$_2$	CCl$_4$	NMR	6.65(20°)	1.1	2.2	3.7		2411
	CH$_3$C(O)NH(CH$_3$)	CCl$_4$	NMR	3.89(20°)	0.78	1.7	3.1		2411
	p-dioxane	CCl$_4$	IR	1.3	0.16			175	2281
		CCl$_4$	NMR	5.6(mf^{-1})		5.0	15.8		2320
	(CH$_2$)$_4$O	CCl$_4$	IR	1.7	0.31			205	2281
	C$_6$H$_5$C(O)CH$_3$	CCl$_4$	IR	2.8	0.61			110	2281
	(C$_6$H$_5$)$_2$CO	CCl$_4$	IR	2.5	0.54			105	2281
	azobenzene	CCl$_4$	IR	0.5	-0.41			170	2281
	CH$_3$NO$_2$	CCl$_4$	IR	1.0	1.4			55	2281
3,4-dichloro-phenol	(C$_2$H$_5$)$_3$N	C$_6$H$_{12}$	UV	556	3.7				674a
		CCl$_4$	IR	292(27°)					1530a
3,5-dichloro-phenol	(C$_2$H$_5$)$_3$N	CCl$_4$	IR	494(27°)	4.0				1530a
		C$_6$H$_{12}$	UV	908					674a
	aniline	CCl$_4$	IR	12(27°)				448	1457a

Compound	Base	Solvent	Method						Ref.
	4-F-aniline	CCl$_4$	IR	8.9 (27°)				433	1457a
	3-F-aniline	CCl$_4$	IR	3.3 (27°)				391	1457a
	4-Cl-aniline	CCl$_4$	IR	6.2 (27°)				411	1457a
	3-Cl-aniline	CCl$_4$	IR	44 (27°)				303	1457a
	4-Br-aniline	CCl$_4$	IR	5.8 (27°)				399	1457a
	3-Br-aniline	CCl$_4$	IR	3.1 (27°)				383	1457a
	4-I-aniline	CCl$_4$	IR	4.6 (27°)				414	1457a
	3-I-aniline	CCl$_4$	IR	3.2 (27°)				402	1457a
2,4,5-trichloro-phenol	(C$_2$H$_5$)$_3$N	CCl$_4$	IR	75 (27°)					1530a
3,5-trifluoro-methylphenol	(C$_2$H$_5$)$_3$N	C$_6$H$_{12}$	UV	2540	4.6				674a
3-trifluoromethyl-4-nitro-phenol									
2,6-dichloro-4-nitro-phenol	(C$_2$H$_5$)$_3$N	C$_6$H$_{12}$	UV	6170	5.2				674a
	C$_5$H$_5$N	CCl$_4$	IR	163 ±2				950	1031a
	3-chloropyridine	CCl$_4$	IR	33.3 ±0.3				680	1031a
	3,5-dichloropyridine	CCl$_4$	IR	8.1 ±0.4				600	1031a
	2,4,6-collidine	CCl$_4$	IR	546 ±8				1200	1031a
	4-picoline	CCl$_4$	IR	315 ±7				1050	1031a
2,4,6-trimethyl-phenol	(C$_2$H$_5$)$_3$N	CCl$_4$	IR	3.2 (27°)				395	1530a
	C$_5$H$_5$N	CCl$_4$	IR	4.61 ±0.12	0.91			330	1032
	3-Cl-pyridine	CCl$_4$	IR	2.01 ±0.10	0.41			271	1032
	3,5-dichloropyridine	CCl$_4$	IR	0.74 ±0.04	-0.18			720	1032
	piperidine	CCl$_4$	IR	9.27 ±0.23	1.32			419	1032
	4-picoline	CCl$_4$	IR	6.44 ±0.13	1.10			481	1032
	2,4,6-collidine	CCl$_4$	IR	6.03 ±0.04	1.06				1032
2-methyl-6-t-butylphenol	ethylene trithiocarbonate	CCl$_4$	IR	0.6	-0.30			131	2281
2,6-di-t-butylphenol	CH$_3$C(O)N(CH$_3$)$_2$	CCl$_4$	NMR	0.87 (20°)	-0.12	2.7	9.5		2411
	CH$_3$C(O)NH(CH$_3$)	CCl$_4$	NMR	0.16 (20°)	-1.1	1.4	8.4		2411
	p-dioxane	CCl$_4$	IR	0.9				130	2281
		CCl$_4$	NMR	0.7 (mf^{-1})	-0.06				2320

APPENDIX (continued)

Acid	Base	Phase or Solvent	Method	$K_{25°C}$, ℓ/mole	$-\Delta G_{25°C}$, kcal/mole	$-\Delta H$, kcal/mole	$-\Delta S_{25°C}$, cal deg^{-1} mole^{-1}	$\Delta\nu_{AH}$, cm^{-1}	Ref.
2,6-di-t-butylphenol	$(CH_2)_4O$	CCl_4	IR	0.5	-0.41	8.0	28.2	85	2281
	$C_6H_5C(O)CH_3$	CCl_4	IR	0.3	-0.71			70	2281
	$(C_6H_5)_2CO$	CCl_4	IR	0.3	-0.71			75	2281
	CH_3NO_2	CCl_4	IR	0.1	-1.4			35	2281
	CH_3CN	C_2Cl_4	IR	0.41	-0.53	0.8 ±0.5	4.4		1496
	C_6H_5CN	C_2Cl_4	IR	0.43	-0.50	0.9 ±0.5	4.6		1496
2,4,6-tri-t-butylphenol	$(CH_2)_4O$	CCl_4	IR	0.5	-0.41			80	2281
4-methyl-2,6-di-t-butylphenol	$(CH_2)_4O$	CCl_4	IR					105	2281
	C_5H_5N	CCl_4	CAL			3.3			1779
4-carbethoxy-2,6-di-t-butylphenol	$(CH_2)_4O$	CCl_4	IR	1.1	0.06			95	2281
4-cyano-2,6-di-t-butylphenol	$(CH_2)_4O$	CCl_4	IR	1.3	0.16			105	2281
4-nitro-2,6-di-t-butylphenol	$(CH_2)_4O$	CCl_4	IR	1.3	0.16			100	2281
4-formyl-2,6-di-t-butylphenol	$(CH_2)_4O$	CCl_4	IR	0.8	-0.13			85	2281
2,6-xylenol	$(CH_2)_4O$	CCl_4	IR						2281
	CH_3CN	CCl_4	IR		0.33	2.88 ±0.10	8.6	112	2654
	mesitylene	CCl_4	IR		-0.92	1.26 ±0.20	7.3	44	2654
	AMIDES								
C_6Cl_5OH	$HC(O)N(CH_3)_2$	CCl_4	NIR	54(20°)	2.3	4.6	7.7	364	914
	$HC(O)N(CH_3)_2$	CS_2	NIR	74(20°)	2.5	5.8	11.1	398	914
	$CH_3C(O)N(C_2H_5)_2$	CCl_4	NIR	126(20°)	2.8	5.6	9.4	435	914

Compound	Solvent	Method						
CH₃C(O)N(C₂H₅)₂	CS₂	NIR	173(20°)	3.0	5.7	9.1	440	914
CH₂ClC(O)N(C₂H₅)₂	CCl₄	NIR	32(20°)	2.0	5.0	10.1	330	914
CH₃C(O)N(C₆H₁₁)₂	CCl₄	NIR	174(20°)	3.0	5.0	6.7	464	914
CH₂ClC(O)N(C₆H₁₁)₂	CS₂	NIR	243(20°)	3.1	6.4	11.1	457	914
CH₃C(O)N(CH₂)₅	CCl₄	NIR	31(20°)	2.0	4.2	7.4	327	914
CH₂ClC(O)N(CH₂)₅	CCl₄	NIR	113(20°)	2.7	5.9	10.7	415	914
CH₃C(O)N(C₆H₅)₂	CCl₄	NIR	32(20°)	2.0	4.5	8.4	323	914
CH₂ClC(O)N(C₆H₅)₂	CCl₄	NIR	42(20°)	2.1	5.1	10.1	337	914
C₂H₅C(O)N(C₂H₅)₂	CCl₄	NIR	15(20°)	1.5	3.9	7.7	270	914
C₂H₅C(O)N(C₆H₁₁)₂	CCl₄	NIR	94(20°)	2.6	5.5	9.7	415	914
C₂H₅C(O)N(CH₂)₅	CCl₄	NIR	101(20°)	2.5	5.5	9.7	415	914
C₂H₅C(O)N(C₆H₅)₂	CCl₄	NIR	79(20°)	2.0	4.8	10.1	404	914
C₃H₇C(O)N(C₂H₅)₂	CCl₄	NIR	33(20°)	2.5	5.2	9.4	328	914
C₃H₇C(O)N(C₆H₁₁)₂	CCl₄	NIR	80(20°)			9.1	408	914
C₃H₇C(O)N(CH₂)₅	CCl₄	NIR	82(20°)	2.5	5.3	9.4	409	914
C₃H₇C(O)N(C₆H₅)₂	CCl₄	NIR	35(20°)	2.0	4.7	9.1	330	914
C₆H₅C(O)N(C₂H₅)₂	CCl₄	NIR	82(20°)	2.5	4.9	8.0	390	914
C₆H₅C(O)N(C₆H₁₁)₂	CCl₄	NIR	109(20°)	2.7	5.6	9.7	410	914
C₆H₅C(O)N(CH₂)₅	CCl₄	NIR	85(20°)	2.5	5.6	10.4	408	914
C₆H₅C(O)N(C₆H₅)₂	CCl₄	NIR	30(20°)	1.9	4.2	7.7	318	914
p-NO₂C₆H₄C(O)N(C₆H₁₁)₂	CCl₄	NIR	35(20°)	2.0	4.6	8.7	350	914
p-NO₂C₆H₄C(O)N(C₂H₅)₂	CCl₄	NIR	32(20°)	2.0	5.3	11.1	330	914
p-NO₂C₆H₄C(O)N(CH₂)₅	CCl₄	NIR	30(20°)	2.0	5.1	10.4	325	914
p-NO₂C₆H₄C(O)N(C₆H₅)₂	CCl₄	NIR	12(20°)	1.4	3.0	5.4	262	914
$\overline{CH_3NCH=CHCH=CHC=O}$	CCl₄	NIR	150(20°)	2.9	5.2	7.7	448	914

AMINES

Compound	Solvent	Method						
C₅H₅N	CCl₄	NIR	111(20°)	2.7	5.8	10.4	805	909
C₅H₅N	CS₂	NIR	108(20°)	2.7	7.3	15.4		909
collidine	CCl₄	NIR	225(20°)	3.1	7.6	15.1	1098	909
(C₆H₅CH₂)₃N	CCl₄	NIR	1.3(20°)	0.1	3.0	9.7		909

ETHERS

Compound	Solvent	Method						
(CH₂)₅O	CCl₄	NIR	7.8(20°)	1.1	5.3	14.1	385	908

Acid	Base	Phase or Solvent	Method	$K_{25°C}$, ℓ/mole	$-\Delta G_{25°C}$, kcal/mole	$-\Delta H$, kcal/mole	$-\Delta S_{25°C}$, cal deg^{-1} mole^{-1}	$\Delta\nu_{AH}$ cm^{-1}	Ref.
C_6Cl_5OH	**ESTERS**								
	$CH_3C(O)OC_2H_5$	CCl_4	NIR	5.4(20°)	0.9	4.5	12.1	185	908
	KETONES								
	$(CH_3)_2CO$	CCl_4	NIR	9.1(20°)	1.2	4.7	11.7	229	908
	cyclohexanone	CCl_4	NIR	10(20°)	1.3	4.6	11.1	245	908
	2,6-dimethyl-γ-pyrone	CCl_4	NIR	186(20°)	3.0	6.2	10.7	459	905
	ORGANOPHOSPHORUS COMPOUNDS								
	$HP(O)(OCH_3)_2$	CCl_4	NIR	61(20°)	2.4	4.7	7.7	400	910
	$HP(O)(OC_2H_5)_2$	CCl_4	NIR	86(20°)	2.6	4.7	7.0	425	910
	$HP(O)[OCH(CH_3)_2]_2$	CCl_4	NIR	107(20°)	2.7	5.0	7.7	450	910
	$(CH_3)_3PO$	CCl_4	NIR	1136(20°)	4.1	5.9	6.0	623	910
	$CH_3P(O)(OC_2H_5)_2$	CCl_4	NIR	214(20°)	3.1	5.3	7.4	500	910
	$CH_3PO(OCH_2CH_2CH_2Cl)_2$	CCl_4	NIR	145(20°)	2.9	4.8	6.4	473	910
	$ClCH_2P(O)(OC_2H_5)_2$	CCl_4	NIR	110(20°)	2.7	4.9	7.4	440	910
	$Cl_3CP(O)(OC_2H_5)_2$	CCl_4	NIR	30(20°)	1.9	4.0	7.0	330	910
	$(CH_3O)_3PO$	CCl_4	NIR	103(20°)	2.7	5.0	7.7	433	910
	$(C_2H_5)_3PO$	CCl_4	NIR	2096(20°)	4.4	6.6	7.4	685	910
	$C_2H_5P(O)(OCH_3)_2$	CCl_4	NIR	187(20°)	3.0	5.3	7.7	488	910
	$C_2H_5P(O)(OC_2H_5)_2$	CCl_4	NIR	235(20°)	3.1	5.3	7.4	511	910
	$(C_2H_5O)_3PO$	CCl_4	NIR	153(20°)	2.9	5.2	7.7	481	910
	$(C_2H_5)_2NP(O)(OC_2H_5)_2$	CCl_4	NIR	258(20°)	3.2	5.3	7.0	519	910
	$Cl_2CHP(O)(OC_2H_5)_2$	CCl_4	NIR	56(20°)	2.3	5.2	9.7	374	910
	$(C_6H_5)_3PO$	CS_2	NIR	674(20°)	3.8	5.7	6.4	578	910
	$(C_6H_5)_3PO$	CCl_4	NIR	1072(20°)	4.0	7.8	12.7	620	910
C_6F_5OH	$(C_6H_5)_3PO$	CCl_4	NIR	6000	5.0	5.5	1.1	620	917

α-naphthol

AMINES

Compound	Solvent	Method						
$(C_2H_5)_3N$	CCl_4	NIR	109 (20°)	2.7	6.6	13.1		909
	C_7H_{16}	UV	117	2.8				278
	CH_2Cl_2	UV	61 (20.7°)					1353
$(n\text{-}C_3H_7)_3N$	CCl_4	NIR	31 (20°)	1.8	6.1	14.4	380	909
$n\text{-}C_4H_9NH_2$	C_7H_{16}	UV	146	3.0				278
$(n\text{-}C_4H_9)_2NH$	C_7H_{16}	UV	162	3.0				278
$(n\text{-}C_4H_9)_3N$	CCl_4	NIR	44 (20°)	2.1	5.9	12.7	383	909
$(C_6H_5)_3N$	CCl_4	NIR	1.9 (20°)	0.4	1.3	3.0		909
C_5H_5N	CCl_4	NIR	82 (20°)	2.5	5.7	10.7	530	909
	CCl_4	CAL			7.0			1779
collidine	CCl_4	NIR	201 (20°)	3.0	6.7	12.4	535	909

AMINE OXIDES

Compound	Solvent	Method						
$(CH_3)_3NO$	CH_2Cl_2	UV	6310 (20.7°)					1353

ETHERS

Compound	Solvent	Method						
$(C_2H_5)_2O$	C_6H_{12}	UV	11 (27°)					238
p-dioxane	C_6H_{12}	UV	13 (27°)					238
	$i\text{-}C_8H_{18}$	UV	21 (20°)					102
$(CH_2)_4O$	C_6H_{12}	UV	22 (27°)	1.7	5.4 ±0.2	12.4		238

ORGANOPHOSPHORUS COMPOUNDS

Compound	Solvent	Method						
$HP(O)(OCH_3)_2$	CCl_4	NIR	168 (20°)	2.9	6.0	10.4	310	910
$HP(O)(OC_2H_5)_2$	CCl_4	NIR	188 (20°)	3.0	5.1	7.0	316	910
$HP(O)[OCH(CH_3)_2]_2$	CCl_4	NIR	236 (20°)	3.1	5.8	9.1	336	910
$CH_3P(O)(OC_2H_5)_2$	CCl_4	NIR	435 (20°)	3.5	5.6	7.0	376	910
$CH_3P(O)(OCH_2CH_2CH_2Cl)_2$	CCl_4	NIR	308 (20°)	3.3	6.2	9.7	356	910
$ClCH_2P(O)(OC_2H_5)_2$	CCl_4	NIR	217 (20°)	3.1	5.7	8.7	330	910
$Cl_2CHP(O)(OC_2H_5)_2$	CCl_4	NIR	118 (20°)	2.7	5.7	10.1	283	910
$Cl_3CP(O)(OC_2H_5)_2$	CCl_4	NIR	82 (20°)	2.5	5.1	8.7	273	910
$(CH_3)_3PO$	CCl_4	NIR	2575 (20°)	4.5	7.3	9.4	506	910
$(CH_3O)_3PO$	CCl_4	NIR	239 (20°)	3.2	5.6	8.0	323	910
$C_2H_5P(O)(OCH_3)_2$	CCl_4	NIR	482 (20°)	3.6	6.2	8.7	383	910

APPENDIX (continued)

Acid	Base	Phase or Solvent	Method	$K_{25°C}$, ℓ/mole	$-\Delta G_{25°C}$, kcal/mole	$-\Delta H$, kcal/mole	$-\Delta S_{25°C}$, cal deg^{-1} mole^{-1}	$\Delta\nu_{AH}$, cm^{-1}	Ref.
α-naphthol	$C_2H_5P(O)(OC_2H_5)_2$	CCl_4	NIR	476(20°)		6.1	8.4	396	910
	$(C_2H_5)_3PO$	CCl_4	NIR	3763(20°)		7.8	10.4	530	910
	$(C_2H_5O)_3PO$	CCl_4	NIR	324(20°)		6.4	10.4	363	910
	$(C_6H_5)_3PO$	CCl_4	NIR	1166(20°)		7.0	9.7	443	910
	$(C_2H_5)_2NP(O)(OC_2H_5)_2$	CCl_4	NIR	582(20°)		6.1	8.0	406	910
	π−BASES								
	benzene	CCl_4	IR		-0.69(26°)			52	2655
		CCl_4	IR	0.26(40°)				50	2398
	toluene	CCl_4	IR		-0.58(26°)			59	2655
		CCl_4	IR	0.38(40°)				61	2398
	m-xylene	CCl_4	IR		-0.45(26°)			69	2655
	o-xylene	CCl_4	IR		-0.44(26°)			69	2655
	p-xylene	CCl_4	IR		-0.40(26°)			70	2655
	mesitylene	CCl_4	IR		-0.26(26°)			61	2398
		CCl_4	IR	0.43(40°)				80	2655
		CCl_4	IR	0.52(40°)				78	2398
	hexamethylbenzene	CCl_4	IR		-0.11(26°)			109	2655
	durene	CCl_4	IR	0.46(40°)				87	2398
	naphthalene	CCl_4	IR		-0.26(26°)			47	2655
	phenanthrene	CCl_4	IR		-0.22(26°)			48	2655
β-naphthol	$(C_2H_5)_3N$	C_7H_{14}	UV	102	2.7				278
	$n\text{-}C_4H_9NH_2$	CH_2Cl_2	UV	54(20.7°)					1353
	$(n\text{-}C_4H_9)_2NH$	C_7H_{14}	UV	128	2.9				278
	C_5H_5N	C_7H_{14}	UV	138	2.9				278
	$2\text{-}CH_3\text{-}C_5H_4N$	C_6H_{12}	UV	66.5(24°)	2.5	5.80 ±0.17	11.15 ±0.55		236
	$3\text{-}CH_3\text{-}C_5H_4N$	C_6H_{12}	UV	86.9(24°)	2.6	6.90 ±0.15	14.30 ±0.50		236
		C_6H_{12}	UV	79.3(24°)	2.6	6.35 ±0.07	12.62 ±0.27		236

Compound	Reagent	Solvent	Method						Ref.
4-CH₃-C₅H₄N		C_6H_{12}	UV	91.2(24°)	2.7	7.14 ±0.08	15.03 ±0.29		236
$(CH_3)_3NO$		CH_2Cl_2	UV	6560(20.7°)		5.9 ±0.3	14.4		1353
$(C_2H_5)_2O$		C_6H_{12}	UV	11(27°)					238
p-dioxane		C_6H_{12}	UV	13(27°)					238
$(CH_2)_4O$		$i\text{-}C_8H_{18}$	UV	18(20°)	1.6				102
$(CH_2)_4O$		C_6H_{12}	UV	20(27°)	1.8				238
benzene		C_6H_{12}	UV	0.16(24°)	-1.1				236
benzene		CCl_4	IR		-0.71(26°)	1.57 ±0.10	8.96 ±0.33	50	2655
toluene		C_6H_{12}	UV	0.24(24°)	-0.8				236
toluene		CCl_4	IR		-0.62(26°)	1.96 ±0.19	9.47 ±0.64	58	2655
o-xylene		C_6H_{12}	UV	0.50(24°)	-0.4				236
o-xylene		CCl_4	IR		-0.46(26°)	2.61 ±0.06	10.15 ±0.22	68	2655
m-xylene		C_6H_{12}	UV	0.53(24°)	-0.4				236
m-xylene		CCl_4	IR		-0.47(26°)	2.65 ±0.12	10.12 ±0.40	68	2655
p-xylene		C_6H_{12}	UV	0.48(24°)	-0.4				236
p-xylene		CCl_4	IR		-0.43(26°)	2.52 ±0.10	9.97 ±0.30	70	2655
mesitylene		C_6H_{12}	UV	0.97(24°)	0.0				236
mesitylene		CCl_4	IR		-0.28(26°)	3.28 ±0.11	11.10 ±0.38	79	2655
hexamethylbenzene		CCl_4	IR		-0.10(26°)			105	2655
naphthalene		CCl_4	IR		-0.27(26°)			47	2655
phenanthrene		CCl_4	IR		-0.23(26°)			47	2655
catechol	$(CH_3)_2CHCN$	CCl_4	IR	9.1	1.3	4.6	11.1	194	2542
catechol	isopropylbenzene	CCl_4	IR	0.10	-1.4	1.6	10.1	72	2542
catechol	cyclohexene	CCl_4	IR	0.23	-0.87	3.8	15.7	129	2542
vanillin	$(CH_3)_2CHCN$	CCl_4	IR	0.87	-0.08	3.5	12.0	143	2542
vanillin	isopropylbenzene	CCl_4	IR	0.11	-1.3	1.4	9.1	33	2542
vanillin	cyclohexene	CCl_4	IR	0.08	-1.5	1.6	10.4	50	2542
CH_3COOH	$(CH_3)_2CO$ (2:1)		NMR	14(mf⁻¹)(30°)					1727
CH_3COOH	$(CH_3CO)_2O$ (2:1)		NMR	34(mf⁻¹)(30°)					1727
CH_3COOH	p-dioxane (2:1)		NMR	100(mf⁻¹)(30°)					1727
$ClCH_2COOH$	CH_3CN		IR	5.3	1.0	5.6	15.4		1899a
$ClCH_2COOH$	$(C_6H_5)_2SO$	CCl_4	IR	1548 ±34	4.35 ±0.01	6.47 ±0.85	6.9 ±2.9		976a
$ClCH_2COOH$	$(C_6H_5CH_2)_2SO$	CCl_4	IR	4105 ±76	4.94 ±0.01	7.72 ±0.75	9.4 ±2.5		976a
$ClCH_2COOH$	$(CH_3)_2SO$	CCl_4	IR	5631 ±88	5.12 ±0.01	8.65 ±0.39	11.8 ±1.3		976a
$ClCH_2COOH$	$(C_6H_5)_3PO$	CCl_4	IR	7161 ±101	4.21 ±0.01	9.23 ±0.48	16.6 ±1.5		976a
$ClCH_2COOH$	$(C_6H_5)_2SeO$	CCl_4	IR	25,690 ±890	6.03 ±0.02	11.77 ±0.73	19.3 ±2.4		976a

APPENDIX (continued)

Acid	Base	Phase or Solvent	Method	$K_{25°C}$, ℓ/mole	$-\Delta G_{25°C}$, kcal/mole	$-\Delta H$, kcal/mole	$-\Delta S_{25°C}$, cal deg^{-1} mole^{-1}	$\Delta\nu_{AH}$ cm^{-1}	Ref.
ClCH₂COOH	CH₃CN	CCl₄	IR	11.2	1.4	7.0	18.8		1899a
Cl₂CHCOOH	(C₆H₅)₂SO	CCl₄	IR	2375 ±108	4.61 ±0.03	6.91 ±0.03	7.6 ±1.6		976a
	(C₆H₅CH₂)₂SO	CCl₄	IR	12,480 ±160	5.60 ±0.01	9.10 ±0.36	11.8 ±1.2		976a
	(CH₃)₂SO	CCl₄	IR	19,640 ±1200	5.87 ±0.04	9.44 ±0.35	12.0 ±1.2		976a
	(C₆H₅)₃PO	CCl₄	IR	43,610 ±980	6.33 ±0.01	12.24 ±0.60	19.9 ±2.0		976a
	(C₆H₅)₂SeO	CCl₄	IR	148,700 ±6700	7.07 ±0.03	14.22 ±0.50	23.9 ±1.7		976a
	CH₃CN			20.9	1.8	7.3	18.5		1899a
Cl₃CCOOH	(C₆H₅)₂SO	CCl₄	IR	4772 ±232	5.02 ±0.03	7.45 ±0.31	8.1 ±1.0		976a
	(C₆H₅CH₂)₂SO	CCl₄	IR	35,010 ±670	6.21 ±0.01	9.72 ±0.51	11.7 ±1.7		976a
	(CH₃)₂SO	CCl₄	IR	45,300 ±1130	6.36 ±0.01	10.13 ±0.30	12.7 ±1.0		976a
	(C₆H₅)₃PO	CCl₄	IR	70,030 ±1540	6.62 ±0.01	13.98 ±1.22	24.5 ±4.1		976a
	(C₆H₅)₂SeO	CCl₄	IR	241,100 ±6800	7.35 ±0.02	15.95 ±1.31	28.7 ±4.4		976a
	CH₃CN	CCl₄	IR	27.4	2.0	8.1	20.4		1899a
F₃CCOOH	CH₃CN		IR	36.3	2.1	8.8	22.5		1899a
C₆H₅COOH	(C₂H₅)₃N	C₆H₆	UV	3760	4.9	11.0	20.7		585
	(C₆H₅NH)₂C=NH	C₆H₆	UV	2.25 x 10⁵	7.3	14.1	22.9		585
(CH₃)₃CNH₂	CH₃C(O)NH(CH₃)	CHCl₃	NMR	3.9(36°)	0.9	1.8	3.0		2410
	benzil	base	GLC			0.8			2521
	N-nitrosomethylaniline		GLC			0.6			2521
i-C₃H₇SH	HC(O)N(CH₃)₂	CCl₄	NMR	0.031(32°)	-2.0	0.9	10		1096a
	CH₃C(O)OC₂H₅	CCl₄	NMR	0.057(31°)	-1.7	1.1	9		1096a
	(CH₂)₄O	CCl₄	NMR	0.061(30°)	-1.6	1.0	9		1096a
	(CH₃)₂CO	CCl₄	NMR	0.047(32°)	-1.8	0.9	9		1096a
	CH₃CN	CCl₄	NMR	0.13(31°)	-1.2	0.6	7		1096a
	(C₂H₅O)₃PO	CCl₄	NMR	0.065(31°)	-1.6	1.0	9		1096a
	CH₂(CH₃O)₂PO	CCl₄	NMR	0.18(31°)	-1.0	1.0	7		1096a
	[(CH₃)₂N]₃PO	CCl₄	NMR	0.02(44°)	-2.2	1.1	11		1096a
	(CH₃O)₂SO	CCl₄	NMR	0.12(30°)	-1.2	0.9	7		1096a

Compound	Base	Method	Solvent						Ref.
$n\text{-}C_4H_9SH$	$(CH_3)_2S$	NMR	CCl_4	0.03 (32°)	-2.1	0.9	10		1096a
	$(CH_3)_4S$	NMR	CCl_4	0.031 (32°)	-2.0	0.8	9		1096a
	$(CH_3)_2S_2$	NMR	CCl_4	0.037 (32°)	-1.9	0.5	8		1096a
	$[(CH_3)_2N]_2CO$	NMR	CCl_4	0.059 (35°)	-1.6	1.1	9		1096a
	p-dioxane	NMR	CCl_4	0.047 (35°)	-1.8	0.9	9		1096a
	$(C_2H_5O)_3PO$	NMR	CCl_4	0.12 (34°)	-1.2	1.0	7		1096a
	$(CH_3)_2SO$	NMR	CCl_4	0.17 (34°)	-1.0	0.9	6		1096a
$t\text{-}C_4H_9SH$	$HC(O)N(CH_3)_2$	NMR	CCl_4	0.084 (36°)	-1.4	0.9	8		1096a
	$CH_3C(O)OC_2H_5$	NMR	CCl_4	0.073 (35°)	-1.5	0.7	8		1096a
	p-dioxane	NMR	CCl_4	0.055 (32°)	-1.7	0.6	8		1096a
	$(CH_3)_2CO$	NMR	CCl_4	0.064 (35°)	-1.6	0.9	8		1096a
	$(C_2H_5O)_3PO$	NMR	CCl_4	0.069 (33°)	-1.6	1.0	9		1096a
	$(CH_3)_2SO$	NMR	CCl_4	0.30 (32°)	-0.7	0.9	5		1096a
	$(CH_3)_4S$	NMR	CCl_4	0.14 (36°)	-1.1	0.7	6		1096a
	$(CH_3)_2S_2$	NMR	CCl_4	0.031 (36°)	-2.0	0.6	9		1096a
C_6H_5SH	$CH_3C(O)NH(CH_3)$	NMR	CCl_4	0.14 (22°)	-1.3	0.9	7.4		1624
	$HC(O)N(CH_3)_2$	NMR	CCl_4	0.25 (26°)	-0.85	1.8	8.9		1623
	C_5H_5N	NMR	CCl_4	0.22 (26°)	-0.90	2.4	11.1		1623
	N-methylpyrazole	NMR	CCl_4	0.14 (26°)	-1.2	2.1	11.1		1623
	$(n\text{-}C_4H_9O)_3PO$	NMR	CCl_4	0.43 (26°)	-0.50	2.0	8.4		1623
	benzene	NMR	CCl_4	0.039 (26°)	-1.9	0.5	8.1		1623
p-thiocresol	ethylene trithiocarbonate	UV	CCl_4	0.64 (35°)					1723
$p\text{-}Br\text{-}C_6H_4SH$	ethylene trithiocarbonate	UV	CCl_4	0.75 (35°)					1723
$CHCl_3$	ethylene trithiocarbonate	UV	CCl_4	0.87 (35°)					1723
	$CH_3C(O)NH(CH_3)$	NMR	CCl_4	0.48 (36°)	-0.4	-0.4	0.2		2410
	$HC(O)N(CH_3)_2$	NMR	CCl_4	0.91 (20°)	-0.08	1.4	5.0		917a
		NMR	CCl_4	0.89	-0.1	1.1	4.0		2244
	$CH_3C(O)N(CH_3)_2$	NMR	CCl_4	0.82 (36°)	-0.2	1.1	4.4	8	2410
		NMR	CCl_4			3.1 ±0.15	10.6 ±0.5		20
	$CH_3C(O)N(C_2H_5)_2$	NMR	CCl_4	1.1 (20°)	0.03	1.6	5.4		917a
	$CH_2ClC(O)N(C_2H_5)_2$	NMR	CCl_4	0.75 (20°)	-0.19	1.2	4.7		917a
	$CH_3C(O)N(CH_2)_5$	NMR	CCl_4	1.1 (20°)	0.01	1.5	5.4		917a
	N-methyl-2-pyridone	NMR	CCl_4	1.2 (20°)	0.11	1.6	5.0		917a

APPENDIX (continued)

Acid	Base	Phase or Solvent	Method	$K_{25°C}$, ℓ/mole	$-\Delta G_{25°C}$, kcal/mole	$-\Delta H$, kcal/mole	$-\Delta S_{25°C}$, cal deg^{-1} mole^{-1}	$\Delta\nu_{AH}$ cm^{-1}	Ref.
CHCl$_3$	N-methyl-2-pyrrolidone	C$_6$H$_{12}$	NMR	3.22(28°)	0.73	3.99 ±0.02	10.9 ±0.1		2593
	(C$_2$H$_5$)$_3$N	C$_6$H$_{12}$	NMR	4.70(mf^{-1})					1089
		C$_6$H$_{12}$	NMR	0.42(28°)	-0.47	4.05 ±0.03	15.2 ±0.1		2593
		C$_6$H$_{12}$	DEC	5.0(mf^{-1})					1583
		C$_6$H$_{12}$	NMR	4.2(mf^{-1})		4.15 ±0.20	11.0		536
		C$_6$H$_{12}$	NMR	0.43(27°)		4.5 ±0.3	16.8	79	2291
	(n-C$_4$H$_9$)$_3$N	C$_6$H$_{12}$	NMR	0.28(22°)	-0.5	3.6	14.8		1802
	(C$_6$H$_{13}$)$_3$N	C$_6$H$_{12}$	NMR	0.17(22°)	-0.8	3.6	15.8		1802
	(C$_8$H$_{17}$)$_3$N	C$_6$H$_{12}$	NMR	0.15(22°)	-1.1	3.4	15.3		1802
	quinuclidine	C$_6$H$_{12}$	NMR	1.21(37°)	-1.2	4.1 ±0.1	13.9		2291
	C$_5$H$_5$N	CCl$_4$	NMR	1.90(mf^{-1})	-0.04				1089
		CCl$_4$	NMR	1.67(mf^{-1})		2.4			762
		CCl$_4$	NMR	0.38(27°)				33	932
		C$_6$H$_{12}$	NMR	1.4(9.5°)				46	2291
	2,6-dimethylpyridine	C$_6$H$_{12}$	NMR	0.80(27°)				59	2291
	(CH$_3$)$_2$NH	gas	VP		-0.9	5.8 ±1.1	22.5		2145
	aniline	C$_6$H$_{12}$	NIR	0.51	0.41	1.7	7.1		2588
	C$_6$H$_{11}$NH$_2$	C$_6$H$_{12}$	NIR	1.1	0.06	3.6	11.9		2588
	N-nitrosomethylaniline	base	GLC			1.8			2521
	CH$_3$C(O)OC$_2$H$_5$	C$_6$H$_{12}$	NMR	0.67(28°)	-0.21	2.5 ±0.1	9.0 ±0.3		2593
		C$_6$H$_{12}$	NMR	1.0(27°)					2291
	(C$_2$H$_5$)$_2$O	C$_6$H$_{12}$	DEC	5.6(mf^{-1})				12	1583
		C$_6$H$_{12}$	NMR	3.8(mf^{-1})					1089
		CCl$_4$	NMR	1.46(mf^{-1})					1089
	(CH$_2$)$_4$O	C$_6$H$_{12}$	NMR	5.4(mf^{-1})					536
		C$_6$H$_{12}$	NMR	0.54(27°)	-0.34	3.6 ±0.5	13.2		2291
		CCl$_4$	NMR	0.34(27°)				5	2291
	(n-C$_3$H$_7$)$_2$O	CHCl$_3$	NMR	1.2(mf^{-1})		2.6		5	123

Compound	Solvent	Method						Ref.
[(CH₃)₂CH]₂O	CCl₄	NMR	2.06 (mf⁻¹)		2.8			1089
	CHCl₃	NMR	2.4 (mf⁻¹)					123
(n-C₄H₉)₂O	C₆H₁₂	NMR	0.243 (28°)		2.35 ±0.12	5.32 ±0.20		2593
	C₆H₁₂	NMR	1.33 (mf⁻¹)		1.9			2291
	CHCl₃	NMR	1.0 (mf⁻¹)		2.4			123
(n-C₅H₁₁)₂O	base	GLC	0.39 (30°)	-0.5	2.38	9.69		2251
(n-octyl)₂O	CCl₄	NMR	0.50 (40°)	-0.3	2.2	8.4		1184
p-dioxane	C₆H₁₂	NMR	0.582 (28°)	-0.3	2.56 ±0.09	9.64 ±0.32		2593
(CH₃)₂CO	CCl₄	NMR	2.07 (mf⁻¹)					1089
	C₆H₁₂	IR	1.2 (30°)					2584
	C₆H₁₂	DEC	4.1 (mf⁻¹)					1583
	C₆H₁₂	UV	0.8	-0.13	2.34 ±0.13	8.4 ±0.5		2278
	C₆H₁₂	NMR	0.75 (28°)	-0.15	3.6	15.1		2593
	C₆H₁₂	NMR	0.21	-0.9	2.4	8.1		2291
	CHCl₃	CAL	1.0	0.0				1249
	C₆H₁₂	NMR	2.1 (mf⁻¹)					1583
	C₆H₁₂	NMR	0.67 (27°)					2291
(C₂H₅)₂CO	C₆H₁₂	NMR	0.73 (20°)	-0.18	2.0	7.5		1590a
(C₃H₇)₂CO	C₆H₁₂	NMR	0.76 (20°)	-0.16	2.1	7.8		1590a
(C₄H₉)₂CO	C₆H₁₂	NMR	0.81 (20°)	-0.12	2.3	8.1		1590a
(C₅H₁₁)₂CO	C₆H₁₂	NMR	0.86 (20°)	-0.09	2.4	8.6		1590a
(C₆H₁₃)₂CO	C₆H₁₂	NMR	0.86 (20°)	-0.09	2.5	8.7		1590a
cyclohexanone	C₆H₁₂	UV	1.2 (30°)					2584
	C₆H₁₂	NMR	1.02 (28°)	0.17	2.4 ±0.1	8.1 ±0.4		2593
CH₃C(O)C₆H₅	base	GLC			1.85			2521
(C₆H₅)₂CO	base	GLC			1.96			2521
benzil	base	GLC			1.10			2521
CH₃CN	CCl₄	NMR	3.2 (mf⁻¹)					210
	CCl₄	NMR	1.1	0.08				1089
[(CH₃)₂N]₃PO	C₆H₁₂	NMR	13.4 (20°)	1.7	4.9 ±0.3	10.7	31	1850
	C₆H₁₂	NMR	15.5 (29°)	1.5	4.1	8.8	33	2291
	C₆H₁₂	NMR	13.4 (20°)	0.61	2.7	7.0	30	914a
(C₆H₅)₃PO	CCl₄	NMR	3.0 (20°)	0.45	2.5	6.9	24	914a
	CCl₄	NMR	2.3 (20°)	0.96	3.6	8.9	11	914a
(C₂H₅O)₃PO	C₆H₁₂	NMR	5.6 (20°)					914a
	C₆H₁₂	NMR	4.64 (28°)	0.95	3.81 ±0.04	9.6 ±0.1		2593

APPENDIX (continued)

Acid	Base	Phase or Solvent	Method	$K_{25°C}$, ℓ/mole	$-\Delta G_{25°C}$, kcal/mole	$-\Delta H$, kcal/mole	$-\Delta S_{25°C}$, cal deg^{-1} mole^{-1}	$\Delta\nu_{AH}$ cm^{-1}	Ref.
$CHCl_3$	$(C_2H_5O)_3PO$	CCl_4	NMR	1.3(20°)	0.10	2.0	6.4	8	914a
	$(C_8H_{17})_3PO$	C_6H_{12}	NMR	13.9(21°)	1.5	5.0	11.7		1802
	$(n\text{-}C_4H_9O)_3PO$	C_6H_{12}	NIR	4.8(29°)	1.0	4.2	10.7		1803
		C_6H_{12}	NMR	5.2(21°)	0.9	4.3	11.4		1802
	$(C_2H_5O)_2P(O)CH_3$	CCl_4	NMR	1.5(20°)	0.19	2.1	6.4	12	914a
	$(C_2H_5O)_2P(O)CCl_3$	CCl_4	NMR	0.66(20°)	-0.27	1.7	6.6		914a
		C_6H_{12}	NMR	2.3(20°)	0.45	2.9	8.2	3	914a
	$(C_2H_5O)_2P(O)(i\text{-}C_3H_7)$	CCl_4	NMR	1.4(20°)	0.16	2.3	7.2	18	914a
		C_6H_{12}	NMR	6.1(20°)	1.0	4.0	10.1	18	914a
	$C_2H_5C(CH_2O)_3P$	C_6H_{12}	NMR	2.8(32°)	0.7	2.7 ±0.3	6.7		2291
	$(CH_3O)_3P$	C_6H_{12}	NMR	3.6(31°)	0.8	2.7 ±0.2	6.4		2291
	benzene	CCl_4	NMR	0.13(40°)	-1.1	1.7	9.4		1184
		C_6H_{12}	NMR	1.31(mℓ$^{-1}$)		2.0 ±0.1			2291
		C_6H_{12}	NMR	1.06(mℓ$^{-1}$)		2.0	6.5 ±0.5		536
	toluene	CCl_4	NMR	0.17(40°)	-0.9	2.2	10.4		1184
	mesitylene	CCl_4	NMR	0.21(40°)	-0.8	2.4	10.7		1184
	α-methylnaphthalene	CCl_4	NMR	0.22(40°)	-0.8	1.6	8.0		1184
	$(CH_3)_2SO$	CCl_4	NMR	1.2(20°)	0.11	1.6	5.0		917a
		CCl_4	NMR	1.1	0.03	1.6	5.3		2244
		$CHCl_3$	NMR	0.3(mℓ$^{-1}$)		3.3		13	1551
	$(i\text{-}C_3H_7)_2SO$	CCl_4	NMR	1.4(20°)	0.18	1.9	5.8		917a
	$(n\text{-}C_4H_9)_2SO$	CCl_4	NMR	1.3(20°)	0.12	1.8	5.6		917a
	$(CH_2)_4SO$	CCl_4	NMR	1.2(20°)	0.09	1.7	5.4		917a
	$(C_2H_5O)_2SO$	CCl_4	NMR	0.32(20°)	-0.69	0.9	5.3		917a
	$(CH_3)_2S$	C_6H_{12}	NMR	0.17(28°)	-1.0	0.9 ±0.3	6.6 ±0.9		2593
	$(CH_2)_4S$	C_6H_{12}	NMR	0.28(28°)	-0.7	2.3 ±0.7	10.2 ±0.6		2593
		C_6H_{12}	NMR	0.32(27°)	-0.7	2.4 ±0.1	10.4	22	2291
	$(C_2H_5)_2S$	C_6H_{12}	NMR	0.22(28°)	-0.9	1.7 ±0.02	8.65 ±0.08		2593

Solute	Base	Solvent	Method						
	(n-octyl)₂S	C₆H₁₂	NMR	0.28 (27°)	-0.48	1.9	8.1	21	2291
	C₂H₅NO₂	base	GLC	0.42 (30°)	-0.59	1.53 ±0.03	7.1 ±0.1		2251
	ethylene carbonate	C₆H₁₂	NMR	0.36 (28°)	-1.7				2593
	γ-thiobutyrolactone	CHCl₃	NMR	0.06	-0.3				1551
	CHCl₃	C₆H₁₂	NMR	0.59 (31.5°)		2.8 ±0.2	10.4		2291
	propylene carbonate	C₆H₁₂	NMR	0.013 (28°)	-1.4				1089
		CHCl₃	NMR	0.09					1551
CHF₃	Cl⁻	CCl₄	NMR	2.51 (27°)	0.1				932
	Br⁻	CH₃CN	NMR	1.18 (27°)		1.38 ±0.04	4.2 ±0.1		932
	I⁻	CCl₄	NMR	1.78 (27°)	-0.2				932
	Cl⁻	CH₃CN	NMR	0.73 (27°)		0.93 ±0.04	3.7 ±0.1		932
	Br⁻	CCl₄	NMR	1.07 (27°)	-0.4				932
	I⁻	CH₃CN	NMR	0.48 (27°)		0.84 ±0.01	4.3 ±0.1		932
CHBr₃	Cl⁻	CCl₄	NMR	2.43 (27°)					932
	Br⁻	CCl₄	NMR	1.38 (27°)					932
	I⁻	CCl₄	NMR	0.55 (27°)					932
	C₅H₅N	CCl₄	NMR	2.45 (27°)	0.06				932
	(n-octyl)₂O	CH₃CN	NMR	1.10 (27°)		1.10 ±0.04	3.5 ±0.2		932
	(n-octyl)₂S	CCl₄	NMR	1.86 (27°)	0.06				932
	Cl⁻	CH₃CN	NMR	1.11 (27°)		1.07 ±0.05	3.4 ±0.2		932
	Br⁻	CCl₄	NMR	1.95 (27°)	0.09				932
	I⁻	CH₃CN	NMR	1.17 (27°)		0.71 ±0.09	2.0 ±0.3		932
	C₅H₅N	CCl₄	NMR	0.34 (27°)					932
CHI₃	(n-octyl)₂O	base	GLC	0.41 (30°)	-0.49	2.2	8.9		2251
	(n-octyl)₂S	base	GLC	0.73 (30°)	-0.14	2.3	8.1		2251
	Cl⁻	CH₃CN	NMR	3.30 (27°)					932
	Br⁻	CH₃CN	NMR	3.76 (27°)					932
	I⁻	CH₃CN	NMR	3.56 (27°)					932
	C₅H₅N	CCl₄	NMR	0.39 (27°)					932
CHCl₂Br	(n-octyl)₂O	base	GLC	0.41 (30°)	-0.48	2.3	9.4		2251
	(n-octyl)₂S	base	GLC	0.50 (30°)	-0.37	2.0	7.8		2251
CHBr₂Cl	(n-octyl)₂O	base	GLC	0.42 (30°)	-0.46	2.8	10.8		2251
	(n-octyl)₂S	base	GLC	0.62 (30°)	-0.24	2.0	7.5		2251
CHCl₂CN	benzene	C₇H₁₆	NMR	0.31 (40°)	-0.8	1.7	8.4		1184
	toluene	C₇H₁₆	NMR	0.38 (40°)					1184
	mesitylene	C₇H₁₆	NMR	0.70 (40°)	-0.3	2.3	8.7		1184

Acid	Base	Phase or Solvent	Method	$K_{25°C}$, ℓ/mole	$-\Delta G_{25°C}$, kcal/mole	$-\Delta H$, kcal/mole	$-\Delta S_{25°C}$, cal deg^{-1} mole^{-1}	$\Delta\nu_{AH}$ cm^{-1}	Ref.
$CHCl_2CN$	pentamethylbenzene	C_7H_{16}	NMR	1.5(40°)	0.03	3.5	11.6		1184
	p-dioxane	C_7H_{16}	NMR	1.4(40°)	-0.4	1.5	6.4		1184
	α-methylnaphthalene	C_7H_{16}	NMR	0.53(40°)					1184
$CDCl_3$	$(CH_3)_2CO$	C_6H_{14}	IR	0.90(31°)	0.0	3.5 ±0.2	11.7		1216
$CDCl_2Br$	$(CH_3)_2CO$	C_6H_{14}	IR	0.80(31°)	-0.07	3.3 ±0.2	11.3		1216
$CDBr_3$	$(CH_3)_2CO$	C_6H_{14}	IR	0.45(31°)	-0.42	2.7 ±0.2	10.5		1216
CH_2Cl_2	benzene	C_7H_{16}	NMR	0.12(40°)	-1.3	1.7	10.1		1184
	toluene	C_7H_{16}	NMR	0.15(40°)	-1.2	2.0	10.0		1184
	mesitylene	C_7H_{16}	NMR	0.17(40°)	-1.2	2.1	11.1		1184
	α-methylnaphthalene	C_7H_{16}	NMR	0.22(40°)	-1.0	1.4	8.0		1184
	(n-octyl)$_2$O	base	GLC	0.28(30°)	-0.7	1.5	7.5		2251
	p-dioxane	C_7H_{16}	NMR	0.20(40°)	-1.0	1.3	7.7		1184
	(n-octyl)$_2$S	base	GLC	0.36(30°)	-0.6	1.3	6.5		2251
	$[(CH_3)_2N]_3PO$	C_6H_{12}	NMR	1.4(20°)					1850
CH_2Br_2	(n-octyl)$_2$O	base	GLC	0.30	-0.7	1.8	8.3		2251
	(n-octyl)$_2$S	base	GLC	0.46	-0.5	1.4	6.1		2251
CH_2BrCl	(n-octyl)$_2$O	base	GLC	0.29	-0.7	1.9	8.7		2251
	(n-octyl)$_2$S	base	GLC	0.41	-0.5	1.5	6.7		2251
$CH(NO_2)_3$	benzene	C_6H_{12}	NMR	0.87(33°)					1077
	toluene	C_6H_{12}	NMR	1.6(33°)					1077
	xylene	C_6H_{12}	NMR	2.8(33°)					1077
	mesitylene	C_6H_{12}	NMR	4.2(33°)					1077
	durene	C_6H_{12}	NMR	8.8(33°)					1077
	pentamethylbenzene	C_6H_{12}	NMR	25(33°)					1077
	hexamethylbenzene	C_6H_{12}	NMR	32(33°)					1077
$C_7F_{15}H$	$(C_2H_5)_3N$	base	NMR	1.5 (mf^{-1})(30°)	4.3				42
	C_5H_5N	base	CAL			4.9 ±0.9			653
	C_5H_5N	C_6H_{12}	NMR	1.47(9°)		4.2 ±0.3			2290a

Donor	Base	Solvent	Method						Ref.
	quinuclidine	base	CAL			4.7 ±0.2			2290
		C₆H₁₂	NMR	1.5(9°)					2290a
	(C₂H₅)₂O	base	NMR	0.6(mf⁻¹)(30°)		1.7 ±0.4			42
	(CH₃)₂CO	base	NMR	1.2(30°)		2.1 ±0.2			2290a
		base	NMR	2.0(mf⁻¹)(30°)		2.5 ±0.5			42
		base	NMR	3.73(30°)		3.3 ±0.3			2290a
	(C₂H₅)₂CO	base	NMR	2.2(mf⁻¹)(30°)		1.8 ±0.4			42
	3-pentanone	base	NMR	4.4(30°)		2.4 ±0.2			2290a
	C₂H₅C(CH₂O)₃P	C₆H₁₂	NMR	2.0(7.5°)					2290a
	[(CH₃)₂N]₃PO	base	NMR	11.3	1.4	3.7 ±0.3		7.7	2691
C₆H₄C≡CH	(CH₃)₂CO	C₆H₁₂	IR	0.44	-0.49				354
		C₆H₁₄	IR	0.45	-0.47				354
		decane	IR	0.43	-0.50				354
		tetradecane	IR	0.43	-0.50				354
CH₃(CH₂)₄C≡CH	(CH₃)₂CO	CCl₄	IR	0.14	-1.2	1.5 ±0.5		1.0	881
	CH₃CN	CCl₄	IR	0.16	-1.1	1.8 ±0.5		2.3	881
	C₅H₅N	CCl₄	IR	0.21	-0.9	2.0 ±0.5		3.7	881
CH₃(CH₂)₅C≡CH	CH₃C(O)NH(CH₃)₂	CHCl₃	NMR	5.8(36°)	1.1	4.85 ±0.15	36	1.3	2410
	CH₃C(O)N(CH₃)₂	C₆H₁₂	NIR	10(22°)	1.3	3.3		11.9	1401
		CHCl₃	NMR	6.0(36°)	1.2			7.0	2410
C₆H₅NH₂	CH₃C(O)OC₂H₅	C₆H₁₂	NIR	1.6(22°)	0.2	3.13 ±0.12	11	9.7	1401
	CH₃C(O)OC₄H₉	C₆H₁₂	NIR	1.5(22°)	0.2	2.99 ±0.05	10	9.4	1401
	(n-C₃H₇)₂O	C₆H₁₂	NIR	0.50(22°)	-0.4	2.77 ±0.03	26	10.6	1401
	(CH₂)₄O	C₆H₁₂	NIR	1.2(22°)	0.1	3.04 ±0.05	29	9.9	1401
	CH₃OC₆H₅	C₆H₁₂	NIR	0.73(22°)	-0.2	1.94 ±0.09	8	7.2	1401
	(n-C₄H₉O)₃PO	C₆H₁₂	NIR	9.65(23°)	1.3	3.83		8.5	1803
	(n-C₈H₁₇O)₃PO	C₆H₁₂	NIR	23.2(23°)	1.8	4.52		9.1	1803
	(C₆H₅)₂CO	C₆H₁₂	UV	0.9	-0.06				239
	benzene	C₆H₁₂	NIR	0.27	-0.77	1.64		8.1	2587
	C₆H₅N(CH₃)₂	C₆H₁₂	NIR	0.71	-0.20	1.63		6.1	2587
	C₅H₅N	C₆H₁₂	NIR	1.65	0.30	3.43		10.5	2587
	(C₆H₁₁)N(CH₃)₂	C₆H₁₂	NIR	0.69	-0.20	3.35		12.0	2587
	C₅H₅N	C₆H₁₂	CAL	0.80 ±0.03	-0.13	3.75 ±0.07		13.0	2
	C₅H₅N	CCl₄	NMR	0.5(mf⁻¹)					1616
m-Cl-C₆H₄NH₂	HC(O)N(CH₃)₂	CCl₄	IR	2.5(20°)					1419
p-Br-C₆H₄NH₂	C₅H₅N	CCl₄	IR	1.0(20°)					1419

371

APPENDIX (continued)

Acid	Base	Phase or Solvent	Method	$K_{25°C}$, ℓ/mole	$-\Delta G_{25°C}$, kcal/mole	$-\Delta H$, kcal/mole	$-\Delta S_{25°C}$, cal deg^{-1} mole^{-1}	$\Delta\nu_{AH}$ cm^{-1}	Ref.
p-Br-$C_6H_4NH_2$	$2,6$-CH_3-C_5H_3N	CCl_4	IR	$1.1(20°)$					1419
	$(n$-$C_4H_9)_2O$	CCl_4	IR	$0.35(20°)$					1419
	p-dioxane	CCl_4	IR	$0.6(20°)$					1419
	$(CH_2)_5O$	CCl_4	IR	$0.5(20°)$					1419
	eucalyptol	CCl_4	IR	$0.4(20°)$					1419
	$CH_3C(O)C_6H_5$	CCl_4	IR	$1.5(20°)$					1419
	cyclohexanone	CCl_4	IR	$1.5(20°)$					1419
	cyclopentanone	CCl_4	IR	$1.3(20°)$					1419
	CH_3CN	CCl_4	IR	$1.0(20°)$					1419
	N-methylpyrrolidone	CCl_4	IR	$3.2(20°)$					1419
	$(CH_3)_2SO$	CCl_4	IR	$4.5(20°)$					1419
$C_6H_{11}NH_2$	$C_6H_{11}N(CH_3)_2$	C_6H_{12}	CAL	0.08 ± 0.01	-1.50	2.68 ± 0.05	14.0		2
	$(CH_2)_4O$	C_6H_{12}	CAL	0.11 ± 0.01	-1.31	3.06 ± 0.05	14.7		2
$C_6H_{13}NH_2$	$C_6H_{11}N(CH_3)_2$	C_6H_{12}	CAL	0.10 ± 0.01	-1.36	3.05 ± 0.05	14.8		2
	$(CH_2)_4O$	C_6H_{12}	CAL	0.13 ± 0.01	-1.21	3.34 ± 0.05	15.3		2
$NH_2CH_2CH_2NH_2$	C_5H_5N	CCl_4	NMR	$6.6(mf^{-1})$					1616
$(CH_3)_2NH$	C_5H_5N	CCl_4	NMR	$4.2(mf^{-1})$					1616
HNCS	$(C_2H_5)_2O$	CCl_4	IR	41	2.2	6.3	13.8	530	139
	$(n$-$C_3H_7)_2O$	CCl_4	IR	26	1.9	6.7	16.1	554	139
	$(n$-$C_4H_9)_2O$	CCl_4	IR	27	2.0	6.4	14.8	552	139
	$CH_3OC_6H_5$	CCl_4	IR	2.5	0.54	3.7	10.6	350	139
	$(C_6H_5)_2O$	CCl_4	IR	1.1	0.06	2.2	7.2	237	139
	$(C_6H_5CH_2)_2O$	CCl_4	IR	12	1.5	5.4	13.1	440	139
	$CH_3O(t$-$C_4H_9)$	CCl_4	IR	56	2.4	7.2	16.1	562	139
	$C_2H_5O(t$-$C_4H_9)$	CCl_4	IR	48	2.3	7.5	17.4	562	139
	$(CH_2)_3O$	CCl_4	IR	86	2.6	6.0	11.4	550	139
	$(CH_2)_4O$	CCl_4	IR	54	2.4	6.2	12.8	545	139
	2-methyl-tetrahydrofuran	CCl_4	IR	74	2.5	6.7	14.1	562	139

Base	Compound	Solvent	Method					n	Ref.
	2,5-dimethyl tetrahydrofuran	CCl_4	IR	88	2.7	7.4	15.8	570	139
	p-dioxane	CCl_4	IR	24	1.9	5.3	11.4	448	139
	$(CH_2)_6O$	CCl_4	IR	39	2.2	6.2	13.4	540	139
	$(CH_3)_2S$	CCl_4	IR		0.61 ±0.02	3.48 ±0.20	9.65 ±0.7	380	138
	$CH_3SC_2H_5$	CCl_4	IR		0.57 ±0.02	3.47 ±0.20	9.74 ±0.7	390	138
	$(C_2H_5)_2S$	CCl_4	IR		0.67 ±0.02	3.49 ±0.20	9.45 ±0.7	400	138
	$(n\text{-}C_4H_9)_2S$	CCl_4	IR		0.65 ±0.02	3.59 ±0.20	9.85 ±0.7	400	138
	$(t\text{-}C_4H_9)_2S$	CCl_4	IR		0.86 ±0.02	4.48 ±0.20	12.18 ±0.7	440	138
	$(CH_2)_5S$	CCl_4	IR		0.58 ±0.02	3.69 ±0.20	10.43 ±0.7	400	138
	$(CH_2)_4S$	CCl_4	IR		0.56 ±0.02	3.62 ±0.20	9.95 ±0.7	420	138
	$(CH_2)_3S$	CCl_4	IR		0.42 ±0.02	3.28 ±0.20	9.60 ±0.7	383	138
	$(n\text{-}C_4H_9)_2S$	CCl_4	IR		0.35 ±0.02	3.67 ±0.20	11.15 ±0.7	410	138
	$(n\text{-}C_3H_7)_2S$	CCl_4	IR		0.44 ±0.02	3.54 ±0.20	10.40 ±0.7	400	138
	CH_3CN	CCl_4	IR		1.46 ±0.03	4.55 ±0.25	10.54 ±0.7	280	138
	$C_6H_5CH_2CN$	CCl_4	IR		1.35 ±0.03	4.53 ±0.25	10.45 ±0.7	284	138
	C_2H_5SCN	CCl_4	IR		1.21 ±0.03	4.32 ±0.25	7.52 ±0.7	275	138
	C_2H_5NCS	CCl_4	IR		0.04 ±0.03	2.28 ±0.25	10.2 ±0.7	216	138
	hexaethylbenzene	CCl_4	IR		0.52 ±0.02	3.57 ±0.20	9.36 ±0.7	170	138
	hexamethylbenzene	CCl_4	IR		0.26 ±0.02	3.05 ±0.20	7.6 ±0.7	170	138
	mesitylene	CCl_4	IR		-0.12 ±0.02	2.30 ±0.20	9.03 ±0.7	124	138
	CH_3NO_2	CCl_4	IR		0.49 ±0.02	3.15 ±0.20		145	138
$(C_2H_5)_2NH$	benzil	base	GLC			0.8			2521
	N-nitrosomethylaniline		GLC			0.5			2521
$(C_6H_5)_2NH$	$CH_3C(O)N(CH_3)_2$	CCl_4	IR	4.3(20°)	0.54			112	1419
	$HC(O)N(C_2H_5)_2$	CCl_4	IR		0.75			93	1996
	$CH_3C(O)N(C_2H_5)_2$	CCl_4	IR		0.47			106	1996
	$CClH_2C(O)N(C_2H_5)_2$	CCl_4	IR		0.58			85	1996
	$CH_3C(O)N(C_6H_5)_2$	CCl_4	IR		0.57			77	1996
	$C_2H_5C(O)N(C_2H_5)_2$	CCl_4	IR					98	1996
	$(C_2H_5)_3N$	CCl_4	IR	0.4(20°)	0.18				1419
	C_5H_5N	CCl_4	IR	1.5(20°)	-0.3			149	1419
	C_5H_5N	CCl_4	IR	1.35 ±0.02					1033
	$CH_3C(O)OCH_3$	CCl_4	IR		0.49			41	1996
	$CH_3C(O)OC_2H_5$	CCl_4	IR					56	1033
	$CH_3C(O)OC_2H_5$	CCl_4	IR	2.27 ±0.20	-0.4			44	1419

APPENDIX (continued)

Acid	Base	Phase or Solvent	Method	$K_{25°C}$, ℓ/mole	$-\Delta G_{25°C}$, kcal/mole	$-\Delta H$, kcal/mole	$-\Delta S_{25°C}$, cal deg^{-1} mole^{-1}	$\Delta\nu_{AH}$ cm^{-1}	Ref.
$(C_6H_5)_2NH$	$(C_2H_5)_2O$	CCl_4	IR	0.5(20°)				94	1419
	p-dioxane	C_6H_{12}	UV	0.28(27°)					2169
		C_6H_{12}	UV	0.58(27°)					2169
		CCl_4	IR	0.88(20°)				77	1419
		CCl_4	IR	0.76 ±0.04				77	1033
	$(CH_2)_4O$	C_6H_{12}	UV	0.4(27°)	-0.16				2169
		TMP	IR	1.53(22°)					2469
		TMP	UV	1.38(22°)					2469
	$(CH_2)_5O$	CCl_4	IR	0.75(20°)				94	1419
	$(n\text{-}C_4H_9)_2O$	CCl_4	IR	0.3(20°)				92	1419
	$(CH_3)_2CO$	CCl_4	IR	2.2(20°)				48	1419
	$(C_2H_5)_2CO$	CCl_4	IR		-0.19			50	1996
	$CH_3C(O)(i\text{-}C_4H_9)$	CCl_4	IR		-0.07			54	1996
	cyclohexanone	CCl_4	IR	1.8(20°)					1419
	cyclopentanone	CCl_4	IR	1.9(20°)					1419
	piperitone	CCl_4	IR		0.21			68	1996
	mesityl oxide	CCl_4	IR		0.11			65	1996
	CH_3CN	CCl_4	IR	1.1(20°)				47	1419
	C_6H_5CN	CCl_4	IR	0.97(20°)					1419
	eucalyptol	CCl_4	IR	0.64(20°)					1419
	$(CH_3)_2SO$	CCl_4	IR	7.1(20°)					1419
	$(CH_3)_2NC(O)N(CH_3)_2$	CCl_4	IR	4.7(20°)					1419
	piperidine	CCl_4	IR	0.85 ±0.05	-0.10			176	1032
	3,5-lutidine	CCl_4	IR	2.06 ±0.02	0.43			160	1032
	4-picoline	CCl_4	IR	1.76 ±0.02	0.34			155	1032
	3,5-dichloropyridine	CCl_4	IR	0.31 ±0.04	-0.69			113	1032
	$HC(O)N(CH_3)_2$	CCl_4	IR		0.09				1996
$C_6H_5NH(CH_3)$	$HC(O)N(CH_3)_2$	CCl_4	IR	0.58	-0.32	4.9	17.5	58	2278

374

CH$_3$C(O)N(CH$_3$)$_2$	C$_6$H$_{12}$	NIR	8.0(22°)	1.2	5.16 ±0.07	13.3	60	1401
CH$_3$C(O)N(C$_2$H$_5$)$_2$	CCl$_4$	IR		0.30			67	1996
CH$_3$C(O)N(C$_6$H$_5$)$_2$	CCl$_4$	IR		-0.04			49	1996
CClH$_2$C(O)N(C$_2$H$_5$)$_2$	CCl$_4$	IR		0.08			53	1996
C$_5$H$_5$N	CCl$_4$	IR	1.3	0.16	3.8			2278
CH$_3$C(O)OC$_2$H$_5$	C$_6$H$_{12}$	NIR	0.96(22°)	-0.06	3.49 ±0.07	12.2	20	1401
CH$_3$C(O)O(n-C$_4$H$_9$)	C$_6$H$_{12}$	NIR	1.0(22°)	-0.04	3.28 ±0.05	11.5	20	1401
(n-C$_3$H$_7$)$_2$O	C$_6$H$_{12}$	NIR	0.31(22°)	-0.7	3.07 ±0.04	11.1	51	1401
(CH$_2$)$_4$O	C$_6$H$_{12}$	NIR	0.78(22°)	-0.18	3.30 ±0.10	12.6	52	1401
CH$_3$OC$_6$H$_5$	C$_6$H$_{12}$	NIR	0.36(22°)	-0.6	1.82 ±0.06	11.7	13	1401
(CH$_3$)$_2$CO	C$_6$H$_{12}$	NIR	1.2(22°)	0.07	3.02 ±0.22	8.0	22	1401
CH$_3$C(O)C$_6$H$_5$	C$_6$H$_{12}$	NIR	1.1(22°)	0.02	3.00 ±0.04	9.9	22	1401
(C$_6$H$_5$)$_2$CO	C$_6$H$_{12}$	UV	1.4	0.07		10.0		239
cyclohexanone	C$_6$H$_{12}$	NIR	1.2(22°)	0.6	3.22 ±0.03	10.6	25	1401
quinoxaline	C$_6$H$_{12}$	UV	2.7	0.85				239
(C$_2$H$_5$)$_2$NNO	C$_6$H$_{12}$	UV	4.2	0.2				239
thiocamphor	CCl$_4$	IR	1.4	1.15				2278
(n-C$_4$H$_9$O)$_3$PO	C$_6$H$_{12}$	NIR	7.3(23°)	1.65	4.15	10.1		1803
(n-C$_8$H$_{17}$O)$_3$PO	C$_6$H$_{12}$	NIR	17.1(23°)	-1.15	4.80	10.6		1803
benzene	C$_6$H$_{12}$	NIR	0.14	-0.46	1.54	9.0		2587
C$_6$H$_5$N(CH$_3$)$_2$	C$_6$H$_{12}$	NIR	0.46	-0.16	1.83	7.7		2587
C$_5$H$_5$N	C$_6$H$_{12}$	CAL	1.31	-0.34	3.77	12.1		2587
C$_6$H$_{11}$N(CH$_3$)$_2$	C$_6$H$_{12}$	NIR	0.56 ±0.02	0.33	3.70 ±0.05	13.5		2
C$_6$H$_{11}$N(CH$_3$)$_2$	C$_6$H$_{12}$	CAL	0.57	-1.66	3.76	13.7	145	2587
HC(O)N(CH$_3$)$_2$	CCl$_4$	NIR	0.06 ±0.01	1.18	2.68 ±0.05	14.6	167	2
CH$_3$C(O)N(C$_2$H$_5$)$_2$	CCl$_4$	IR		1.33			134	1996
CCl$_1$H$_2$C(O)N(C$_2$H$_5$)$_2$	CCl$_4$	IR		0.97			112	1996
CCl$_2$HC(O)N(C$_2$H$_5$)$_2$	CCl$_4$	IR		0.61			128	1996
CH$_3$C(O)N(C$_6$H$_5$)$_2$	CCl$_4$	IR		1.11			166	1996
C$_2$H$_5$C(O)N(C$_2$H$_5$)$_2$	CCl$_4$	IR		1.25			331	1996
(C$_2$H$_5$)$_3$N	C$_6$H$_{12}$	CAL	4.4 ±0.1	0.9	5.9 ±0.2	16.8		1811
	C$_6$H$_{12}$	NMR	23(mf^{-1})(33°)		4.3	8.0		1006
	C$_6$H$_{12}$	MP	3.2 ±0.3	0.7				1525
C$_5$H$_5$N	C$_6$H$_{12}$	CAL	5.5 ±0.8	1.0	5.0 ±0.3	13.4	252	1811
	C$_6$H$_{12}$	NMR	23(mf^{-1})(33°)		4.3	8.0		1006

Row-group labels (left column): C$_6$H$_{11}$NH(CH$_3$), pyrrole, C$_5$H$_5$N

APPENDIX (continued)

Acid	Base	Phase or Solvent	Method	$K_{25°C}$, ℓ/mole	$-\Delta G_{25°C}$, kcal/mole	$-\Delta H$, kcal/mole	$-\Delta S_{25°C}$, cal deg^{-1} mole^{-1}	$\Delta\nu_{AH}$ cm^{-1}	Ref.
pyrrole	C_5H_5N	CCl_4	IR	2.7(20°)	0.53	3.2	8.9		2606
		C_6H_{12}	MP	3.4(33°)					1525
		CCl_4	MP	4.0(33°)					1525
	2-CH_3-C_5H_4N	CCl_4	IR	3.0(20°)	0.58	3.8	10.8		2606
	2,6-CH_3-C_5H_3N	CCl_4	IR	3.7(18°)	0.68	3.4	9.2		2606
	quinuclidine	C_6H_{12}	CAL			5.55 ±0.18			2290
	$HC(O)OCH_3$	CCl_4	IR		-0.04			61	1996
	$CH_3C(O)OCH_3$	CCl_4	IR		0.25			76	1996
	$(C_2H_5)_2O$	C_6H_{12}	UV	0.5(27°)					2169
	p-dioxane	C_6H_{12}	UV	0.8(27°)					2169
		C_6H_{12}	MP	1.7 ±0.3	0.3				1525
	$(CH_2)_4O$	C_6H_{12}	UV	0.4(27°)					2169
	$(CH_2)_5O$	C_6H_{12}	MP	4.8(20°)					1525
	$CH_3C(O)i$-C_4H_9	CCl_4	IR		0.34			93	1996
	piperitone	CCl_4	IR		0.65			108	1996
	mesityl oxide	CCl_4	IR		0.45			94	1996
	CH_3CN	CCl_4	IR		1.9			72	1675
	CH_3CH_2CN	C_6H_{12}	MP	10.8 ±1	1.4				1525
	$(CH_3)_2SO$	CCl_4	CAL	11.3 ±0.4	1.4	4.2 ±0.1	9.4	184	1811
		base	NMR	8.69(mf^{-1})(33°)		3.0 ±0.5			1976
piperidine	C_5H_5N	CCl_4	NMR	3.9(mf^{-1})					1616
pyrrolidine	C_5H_5N	CCl_4	NMR	1.0(mf^{-1})					1616
indole	$HC(O)N(CH_3)_2$	CCl_4	IR	9.8(30°)	1.4	3.4	7.1	166	669
		CCl_4	IR		1.48			158	1996
	$HC(O)N(C_2H_5)_2$	CCl_4	IR		1.49			166	1996
	$CH_3C(O)N(C_2H_5)_2$	CCl_4	IR		1.69			187	1996
	$CH_3C(O)N(C_2H_5)_2$	CCl_4	IR		1.42			145	1996
	$CClH_2C(O)N(C_2H_5)_2$	CCl_4	IR		1.41			158	1996

376

Substrate	Base	Solvent	Method					No.	Ref.
	CCl₂HC(O)N(C₂H₅)₂	CCl₄	IR		1.02			132	1996
	CCl₃C(O)N(C₂H₅)₂	CCl₄	IR		0.69			108	1996
	CBrH₂C(O)N(C₂H₅)₂	CCl₄	IR		1.32			160	1996
	(C₆H₅)CH₂C(O)N(C₂H₅)₂	CCl₄	IR		1.67			182	1996
	C₂H₅C(O)N(C₂H₅)₂	CCl₄	IR		1.62			183	1996
	C₂H₅C(O)N(n-C₄H₉)₂	CCl₄	IR		1.67			188	1996
	C₃H₇C(O)N(C₂H₅)₂	CCl₄	IR		1.71			192	1996
	CH₃(CH₂)₈C(O)N(C₂H₅)₂	CCl₄	IR		1.68			184	1996
	C₆H₅C(O)N(C₂H₅)₂	CCl₄	IR		1.71			190	1996
	C₅H₅N	CCl₄	IR	9.8(30°)	1.4	3.4	7.1	166	669
	HC(O)OCH₃	CCl₄	IR		0.12			61	1996
	HC(O)OC₂H₅	CCl₄	IR		0.29			68	1996
	CH₃C(O)OCH₃	CCl₄	IR		0.39			85	1996
	CH₃C(O)OC₂H₅	CCl₄	IR		0.45			90	1996
	CH₃C(O)OCH=CH₂	CCl₄	IR		0.18			62	1996
	p-dioxane	CCl₄	IR		0.55	1.8	4.3	136	669
	(CH₂)₄O	CCl₄	IR	1.8(30°)	0.4	1.6	4.0	164	669
	(CH₃)₂CO	CCl₄	IR	2.5(30°)	0.6	2.6	6.7	104	669
	CH₃C(O)(C₂H₅)	CCl₄	IR		0.57			93	1996
	(C₂H₅)₂C(O)	CCl₄	IR		0.65			94	1996
	CH₃C(O)(i-C₄H₉)	CCl₄	IR		0.62			96	1996
	(C₆H₅)₂CO	CCl₄	IR	2.7(30°)	0.64	2.0	4.6	97	669
	cyclopentanone	CCl₄	IR		0.63			91	669
	piperitone	CCl₄	IR		0.75			95	1996
	mesityl oxide	CCl₄	IR		1.02			126	1996
	carvone	CCl₄	IR		0.74			109	1996
	β-ionone	CCl₄	IR		0.69			107	1996
	[(CH₃)(C₆H₅)N]₂C(O)	CCl₄	IR		0.79			107	1996
	CH₃CN	CCl₄	IR		1.30			165	1996
	CH₃CN	CCl₄	IR	1.7(30°)	0.28	1.5	4.1	86	669
2-methyl-indole	diethyl ether	C₆H₁₂	UV	0.70(27°)					2169
	p-dioxane	C₆H₁₂	UV	1.9(27°)					2169
	(CH₂)₄O	C₆H₁₂	UV	0.56(27°)					2169
5-fluoro-indole	HC(O)N(CH₃)₂	CCl₄	NMR	19.5	1.76				1676
	HC(O)N(CH₃)₂	CCl₄	IR	17.5	1.70				1676
	CH₃C(O)N(CH₃)₂	CCl₄	NMR	30	2.02				1676

APPENDIX (continued)

Acid	Base	Phase or Solvent	Method	$K_{25°C}$, ℓ/mole	$-\Delta G_{25°C}$, kcal/mole	$-\Delta H$, kcal/mole	$-\Delta S_{25°C}$, cal deg^{-1} mole^{-1}	$\Delta\nu_{AH}$ cm^{-1}	Ref.
5-fluoro-indole	$CH_3C(O)N(CH_3)_2$	CCl_4	IR	29.3	2.00				1676
	$CH_2ClC(O)N(CH_3)_2$	CCl_4	NMR	9.0	1.30				1676
		CCl_4	IR	12.1	1.48				1676
	C_6H_5CHO	CCl_4	IR	2.7	0.59				1676
	$(C_2H_5)_3N$	CCl_4	NMR	3.7	0.78				1676
		CCl_4	IR	3.3	0.71				1676
	$(CH_2=CHCH_2)_3N$	CCl_4	IR	1.4	0.20				1676
	$(n-C_4H_9)_3N$	CCl_4	NMR	2.6	0.57				1676
		CCl_4	IR	2.7	0.59				1676
	C_5H_5N	CCl_4	NMR	5.7	1.03				1676
		CCl_4	IR	5.6	1.02				1676
	$3-Br-C_5H_4N$	CCl_4	NMR	2.0	0.41				1676
		CCl_4	IR	2.7	0.59				1676
	$4-N(CH_3)_2-C_5H_4N$	CCl_4	NMR	29	2.00				1676
		CCl_4	IR	25.1	1.91				1676
	$3,5-Cl-C_5H_3N$	CCl_4	IR	0.97	-0.02				1676
	$C_6H_5N(CH_3)_2$	CCl_4	IR	1.47	0.23				1676
	$C_6H_{11}N(CH_3)_2$	CCl_4	IR	5.4	1.00				1676
	quinuclidine	CCl_4	NMR	6.1	1.07				1676
	$(C_2H_5)_2O$	CCl_4	IR	1.71	0.32				1676
	$(CH_2)_4O$	CCl_4	NMR	2.9	0.63				1676
		CCl_4	IR	2.2	0.47				1676
	cyclohexanone	CCl_4	NMR	5.3	0.99				1676
		CCl_4	IR	5.6	1.02				1676
	$p-CH_3O-C_6H_4C(O)CH_3$	CCl_4	IR	5.3	0.99				1676
	C_6H_5CN	CCl_4	IR	2.2	0.47				1676
	$(C_6H_5)_3PO$	CCl_4	IR	112	2.80				1676
	$(CH_3O)_3PO$	CCl_4	NMR	32	2.05				1676

						Ref.
carbazole	[(CH₃)₂N]₃PO	CCl₄	IR	31.1	2.04	1676
		CCl₄	NMR	225	3.21	1676
	(CH₃)₂SO	CCl₄	IR	213	3.18	1676
		CCl₄	NMR	32	2.05	1676
	(C₆H₅)₂SO	CCl₄	IR	30.6	2.03	1676
	[(CH₃)₂N]₂C(O)	CCl₄	IR	15.5	1.62	1676
		CCl₄	NMR	30	2.02	1676
	[(CH₃)₂N]₂C(NH)	CCl₄	IR	26.3	1.94	1676
		CCl₄	NMR	46	2.27	1676
		CCl₄	IR	45.2	2.26	1676
	p-dioxane	C₆H₁₂	UV	2.5(27°)		2169
	(CH₂)₄O	C₆H₁₂	UV	1.5(27°)		2169
maleimide	CH₃C(O)N(CH₃)₂	CCl₄	IR	32	2.1	1433
	C₅H₅N	CCl₄	IR	18	1.7	1433
	4-picoline	CCl₄	IR	22	1.8	1433
	(C₂H₅)₂O	CCl₄	IR	4.0	0.8	1433
	(i-C₃H₇)₂O	CCl₄	IR	4.0	0.8	1433
	p-dioxane	CCl₄	IR	5.5	1.0	1433
	eucalyptol	CCl₄	IR	4.5	0.9	1433
	CH₃CN	CCl₄	IR	2.5	0.5	1433
	(n-C₄H₉O)₃PO	CCl₄	IR	55	2.4	1433
	[(CH₃)₂N]₃PO	CCl₄	IR	255	3.3	1433
	(CH₃)₂SO	CCl₄	IR	60	2.4	1433
succinimide	CH₃C(O)N(CH₃)₂	CCl₄	IR	22	1.8	1433
	C₅H₅N	CCl₄	IR	22	1.8	1433
	4-picoline	CCl₄	IR	20.5	1.8	1433
	(C₂H₅)₂O	CCl₄	IR	4.0	0.8	1433
	(i-C₃H₇)₂O	CCl₄	IR	3.0	0.7	1433
	p-dioxane	CCl₄	IR	7.0	1.2	1433
	eucalyptol	CCl₄	IR	7.0	1.2	1433
	CH₃CN	CCl₄	IR	2.0	0.4	1433
	(n-C₄H₉O)₃PO	CCl₄	IR	42	2.2	1433
	[(CH₃)₂N]₃PO	CCl₄	IR	168	3.0	1433
	(CH₃)₂SO	CCl₄	IR	65	2.5	1433
CH₃C(O)NH(CH₃)	HC(O)N(CH₃)₂	CCl₄	IR	105	0.96	1996
	HC(O)N(C₂H₅)₂	CCl₄	IR	105	1.10	1996

APPENDIX (continued)

Acid	Base	Phase or Solvent	Method	$K_{25°C}$, ℓ/mole	$-\Delta G_{25°C}$, kcal/mole	$-\Delta H$, kcal/mole	$-\Delta S_{25°C}$, cal deg^{-1} mole^{-1}	$\Delta\nu_{AH}$ cm^{-1}	Ref.
CH₃C(O)NH(CH₃)	CH₃C(O)N(CH₃)₂	CCl₄	NMR	10	1.36			122	2244
	CH₃C(O)N(C₂H₅)₂	CCl₄	IR		1.30			90	1996
	CH₃C(O)N(C₆H₅)₂	CCl₄	IR		0.93			95	1996
	CClH₂C(O)N(C₂H₅)₂	CCl₄	IR		0.89			112	1996
	C₂H₅C(O)N(C₂H₅)₂	CCl₄	IR		1.10			145	1996
	C₅H₅N	CCl₄	IR	1.8(26°)	0.35	4.6	14.3	50	233
	HC(O)OCH₃	CCl₄	IR		0.16			41	1996
	CH₃C(O)OCH₃	CCl₄	IR		-0.26			60	1996
	CH₃C(O)(i-C₄H₉)	CCl₄	IR		0.17			68	1996
	mesityl oxide	CCl₄	IR		0.42			70	1996
	piperitone	CCl₄	IR		0.52			60	1996
	(C₆H₅)₂C(O)	CCl₄	IR	1.9(26°)	0.38	2.9	8.5	100	233
	benzene	CCl₄	IR	0.25(26°)	-0.82				233
	CH₃SC₂H₅	CCl₄	IR	0.30(26°)	-0.72	1.4	7.1		233
	(CH₃)₂SO	CCl₄	NMR	12	1.47				2244
	ethylene trithiocarbonate	CCl₄	IR	0.65(26°)	-0.26	1.0	4.2	90	233
C₆H₅NHCO₂C₂H₅	C₅H₅N	CCl₄	IR	2.5(26°)	0.54	5.1	15.3	178	233
	(C₆H₅)₂CO	CCl₄	IR	1.3(26°)	0.16	3.2	10.2	83	233
C₆H₅CH₂NOH	(C₂H₃)₃N	CCl₄	NIR	14	1.6	6.3 ±0.8	15.4	380	754
	(C₂H₅)₂O	CCl₄	NIR	2.3	0.5	4.1 ±0.8	12.1	220	754
	benzene	CCl₄	NIR	0.5	-0.4	1.8 ±0.8	7.4	55	754
	(CH₃)₂SO	CCl₄	NIR	11	1.4	5.2 ±0.8	12.7	300	754
C₆H₅CH₂NHOCH₂-C₆H₅	(C₂H₅)₃N	CCl₄	NIR	4.1	0.8	4.3 ±0.8	11.7	190	754
	(C₂H₅)₂O	CCl₄	NIR	1.6	0.3	2.4 ±0.8	7.0	70	754
	(CH₃)₂SO	CCl₄	NIR	3.2	0.7	3.4 ±0.8	9.1	130	754
HN₃	C₅H₅N	CCl₄	IR	4.75 ±0.17	0.92	3.72 ±0.17	9.5	418 ±15	1781

Acid	Base	Solvent	Method						Ref.
HNCO	$(C_2H_5)_2O$	CCl_4	IR	2.03 ±0.05	0.42	3.40 ±0.07	10.0	222 ±10	1781
	$(CH_2)_3O$	CCl_4	IR	2.17 ±0.14	0.46	2.61 ±0.14	7.2	121 ±7	1781
	$(CH_2)_4O$	CCl_4	IR	2.76 ±0.08	0.60	3.10 ±0.08	8.4	207 ±10	1781
	CH_3CN	CCl_4	IR	2.52 ±0.12	0.55	2.43 ±0.22	6.3	110 ±7	1781
	$(t\text{-}C_4H_9)_2S$	CCl_4	IR	0.99 ±0.16	-0.06	2.64 ±0.05	9.1	222 ±10	1781
	C_5H_5N	CCl_4	IR	33.9 ±0.07	2.09	5.42 ±0.14	11.2	624	1782
	$(C_2H_5)_2O$	CCl_4	IR	6.68 ±0.27	1.12	3.92 ±0.08	9.4	346 ±15	1782
	$(CH_2)_3O$	CCl_4	IR	4.76 ±0.25	0.93	3.49 ±0.14	8.6	226	1782
	$(CH_2)_4O$	CCl_4	IR	9.10 ±0.06	1.31	4.24 ±0.08	9.8	361	1782
	CH_3CN	CCl_4	IR	4.60 ±0.4	0.90	3.21 ±0.08	7.7	182	1782
	hexaethylbenzene	CCl_4	IR	1.36 ±0.05	0.18	2.41 ±0.24	7.5	133	1782
	$(t\text{-}C_4H_9)_2S$	CCl_4	IR	1.43 ±0.07	0.21	2.84 ±0.08	8.8	318	1782
HF	$(CH_3)_4N^+F^-$ (s)	vapor	VP			37			1009
	$(CH_3)_2O$	vapor	IR	14 (atm^{-1}, 30°)		10.3		505	2444
	$CH_3OC_2H_5$	vapor	IR	20 (atm^{-1}, 30°)		8.8		535	2444
	$(C_2H_5)_2O$	vapor	IR	10 (atm^{-1}, 30°)		8.2		575	2444
HCl	$(CH_3)_2O$	vapor	NMR			7.1 ±0.8			896
	$(C_2H_5)_2O$	vapor	VP			7.6			871
	$(CH_3)_2O$	vapor	IR			5.6		316	197
	$(C_2H_5)_2O$	vapor	VP			7.5 ±0.3		336	197
HBr	$(n\text{-}C_4H_9)_4N^+Cl^-$ (S)	vapor	VP			14.2			1557
	$(n\text{-}C_4H_9)_4N^+Br^-$ (S)	vapor	VP			9.1			1557
HI	$(n\text{-}C_4H_9)_4N^+Br^-$ (S)	vapor	VP			12.8			1557
	$(n\text{-}C_4H_9)_4N^+I^-$ (S)	vapor	VP			12.4			1557

ANNOTATED BIBLIOGRAPHY

Our intention in assembling this annotated bibliography was to update the comprehensive bibliography published by Pimentel and McClellan. Therefore, we have tried to include all pertinent references on theory, spectroscopic measurements, thermodynamics, kinetics, and biological applications which have appeared in the literature between 1960 and January, 1974. General references and representative research papers are also included for other areas of hydrogen bonding with the hope that readers can find the information they seek with a minimum of literature searching.

Journal abbreviations are those used by *Chemical Abstracts*. The Russian journal citations are followed by either the corresponding page in the English translation or the *Chemical Abstract* citation. Abbreviations for experimental methods are the same as those used in the text. Other abbreviations used are BIOL, biological applications; CAL, calorimetric method; poly A, polyadenylic acid; poly C, polycytidylic acid; poly G, polyguanylic acid; poly I, polyinosinic acid; poly U, polyuridylic acid.

1. E. W. Abel, D. A. Armitage, and S. P. Tyfield, *J. Chem. Soc.*, *A*, 554 (1967). NMR: Adducts of $CHCl_3$ with organometallic bases.

2. L. Abello, M. Kern, D. Caceres, and G. Pannetier, *Bull. Soc. Chim. Fr.*, 94 (1970). CAL: Adducts of cyclohexylamines, anilines, with N,N-dimethylcyclohexylamine and tetrahydrofuran.

3. L. Abello, B. Servais, M. Kern, and G. Pannetier, *Bull. Soc. Chim. Fr.*, 4038 (1968). CAL: Self-association, $PhNH_2$, PhNHMe, and $PhNMe_2$ and saturated counterparts.

4. L. Abello, B. Servais, M. Kern, and G. Pannetier, *Bull. Soc. Chim. Fr.*, 4360 (1968). Self-association, anilines in C_6H_6, CCl_4, cyclohexane.

5. R. J. Abraham, *Mol. Phys.*, $\underline{4}$, 369 (1961). NMR: CH_3CN and CH_4 in various solvents.

6. M. A. Abramovich, I. M. Ginzburg, and D. V. Ioffe, *Teor. Eksp. Khim.*, $\underline{7}$, 225 (1971); *Chem. Abstr.*, $\underline{75}$, 42672z (1971). IR: Intramolecular, hydroxycarboxylic acids.

6a. P. Acharya, P. Kumara, and P. S. Narayanan, *Pramana*, $\underline{1}$, 161 (1973); *Chem. Abstr.*, $\underline{79}$, 130960e (1973). IR: Dynamics, H_2O in ferroelectric $K_4[Fe(CN)_6] \cdot 3H_2O$.

6b. P. K. Acharya and P. S. Narayanan, *Indian J. Pure Appl. Phys.*, $\underline{10}$, 827 (1972). IR, Raman: Potential well in $KH_3(SeO_3)_2$.

383

6c. P. K. Acharya and P. S. Narayanan, *Spectrochim. Acta*, 29A, 925 (1973). Raman, IR: Ferroelectric alkali trihydrogen selenites.

7. W. Adam, A. Grimison, R. Hoffmann, and C. Zuazaga de Ortiz, *J. Amer. Chem. Soc.*, 90, 1509 (1968). Theory: Extended Hückel treatment of pyridine-water and pyridine-methanol.

8. P. Adamek, P. Endrle, and Z. Ksandr, *Collect. Czechoslov. Chem. Commun.*, 36, 3539 (1971). IR: Adducts of substituted phenols with n-C_4H_9CN, thermodynamic data.

9. P. Adamek and Z. Ksandr, *Collect. Czechoslov. Chem. Commun.*, 33, 3053 (1968). IR: Accuracy of equilibrium constant.

10. P. Adamek and Z. Ksandr, *Collect. Czechoslov. Chem. Commun.*, 35, 1587 (1970). IR: Optimum conditions for determining K.

11. H. E. Affsprung, S. D. Christian, and A. M. Melnick, *Spectrochim. Acta*, 20, 285 (1964). IR: Hetero-dimerization of acetic and trichloroacetic acids in CCl_4.

12. H. E. Affsprung, S. D. Christian, and J. D. Worley, *Spectrochim. Acta*, 20, 1415 (1964). Near-IR: Self-association, γ-butyrolactam.

13. C. Agami and M. Caillot, *Bull. Soc. Chim. Fr.*, 1990 (1969). NMR, IR: Δv_{CH}, chemical shifts for $C_6H_5C \equiv CH$ in organic solvents.

14. M. Ageno, *Theor. Chim. Acta*, 17, 334 (1970). Theory: Model for anomalous water.

15. M. Ageno, *Atti Accad. Naz. Lincei, Cl. Sci. Fis., Mat. Natur. Rend.*, 47, 514 (1969); *Chem. Abstr.*, 73, 48668h (1970). Theory: Anomalous water and the collective model of the hydrogen bond.

16. M. Ageno, *Proc. Nat. Acad. Sci. U. S.*, 57, 567 (1967). Theory: Cyclic aggregates in the structure of water.

17. M. Ageno, E. Dore, and C. Frontali, *Biopolymers*, 9, 116 (1970). BIOL: Denaturation of DNA.

18. M. Ageno, E. Dore, and C. Frontali, *Biophys. J.*, 9, 1281 (1969). BIOL: Alkaline denaturation of DNA.

20. J. Ahlf and D. Platthaus, *Ber. Bunsenges, Physik. Chem.*, 74, 204 (1970). NMR: Thermodynamic data, self-association, phenol, N-methylacetamide, N-ethylacetamide, N-methylpropionamide.

20a. A. Aihara and Y. Shirota, *Bull. Chem. Soc. Jap.*, 45, 935 (1972). IR: Bending vibrations, crystalline p-hydroxybenzaldehyde.

21. G. Aksnes, *Acta Chem. Scand.*, 14, 1475 (1960). IR: K, adducts of phenol with triethylphosphate.

22. G. Aksnes and P. Albriktsen, *Acta Chem. Scand.*, 22, 1866 (1968). IR: Thermodynamic data, adducts of organophosphorus compounds with phenol, v_{OH} correlations.

23. G. Aksnes and P. Albriktsen, *Acta Chem. Scand.*, 20, 1330 (1966). IR: Intramolecular, monomethyl ethers of diols.

24. G. Aksnes and K. Bergesen, *Acta Chem. Scand.*, 18, 1586 (1964). IR: Intramolecular, hydroxyalkyl diphenylphosphine oxides $HO(CH_2)_n P(O)Ph_2$ with n = 1-4.

25. G. Aksnes and T. Gramstad, *Acta Chem. Scand.*, 14, 1485 (1960). Near-IR: Thermodynamic data, phenol adducts of organophosphorus bases, $\Delta\nu$-ΔH relationship.

26. H. Akutsu and M. Tsuboi, *Bull. Chem. Soc. Jap.*, 43, 3391 (1970). BIOL: Helical structure, poly A, G with poly U.

27. L. Al-Adhami and D. J. Millen, *Nature*, 211, 1291 (1966). IR: Gas phase, HNO_3-$(C_2H_5)_2O$ adduct.

27a. G. Alagona, R. Cimiraglia, and U. Lamanna, *Theor. Chim. Acta*, 29, 93 (1973). Theory: LCAO calculations on $H_5O_2^+$.

27b. G. Alagona, R. Cimiraglia, E. Scrocco, and J. Tomasi, *Theor. Chim. Acta*, 25, 103 (1972). Theory: Electrostatic model.

28. W. J. Albery in *Prog. Reaction Kinetics*, G. Porter, ed., 4, 353 (1967). Kinetics: Review, role of solvent in aqueous proton transfer reactions.

29. M. Alei, Jr. and A. E. Florin, *J. Phys. Chem.*, 73, 863 (1969). NMR: ^{17}O, $(H_2O)_2$ in NH_3.

30. D. O. Alford, A. Menefee, and C. B. Scott, *Chem. Ind. (London)*, 514 (1959). IR: Phosphorotetrathioic acids.

30a. T. D. Alger, D. M. Grant, and J. R. Lyerla, Jr., *J. Phys. Chem.*, 75, 2539 (1971). NMR: Self-association, ^{13}C spin-lattice relaxation studies, HCOOH, CH_3COOH.

31. E. A. Allan and L. W. Reeves, *J. Phys. Chem.*, 67, 591 (1963). NMR: Intramolecular, *cis-trans* isomers of *o*-halophenols.

32. E. A. Allan and L. W. Reeves, *J. Phys. Chem.*, 66, 613 (1962). NMR: Intramolecular, thermodynamic data, *o*-halophenols.

33. G. A. Allen and M. A. Caldin, *Quart. Rev. (London)*, 7, 255 (1953). Review: Self-association, carboxylic acids.

34. G. Allen, J. G. Watkinson, and K. H. Webb, *Spectrochim. Acta*, 22, 807 (1966). IR: Self-association, benzoic acid in vapor phase and in inert solvents.

35. L. C. Allen and P. A. Kollman, *Science*, 167, 1443 (1970). Theory: CNDO/2 calculations on anomalous water.

36. L. C. Allen and P. A. Kollman, *J. Amer. Chem. Soc.*, 92, 4108 (1970). Theory: CI calculations on cyclic HF and H_2O polymers.

37. A. Allerhand and P. von R. Schleyer, *J. Amer. Chem. Soc.*, 85, 1715 (1963). IR: C—H bonds as proton donors.

38. A. Allerhand and P. von R. Schleyer, *J. Amer. Chem. Soc.*, 85, 1233 (1963). IR: Effect of solvents on ν_{X-H}.

39. A. Allerhand and P. von R. Schleyer, *J. Amer. Chem. Soc.*, 85, 866 (1963). IR: $\Delta\nu_{OH}$, phenol and MeOH adducts of RCN, RNC.

40. A. Allerhand and P. von R. Schleyer, *J. Amer. Chem. Soc.*, 85, 371 (1963). IR: Solvent effects on O—H...O complexes, concentration dependence of $\Delta\nu$.

41. A. Allerhand and P. von R. Schleyer, *J. Amer. Chem. Soc.*, 84, 1322 (1962). IR: Phenol and MeOH adducts of RCN, RNC, $\Delta\nu_{OH}$.

42. S. K. Alley, Jr. and R. L. Scott, *J. Phys. Chem.*, 67, 1182 (1963). NMR: Thermodynamic data, $C_7F_{15}H$ adducts with bases.

43. A. L. Allred and R. N. Wendricks, *J. Chem. Soc. A*, 778 (1966). NMR: Effect of electrolytes on ^1H spectrum of liquid NH_3.

44. J. Almlöf, *Acta Crystallogr., Sect. B*, 28, (Pt. 2), 481 (1972). X-Ray: $HClO_4 \cdot 3H_2O$ structure at -188°C.

44a. J. Almlöf, *Chem. Phys. Lett.*, 17, 49 (1972). Theory: Gaussian calculation, HF_2^- and DF_2^-.

44b. J. Almlöf, Å. Kvick, and J. O. Thomas, *J. Chem. Phys.*, 59, 3901 (1973). Theory: LCAO calculations on glycine.

44c. J. Almlöf, L. Lindgren, and J. Tegenfeldt, *J. Mol. Struct.*, 14, 427 (1972). Theory: *Ab initio* calculations, hydrates.

45. J. Almlöf, J. O. Lundgren, and I. Olovsson, *Acta Crystallogr., Sect. B*, 27, 898 (1971). X-Ray: $HClO_4 \cdot 2.5H_2O$.

46. J. Almlöf and O. Mårtensson, *Int. J. Quantum Chem.*, 6, 491 (1972). Theory: Extended Hückel calculations on pyrrole-pyridine hydrogen-bonded complex.

47. J. Almlöf and O. Mårtensson, *Acta Chem. Scand.*, 25, 1413 (1971). Theory: Extended Hückel calculations on formic and acetic acid dimers.

48. J. Almlöf and O. Mårtensson, *Acta Chem. Scand.*, 25, 355 (1971). Theory: Extended Hückel calculations on formamide dimer.

49. G. Altardi and F. Amaldi, *Ann. Rev. Biochem.*, 39, 183 (1970). BIOL: Review, rRNA.

50. A. C. P. Alves and J. M. Hollas, *Mol. Phys.*, 23, 927 (1972). UV: Intramolecular, tropolone vapor.

51. E. S. Amis, *Solvent Effects on Reaction Rates and Mechanisms*, Academic Press, New York, 1966, p. 200. Kinetics: Review.

52. A. V. Amosov, G. T. Petrovskii, and D. M. Yudin, *Teor. Eksp. Khim.*, 6, 271 (1970); *Chem. Abstr.*, 73, 61113t (1970). ESR: Hydrogen bonding to AlO_4 and SiO_4 in γ-irradiated silica.

53. V. Ananthanarayanan, *Spectrochim. Acta*, 20, 197 (1964). Raman: Self-association, adipic acid.

54. C. M. W. Anderson, J. L. Duncan, and F. J. C. Rossotti, *J. Chem. Soc.*, 140, 2165, 4201 (1961). IR: Self-association, pyrazole derivatives, imidazole.

54a. G. R. Anderson and G. J. Jiang, *J. Phys. Chem.*, 77, 2560 (1973). Theory: Potential function for $H_5O_2^+$.

54b. G. R. Anderson and E. R. Lippincott, *J. Chem. Phys.*, 55, 4077 (1971). Theory: Vibronic effects in hydrogen bonding.

55. R. G. Anderson and M. C. R. Symons, *Trans. Faraday Soc.*, 65, 2550 (1969). NMR: t-Butanol-H_2O mixtures.

56. B. D. Andrews, A. J. Poynton, and I. D. Rae, *Aust. J. Chem.*, 25, 639 (1972). NMR: Intramolecular, o-substituted acetanilides.

57. B. D. Andrews and I. D. Rae, *Aust. J. Chem.*, 24, 413 (1971). NMR: Intramolecular, 2-substituted 1,3-phenylenediamines.

58. B. D. Andrews, I. D. Rae, and B. E. Reichert, *Tetrahedron Letters*, 1859 (1969). NMR: Intramolecular, *o*-substituted anilides.

58a. P. R. Andrews, *Aust. J. Chem.*, <u>25</u>, 2243 (1972). IR: Self-association, *cis* isomers of *o*-R$\overline{C_6H_4}$NHC(O)H.

59. F. A. L. Anet and J. M. Muchowski, *Proc. Chem. Soc.*, 219 (1962). NMR: Self-association, *m*-substituted phenols.

60. J. Applequist and V. Damle, *J. Amer. Chem. Soc.*, <u>88</u>, 3895 (1966). BIOL: Thermodynamics, one-stranded helix-coil equilibrium in poly A.

61. J. M. Appleton, B. D. Andrews, I. D. Rae, and B. E. Reichert, *Aust. J. Chem.*, <u>23</u>, 1667 (1970). NMR: Intramolecular, *o*-substituted anilides and anilines.

62. K. Arakawa, K. Sasaki, and Y. Endo, *Bull. Chem. Soc. Jap.*, <u>42</u>, 2079 (1969). Structure theory of H_2O.

63. R. Arav and K. G. Wagner, *Eur. J. Biochem.*, <u>13</u>, 267 (1970). BIOL: Binding of ribonucleotides to basic polyamino acids.

64. R. B. Archibald and A. D. E. Pullin, *Spectrochim. Acta*, <u>12</u>, 34 (1958). IR: Solvent effects on ν_{AH}.

65. I. M. Arefev and V. I. Malyshev, *Opt. Spektrosk.*, <u>13</u>, 206 (1962). [English, 112]. IR: HX adducts with acetone, dioxane, ether.

66. J. Arient, J. Knizek, J. Marhan, and V. Slavik, *Collect. Czechoslov. Chem. Commun.*, <u>33</u>, 3280 (1968). IR: Intramolecular, dianthrimides and diphthaloylcarbazoles.

67. Y. Armand and P. Arnaud, *Ann. Chim. (Treizieme Serie)*, <u>9</u>, 433 (1964). IR: Intramolecular, unsaturated alcohols.

67a. M. Arnaudov, A. Dobrev, L. Shishkova, and Khr. Ivanov, *Zh. Prikl. Spektrosk.*, <u>18</u>, 242 (1973); *Chem. Abstr.*, <u>79</u>, 41571m (1973). IR: Intramolecular, 3-(benzoylamino) propanoic acids.

68. E. M. Arnett in *Prog. Phys. Org. Chem.*, S. G. Cohen, A. Streitwieser, Jr., R. W. Taft, eds., <u>1</u>, 223, Interscience, 1963. Review: Quantitative comparisons of weak organic bases, $\Delta\nu$-pKa correlation.

69. E. M. Arnett, W. G. Bentrude, J. J. Burke, and P. McC. Duggleby, *J. Amer. Chem. Soc.*, <u>87</u>, 1541 (1965). CAL: Design of calorimeter apparatus for hydrogen bonding studies.

70. E. M. Arnett, L. Joris, E. Mitchell, T. S. S. R. Murty, T. M. Gorrie, and P. von R. Schleyer, *J. Amer. Chem. Soc.*, <u>92</u>, 2365 (1970). Thermodynamic data, *p*-fluorophenol adducts in CCl_4.

71. E. M. Arnett and E. J. Mitchell, *J. Amer. Chem. Soc.*, <u>93</u>, 4052 (1971). Proton transfer versus hydrogen bonding.

71a. E. M. Arnett, T. S. S. R. Murty, P. von R. Schleyer, and J. Joris, *J. Amer. Chem. Soc.*, <u>89</u>, 5955 (1967). CAL: Hydrogen bond enthalpies.

72. G. Arnold and C. Schiele, *Z. Naturforsch.*, B, <u>23</u>, 1192 (1968). IR: Substituted benzohydrazides.

73. J. Arnold, J. E. Bertie, and D. J. Millen, *Proc. Chem. Soc.*, 121 (1961). IR: Gas-phase, adducts of HX with $(CH_3)_2O$.

74. J. Arnold and D. J. Millen, *J. Chem. Soc.*, 510 (1965). IR: Gas-
     phase spectra, HF adducts with carbonyl compounds.

75. J. Arnold and D. J. Millen, *J. Chem. Soc.*, 503 (1965). IR: Gas-
     phase, ether-hydrogen fluoride systems.

76. J. T. Arnold and M. E. Packard, *J. Chem. Phys.*, 19, 1608 (1951).
     NMR: Concentration dependence of OH proton resonance of ethanol.

77. J. T. Arnold and M. E. Packard, *Phys. Rev.*, 83, 210 (1951).
     NMR: Concentration dependence of OH proton resonance of ethanol.

78. K. Arnold and G. Klose, *Tetrahedron*, 25, 3775 (1969). NMR:
     Intramolecular, β-thioxo ketones.

79. K. Arnold and G. Klose, *Mol. Phys.*, 13, 391 (1967). Theory:
     Computed bond length versus bond energy curves.

80. K. Arnold, G. Klose, P. Thomas, and E. Uhlemann, *Tetrahedron*, 25,
     2957 (1969). NMR: Intramolecular, β-thioxo ketones.

81. S. Arnott, *Science*, 167, 1694 (1970). BIOL: Crystallography of
     DNA, difference synthesis supports Watson-Crick base pairing.

82. S. Arnott, S. D. Dover, and A. J. Wonacott, *Acta Crystallogr.*,
     Sect. B, 25, 2192 (1969). BIOL: Structures of DNA and RNA.

83. S. Arnott, W. Fuller, A. Hodgson, and I. Prutton, *Nature*, 220, 561
     (1968). X-Ray: Double helices formed by poly (A + U), poly
     (I + C), and poly (G + C).

84. S. Arrhenius, *J. Macromol. Sci. Chem.*, 4, 243 (1970). Theory:
     Oscillating ion bond.

84a. M. Arsic-Eskinja, *Ber. Kernforschungsanlage Juelich*, 1972, Juel
     872-FF, 36 pp.; *Chem. Abstr.*, 78, 102967y (1973). Neutron
     scattering: $KHCO_3$, KH_2PO_4.

85. M. Asselin, G. Belanger, and C. Sandorfy, *J. Mol. Spectrosc.*, 30,
     96 (1969). IR: Anharmonicity of OH bands in associated alcohols.

86. M. Asselin and C. Sandorfy, *Chem. Phys. Lett.*, 8, 601 (1971).
     IR: Self-association, band width and frequency in alcohols.

87. M. Asselin and C. Sandorfy, *Can. J. Chem.*, 49, 1539 (1971). IR:
     Anharmonicity, in-plane O—H bending combination with v_{OH} alcohols.

88. M. Asselin and C. Sandorfy, *J. Chem. Phys.*, 52, 6130 (1970).
     Near-IR: Self-association, alcohols, band width, double excita-
     tion.

89a. P. W. Atkins and M. C. R. Symons, *Mol. Phys.*, 23, 831 (1972).
     IR: $(H_2O)_2$ in CCl_4.

90. G. Atkinson, U. S. Clearinghouse Fed. Sci. Tech. Inform., AD 1969,
     AD-695708, 9 pp. Kinetics: Ultrasonic absorption, *t*-butanol,
     phenol, ethanol, ε-caprolactam.

90a. D. H. Aue, H. M. Webb, and M. T. Bowers, *J. Amer. Chem. Soc.*, 95,
     2699 (1973). Intramolecular: Gas-phase thermodynamic data,
     diaminoalkanes.

91. E. Augdahl and T. Ledaal, *Spectrochim. Acta*, 26A, 1173 (1970).
     IR: Δv_{OH}, phenol adducts of cyclic ketones.

92. B. S. Ault and G. C. Pimentel, *J. Phys. Chem.*, 77, 57 (1973). IR: 1:1 complex of HCl and H_2O and deuterated analogues in nitrogen matrix at 15°K.

93. B. S. Ault and G. C. Pimentel, *J. Phys. Chem.*, 77, 1649 (1973). IR: NH_3-HCl adduct in nitrogen matrix at 15°K.

94. L. M. Avkhutskii, S. A. Polishchuk, and S. P. Gabuda, *Isv. Sib. Otd. Akad. Nauk SSSR, Ser. Khim. Nauk*, 128 (1968); *Chem. Abstr.*, 70, 52867c (1969). NMR: ^1H and ^{19}F NMR of polycrystalline $\overline{AgF}$•$4H_2O$.

96. T. Axenrod and M. J. Wieder, *J. Amer. Chem. Soc.*, 93, 3541 (1971). NMR: Intramolecular, $^1J(^{15}NH)$ in *o*-substituted anilines.

97. A. Ažman, B. Borstnik, and D. Hadži, *J. Mol. Struct.*, 8, 315 (1971). IR: Intensity, band width, ν_{AH}.

98. A. Ažman, J. Koller, and D. Hadži, *Chem. Phys. Lett.*, 5, 157 (1970). Theory: CNDO calculations on ring models of polywater.

99. R. G. Azrak and E. B. Wilson, *J. Chem. Phys.*, 52, 5299 (1970). Microwave: Intramolecular, 2-haloethanols.

100. H. Baba, A. Matsuyama, and H. Kokubun, *Spectrochim. Acta*, 25A, 1709 (1969). UV: Proton transfer in *p*-nitrophenol-triethylamine adduct.

101. H. Baba, A. Matsuyama, and H. Kokubun, *J. Chem. Phys.*, 41, 895 (1964). UV: Proton transfer in *p*-nitrophenol-triethylamine adducts in dichloroethane.

102. H. Baba and S. Suzuki, *J. Chem. Phys.*, 35, 1118 (1961). UV: Thermodynamic data, adducts of phenol with *p*-dioxane in isooctane.

103. J. Bacon and D. P. Santry, *J. Chem. Phys.*, 56, 2011 (1972). Theory: Perturbation and CNDO calculations on HF polymers.

104. J. Bacon and D. P. Santry, *J. Chem. Phys.*, 55, 3743 (1971). Theory: Perturbation CNDO treatment of $(HF)_2$ and $(HCN)_2$.

105. R. F. W. Bader, *Can. J. Chem.*, 42, 1822 (1964). Theory: Model for hydrogen bond reactions.

106. R. M. Badger, *J. Chem. Phys.*, 8, 288 (1940). IR: $\Delta\nu$-ΔH correlation.

107. R. M. Badger and S. H. Bauer, *J. Chem. Phys.*, 5, 839 (1937). IR: $\Delta\nu$-ΔH correlation.

108. S. Badilescu and I. I. Badilescu, *Rev. Roum. Chim.*, 14, 329 (1969); *Chem. Abstr.*, 71, 38165m (1969). IR: Intramolecular, dihydroxyacetophenones.

109. G. D. Bagratishivili, G. V. Tsitsishvili, and K. A. Bezhashvili, *J. Phys. Chem. USSR*, 36, 1091 (1962). IR: Intramolecular, *o*-anilines.

110. J. Bailey and S. M. Walker, *Eur. Polym. J.*, 8, 339 (1972); *Chem. Abstr.*, 77, 6866y (1972). NMR, Kinetics: N-ethyl urethane in CCl_4, ethyl propionate, and 1,2-dimethoxyethane.

111. W. F. Baitinger, P. von R. Schleyer, and K. Mislow, *J. Amer. Chem. Soc.*, 87, 3168 (1965). IR: Intramolecular, hydroxy derivatives of biphenyl.

112. W. F. Baitinger, P. von R. Schleyer, T. S. S. R. Murty, and L. Robinson, *Tetrahedron*, 20, 1635 (1964). IR: Intramolecular, β-nitroalcohols.

113. A. Bajorek *et al.*, *Inst. Nucl. Phys.*, Cracow, Rep., No. 601 (1968); *Chem. Abstr.*, 70, 15057t (1969). Neutron scattering: Hydrates.

114. A. W. Baker and D. E. Bublitz, *Spectrochim. Acta*, 22, 1787 (1966). IR: Intramolecular, ferrocenyl alcohols and phenols.

115. A. W. Baker and W. W. Kaeding, *J. Amer. Chem. Soc.*, 81, 5904 (1959). IR: Intramolecular, $\Delta\nu_{OH}$ for *o*-halophenols and 2,6-dihalophenols.

116. A. W. Baker, H. O. Kerlinger, and A. T. Shulgin, *Spectrochim. Acta*, 20, 1467 (1964). IR: Intramolecular, ΔH errors, temperature dependency of ν_{OH} band absorptivities.

117. A. W. Baker and A. T. Shulgin, *Spectrochim. Acta*, 22, 95 (1966). IR: ΔH for *cis-trans* equilibrium of *o*-iodophenol in twelve solvents.

118. A. W. Baker and A. T. Shulgin, *Can. J. Chem.*, 43, 650 (1965). IR: Intramolecular, ΔH for *o*-halophenols.

119. A. W. Baker and A. T. Shulgin, *Nature*, 206, 712 (1965). IR: Intramolecular, *o*-trifluoromethyl phenol.

120. A. W. Baker and A. T. Shulgin, *Spectrochim. Acta*, 19, 1611 (1963). IR: Intramolecular, O—H...π, *o*-tritylphenols.

121. A. W. Baker and A. T. Shulgin, *J. Amer. Chem. Soc.*, 80, 5358 (1958). IR: Intramolecular, O—H...π bonds, 2-allylphenol, $\Delta\nu_{OH}$ for *o*-substituted phenols.

122. A. W. Baker and M. D. Yeaman, *Spectrochim. Acta*, 22, 1773 (1966). IR: Band analysis as applied to enthalpy calculations.

123. J. R. Baker, I. D. Watson, and A. G. Williamson, *Aust. J. Chem.*, 24, 2047 (1971). NMR: $CHCl_3$ adducts of ethers.

124. R. Baker and L. K. Dyall, *J. Chem. Soc. B*, 1952 (1971). IR: Intramolecular, bicyclo[2.2.1]heptan-2-ols.

125. Yu. G. Baklagina, M. V. Vol'kenshtein, and Yu. D. Krondraskiv, *Zh. Strukt. Khim.*, 7, 399 (1966). X-Ray: Complex of 9-methyladenine with 1-methyl-5-bromouracil.

126. A. Balasubramanian, *Indian J. Chem.*, 1, 329 (1963). UV: Effect of hydrogen bonding solvents on $n \to \sigma^*$ transitions of *i*-propyl iodide.

127. A. Balasubramanian, J. B. Capindale, and W. F. Forbes, *Can. J. Chem.*, 42, 2674 (1964). UV, IR: Intramolecular, 2,4-dinitrodiphenyl amines.

128. A. Balasubramanian, W. F. Forbes, and J. C. Dearden, *Can. J. Chem.*, 44, 961 (1966). UV: Intramolecular, 3-trifluoromethyl-2-nitrophenol.

129. A. Balasubramanian and C. N. R. Rao, *Spectrochim. Acta*, 18, 1337 (1962). UV: Solvent effects on $n \to \pi^*$ transitions of C=O, C=S, NO_2, N=N chromophores.

130. R. Balasubramanian, R. Chidambaram, and G. N. Ramachandran,
     Biochim. Biophys. Acta, <u>221</u>, 196 (1970). Theory: Modified
     Lippincott-Schroeder potential function.

131. R. E. Ballard and C. H. Park, *Spectrochim. Acta*, 399 (1970).
     Electronic: Self-association, carboxylic acids.

132. B. W. Bangerter and S. I. Chan, *J. Amer. Chem. Soc.*, <u>91</u>, 3910
     (1969). BIOL: NMR, base-stacking interaction in adenylyl-
     (3'→5')-cytidine and cytidylyl-(3'→5')-adenosine, aqueous
     solutions.

133. B. W. Bangerter and S. I. Chan, *Biopolymers*, <u>6</u>, 983 (1968).
     BIOL: NMR, binding of purine to poly U.

134. B. W. Bangerter and S. I. Chan, *Proc. Natl. Acad. Sci. U. S.*,
     <u>60</u>, 1144 (1968). BIOL: NMR, interaction of adenosine with
     poly U.

135. J. Bankovskis *et al.*, *Latv. PSR Zinat. Akad. Vestis, Kim. Ser.*,
     39, 52, 60 (1968); *Chem. Abstr.*, <u>70</u>, 72456w, 72454u (1969).
     UV: Hydrogen bonding of 5-halo derivatives of 8-mercaptoquino-
     line with OH solvents.

136. T. M. Barakat, N. Legge, and A. D. E. Pullin, *Trans. Faraday Soc.*,
     <u>59</u>, 1773 (1963). IR: Δν, hydrogen-bonded HNCS.

137. T. M. Barakat, N. Legge, and A. D. E. Pullin, *Trans. Faraday Soc.*,
     <u>59</u>, 1764 (1963). IR: Self-association, HNCS, DNCS.

138. T. M. Barakat, M. J. Nelson, S. M. Nelson, and A. D. E. Pullin,
     Trans. Faraday Soc., <u>65</u>, 41 (1969). IR: Δν, thermodynamic data,
     adducts of HNCS.

139. T. M. Barakat, M. J. Nelson, S. M. Nelson, and A. D. E. Pullin,
     Trans. Faraday Soc., <u>62</u>, 2674 (1966). IR: Thermodynamic data,
     adducts of HNCS.

140. J. R. Barcelo, *Ion (Madrid)*, <u>32</u>, 461, 594, 603, 609, 658, 712
     (1972); *Chem. Abstr.*, <u>78</u>, 123390v (1973). Review: Chloro- and
     fluoro-acetic acids.

140a. L. Barcza and M. T. Pope, *J. Phys. Chem.*, <u>77</u>, 1795 (1973).
     NMR: K, adducts of 1,2-dihydroxybenzene with X^-, ClO_4^-.

141. L. Bardet, J. Maillols, and H. Maillols, *Compt. Rend. Acad. Sci.*,
     Ser. B, <u>270</u>, 158 (1970). Raman: Succinic acid.

141a. A. J. Barnes, J. B. Davies, H. E. Hallam, and J. D. R. Howells,
     J. Chem. Soc., Faraday 2, 246 (1973). IR: Matrix studies of
     HI complexes at 20°K.

141b. A. J. Barnes, H. E. Hallam, and G. F. Scrimshaw, *Trans. Faraday
     Soc.*, <u>65</u>, 3150, 3159, 3172 (1969). IR: Cryogenic studies of
     hydrogen halide polymers.

141c. A. J. Barnes, H. E. Hallam, and D. Jones, *Proc. Roy. Soc.*,
     London, <u>335A</u>, 97 (1973). IR: Intramolecular, self-association,
     alcohols.

142. J. Barrett and A. L. Mansell, *Nature*, <u>187</u>, 138 (1960). UV: Self-
     association, H_2O, HDO, D_2O.

143. G. M. Barrow, *Spectrochim. Acta*, 16, 799 (1960). Theory: Double-minimum potential energy curves, pKa correlations.

144. C. D. Barry, A. C. T. North, J. A. Glasel, R. J. P. Williams, and A. V. Xavier, *Nature*, 232, 236 (1971). BIOL, NMR: Use of lanthanide shift reagents for determination of nucleotide conformations.

145. J. R. Bartels-Keith and R. F. W. Cieciuch, *Can. J. Chem.*, 46, 2593 (1968). NMR: Intramolecular, *o*-substituted acetanilides.

146. J. Bartsch and V. Prey, *Justus Liebigs Ann. Chem.*, 717, 198 (1968). IR: Intramolecular, substituted gluco- and fructo-pyranoses and furanoses.

147. M. R. Basila, E. L. Saier, and L. R. Cousins, *J. Amer. Chem. Soc.*, 87, 1665 (1965). IR: K, adducts of 2-butanol with π-bases.

148. G. E. Bass, *Proc. Natl. Acad. Sci. U. S.*, 62, 345 (1969). Theory: Resonance stabilization of cyclic groups of hydrogen-bonded water molecules.

149. G. E. Bass and L. J. Schaad, *J. Amer. Chem. Soc.*, 93, 4585 (1971). Theory: Pariser-Parr-Pople calculation of hydrogen bonds in DNA replication plane.

150. S. Basu, *Z. Naturforsch. B*, 23, 562 (1968). BIOL: Effect of temperature, HCHO, and nucleate ion concentration on the denaturation of DNA at low temperature.

151. S. Basu and N. N. Das Gupta, *Biochim. Biophys. Acta*, 174, 74 (1969). BIOL, UV: DNA.

152. S. Basu and L. Loh, *Biochim. Biophys. Acta*, 76, 131 (1963). BIOL: Absorption and fluorescence spectra of mixtures of nucleosides.

153. E. Bauer and M. Magat, *J. Phys. Radium*, 9, 319 (1938). IR: Model for solvent effects on frequency shifts based on dielectric constant.

154. K. Bauge and J. W. Smith, *J. Chem. Soc. A*, 616 (1966). Dipole moments of tri-*n*-butylammonium and triethylammonium salts in benzene.

155. J. W. Bayles and B. Evans, *J. Chem. Soc.*, 6984 (1965). Electronic: Association of 2,4-dinitrophenol with pyrrolidines and piperidines.

156. J. W. Bayles and A. F. Taylor, *J. Chem. Soc.*, 417 (1961). Electronic: Thermodynamic data, adducts of 2,4-dinitrophenol with butyl amines.

157. N. S. Bayliss, A. R. H. Cole, and L. H. Little, *Aust. J. Chem.*, 8, 26 (1955). IR: Solvent effects on stretching frequencies.

159. I. J. Bear and W. G. Mumme, *Spectrochim. Acta*, 26A, 755 (1970). IR: Zirconium sulfate hydrates.

160. K. Beardsley, T. Tao, and C. R. Cantor, *Biochemistry*, 9, 3524 (1970). BIOL: Conformation of the anticodon loop of phenylalanine tRNA.

161. L. A. Beath, A. G. Williamson, *J. Chem. Thermodyn.*, $\underline{1}$, 51 (1969).
      CAL: Enthalpies of mixing, ethers with CCl_4 and $CHCl_3$.

162. J. L. Beauchamp, *Ann. Rev. Phys. Chem.*, $\underline{22}$, 527 (1971). Review:
      Ion cyclotron resonance, study of strong hydrogen bonds in the
      gas phase.

163. J. L. Beauchamp, D. Holtz, S. D. Woodgate, and S. L. Pratt,
      J. Amer. Chem. Soc., $\underline{94}$, 2798 (1972). Ion Cyclotron Resonance:
      Basicities of alkyl halides in the gas phase.

164. J. K. Becconsall and P. Hampson, *Mol. Phys.*, $\underline{10}$, 21 (1966).
      NMR: Solvent effects, ^{1}H and ^{13}C NMR chemical shifts for CH_3I
      and CH_3CN.

165. A. Becker and J. Hurwitz, *Prog. Nucl. Acid Res. Mol. Biol.*, $\underline{11}$,
      423 (1971). BIOL: Review, DNA replication.

166. E. D. Becker, *High Resolution NMR*, Academic Press, New York,
      1969, p. 232. Review.

167. E. D. Becker, *Spectrochim. Acta*, $\underline{17}$, 436 (1961). IR: Thermo-
      dynamic data, methanol adducts with bases.

168. E. D. Becker, *Spectrochim. Acta*, $\underline{15}$, 743 (1959). NMR: Self-
      association of $CHCl_3$.

169. E. D. Becker, *J. Chem. Phys.*, $\underline{31}$, 269 (1959). NMR: *t*-Butanol
      dimer.

170. E. D. Becker, U. Liddel, and J. N. Shoolery, *J. Mol. Spectrosc.*,
      $\underline{2}$, 1 (1958). NMR: Self-association, ethanol in CCl_4.

170a. P. Becker, H. Brusset, and H. Gillier-Pandraud, *Compt. Rend. Acad.
      Sci.*, $\underline{274C}$, 1043 (1972). X-Ray: Pyrogallol at -150°C, intra-
      and intermolecular hydrogen bonds.

171. W. Beckering, *J. Phys. Chem.*, $\underline{65}$, 206 (1961). IR: Intramolec-
      ular, hydrogen bonding to π-electrons in *o*-substituted phenols.

172. B. Behera and P. S. Zacharias, *Spectrochim. Acta*, $\underline{27A}$, 2273
      (1971). IR: Intramolecular, triazene 1-oxides.

173. V. Bekárek, *J. Chem. Soc. D*, 1565 (1971). IR: Correlation of
      ν_{OH} of phenols with Hammett σ.

173a. V. Bekárek, I. Janu, J. Jirkovský, J. Socha, and J. Klicnar,
      Collect. Czech. Chem. Commun., $\underline{37}$, 3447 (1972). IR: Intramo-
      lecular, substituted 2-nitroanilines.

174. V. Bekárek, J. Kavalek, J. Socha, and S. Andrijsek, *Chem. Commun.*,
      630 (1968). NMR: Intramolecular, *o*-nitroanilines.

174a. C. Belin and J. Potier, *J. Chim. Phys. Physicochim. Biol.*, $\underline{69}$,
      1222 (1972); $\underline{70}$, 490 (1973). IR: Perchloric acid-nitromethane
      adduct.

175. C. L. Bell and G. M. Barrow, *J. Chem. Phys.*, $\underline{31}$, 300 (1959).
      Near-IR: Evidence for double-minimum potential, alcohols.

176. R. P. Bell and J. E. Crooks, *J. Chem. Soc.*, 3513 (1962). UV, IR:
      Proton transfer and deuterium isotope effects, *p*-nitrophenol
      adducts with amines.

177. L. J. Bellamy, *Advances in Infrared Group Frequencies*, Methuen and Co., Ltd., Great Britain, 1968, Chapter 8. IR: Review, ν_{AH}.

178. L. J. Bellamy, C. P. Conduit, R. J. Pace, and R. L. Williams, *Trans. Faraday Soc.*, $\underline{55}$, 1677 (1959). IR: Solvent effects on ν_{SO}.

179. L. J. Bellamy, G. Eglinton, and J. F. Morman, *J. Chem. Soc.*, 4762 (1961). IR: K, phenol-ether complexes in CCl_4.

180. L. J. Bellamy and H. E. Hallam, *Trans. Faraday Soc.*, $\underline{55}$, 220 (1959). IR: Solvent effects on ν_{AH}.

181. L. J. Bellamy, H. E. Hallam, and R. L. Williams, *Trans. Faraday Soc.*, $\underline{54}$, 1120 (1958). IR: Solvent effects on ν_{AH}.

182. L. J. Bellamy, R. F. Lake, and R. J. Pace, *Spectrochim. Acta*, $\underline{19}$, 443 (1963). IR: Acetic acid dimerization, cyclic and open structures.

183. L. J. Bellamy and A. J. Owen, *Spectrochim. Acta*, $\underline{25A}$, 329 (1969). IR: Empirical relationship between $\Delta\nu_{AH}$ and A...$\overline{B}$ bond distances.

184. L. J. Bellamy, A. R. Osborn, E. R. Lippincott, and A. R. Bandy, *Chem. Ind. (London)*, 686 (1969). Near-IR, Raman: Anomalous water.

185. L. J. Bellamy, A. R. Osborn, and R. J. Pace, *J. Chem. Soc.*, 3749 (1963). IR: $\Delta\nu_{OH}$ and $\Delta\nu_{OD}$, deuterium isotope effect.

186. L. J. Bellamy and R. J. Pace, *Spectrochim. Acta*, $\underline{27A}$, 705 (1971). IR: $\Delta\nu_{CO}$, $\Delta\nu_{OH}$ for alcohol-ketone adducts.

186a. L. J. Bellamy and R. J. Pace, *Spectrochim. Acta*, $\underline{28A}$, 1869 (1972). IR: Effects of nonequivalent hydrogen bonding on $\overline{\nu_S}$, RNH_2, H_2O.

187. L. J. Bellamy and R. J. Pace, *Spectrochim. Acta*, $\underline{25A}$, 319 (1969). IR: Correlation of $\Delta\nu_{XH}$ with X...Y distances in $\overline{X—H}$...Y.

188. L. J. Bellamy and R. J. Pace, *Spectrochim. Acta*, $\underline{22}$, 525 (1966). IR: Structure of alcohol and phenol dimers.

189. L. J. Bellamy and R. J. Pace, *Spectrochim. Acta*, $\underline{19}$, 435 (1963). IR: Oxalic acid dimerization, cyclic and open structures.

190. L. J. Bellamy and P. E. Rogasch, *Proc. Roy. Soc. (London)*, $\underline{257A}$, 98 (1960). IR: Effect of proton transfer on band width.

191. L. J. Bellamy and R. L. Williams, *Proc. Roy. Soc. (London)*, $\underline{255A}$, 22 (1960). IR: Solvent effects on group frequencies.

192. L. J. Bellamy and R. L. Williams, *Proc. Roy. Soc. (London)*, $\underline{254A}$, 119 (1960). IR: Solvent effects on the ν_{OH} of phenol, 2,6-disubstituted phenols.

193. L. J. Bellamy and R. L. Williams, *Trans. Faraday Soc.*, $\underline{55}$, 14 (1959). IR: Solvent effects on ν_{CO}.

194. A. M. Bellocq, C. Perchard, A. Novak, and M. L. Josien, *J. Chim. Phys.*, $\underline{62}$, 1334 (1965). IR: Spectra of self-associated imidazole.

195. L. Bellon and J. L. Abboud, *Compt. Rend. Acad. Sci.*, $\underline{261}$, 3106 (1965). UV: Adducts of $CHCl_3$ and $CDCl_3$ with bases.

196. L. P. Belozerskaya and D. N. Shchepkin, *Tepl. Dvizhenie Mol.
      Mezhmol. Vzaimodeistire Zhidk. Rastvorakh*, 166 (1969); *Chem.
      Abstr.*, 74, 117782a (1971). IR: ν_{NH}, pyrrole-ether adduct.

197. L. P. Belozerskaya and D. N. Shchepkin, *Opt. Spektrosk.*, supple-
      ment 3, 290 (1967). [English, 146 (1968)]. Far-IR: HX-ether
      adducts.

198. A. A. Belyaeva, M. I. Dvorkin, and L. D. Shcherba, *Opt. Spec-
      trosk.*, 31, 392 (1971). [English, 210]. IR: KOH and $(KOH)_2$
      in argon matrix.

198a. W. S. Benedict, N. Gailar, and E. K. Plyler, *J. Chem. Phys.*, 24,
      1139 (1956). IR: Water vapor.

199. H. A. Benesi and J. H. Hildebrand, *J. Amer. Chem. Soc.*, 71, 2703
      (1949). Equation for calculating equilibrium constant.

200. L. Benjamin and G. C. Benson, *J. Phys. Chem.*, 67, 858 (1963).
      CAL: Deuterium isotope effect.

200a. A. Ben-Naim, *Chem. Phys. Lett.*, 13, 406 (1972). Theory: Solute
      effects, water structure.

200b. A. Ben-Naim and F. H. Stillinger, Jr., *Water Aqueous Solutions:
      Structure, Thermodynamics, Transport Processes*, R. A. Horne, ed.,
      Interscience, New York, 1972, pp. 295-330. Theory: Water
      structure.

201. A. Benoit, *Spectrochim. Acta*, 19, 2011 (1963). Far-IR: $HCoO_2$
      and $DCoO_2$.

202. A. M. Benson and H. G. Drickamer, *J. Chem. Phys.*, 27, 1164 (1957).
      IR: Model for solvent effects on stretching frequencies.

203. H. A. Bent, *Chem. Rev.*, 68, 587 (1968). Review of structural
      chemistry of donor-acceptor interactions.

204. W. E. Bentz and L. D. Colebrook, *Can. J. Chem.*, 47, 2473 (1969).
      NMR: Hydrogen bonding in acetylenes containing oxygen.

205. B. Berglund and J. Tegenfeldt, *Mol. Phys.*, 26, 633 (1973).
      NMR: ^2D, deuterated $NaHC_2O_4 \cdot H_2O$.

206. K. Bergman, M. Eigen, and L. de Maeyer, *Ber. Bunsenges, Physik.
      Chem.*, 67, 819 (1963). Kinetics: Self-association, ε-
      caprolactam.

207. E. D. Bergmann, H. Weiler-Feilchenfeld, and Z. Neiman, *J. Chem.
      Soc. B*, 1334 (1970). BIOL, UV: Adenine derivatives.

208. P. J. Berkeley, Jr. and M. W. Hanna, *J. Amer. Chem. Soc.*, 86,
      2990 (1964). Theory: Calculation of NMR chemical shift due to
      hydrogen bonding; $CHCl_3$ adducts of nitrogen bases.

209. P. J. Berkeley, Jr. and M. W. Hanna, *J. Chem. Phys.*, 41, 2530
      (1964). NMR: $CHCl_3$-amine adducts.

210. P. J. Berkeley, Jr. and M. W. Hanna, *J. Phys. Chem.*, 67, 846
      (1963). NMR: K for $CHCl_3$ with nitrogen donors.

211. I. B. Berlman, *Chem. Phys. Lett.*, 3, 61 (1969). Intramolecular,
      2,5-diphenyl-1,3,4-oxadiazole and 2-phenyl-5-(4-biphenylyl)-
      1,3,4-oxadiazole.

212. M. C. Bernard-Houplain, C. Bourdéron, J. J. Péron, and C. Sandorfy, *Chem. Phys. Lett.*, 11, 149 (1971). IR: Effect of bromine-containing solvents on O—H...O and N—H...N systems.

212a. M. C. Bernard-Houplain, G. Belanger, and C. Sandorfy, *J. Chem. Phys.*, 57, 530 (1972). IR: Self-association, N-methylaniline at low temperature.

212b. M. C. Bernard-Houplain and C. Sandorfy, *Can. J. Chem.*, 51, 1075 (1973). IR: Self-association, pyrrole, indole.

212c. M. C. Bernard-Houplain and C. Sandorfy, *J. Chem. Phys.*, 56, 3412 (1972). IR: Self-association, Me_2NH at low temperature.

213. C. F. Bernasconi, *J. Phys. Chem.*, 75, 3636 (1971). Kinetics: Intramolecular, Meisenheimer complexes.

213a. C. V. Berney, *J. Amer. Chem. Soc.*, 95, 708 (1973). IR: Solid CF_3COOH, CF_3COOD.

214. C. V. Berney, R. L. Redington, and K. C. Lin, *J. Chem. Phys.*, 53, 1713 (1970). IR: Matrix-isolated carboxylic acid dimers at 4°K.

215. H. J. Bernstein, *J. Amer. Chem. Soc.*, 85, 484 (1963). Energy cycle for hydrogen bond energies, frequency shift correlations.

216. H. Berthod and A. Pullman, *Compt. Rend. Acad. Sci.*, 259, 2711 (1964). Theory: Effect of hydrogen bonding on dipole moment of nucleotide base pairs.

216a. H. Berthod and A. Pullman, *Chem. Phys. Lett.*, 14, 217 (1972). Theory: Formamide dimer.

217. H. Berthod and B. Pullman, *Biochim. Biophys. Acta*, 246, 359 (1971). Theory: MO calculations, conformation of nucleic acids.

218. H. Berthod and B. Pullman, *Biochim. Biophys. Acta*, 232, 595 (1971). Theory: MO calculations, conformation of nucleic acids.

219. C. Berthomieu and C. Sandorfy, *J. Mol. Spectrosc.*, 15, 15 (1965). IR: Frequencies and intensities of ν_{NH} for secondary amines.

219a. J. E. Bertie and M. V. Falk, *Can. J. Chem.*, 51, 1713 (1973). Far-IR: Gas phase, Me_2O-HCl.

220. J. E. Bertie and D. J. Millen, *J. Chem. Soc.*, 514 (1965). IR: Gas-phase spectra of DCl adducts with ethers.

221. J. E. Bertie and D. J. Millen, *J. Chem. Soc.*, 497 (1965). IR: Gas-phase spectra of ether-hydrogen chloride adducts.

222. J. E. Bertie and D. J. Millen, *J. Chem. Soc.*, 510 (1964). IR: HF adducts with bases.

223. M. Bertolucci, F. Jantzef, and D. L. Chamberlain, Jr., *Advan. Chem. Ser.*, 87, 124 (1968). IR: Reflection spectra, adsorption of proton donors on trialkyl phosphates and calcium phosphates.

224. G. L. Bertrand, R. D. Beaty, and H. A. Burns, *J. Chem. Eng. Data*, 13, 436 (1968). CAL: Construction of calorimeter.

225. P. Besarini, G. Galloni, and S. Ghersetti, *Spectrochim. Acta*, 20, 267 (1964). IR: $\Delta\nu_{OH}$, K, phenol adducts of sulfones.

226. S. Besnainou, *J. Chim. Phys.*, <u>67</u>, 1825 (1970). Electronic: $\pi \rightarrow \pi^*$, anilines associated with dioxane or methanol.

227. S. Besnainou and C. Benoit, *Compt. Rend. Acad. Sci. Paris*, Ser. C, <u>270</u>, 2124 (1970); Ser. B, <u>274</u>, 478 (1972). Theory: Hydrogen bond in solid imidazole by CNDO/2.

228. S. Besnainou, R. Prat, and S. Bratož, *J. Chim. Phys.*, <u>61</u>, 222 (1964). Theory, UV: Pariser-Parr-Pople calculations on the influence of the hydrogen bond on $\pi \rightarrow \pi^*$ bands of organics.

229. G. Bessis, *Cahiers Phys.*, <u>1961</u>, 105. Theory: VB calculations of FHF⁻.

230. G. Bessis and S. Bratož, *J. Chim. Phys.*, <u>57</u>, 769 (1960). Theory: Early VB studies on FHF⁻.

231. E. Bessler and G. Bier, *Makromol. Chem.*, <u>122</u>, 30 (1969). IR: ν_{NH}, polyamides.

232. N. N. Bessonova *et al.*, *Zh. Obshch. Khim.*, <u>43</u>, 144 (1973). [English, p. 141]. IR: Intramolecular, *o*-aminoacetanilides.

233. K. R. Bhaskar and C. N. R. Rao, *Biochim. Biophys. Acta*, <u>136</u>, 561 (1967). Near-IR: Self-association of N-methylacetamide, N-phenylurethane, thermodynamic data.

234. K. R. Bhaskar and S. Singh, *Spectrochim. Acta*, <u>23A</u>, 1155 (1967). IR: Thermodynamic data, adduct of phenol with 2-aminopyridine.

235. K. R. Bhaskar, S. Singh, S. N. Bhat, and C. N. R. Rao, *Inorg. Nucl. Chem.*, <u>28</u>, 1915 (1966). IR: Thermodynamic data, adduct of phenol with triphenylamine.

236. B. B. Bhowmik, *J. Phys. Chem.*, <u>74</u>, 4442 (1970). Electronic: Adducts of β-naphthol with azaaromatics and aromatic hydrocarbons.

237. B. B. Bhowmik, *Indian J. Chem.*, <u>7</u>, 788 (1969). UV: Association of α- and β-naphthols with azaaromatics, K.

238. B. B. Bhowmik and S. Basu, *Trans. Faraday Soc.*, <u>59</u>, 813 (1963). IR: K, adducts of substituted phenols with ethers.

239. B. B. Bhowmik and S. Basu, *Trans. Faraday Soc.*, <u>58</u>, 48 (1962). UV: K, adducts of amines.

239a. V. V. Bhujle and M. R. Padhye, *Indian J. Pure Appl. Phys.*, <u>10</u>, 867 (1972). IR: Interaction of $CHCl_3$ and $CDCl_3$ with amino-pyridines.

239b. J. Biais, B. Lemanceau, and C. Lussan, *J. Chim. Phys.*, <u>64</u>, 1019 (1967). NMR: Self-association, alcohols.

239c. J. Biais, J. Dos Santos, and B. Lemanceau, *J. Chim. Phys.*, <u>67</u>, 806 (1970). NMR: Self-association, alcohols.

239d. R. Bicca de Alencastroa and C. Sandorfy, *Can. J. Chem.*, <u>51</u>, 985, 1443 (1973). IR: Adducts of thiols with C_5H_5N, Et_3N, self-association of thiols.

240. G. Biczó, J. Ladik, and J. Gergely, *Acta Phys. Hung.*, <u>20</u>, 11 (1966). Theory: Tunneling frequency of hydrogen bond in nucleotide base pairs.

241. G. Biczó, J. Ladik, and J. Gergely, *Phys. Letters*, 13, 317 (1964). Theory: Tunneling frequencies in nucleotide base pairs.

242. G. Biczó, J. Ladik, and J. Gergely, *Magy. Kem. Folyoirat*, 71, 292 (1965); *Chem. Abstr.*, 60, 14055f (1965). Theory: Proton tunneling in nucleotide base pairs.

243. A. I. Biggs and R. A. Robinson, *J. Chem. Soc.*, 388 (1961). pK values for substituted anilines and phenols.

244. F. Bigoli, A. Manotti Lanfredi, A. Tiripicchio, and M. Tiripicchio Camellini, *Acta Crystallogr.*, Sect. B, 26, 1075 (1970). X-Ray: Hydrogen bonds in $[Mg(H_2O)_6][H_3IO_6]$.

244a. J. M. Bijen, *J. Mol. Struct.*, 17, 69 (1973). IR, Electron diffraction: Intramolecular, gaseous glycol monoformate.

245. P. Biscarini, G. Galloni, and S. Ghersetti, *Bull. Sci. Fac. Chim. Ind. Bologna*, 21, 154 (1963). IR: Phenol-sulfoxide adducts, $\Delta\nu$-σ^* correlation.

246. P. Biscarini, G. Galloni, and S. Ghersetti, *Spectrochim. Acta*, 20, 267 (1964). IR: K, $\Delta\nu_{OH}$, phenol-sulfone adducts.

248. J. O. Bishop, *Biochem. J.*, 116, 223 (1970). BIOL: Equilibrium interpretation of DNA:RNA hybridization data.

249. J. O. Bishop, *Nature*, 224, 600 (1969). BIOL: Mechanisms of DNA-RNA hybridization.

250. J. O. Bishop, F. W. Robertson, J. A. Burns, and M. Melli, *Biochem. J.*, 115, 361 (1969). BIOL: Analysis of DNA-RNA hybridization data.

251. R. J. Bishop and L. E. Sutton, *J. Chem. Soc.*, 6100 (1964). K, phenol-pyridine adduct in various solvents.

252. H. D. Bist and V. N. Sarin, *Proc. Nucl. Phys. Solid State Phys. Symp.*, *13th*, 3, 252 (1968). IR: Vapor phase, phenol.

253. J. L. Bjorkstam, *J. Phys. Soc. Jap.*, Suppl., 28, 101 (1969). NMR: Hydrogen-bonded ferroelectrics.

254. P. Bladon, *Annual Rev. NMR Spectrosc.*, E. F. Mooney, ed., 2, 1 (1969), Academic Press. NMR: Review, includes short discussion of keto-enol equilibria, self-association, complex formation.

255. R. D. Blake, J. R. Fresco, and R. Langridge, *Nature*, 225, 32 (1970). X-Ray: Single crystals of mixtures of tRNAs.

256. R. D. Blake, L. C. Klotz, and J. R. Fresco, *J. Amer. Chem. Soc.*, 90, 3556 (1968). UV: Complex formation in equimolar mixtures of poly A and poly U.

257. R. F. Blanks and J. M. Prausnitz, *J. Chem. Phys.*, 38, 1500 (1963). IR: Thermodynamic data, adducts of 2-propanol with ethers.

257a. R. Blinc, *Advances in Magnetic Resonance*, J. S. Waugh, ed., 3, 141 (1968). Review: Magnetic resonance, hydrogen-bonded ferroelectrics.

258. R. Blinc and D. Hadži in *Hydrogen Bonding*, D. Hadži, ed., Pergamon Press, 1959, p. 147. IR, Theory: Double-minimum potential.

258a. R. Blinc and D. Hadži, *Spectrochim. Acta*, 16, 852 (1960). IR,
NMR: Proton tunneling in compounds containing strong hydrogen
bonds.

258b. R. Blinc and D. Hadži, *Nature*, 212, 1307 (1966). NMR: ^{1}D NMR,
ferroelectric crystals.

259. R. Blinc, D. Hadži, and A. Novak, *Z. Elektrochem.*, 64, 567 (1960).
Theory, IR: Semiempirical studies of hydrogen bond potential
curve, lack of correlation between X...Y distance and ν_{X-H}.

259a. R. Blinc, M. Pintar, and I. Zupancic, *J. Phys. Chem. Solids*, 28,
405 (1967). NMR: Deuterated triglycine sulfate.

260. R. Blinc and M. Ribaric, *Phys. Rev.*, 130, 1816 (1963). Theory:
Proton-lattice interactions in hydrogen-bonded ferroelectric
crystals.

261. R. Blinc and S. Svetina, *Phys. Rev.*, 147, 430 (1966). Theory:
Ferroelectrics, KH_2PO_4.

262. R. Blinc, Z. Trontelj, and B. Volavsek, *J. Chem. Phys.*, 44, 1028
(1966). NMR: KH_2F_3, unsymmetrical hydrogen bond.

263. U. Blindheim and T. Gramstad, *Spectrochim. Acta*, 25A, 1105 (1969).
IR: $\Delta\nu_{OH}$, phenol adducts of phosphoryl compounds.

264. U. Blindheim and T. Gramstad, *Spectrochim. Acta*, 21, 1073 (1965).
IR: Thermodynamic data, phenol adducts of phosphoryl compounds.

265. A. C. Blizzard and D. P. Santry, *J. Theor. Biol.*, 25, 461 (1969).
BIOL: CNDO/2 calculations, electronic structure of the G—C
base pair, proton transfer in the excited state.

265a. M. Blume and R. E. Watson, *Proc. Roy. Soc. (London)*, A271, 565
(1963). Spin-orbit coupling.

266. S. M. Blumenfeld and H. Fast, *Spectrochim. Acta*, 24A, 1449 (1968).
Raman: Low-frequency data for HCOOH and HCOOD.

267. R. M. Bock and J. L. Hoffman, *Methods in Enzymology*, Vol. XII,
L. Grossman and K. Moldave, eds., Nucleic Acids, Part B, Academic
Press, New York, 1968, p. 253. BIOL: Solvent perturbation
spectra of nucleic acids.

268. N. Bodor, M. J. S. Dewar, and A. J. Harget, *J. Amer. Chem. Soc.*,
92, 2929 (1970). Theory: Nucleotide bases, SCF-MO π-electron
calculations.

269. H. Boedtker and D. G. Kelling, *Biochem. Biophys. Res. Commun.*,
29, 758 (1967). UV: Helical content of 5s RNA.

270. Yu. S. Bogachev, L. K. Vasianina, N. N. Shapet'ko, and T. L.
Alexeeva, *Org. Magn. Resonance*, 4, 453 (1972). NMR: Self-
association, phenol.

271. V. S. Bogdanov, T. M. Ushakova, A. N. Volkov, and A. V. Bogdanova,
Khim. Atsetilena, 403 (1968); *Chem. Abstr.*, 70, 110378g (1969).
NMR: Intramolecular, ethynyl-vinyl ethers and sulfur- and
nitrogen-containing analogs.

272. W. Bol, *J. Appl. Crystallogr.*, 1, 234 (1968). X-Ray: Liquid
water.

273. R. W. Bolander, J. L. Kassner, Jr., and J. T. Zung, *J. Chem. Phys.*, 50, 4402 (1969). Theory: Semiempirical statistical mechanical treatment of hydrogen-bonded clusters in the vapor phase.

274. J. Bolard, *J. Chim. Phys. Physicochim. Biol.*, 66, 389 (1969). UV, IR: (-) Menthol association with acetone.

274a. I. E. Boldeskul *et al.*, *Zh. Prikl. Spektrosk.*, 16, 859 (1972). IR: Thermodynamic data, tetraalkylphosphonium salt-alcohol systems.

275. T. F. Bolles and R. S. Drago, *J. Amer. Chem. Soc.*, 87, 5015 (1965). CAL: Calorimetric method.

276. R. Bonaccorsi, C. Petrongolo, E. Scrocco, and J. Tomasi, *Theor. Chim. Acta*, 20, 331 (1971). Theory: Electrostatic model for water dimer.

277. A. Bondi, *J. Phys. Chem.*, 68, 441 (1964). Tabulations of van der Waals radii.

277a. O. D. Bonner, *J. Phys. Chem.*, 76, 1228 (1972). IR: HDO.

278. M. Bonnet and A. Julg, *J. Chim. Phys.*, 59, 723 (1962). UV: Thermodynamic data, adducts of phenol and α-naphthol with amines.

278a. G. J. Boobyer, *Spectrochim. Acta*, 23A, 325 (1967). IR: Δν, ΔB, *n*-propylacetylene adducts.

278b. G. J. Boobyer and W. J. Orville-Thomas, *Spectrochim. Acta*, 22, 147 (1966). IR: Model for ν_{AH} intensity changes.

278c. P. Bordewijk, M. Kunst, and A. Rip, *J. Phys. Chem.*, 77, 548 (1973). Self-association, heptanol-1 in CCl_4.

279. V. E. Borisenko and D. N. Shchepkin, *Opt. Spektrosk.*, 29, 46 (1970); [English, 24]. IR: $CHCl_3$ and $CDCl_3$ in gas-phase, aprotic, and electron-donor solvents.

280. V. E. Borisenko and G. N. Volokhina, *Opt. Spektrosk.*, 29, 683 (1970); [English, 365]. IR: Association of *t*-butylacetylene with various bases.

281. E. Ya. Borisova, M. G. Zaitseva, and E. M. Cherkasova, *Reakts. Sposobnost Org. Soedin.*, 6, 135 (1969); *Chem. Abstr.*, 71, 90660h (1969). IR: Intramolecular, N-(ω-aminoalkyl) amides.

282. N. P. Borisova, D. N. Gleborskii, and A. A. Krasheninnikov, *Vestn. Leningrad. Univ., Khim.*, 1970, 134; *Chem. Abstr.*, 73, 123638d (1970). Theory: Hückel calculations on model systems.

283. P. M. Borodin, R. D. Singh, and V. A. Shcherbakov, *Zh. Strukt. Khim.*, 9, 1078 (1968); *Chem. Abstr.*, 70, 72684u (1969). NMR: ^{19}F in DF-D_2O and DF—H_2O—D_2O systems.

284. R. Borsdorf and A. Preiss, *J. Prakt. Chem.*, 311, 36 (1969). IR: Transannular hydrogen bonds in 1,4-cycloheptanediols.

285. L. Borucki, *Ber. Bunsenges, Physik. Chem.*, 71, 504 (1967). Kinetics: Self-association, benzoic acid.

286. A. A. Bothner-By and R. E. Glick, *J. Chem. Phys.*, 26, 1651 (1957). BIOL: NMR, magnetic anisotropy, ring currents.

287. M. Boublik, P. Barth, and J. Sponar, *Collect. Czech. Chem. Commun.*, 32, 2201 (1966). BIOL: Changes in the MW and shape, calf thymus DNA particles.

288. A. D. Boul, D. J. Chadwick, and G. D. Meakins, *J. Chem. Soc. D*, 1624 (1971). IR: Intramolecular, *trans*-2-halogenocyclohexanols.

289. H. Bourassa-Bataille, P. Sauvageau, and C. Sandorfy, *Can. J. Chem.*, 41, 2240 (1963). IR: Intramolecular, *o*-halophenols.

289a. C. Bourdéron, *Can. J. Chem.*, 50, 1199 (1972). IR: Cyclohexanone adduct with 2,6-di-*t*-butyl-*p*-cresol.

289b. C. Bourdéron, J. J. Péron, and C. Sandorfy, *J. Phys. Chem.*, 76, 864 (1972). IR: Self-association, sterically hindered alcohols.

289c. C. Bourdéron, J. J. Péron, and C. Sandorfy, *J. Phys. Chem.*, 76, 869 (1972). Near-IR: Self-association, methanol.

290. C. Bourdéron and C. Sandorfy, *J. Chem. Phys.*, 59, 2527 (1973). Near-IR: Self-association, alcohols.

291. A. J. R. Bourn, D. G. Gillies, and E. W. Randall, *Tetrahedron*, 20, 1811 (1964). NMR: *Cis-trans* isomerism in formanilide.

292. J. Bournay and Y. Marechal, *J. Chem. Phys.*, 55, 1230 (1971). IR: Self-association, propynoic and acrylic acid.

293. K. Bowden, J. G. Irving, and M. J. Price, *Can. J. Chem.*, 46, 3903 (1968). NMR: Association of substituted mesitylenes and durenes with solvents.

294. H. C. Bowen and J. W. Linnett, *J. Chem. Soc. A*, 1675 (1966). Theory: Approximate calculations on HF_2^- and water dimer (treated as a 4-electron problem).

294a. M. J. T. Bowers and R. M. Pitzer, *J. Chem. Phys.*, 59, 163 (1973). Theory: Bond orbital analysis of $(H_2O)_2$.

294b. M. T. Bowers and W. H. Flygare, *J. Mol. Spectrosc.*, 19, 325 (1966). IR: Matrix studies of hydrogen halide dimers.

295. R. S. Bowman, D. R. Stevens, and W. E. Baldwin, *J. Amer. Chem. Soc.*, 79, 87 (1957). IR: Self-association, phenols.

296. D. B. Boyd, *Theor. Chim. Acta*, 14, 402 (1969). BIOL: MO for components of ATP.

296a. S. F. Boys, *Proc. Roy. Soc.*, A200, 542 (1950). Theory: Electronic wavefunctions.

297. P. A. Bozhulin, L. Singurel, and A. V. Chekunov, *Zh. Prikl. Specktrosk.*, 9, 303 (1968). IR, Raman: Solutions of Me_2SO in $CHCl_3$ and H_2O.

298. J. R. Bracelo and C. Otero, *Spectrochim. Acta*, 18, 1231 (1962). IR: Self-association, fluoroacetic acids.

299. E. M. Bradbury and C. Crane-Robinson, *Nature*, 220, 1079 (1968). BIOL: Review, high resolution NMR studies of biopolymers.

300. E. M. Bradbury and A. Elliott, *Spectrochim. Acta*, 19, 995 (1963). IR: Spectra of crystalline N-methyl acetamide.

301. D. F. Bradley and H. A. Nash, *Molecular Associations in Biology*, B. Pullman, ed., Academic Press, 1968, p. 137. Theory: DNA replication.

302. J. Brahms, *J. Chim. Phys. Physicochim. Biol.*, 65, 105 (1968). BIOL: CD, conformation and stability of oligonucleotides and nucleic acids.

303. J. Brahms, J. C. Maurizot, and A. M. Michelson, *J. Mol. Biol.*, 25, 481 (1967). BIOL: Stacking of 3'→5' dinucleoside phosphates.

304. J. Brahms, A. M. Michelson, and K. E. van Holde, *J. Mol. Biol.*, 15, 467 (1966). BIOL: Adenylate oligomers in single- and double-strand conformation.

305. J. Brahms and C. Sadron, *Nature*, 212, 1309 (1966). BIOL: Conformational stability of polynucleotides.

306. J. Brandmüller and K. Seevogel, *Spectrochim. Acta*, 20, 453 (1964). IR: K, adducts of alcohols with pyridine.

307. W. W. Brandt and J. Chojnowski, *Spectrochim. Acta*, 25A, 1639 (1969). NMR, IR: Phenol adducts of alkyl halides, tri-*n*-butyl amine, tri-*n*-butyl phosphine.

308. D. A. Brant, *Macromolecules*, 1, 291 (1968). Theory: Empirical hydrogen bond and van der Waals potential functions used to predict polypeptide conformations.

309. J. W. Brasch and R. J. Jakobsen, *Spectrochim. Acta*, 20, 1644 (1964). Far-IR: Use of polyethylene matrix.

310. J. W. Brasch, R. J. Jakobsen, W. G. Fateley, and N. T. McDevitt, *Spectrochim. Acta*, 24A, 203 (1968). Far-IR: Polymorphic forms of phenol.

311. J. W. Brasch, Y. Mikawa, and R. J. Jakobsen, *Applied Spectroscopy Reviews*, 1, 187 (1968). Review: Far-infrared studies of hydrogen bonding.

312. J. W. Brasch, Y. Mikawa, and R. J. Jakobsen, *Colloq. Spectrosc. Int., Ottawa, 13th*, 174 (1967). IR: Effect of pressure on ν_{OH} band of solid alcohols.

312a. S. Bratan and F. Strohbusch, *Chem. Ber.*, 105, 2284 (1972). NMR: Intramolecular, dibenzoylmethanes.

313. S. Bratož in *Advances in Quantum Chemistry*, P.-O. Löwdin, ed., Academic Press, New York, 1967, Vol. 3, p. 209. Review of theory.

314. S. Bratož and D. Hadži in *Hydrogen Bonding*, D. Hadži, ed., Pergamon Press, New York, 1959, p. 111.

315. S. Bratož and D. Hadži, *J. Chem. Phys.*, 27, 991 (1957). IR: Band broadening, anharmonicity, double-minimum phenomena.

316. S. Bratož, D. Hadži, and G. Rossmy, *Trans. Faraday Soc.*, 52, 464 (1956). IR: Intramolecular, β-diketones.

317. E. A. Braude and F. Sondheimer, *J. Chem. Soc.*, 3754 (1955). UV: Effect of steric conformation on electronic spectra of conjugated systems.

318. J. T. Braunholtz, G. E. Hall, F. G. Mann, and N. Sheppard, *J. Chem. Soc.*, 868 (1959). IR: Studies of XOOH group where X is P, As, S, Se.

319. S. E. Bresler, *Introduction to Molecular Biology*, R. A. Zimmermann, translation ed., Academic Press, 1971, Chapters 3-5. BIOL: Nucleic acids, structures and functions.

320. J. Brickmann, A. Licha, and H. Zimmermann, *Ber. Bunsenges, Phys. Chem.*, 75, 1324 (1971). Theory: Proton states of hydrogen bonds in a two-dimension double-minimum potential.

321. J. Brickmann and H. Zimmermann, *J. Chem. Phys.*, 50, 1608 (1969). Theory: Lingering time of proton in walls of a double-minimum hydrogen bond potential.

322. J. Brickmann and H. Zimmermann, *Ber. Bunsenges, Phys. Chem.*, 71, 160 (1967). Theory: Proton transfer in a symmetric double-minimum hydrogen bond potential.

323. J. Brickmann and H. Zimmermann, *Ber. Bunsenges, Phys. Chem.*, 70, 521 (1966). Theory: Tunneling frequencies in hydrogen bond of imidazole.

324. J. Brickmann and H. Zimmermann, *Ber. Bunsenges, Phys. Chem.*, 70, 157 (1966). Theory: Proton tunneling in a double-minimum potential.

325. G. Briegleb, *Electronen-Donor-Acceptor-Komplexe*, Springer Verlag, Berlin, 1961, Chapter 12. Determination of K.

326. C. Brissette and C. Sandorfy, *Can. J. Chem.*, 38, 34 (1960). IR: Amine hydrohalides.

327. R. J. N. Brits, M. J. De Vries, and H. G. Raubenheimer, *J. S. Afr. Chem. Inst.*, 21, 183 (1968); *Chem. Abstr.*, 70, 86887z (1970). IR: Association of phenol, *o*-cresol, and pyrrole with π-bases.

327a. P. Broadhead and G. A. Newman, *Spectrochim. Acta*, 28A, 1915 (1972). IR: $B(OH)_3$, HBO_2, B_2O_3.

328. A. I. Brodskii, V. D. Pokhodenko, and V. S. Kuts, *Usp. Khim.*, 39, 753 (1970); [English, 347]. NMR: Review of the self-association and hetero-association of alcohols and phenols.

329. W. V. F. Brookes and C. M. Han, *J. Phys. Chem.*, 71, 650 (1967). IR: Intramolecular, normal coordinate analysis, carboxylic and percarboxylic acids.

330. C. J. W. Brooks, G. Eglinton, and L. Hanaineh, *Spectrochim. Acta*, 22, 161 (1966). IR: Solvent effects on group frequencies, $C{=}O$ and O—H.

331. C. J. W. Brooks, G. Eglinton, and J. F. Morman, *J. Chem. Soc.*, 661 (1961). IR: Intramolecular, salicylic acid derivatives.

332. C. J. W. Brooks, G. Eglinton, and J. F. Morman, *J. Chem. Soc.*, 106 (1961). IR: Self-association, intramolecular, frequency shifts of ν_{CO} and ν_{OH} of aryl carboxylic acids.

333. A. D. Broom, M. P. Schweizer, and P. O. P. Ts'o, *J. Amer. Chem. Soc.*, 89, 3612 (1967). BIOL, NMR: Intermolecular associations of fourteen purine nucleosides in aqueous solution.

334. D. G. Brown, R. S. Drago, and T. F. Bolles, *J. Amer. Chem. Soc.*, 90, 5706 (1968). CAL: ΔH for phenol-ethyl acetate adduct.

336. I. Brown, G. Eglinton, and M. Martin-Smith, *Spectrochim. Acta*, 19, 463 (1963). IR: Intramolecular, *o*-bromophenols.

337. I. Brown, G. Eglinton, and M. Martin-Smith, *Spectrochim. Acta*, 18, 1593 (1962). IR: Intramolecular, *o*-bromophenols.

338. M. P. Brown and R. W. Heseltine, *Chem. Commun.*, 1551 (1968). IR: Self-association and hetero-association of coordinated BH_3.

339. M. P. Brown, R. W. Heseltine, P. A. Smith, and P. J. Walker, *J. Chem. Soc. A*, 410 (1970). IR: BH_3 and BH_2 groups as proton acceptors.

340. R. F. C. Brown, L. Radom, S. Sternhell, and I. D. Rae, *Can. J. Chem.*, 46, 2577 (1968). NMR: Intramolecular, *o*-substituted acetanilides.

341. T. L. Brown, R. L. Gerteis, D. A. Bafus, and J. A. Ladd, *J. Amer. Chem. Soc.*, 86, 2135 (1964). NMR: $Et_3N-C_6H_6$.

342. E. J. Browne and J. B. Polya, *J. Chem. Soc. C*, 1056 (1969). IR: Self-association, triazoles.

343. E. J. Browne and J. B. Polya, *J. Chem. Soc. C*, 824 (1968). IR: Proton tunneling in triazole and pyridyl ketones.

344. J. M. Bruce and P. Knowles, *J. Chem. Soc.*, 5900 (1964). NMR: Effect of hydrogen bonding on spin-spin coupling in alcohols.

345. T. C. Bruice and T. H. Fife, *J. Amer. Chem. Soc.*, 84, 1973 (1962). IR: Intramolecular, cyclopentane-1,2-diol monoacetates.

346. F. V. Brutcher, Jr. and W. Bauer, Jr., *J. Amer. Chem. Soc.*, 84, 2236 (1962). IR: Intramolecular, steroids.

347. J. Brynestad and G. P. Smith, *J. Phys. Chem.*, 72, 296 (1968). Interpretation of isosbestic points.

348. J. B. Bryan and B. Curnutte, *J. Mol. Spectrosc.*, 41, 512 (1972). IR: Normal coordinate analysis of H_2O bonded to four nearest neighbors.

349. A. Bryson and R. L. Werner, *Aust. J. Chem.*, 13, 456 (1960). IR: Intramolecular, 2-nitro-1-naphthylamine, 8-nitro-1-naphthylamine.

349a. B. Brzezinski, T. Dziembowska, and M. Szafran, *Rocz. Chem.*, 47, 445 (1973). NMR: Intramolecular, α-carboxy N-oxides.

350. H. Buc, *Ann. Chim.* t. 8, 409 (1963). IR: Intramolecular, β-diols.

351. H. Buc and J. Neel, *Compt. Rend. Acad. Sci.*, 255, 2947 (1962). IR: Intramolecular, ΔH, 1,3-diols.

352. H. Buc and J. Neel, *Compt. Rend. Acad. Sci.*, 252, 1786 (1961). IR: Intramolecular, 1,3-diols.

353. H. Buchowski, *J. Chim. Phys.*, 68, 1138 (1971). Influence of hydrogen bonding on thermodynamic properties of solutions.

354. H. Buchowski, J. Devaure, P. V. Huong, and J. Lascombe, *Bull. Soc. Chim. Fr.*, 2532 (1966). IR: K, adducts of heptyne-1 with acetone in various solvents.

355.   A. D. Buckingham, *Proc. Roy. Soc. (London)*, A248, 169 (1958).
       IR: Model for solvent effects on ν_{AH}.

356.   A. D. Buckingham, T. Schaefer, and W. G. Schneider, *J. Chem.
       Phys.*, 34, 1064 (1961). NMR: Nitriles in neopentane, acetone,
       benzene.

357.   A. D. Buckingham, T. Schaeffer, and W. G. Schneider, *J. Chem.
       Phys.*, 32, 1227 (1960). NMR: Theory, solvent effects on proton
       chemical shifts.

357a. P. Buckley and M. Brochu, *Can. J. Chem.*, 50, 1149 (1972).
       Microwave: Intramolecular, 2-methoxyethanol.

358.   P. Buckley, P. Giguere, and M. Schneider, *Can. J. Chem.*, 47, 901
       (1969). IR: Intramolecular, 2-chloro- and 2-bromoethanol.

359.   P. Buckley, P. Giguere, and D. Yamamoto, *Can. J. Chem.*, 46, 2917
       (1968). IR: Intramolecular, 2-fluoroethanol.

360.   K. S. Buckton and R. G. Azrak, *J. Chem. Phys.*, 52, 5652 (1970).
       Microwave: Intramolecular, 2-fluoroethanol.

361.   C. E. Bugg and U. Thewalt, *J. Amer. Chem. Soc.*, 92, 7441 (1970).
       BIOL, X-Ray: Base stacking in 6-thioguanine.

362.   C. E. Bugg, U. T. Thewalt, and R. E. Marsh, *Biochem. Biophys.
       Res. Commun.*, 33, 436 (1968). BIOL, X-Ray: Base stacking in
       nucleic acid components, crystal structures of guanine, guanosine,
       and inosine.

363.   K. Buijs and G. R. Choppin, *J. Chem. Phys.*, 40, 3120 (1964).
       IR: Self-association, H_2O.

364.   K. Buijs and G. R. Choppin, *J. Chem. Phys.*, 39, 2035, 2042 (1963).
       Near-IR: Self-association, H_2O.

364a. J. T. Bulmer and H. F. Shurvell, *J. Phys. Chem.*, 77, 256 (1973).
       IR: Self-association, band resolution techniques for acetic acid.

364b. J. T. Bulmer and H. F. Shurvell, *J. Phys. Chem.*, 77, 2085 (1973).
       IR: Band resolution, $CDCl_3$-Bu_2O mixtures.

365.   J. E. Bundschuh, F. Takahashi, and N. C. Li, *Spectrochim. Acta*,
       24A, 1639 (1968). NMR: Association of hydroquinone with tetra-
       hydrofuran.

366.   J. L. Burdett and M. T. Rogers, *J. Phys. Chem.*, 70, 939 (1966).
       NMR: Intramolecular, β-diketones and β-ketoesters.

367.   J. L. Burdett and M. T. Rogers, *J. Amer. Chem. Soc.*, 86, 2105
       (1964). NMR: Intramolecular, β-diketones and β-ketoesters.

368.   S. F. Bureiko and G. S. Denisov, *Zh. Prikl. Spektrosk.*, 14, 276
       (1971); *Chem. Abstr.*, 75, 12811b (1971). IR: Thermodynamic data,
       adducts of secondary amines with Et_3N and $H(CF_2)_4CH_2OH$.

369.   A. B. Burg and J. E. Griffiths, *J. Amer. Chem. Soc.*, 83, 4333
       (1961). IR: Phosphinic acid derivatives, CF_3HPOOH.

370.   J. J. Burke, G. G. Hammes, and T. B. Lewis, *J. Chem. Phys.*, 42,
       3520 (1965). Kinetics: Apparatus for ultrasonic relaxation
       method.

371. A. Burneau and J. Corset, *J. Chem. Phys.*, 56, 662 (1972). IR: Simultaneous vibrational transitions, combination bands between X—H and A—B stretching vibrations in RXH...BAR'.

372. A. Burneau and J. Corset, *Can. J. Chem.*, 51, 2059 (1973). Near-IR: Simultaneous vibrational transitions for adducts of RXH with R'AB.

372a. A. Burneau and J. Corset, *J. Phys. Chem.*, 76, 449 (1972). Near-IR: Association of water with acetone.

372b. A. Burneau and J. Corset, *J. Chem. Phys.*, 58, 5188 (1973). Near-IR: Polemic to Ref. [448a] on water-acetone mixtures.

373. A. Burneau and J. Corset, *Chem. Phys. Lett.*, 9, 99 (1971). IR: Anharmonicity constants, adducts of H_2O, D_2O, HOD.

374. R. E. Burton and J. Daly, *Trans. Faraday Soc.*, 67, 1219 (1971). Theory: CNDO/2 study of influence of dissolved ions on hydrogen bonding in water.

374a. W. K. Busfield, M. P. Ennis, and I. J. McEwen, *Spectrochim. Acta*, 29A, 1259 (1973). IR: Intramolecular, α, ω-diols.

375. W. R. Busing and D. F. Hornig, *J. Phys. Chem.*, 65, 284 (1961). Raman: Effect of KBr, KOH, KCl on H_2O spectrum.

376. R. N. Butler, *J. Chem. Soc. B*, 680 (1969). NMR: Intramolecular, methyl-substituted 5-aminotetrazoles.

377. R. N. Butler and M. C. R. Symons, *Chem. Commun.*, 71 (1969). NMR: ^1H spectra of perchlorate salts in methanol.

378. V. F. Bystrov, K. M. Dyumaev, V. P. Lezina, and G. A. Nikiforov, *Doklady Akad. Nauk SSSR*, 148, 1077 (1963). NMR: Adducts of 2,6-alkylphenols with ether, acetone, triethylamine.

379. V. F. Bystrov, V. V. Ershov, and V. P. Lezina, *Opt. Spectrosk.*, 17, 538 (1964); [English, 290]. NMR: o-Alkylphenols.

380. V. F. Bystrov, I. I. Grandberg, and G. I. Sharova, *Opt. Spektrosk.*, 17, 63 (1964); [English, 31]. NMR: Self-association, methyl derivatives of pyrazole.

381. V. F. Bystrov and V. P. Lezina, *Opt. Spektrosk.*, 16, 790 (1964); [English, 430]. NMR: K_n for self-association of saturated amines.

382. V. F. Bystrov and V. P. Lezina, *Opt. Spektrosk.*, 16, 1004 (1964); [English, 542]. NMR: K_n for self-association of phenols and pyrazole derivatives in CCl_4.

383. M. Cabanetos and B. Wojtkowiak, *Compt. Rend. Acad. Sci.*, Paris, Ser. C, 268, 751 (1969). IR: Self-association of ethynylcarbinols.

384. S. Cabani and P. Gianni, *J. Chem. Soc. A*, 547 (1968). CAL: Limits of calorimetry.

385. D. F. Cadogan and J. H. Purnell, *J. Phys. Chem.*, 73, 3849 (1969). Chromatography: Gas-liquid chromatographic method for determining thermodynamic data, alcohol adducts of didecyl sebacate in squalane.

386. J. Caillet, *Compt. Rend. Acad. Sci.*, Paris, Ser. D, <u>266</u>, 288
     (1968). Theory: Hydrogen bonds in DNA base pairs.

387. T. Cairns and G. Eglinton, *Nature*, <u>196</u>, 535 (1962). IR: Intra-
     molecular, polyphenols.

388. T. Cairns, G. Eglinton, and D. T. Gibson, *Spectrochim. Acta*, <u>20</u>,
     31 (1964). IR: Association of *p*-cresol with Me_2SO.

389. E. F. Caldin, *Chem. Rev.*, <u>69</u>, 135 (1969). Kinetics: Tunneling
     in proton-transfer reactions.

390. E. F. Caldin, *Fast Reactions in Solution*, Wiley & Sons, New York,
     1964, Chapter 4, 5. Kinetics: Temperature-jump and ultrasonic
     relaxation methods.

391. E. F. Caldin and J. E. Crooks, *J. Sci. Instr.*, <u>44</u>, 449 (1967).
     Kinetics: Temperature-jump apparatus.

391a. E. F. Caldin, J. E. Crooks, and D. O'Donnell, *J. Chem. Soc.*,
     Faraday Trans., *1*, <u>69</u>, 1000 (1973). Kinetics: Tetrabromophenol-
     phthalein ethyl ester-amine reactions.

392. G. L. Caldow and H. W. Thompson, *Proc. Roy. Soc. (London)*, <u>A254</u>,
     1 (1960). IR: Solvent effects on stretching frequencies.

393. E. S. Campbell, G. Gelernter, H. Heinen, and V. R. G. Moorti,
     J. Chem. Phys., <u>47</u>, 2690 (1967). Theory: Multipole contribu-
     tions to hydrogen bond in ice.

394. J. Cantacuzene, *Compt. Rend. Acad. Sci.*, Paris, <u>250</u>, 2356 (1960).
     UV: $n \rightarrow \pi^*$ band position of cyclopentanone in chlorinated
     ethanols.

395. J. Cantacuzene, J. Gassier, Y. Lhermitte, and M. Martin, *Compt.*
     Rend. Acad. Sci., Paris, <u>251</u>, 866 (1960). NMR: Chloro-
     substituted ethanols in $CHCl_3$, tetrahydrofuran, pyridine and
     triethylamine.

396. J. Cantacuzene, J. Gassier, Y. Lhermitte, and M. Martin, *Compt.*
     Rend. Acad. Sci., Paris, <u>250</u>, 1474 (1960). NMR: Dilution curves
     for alcohols in various solvents.

397. C. R. Cantor, *Nature*, <u>216</u>, 513 (1967). BIOL: ORD, possible con-
     formation for 5s ribosomal RNA.

398. C. R. Cantor, S. R. Jaskunas, and I. Tinoco, Jr., *J. Mol. Biol.*,
     <u>20</u>, 39 (1966). BIOL: ORD, conformation of polynucleotides.

399. C. R. Cantor, M. M. Warshaw, and H. Shapiro, *Biopolymers*, <u>9</u>, 1059
     (1970). BIOL: CD, conformation of deoxyoligonucleotides.

400. J. M. Cantril and H. A. Pohl, *Int. J. Quantum Chem. Symp.*, <u>4</u>, 165
     (1971). Theory: Simplified SCF calculations for hydrogen bonds
     with second row atoms.

401. R. Cardinaud, *Bull. Soc. Chim. Fr.*, 629 (1960). IR: Self-
     association of alcohol-OD.

402. R. Cardinaud, *Compt. Rend. Acad. Sci.*, Paris, <u>249</u>, 1641 (1959).
     IR: Self-association of alcohol-OD.

402a. G. L. Carlson, W. G. Fateley, A. S. Manocha, and F. F. Bentley,
     J. Phys. Chem., <u>76</u>, 1553 (1972). IR: Intramolecular, *o*-
     halophenols.

402b. G. L. Carlson and W. G. Fateley, *J. Phys. Chem.*, **77**, 1157 (1973). Far-IR: Intramolecular, *o*-substituted phenols.

403. G. L. Carlson, R. E. Witkowski, and W. G. Fateley, *Nature*, **211**, 1289 (1966). Far-IR: Gas-phase spectra of $HNO_3-O(CH_3)_2$ and $CH_3OH-N(CH_3)_3$.

404. G. L. Carlson, R. E. Witkowski, and W. G. Fateley, *Spectrochim. Acta*, **22**, 1117 (1966). Far-IR: Dimers of formic acid, formic acid-d, acetic acid, CD_3COOH in gas phase and solid state.

405. K. D. Carlson, D. Weisleder, and M. E. Daxenbichler, *J. Amer. Chem. Soc.*, **92**, 6232 (1970). IR, NMR: Intramolecular, absolute configurational assignments of diastereomeric β-hydroxy(-ace-toxy)episulfides.

406. J. A. Carrabine and M. Sundaralingam, *Biochemistry*, **10**, 292 (1971). BIOL, X-Ray: Hydrogen bonds in 2:1 complexes of uracil-mercuric chloride.

407. G. R. Cassani and F. J. Bollum, *Biochemistry*, **8**, 3928 (1969). BIOL: Oligodeoxynucleotides with polydeoxynucleotides.

408. G. R. Cassani and F. J. Bollum, *J. Amer. Chem. Soc.*, **89**, 4798 (1967). BIOL: Oligodeoxynucleotide-polydeoxynucleotide inter-actions, adenine-thymine base pairs, melting studies.

409. B. Castagna *et al.*, *Ann. Chim. (Paris)*, **7**, 5 (1972). Dielectric polarization, hydrogen-bonded complexes.

410. B. Casu, M. Reggiani, G. G. Gallo, and A. Vigevani, *Tetrahedron*, **22**, 3061 (1966). NMR: Conformation of glucose in Me_2SO.

411. F. Caujolle, Dang Quoc Quan, and F. Fauran, *Compt. Rend. Acad. Sci.*, Paris, Ser. C, **267**, 1499 (1968). IR: Self-association, alkylphenols.

413. R. Černý, E. Černá, and J. H. Spencer, *J. Mol. Biol.*, **46**, 145 (1969). BIOL: Nucleotide clusters in DNAs.

414. R. W. Chambers, *Prog. Nucl. Acid Res. Mol. Biol.*, **11**, 489 (1971). BIOL: Review, recognition of tRNA by its aminoacyl-tRNA ligase.

415. S. I. Chan, B. W. Bangerter, and H. H. Peter, *Proc. Natl. Acad. Sci. U. S.*, **55**, 720 (1966). BIOL, NMR: Purine binding to di-nucleotides, base stacking.

416. S. I. Chan and G. P. Kreishman, *J. Amer. Chem. Soc.*, **92**, 1102 (1970). BIOL, NMR: Purine protons in the presence of poly U.

417. S. I. Chan and J. H. Nelson, *J. Amer. Chem. Soc.*, **91**, 168 (1969). BIOL, NMR: Base stacking interaction in adenylyl (3'→5)adenosine.

418. S. I. Chan, M. P. Schweizer, P. O. P. Ts'o, and G. K. Helmkamp, *J. Amer. Chem. Soc.*, **86**, 4182 (1964). NMR: Self-association, purine and 6-methyl purine in H_2O.

419. W. L. Chandler and R. H. Dinius, *J. Phys. Chem.*, **73**, 1596 (1969). NMR: Self-association, ethanol.

420. A. K. Chandra and S. Banerjee, *J. Phys. Chem.*, **66**, 952 (1962). UV: Thermodynamic data, phenol-pyridine adduct in heptane.

421. A. K. Chandra and S. Basu, *Trans. Faraday Soc.*, **56**, 632 (1960). UV: K, alcohol adducts of pyridazine and diethyl nitrosoamine.

422. A. K. Chandra and A. B. Sannigrahi, *J. Phys. Chem.*, 69, 2494 (1965). UV: K, adducts of 1-butanol with benzophenone, cyclohexanone.

423. S. S. Chang and E. F. Westrum, Jr., *J. Chem. Phys.*, 36, 2571 (1962). Thermal properties of C1HC1⁻.

424. G. W. Chantry, H. A. Gebbie, and H. N. Mirza, *Spectrochim. Acta*, 23A, 2749 (1967). Far-IR: $CHCl_3$ and CCl_4 complexes with C_6H_6.

425. A. C. Chapman and L. E. Thirlwell, *Spectrochim. Acta*, 20, 937 (1964). IR: Spectra of phosphates.

426. D. Chapman, D. R. Lloyd, and R. H. Prince, *J. Chem. Soc.*, 550 (1964). IR: Aqueous solutions of dicarboxylic acid salts.

427. D. Chapman, D. R. Lloyd, and R. H. Prince, *J. Chem. Soc.*, 3645 (1963). NMR: Intramolecular, α-aminopolycarboxylic acids in H_2O.

428. O. L. Chapman and R. W. King, *J. Amer. Chem. Soc.*, 86, 1256 (1964). NMR: Alcohol-Me_2SO mixtures.

429. E. Chargaff, *Prog. Nucl. Acid Res. Mol. Biol.*, 8, 297 (1968). BIOL: "What really is DNA?"

430. E. Chargaff, *Experientia*, 6, 201 (1950). BIOL: Constancies in base concentration ratios in DNA.

431. G. Chedd, *New Science*, 46, 426 (1970). BIOL: Review of studies questioning role of DNA polymerase.

432. E. Chemouni, M. Fournier, J. Roziere, and J. Potier, *J. Chim. Phys. Physicochim. Biol.*, 67, 517 (1970). IR: Vibrational analysis of $H_5O_2^+$ spectrum.

433. C. C. Chen and W. C. Lin, *J. Chinese Chem. Soc. (Taiwan)*, 13, 49 (1966). NMR: Self-association of butyl alcohols.

434. P. Cheng, *Biochemistry*, 7, 3367 (1968). BIOL: Use of Cotton effect of RNA as probe for base pairing.

435. M. T. Chenon and N. Lumbroso-Bader, *J. Chim. Phys.*, 62, 1208 (1965). NMR: K, N-methylaniline adduct with acetone.

436. M. F. Chernyshova and N. P. Lushina, *Zh. Prikl. Spektrosk.*, 11, 843 (1969); *Chem. Abstr.*, 72, 60900q (1970). IR: Benzanilide and benzenesulfanilimide.

437. P. C. Cherry, W. R. T. Cottrell, G. D. Meakins, and E. E. Richards, *J. Chem. Soc. C*, 459 (1968). IR: Intramolecular, *syn-, anti*-α-hydroximino ketones.

437a. N. O. Cherskaya, V. I. Shilenko, V. A. Shlyapochnikov, and S. S. Novikov, *Izv. Akad. Nauk SSSR, Ser. Khim.*, 620 (1972); *Chem. Abstr.*, 77, 74358j (1972). IR: Thermodynamic data, CCl_4-C_6H_5OH-RNO_2.

438. V. F. Chesnokov, I. M. Bokhovkin, and I. V. Khazova, *Zh. Obshch. Khim.*, 39, 500 (1969); *Chem. Abstr.*, 71, 25203x (1969). IR: Adducts of Me_2SO with phenols.

439. Chi-Yu Chiang, *Diss. Abstr. Int. B*, 31, 6530 (1971). Hydrogen bond formation by nitrosamines.

439a. T. C. Chiang and R. M. Hammaker, *J. Phys. Chem.*, 69, 2715 (1965).
      IR: Self-association, trimethylacetic acid.

440. T. C. Chiang and R. M. Hammaker, *J. Mol. Spectrosc.*, 18, 110
      (1965). NMR: Self-association, trimethylacetic acid in CCl_4.

440a. T. Chiba, *J. Chem. Phys.*, 41, 1352 (1964). NMR: [1]D of ferro-
      electrics.

441. R. Chidambaram, R. Balasubramanian, and G. N. Ramachandran,
      Biochem. Biophys. Acta, 221, 182 (1970). Theory: Semiempirical
      potential functions for hydrogen bond interactions.

442. P. Chioboli, B. Fortunato, and A. Rastelli, *Ric. Sci. Rend.*, *Sez.*
      A., 8, 985 (1965). UV: Adducts of chlorophenols with dioxane
      and Et_3N.

442a. R. Chiron, Y. Graft, and R. Ramachandran, *Bull. Soc. Chim. Fr.*,
      3396 (1972). IR: Intramolecular, γ-ketoamides.

443. D. P. Chock, *Phys. Lett. A*, 34, 17 (1971). Theory: Hydrogen-
      bonded ferroelectrics.

444. H. Chojnacki, *Theor. Chim. Acta*, 22, 309 (1971). Theory: $(H_2O)_2$
      and $(H_2O)_3$, comparison of CNDO/2, INDO, and MINDO/1 results.

445. H. Chojnacki, *Theor. Chim. Acta*, 17, 244 (1970). Theory: Elec-
      trical conductivity of hydrogen-bonded biological systems.

446. H. Chojnacki, *Theor. Chim. Acta*, 12, 373 (1968). Theory:
      Pariser-Parr-Pople calculations on the hydrogen bond in solid
      imidazole.

447. J. Chojnowski, *Bull. Acad. Pol. Sci.*, *Ser. Sci. Chim.*, 18, 317
      (1970); *Chem. Abstr.*, 73, 130515q (1970). Adducts of phenol and
      methanol with Et_3P, Et_3Sb, Et_3As.

447a. J. Chojnowski, *Zeoz. Nauk Politech. Łodz.*, *Chem.*, 77 (1971);
      Chem. Abstr., 77, 145737v (1973). IR, NMR: Thermodynamic data,
      adducts of N, P, As, Sb, O, S, Se, Te, Cl, Br donor sites.

448. J. Chojnowski, *Bull. Acad. Pol. Sci.*, *Ser. Sci. Chim.*, 18, 309
      (1970). Adducts of pyrrole with Et_3P, Et_3As, and Et_3Sb.

448a. G. R. Choppin and M. R. Violante, *J. Chem. Phys.*, 56, 5890 (1972).
      Near-IR: H_2O-$(CH_3)_2CO$, H_2O-dioxane.

448b. G. R. Choppin and M. R. Violante, *J. Chem. Phys.*, 58, 5189 (1973).
      Reply to polemic, Ref. [372b].

449. H. A. Christ, P. Diehl, H. R. Schneider, and H. Dahn, *Helv. Chim.*
      Acta, 44, 856 (1961). NMR: [17]O resonance for acetic acid.

450. H. A. Christ and P. Diehl, *Helv. Phys. Acta*, 36, 170 (1963).
      NMR: [17]O chemical shifts, carbonyl compounds.

451. S. D. Christian, H. E. Affsprung, and S. A. Taylor, *J. Phys.*
      Chem., 67, 187 (1963). Lack of reliability of using partition
      method to determine association constants.

452a. S. D. Christian, S. R. Johnson, H. E. Affsprung, and P. J. Kil-
      patrick, *J. Phys. Chem.*, 70, 3376 (1966). Effect of solvent on
      thermodynamic data.

452b. S. D. Christian and T. L. Stevens, *J. Phys. Chem.*, 76, 2039
      (1972). IR: Self-association, trifluoroacetic acid.

453. S. D. Christian, A. A. Taha, and B. W. Gash, *Quart. Rev.*, 24, 20
     (1970). Review of molecular complexes of water.

454. S. D. Christian and E. E. Tucker, *J. Phys. Chem.*, 74, 214 (1970).
     Solvent effects on formation constants.

455. I. I. Chugnev, V. D. Aksel'rod, and A. A. Baev, *Biochem. Biophys.
     Res. Commun.*, 34, 348 (1969). BIOL: Role of anticodon in the
     acceptor function of tRNA.

456. L. V. Chulpanova and Ya. I. Yashin, *Zh. Fiz. Khim.*, 44, 773
     (1970); *Chem. Abstr.*, 73, 18815m (1970). Gas chromatography:
     ΔH estimated by temperature dependence of retention volumes.

457. M. Cignitti and L. Paoloni, *Theor. Chim. Acta*, 22, 250 (1971).
     Theory: Chains of H₂O molecules, CNDO/2.

458. A. Ciures, *Rev. Roum. Biol.*, *Ser. Zool.*, 13, 211 (1968). BIOL:
     Possible role of the phosphate charges in the winding and packing
     of the DNA molecule.

459. A. Ciures, *Rev. Roum. Inframicrobiol.*, 6, 251 (1969); *Chem.
     Abstr.*, 73, 98v (1970). BIOL: Topological model for circular
     DNA replication.

460. A. D. H. Clague and H. J. Bernstein, *Spectrochim. Acta*, 25A, 593
     (1969). IR: Self-association of carboxylic acids in the gas
     phase.

461. A. D. H. Clague, G. Govil, and H. J. Bernstein, *Can. J. Chem.*,
     47, 625 (1969). NMR: Vapor phase, MeOH-Me₃N adduct.

462. D. Clague and A. Novak, *J. Mol. Struct.*, 5, 149 (1970). Far-IR:
     Self-association, carboxylic acids.

463. P. A. Clarke and N. F. Hepfinger, *J. Chem. Educ.*, 48, 193 (1971).
     NMR: Use of three-dimensional models for chemical shift data of
     hydrogen-bonded adducts.

464. P. A. Clarke and N. F. Hepfinger, *J. Org. Chem.*, 35, 3249 (1970).
     NMR: Nitrile adducts of methanol.

465. J. Clauwaert, *Z. Naturforsch, B*, 23, 454 (1968). BIOL, UV:
     Poly A-poly U complexes.

466. P. Claverie, *J. Chim. Phys. Physicochim. Biol.*, 65, 57 (1968).
     Theory: Calculation of vertical interaction energies between
     stacked base pairs.

467. P. Claverie, B. Pullman, and J. Caillet, *J. Theor. Biol.*, 12,
     419 (1967). BIOL: van der Waals-London interactions between
     stacked purines and pyrimidines.

468. T. A. Claxton and M. C. R. Symons, *J. Chem. Soc. D*, 379 (1970).
     Theory: Extended Hückel treatment of hydrated hydrogen atom.

469. M. F. Claydon and N. Sheppard, *Chem. Commun.*, 1431 (1969). IR:
     Nature of "A, B, C"-type spectra of strongly hydrogen-bonded
     systems.

470. E. Clementi, *J. Chem. Phys.*, 46, 3851 (1967). Theory: Gaussian
     calculations on the reaction $\overline{NH_3}$ + HCl $\rightleftarrows$ NH₄⁺Cl⁻.

411

471.   E. Clementi, *J. Chem. Phys.*, $\underline{47}$, 2323 (1967). Theory: Gaussian
       calculations on the reaction $\overline{NH}_3$ + HCl $\rightleftarrows$ $NH_4{}^+Cl^-$.

473.   E. Clementi and J. N. Gayles, *J. Chem. Phys.*, $\underline{47}$, 3837 (1967).
       Theory: Gaussian study of inner and outer complexes in formation
       of NH_4Cl from HCl and NH_3.

474.   E. Clementi and A. D. McLean, *J. Chem. Phys.*, $\underline{36}$, 745 (1962).
       Theory: FHF^-, SCF calculations with Slater basis.

474a. E. Clementi, *J. Chem. Phys.*, $\underline{34}$, 1468 (1961). Theory: SCF cal-
       culations, $HF_2{}^-$, Slater basis.

475.   E. Clementi, J. Mehl, and W. von Niessen, *J. Chem. Phys.*, $\underline{54}$, 508
       (1971). Theory: Gaussian calculations on guanine-cytosine and
       on formic acid dimer.

476.   R. Clements, R. L. Dean, and J. L. Wood, *J. Chem. Soc. D*, 1127
       (1971). IR: Proton double-minimum potentials in asymmetric
       hydrogen bonds.

476a. R. Clements, R. L. Dean, and J. L. Wood, *J. Mol. Struct.*, $\underline{17}$, 291
       (1973). IR: Asymmetric $(B^1HB^2)^+$ where B^1 and B^2 are pyridine
       derivatives.

477.   R. Clements, R. L. Dean, T. R. Singh, and J. L. Wood, *J. Chem.
       Soc. D*, 1125 (1971). IR: Proton double-minimum potentials in
       symmetric hydrogen bonds.

478.   R. Clements, F. N. Masri, and J. L. Wood, *J. Chem. Soc. D*, 1530
       (1971). IR: Hydrogen-bonded cations $[Me_3NHNMe_3]^+$ and $[Me_3NHpy]^+$.

478a. R. Clements and J. L. Wood, *J. Mol. Struct.*, $\underline{17}$, 265, 283 (1973).
       IR: Symmetric B_2H^+ where B = pyridine derivative.

479.   C. J. Clemett, *J. Chem. Soc. A*, 455 (1969). NMR: H_2O in cyclic
       ethers.

480.   Th. Clerbaux and Th. Zeegers-Huyskens, *Bull. Soc. Chim. Belges*,
       $\underline{75}$, 366 (1966). Self-association, alcohols in cyclohexane.

481.   D. Clotman, D. Van Lerberghe, and Th. Zeegers-Huyskens, *Spectro-
       chim. Acta*, $\underline{26A}$, 1621 (1970). IR: Phenol-Et_3N adducts, σ
       correlations.

481a. D. Clotman and Th. Zeegers-Huyskens, *Spectrochim. Acta*, $\underline{26A}$, 1627
       (1970). IR: Adducts of phenols with amines.

481b. J. F. Coetzee and R. Mei-Shun Lok, *J. Phys. Chem.*, $\underline{69}$, 2690
       (1965). Self-association, phenols, benzoic acids, amines, amides,
       vapor pressure study.

482.   J. F. Coetzee and G. P. Cunningham, *J. Amer. Chem. Soc.*, $\underline{87}$, 2534
       (1965). Reaction of *o*-substituted benzoic acids with amines.

482a. E. R. Cohen and J. W. DuMond, *Rev. Mod. Phys.*, $\underline{37}$, 537 (1965).
       Values of fundamental constants.

482b. N. V. Cohan and M. Weissman, *Chem. Phys. Lett.*, $\underline{22}$, 287 (1973).
       Theory: Polaron model of hydrated electron.

483.   D. R. Cogley, M. Falk, J. N. Butler, and E. Grunwald, *J. Phys.
       Chem.*, $\underline{76}$, 855 (1972). NMR, IR: Self-association, water in
       propylene carbonate.

484. C. Colby, Jr., B. D. Stollar, and M. I. Simon, *Nature, New Biol.*, **229**, 172 (1971). BIOL: DNA—RNA hybrids.

485. A. R. H. Cole, L. H. Little, A. J. Michell, *Spectrochim. Acta*, **21**, 1169 (1965). IR: Solvent effects on ν_{OH} and ν_{SH}.

486. A. R. H. Cole and F. Macritchie, *Spectrochim. Acta*, **15**, 6 (1959). IR: Thermodynamic data, *trans*-dihydrocryptol-dioxane adduct.

487. A. R. H. Cole and A. J. Michell, *Aust. J. Chem.*, **18**, 102 (1965). IR: Effect of solvent on IR measurements of K, ΔH.

488. L. D. Colebrook and D. S. Tarbell, *Proc. Natl. Acad. Sci. U. S.*, **47**, 993 (1961). NMR: Self-association, *n*-propyl mercaptan, benzyl mercaptan, thiophenol.

489. D. C. Colinese, *J. Chem. Soc. B*, 864 (1971). IR, Dipole Moment: N-(4-substituted benzylidene)-4-hydroxyanilines in C_6H_6, CCl_4, dioxane.

490. D. C. Colinese, *J. Chem. Soc. B*, 857 (1971). IR, Dipole Moment: Linear correlation of ν_{OH} and μ_{OH} with Hammett σ for N-(4-hydroxybenzylidene)-4-substituted anilines in C_6H_6, CCl_4, dioxane.

491. D. C. Colinese, D. A. Ibbitson, and C. W. Stone, *J. Chem. Soc. B*, 570 (1971). IR: Correlation of $\Delta\nu_{OH}$ with σ, dipole moments for 4'-substituted 4-hydroxyazobenzenes.

492. C. J. Collins and B. M. Benjamin, *J. Amer. Chem. Soc.*, **89**, 1652 (1967). NMR: Intramolecular, π-cloud, norbornyl tosylates.

493. M. F. Collins, B. C. Haywood, and G. C. Stirling, *J. Chem. Phys.*, **52**, 1828 (1970). Neutron spectrometry as applied to hydrogen bonding.

494. L. Colombo, *J. Chem. Phys.*, **49**, 4688 (1968). Raman: Self-association, single crystal spectrum of imidazole.

495. L. Colombo and K. Furic, *Spectrochim. Acta*, **27A**, 1773 (1971). Raman: Benzoic acid single crystals.

496. P. Combelas, F. Cruege, J. Lascombe, C. Zuivoron, and M. Rey-Lafon, *J. Chim. Phys. Physicochim. Biol.*, **66**, 668 (1969). IR: Amide-carboxylic acid systems.

497. P. Combelas, C. Garrigou-Lagrange, and J. Lascombe, *Ann. Chim.*, **5**, 315 (1970). IR: Review, carbonyl group.

497a. P. Combelas, C. Garrigou-Lagrange, and J. Lascombe, *Biopolymers*, **12**, 611 (1973). IR: Interaction of poly-L-alanine with CF_3COOH.

498. T. M. Connor and C. Reid, *J. Mol. Spectrosc.*, **7**, 32 (1961). NMR: Dilution shift studies of alcohols.

499. P. G. Connors, M. Labanauskas, and W. W. Beeman, *Science*, **166**, 1528 (1969). BIOL: Small angle X-ray scattering, tRNA in solution.

500. T. J. Conocchioli, M. I. Tocher, and R. M. Diamond, *J. Phys. Chem.*, **69**, 1106 (1965). Solvent extraction, adduct of trioctyl phosphine oxide with $HClO_4$.

501. K. Conrow, G. D. Johnson, and R. E. Bowen, *J. Amer. Chem. Soc.*, **86**, 1025 (1964). Statistical analysis and error limits.

502. J. S. Cook and I. H. Reece, *Aust. J. Chem.*, 14, 211 (1961).
      IR: Self-association, hindered alcohols.

503. R. L. Cook and J. E. House, Jr., *Trans. Ill. State Acad. Sci.*,
      62, 223 (1969). IR: $\Delta\nu_{OH}$ alcohol adducts with substituted
      pyridines.

504. R. D. G. Cooper, P. V. De Marco, J. C. Cheny, and N. D. Jones,
      J. Amer. Chem. Soc., 91, 1408 (1969). NMR: Intramolecular,
      configuration of phenoxymethyl penicillin sulfoxide.

505. P. Coppens and T. M. Sabine, *Acta Crystallogr.*, *Sect. B*, 25, 2442
      (1969). Neutron Diffraction: α- and β-deuterated oxalic acid
      dihydrate.

506. Marcia Cordes de N. D. and J. L. Walter, *Spectrochim. Acta*, 24A,
      237 (1968). IR, Raman: Imidazole.

507. R. D. Corsaro and G. Atkinson, *J. Chem. Phys.*, 54, 4090 (1971).
      Kinetics: Ultrasonic, self-association of acetic acid.

508. R. D. Corsaro, G. Atkinson, and B. R. Perlis, *Chem. Instr.*, 2,
      389 (1970). Kinetics: Apparatus for ultrasonic absorption
      measurements.

509. J. Corset, P. V. Huong, and J. Lascombe, *Spectrochim. Acta*, 24A,
      2045 (1968). IR: NH_4^+ salts in liquid NH_3.

510. C. C. Costain and G. P. Srivatsava, *J. Chem. Phys.*, 41, 1620
      (1964). Microwave: Adducts of trifluoroacetic acid.

511. C. C. Costain and G. P. Srivatsava, *J. Chem. Phys.*, 35, 1903
      (1961). Microwave: Gas phase, $CF_3COOH/HCOOH$.

512. R. I. Cotter and W. B. Gratzer, *Eur. J. Biochem.*, 8, 352 (1969).
      IR: Conformation of RNA and proteins in the ribosome.

513. R. I. Cotter and W. B. Gratzer, *Nature*, 221, 154 (1969).
      IR: Conformation of rRNA.

514. F. A. Cotton, V. W. Day, E. E. Hazen, Jr., and S. Larsen, *J. Amer.
      Chem. Soc.*, 95, 4834 (1973). X-Ray: Methylguanidinium dihydrogen
      orthophosphate.

515. C. A. Coulson in *Hydrogen Bonding*, D. Hadži, ed., Pergamon Press,
      New York, 1959, p. 339. Review of theory.

516. C. A. Coulson, *Research*, 10, 149 (1957). Review of theory.

516a. C. A. Coulson, *Proc. Cambridge Phil. Soc.*, 34, 204 (1938).
      LCAO SCF method.

517. C. A. Coulson and U. Danielsson, *Arkiv Fysik*, 8, 239, 245 (1954).
      Theory: Ionic and covalent contributions to the hydrogen bond.
      Classic papers using approximate calculations on model systems.

518. C. A. Coulson and D. Eisenberg, *Proc. Roy. Soc.*, A291, 445 (1966).
      Theory: Dipole moment of water molecule in ice calculated to be
      almost double that of an isolated water molecule.

518a. C. A. Coulson and I. Fischer, *Phil. Mag.*, 40, 386 (1949).
      Equivalence of MO and VB methods.

519. C. L. Coulter and S. W. Hawkinson, *Proc. Natl. Acad. Sci. U. S.*,
      63, 1359 (1969). BIOL: Base pairing in 5-chlorouridine.

520. Y. Courtois, P. Fromageot, and W. Guschlbauer, *Eur. J. Biochem.*,
      6, 493 (1968). BIOL: ORD, protonation of DNA.

521. Y. Courtois, D. Thiele, and W. Guschlbauer, *Bull. Soc. Chem.
      Biol.*, 50, 2373 (1968). BIOL: ORD, CD, protonation of nucleic
      acids and polynucleotides.

522. S. M. Coutts, *Biochim. Biophys. Acta*, 232, 94 (1971). BIOL:
      Thermodynamics and kinetics of guanine-cytosine base pairing.

523. M. Couzi, J. le Calve, P. V. Huong, and J. Lascombe, *J. Mol.
      Struct.*, 5, 363 (1970). IR: Gas phase, HF-ether mixtures.

524. M. Couzi and P. V. Huong, *Spectrochim. Acta*, 26A, 49 (1970).
      IR: Spectra of phenol in forty solvents.

525. A. K. Covington and P. Jones, eds., *Hydrogen-Bonded Solvent
      Systems*, Taylor and Francis, London, 1968.

526. D. J. Cowley and L. H. Sutcliffe, *Spectrochim. Acta*, 25A, 1663
      (1969). NMR: Association of nitrobenzenes with tetrahydrofuran.

527. D. J. Cowley and L. H. Sutcliffe, *Spectrochim. Acta*, 25A, 989
      (1969). UV: Hydrogen bonding between the aromatic ring hydrogen
      of chloronitrobenzenes and tetrahydrofuran.

528. B. G. Cox and P. T. McTigue, *Aust. J. Chem.*, 22, 1637 (1969).
      Electronic: Aldehyde adducts of anions.

530. R. A. Cox and K. Kanagalingom, *Biochem. J.*, 108, 599 (1968).
      BIOL: Denaturation of DNA in presence of urea or formaldehyde.

532. T. Cramer, *Prog. Nucl. Acid Res. Mol. Biol.*, 11, 391 (1971).
      BIOL: Review, three-dimensional structure of tRNA.

533. F. Cramer, H. Doepner, F. Von der Haar, E. Schlimme, and H. Seidel,
      Proc. Natl. Acad. Sci. U. S., 61, 1384 (1968). BIOL: Conforma-
      tion of tRNA.

534. B. M. Craven, *Acta Crystallogr.*, 23, 376 (1967). BIOL, X-Ray:
      5-Nitrouracil monohydrate.

534a. B. M. Craven, C. Cusatis, G. L. Gartland, and E. A. Vizzini, *J.
      Mol. Struct.*, 16, 331 (1973). X-Ray: N—H...O bonds in bar-
      biturates.

535. C. J. Creswell and A. L. Allred, *J. Amer. Chem. Soc.*, 85, 1723
      (1963). NMR: CHX_3 adduct of tetrahydrofuran in C_6H_{12}.

536. C. J. Creswell and A. L. Allred, *J. Phys. Chem.*, 66, 1469 (1962).
      NMR: Thermodynamic data, $CHCl_3$ with C_6H_6 in C_6H_{12}.

537. C. J. Creswell and A. L. Allred, *J. Amer. Chem. Soc.*, 84, 3966
      (1962). CHF_3, CDF_3 association with tetrahydrofuran.

538. C. J. Creswell and G. M. Barrow, *Spectrochim. Acta*, 22, 839
      (1966). IR: K, acetylene-acetone adduct.

539. F. H. C. Crick, *Nature*, 227, 561 (1970). BIOL: Review.

540. F. H. C. Crick, *Science*, 167, 1694 (1970). BIOL: Structure of
      DNA.

541. F. H. C. Crick, *J. Mol. Biol.*, 19, 548 (1966). BIOL: Codon-
      anticodon, the wobble hypothesis.

542.   W. Croalto and A. Fava, *Ann. Chim. (Rome)*, 54, 735 (1964).
       IR: Self-association, vicinal triazoles.

543.   J. R. Crook and K. Schug, *J. Amer. Chem. Soc.*, 86, 4271 (1964).
       NMR: K_n, self-association of hydrazine derivatives.

544.   J. E. Crooks and B. H. Robinson, *Trans. Faraday Soc.*, 66, 1436
       (1970). UV: Kinetics, proton transfer.

544a. J. E. Crooks and B. H. Robinson, *Trans. Faraday Soc.*, 67, 1707
       (1971). Kinetics: Proton transfer, Bromophenol Blue adducts.

545.   J. E. Crooks and B. H. Robinson, *Chem. Commun.*, 979 (1970).
       UV: Kinetics, proton transfer in aprotic solvents.

546.   A. D. Cross and D. M. Crothers, *Biochemistry*, 10, 4015 (1971).
       BIOL, NMR: Single- and double-stranded deoxyribooligonucleotides.

547.   J. Crossley, *Advan. Mol. Relaxation Processes*, 2, 69 (1970).
       Review: Dielectric relaxation.

548.   J. Crossley and C. P. Smyth, *J. Amer. Chem. Soc.*, 91, 2482 (1969).
       Dielectric rotational lifetimes.

549.   D. M. Crothers, *Biopolymers*, 6, 1391 (1968). BIOL: ΔG, base
       pair formation.

550.   D. M. Crothers and D. I. Ratner, *Biochemistry*, 7, 1823 (1968).
       BIOL: Complex formation between actinomycin and deoxyguanosine.

551.   A. J. Cunningham, J. D. Payzant, and P. Kebarle, *J. Amer. Chem.
       Soc.*, 94, 7627 (1972). Gas Phase: Ion cyclotron resonance,
       $H^+(H_2O)n$.

551a. M. Currie and J. C. Speakman, *J. Chem. Soc. A*, 1923 (1970).
       Neutron Diffraction: Potassium hydrogen malonate.

552.   E. A. Cutmore and H. E. Hallam, *Trans. Faraday Soc.*, 58, 40
       (1962). IR: Solvent effects on group frequencies.

553.   E. Czarnecka and A. Tramer, *Acta Phys. Pol.*, 36, 133 (1969);
       Chem. Abstr., 72, 37345d (1970). IR: Intramolecular, salicylic
       acid.

553a. U. Dabrowska, B. Hintze, and C. Belzecki, *Bull. Acad. Pol. Sci.,
       Ser. Sci. Chim.*, 20, 641 (1972). IR: Intramolecular, 1,1-
       substituted-2-hydroxyguanidines and N,N-substituted-1-
       aminooxyformamidines.

554.   U. Dabrowska, R. Stachura, and J. Dabrowski, *Rocz. Chem.*, 44,
       1751 (1970); *Chem. Abstr.*, 74, 93035f (1971). IR: Intramolecular,
       o-hydroxy and *o*-amino sulfones.

555.   U. Dabrowska and T. Urbanski, *Spectrochim. Acta*, 21, 1765 (1965).
       IR: Intramolecular, *o*-nitrophenol and derivatives.

556.   U. Dabrowska and T. Urbanski, *Rocz. Chem.*, 37, 805 (1963).
       IR: Intramolecular, *o*-nitrophenols.

557.   U. Dabrowska, T. Urbanski, M. Witanowski, and L. Sletaniak, *Rocz.
       Chem.*, 38, 1323 (1964). NMR: Intramolecular, nitrophenols.

558.   J. Dabrowski, *Spectrochim. Acta*, 19, 475 (1963). IR: Intramo-
       lecular, enamino ketones.

559.   J. Dabrowski and Z. Swistun, *J. Chem. Soc. B*, 818 (1971);
       Tetrahedron, 29, 2261 (1973). IR, NMR: Bifurcate hydrogen bond
       in *bis*(β-acylvinyl)amines.

559a. J. Dabrowski, Z. Swistun, and U. Dabrowska, *Tetrahedron*, 29,
       2257 (1973). IR, NMR: Bifurcate bond in (*sym-o,o'*-diacyl)-
       diphenylamines.

560.   G. Dahlgren and F. A. Long, *J. Amer. Chem. Soc.*, 82, 1303 (1960).
       IR: Deuterium isotope effects.

561.   A. J. Dale and T. Gramstad, *Spectrochim. Acta*, 28A, 639 (1972).
       NMR: Self-association, phenol, mono- and polyfluorophenols.

562.   J. Daly and R. E. Burton, *Trans. Faraday Soc.*, 66, 2408 (1970).
       Theory: CNDO/2 calculations on H_3O^+ and $H_9O_4^+$.

563.   V. K. Damle, *Biopolymers*, 9, 353 (1970). BIOL: Helix-coil
       equilibrium in two- and three-stranded complexes involving com-
       plementary poly- and oligonucleotides.

564.   S. H. Dandegaonker, *J. Indian Chem. Soc.*, 46, 148 (1969). UV:
       Intramolecular, substituted benzaldehydes and hydroxyhaloaceto-
       phenones.

565.   W. B. Dandliker and V. A. de Saussure, *The Chemistry of Biosur-
       faces*, Vol. 1, M. L. Hair, ed., Marcel Dekker, New York, 1971,
       p. 1. BIOL: Role of water in stabilization of macromolecules
       by hydrophobic bonding.

566.   V. I. Danilov, V. A. Kuprievich, and O. V. Shramko, *Dokl. Akad.
       Nauk SSSR*, 177, 1465 (1967). BIOL: Electronic structure of DNA
       base pairs.

567.   V. I. Danilova and S. Ya. Belomyttsev, *Isv. Vyssh. Uched. Zaved.
       Fiz.*, 11, 126 (1968). UV: $PhNH_2$-$PhNO_2$ and $PhOH$-$PhNO_2$ systems.

568.   S. S. Danyluk and F. E. Hruska, *Biochemistry*, 7, 1038 (1968).
       BIOL, NMR: Effect of pD, purine and pyrimidine bases in D_2O.

569.   B. K. Das, *J. Indian Chem. Soc.*, 47, 170 (1970). UV: Association
       of ethers with isomeric methyl carbazoles.

569a. R. Daudel and C. Sandorfy, *Semiempirical Wave-Mechanical Calcula-
       tions on Polyatomic Molecules*, Yale University Press, New Haven,
       1971.

570.   J. G. David and H. E. Hallam, *Spectrochim. Acta*, 21, 841 (1965).
       IR: Self-association, thiophenols.

571.   J. G. David and H. E. Hallam, *Trans. Faraday Soc.*, 60, 2013
       (1964). IR: Solvent effects on ν_{SH}.

572.   G. Davidovics and J. Chouteau, *Spectrochim. Acta*, 22, 703 (1966).
       IR: Self-association of 2-aminothiazole.

573.   R. S. Davidson and M. Santhanam, *J. Chem. Soc. B*, 1151 (1971).
       Fluorescence and phosphorescence spectra of aromatic amines.

574.   D. R. Davies, *Ann. Rev. Biochem.*, 36, 321 (1967). BIOL, X-Ray:
       Review, X-ray diffraction studies on macromolecules.

575.   D. R. Davies and G. Felsenfeld in *Structural Chemistry and Molec-
       ular Biology*, A. Rich and N. Davidson, eds., W. F. Freeman, San
       Francisco, 1968, p. 423. BIOL: Discussion, RNA structure.

575a. J. B. Davies and H. E. Hallam, *Trans. Faraday Soc.*, 67, 3176 (1971). IR: Matrix studies, $(DCl)_x$.

576. M. Davies, *J. Chem. Educ.*, 46, 17 (1969). Dielectric aspects of hydrogen bond interaction.

577. M. Davies and D. M. L. Griffiths, *J. Chem. Soc.*, 132 (1955). IR: Intramolecular, *o*-methoxybenzoic acid.

578. J. C. Davis, Jr., Ph.D. Thesis, University of California, Berkeley, California, 1959 (AEC Rept. UCRL8909). NMR: K_n, self-association of ethanol in CCl_4.

579. J. C. Davis, Jr. and K. K. Deb in *Advances in Magnetic Resonance*, Vol. 4, J. S. Waugh, ed., Academic Press, New York, 1970, p. 201. Review: NMR studies of hydrogen bonding.

580. J. C. Davis, Jr. and K. S. Pitzer, *J. Phys. Chem.*, 64, 886 (1960). NMR: Chemical shifts of formic, acetic, benzoic acids in C_6H_6 and CCl_4.

581. J. C. Davis, Jr., K. S. Pitzer, and C. N. R. Rao, *J. Phys. Chem.*, 64, 1744 (1960). NMR: Self-association, MeOH, EtOH, 2-propanol, *t*-BuOH.

582. M. M. Davis in *The Chemistry of Nonaqueous Solvents*, Vol. 3, J. J. Lagowski, ed., Academic Press, 1970, p. 1. Effect of hydrogen bonding on solvent properties.

583. M. M. Davis, *Acid-Base Behavior in Aprotic Organic Solvents*, NBS Monograph 105, 1968, U. S. Government Printing Office, Washington, D. C. Role of hydrogen bonding in aprotic solvents.

584. M. M. Davis and M. Paabo, *J. Org. Chem.*, 31, 1804 (1966). UV: 1:1 adducts of aliphatic acids with 1,3-diphenylguanidine.

585. M. M. Davis and M. Paabo, *J. Amer. Chem. Soc.*, 82, 5081 (1960). UV: Thermodynamic data, benzoic acid adducts with triethylamine and diphenylguanidine.

587. T. S. Davis and J. P. Fackler, *Inorg. Chem.*, 5, 242 (1966). IR: Hydrogen bonding of $CDCl_3$, H_2O, MeOH to metal β-ketoenolates.

588. S. K. De and S. R. Palit, *J. Phys. Chem.*, 71, 444 (1967). UV: K, adducts of alcohols with ethylene trithiocarbonate.

589. J. C. Dearden, *Sci. Prog., Nature (Paris)*, 327 (1968). BIOL: Review, effect of hydrogen bonds on properties of H_2O and biological compounds.

590. J. C. Dearden, *Nature*, 206, 1147 (1965). IR: Intramolecular, correlation of ν_{OH} and $O-H...X$ distance.

591. J. C. Dearden and W. F. Forbes, *Can. J. Chem.*, 38, 1837, 1852 (1960). UV: Intramolecular, phenol and aniline derivatives.

592. J. C. Dearden and W. F. Forbes, *Can. J. Chem.*, 38, 896 (1960). UV: Self-association, anilines and phenols.

593. Z. Dega-Szafran, M. Z. Naskret-Barciszewska, and M. Szafran, *Rocz. Chem.*, 47, 749 (1973). NMR: Quinoline N oxide adducts of halogenated acids.

593a. Z. Dega-Szafran, E. Grech, and M. Szafran, *J. Chem. Soc., Perkin Trans. 2*, 1839 (1972). NMR: Lepidine N-oxide adducts of trifluoro- and monochloro-acetic acid.

594. P. G. De Gennes, *Solid State Commun.*, **1**, 132 (1963). Theory: Calculation of collective motions of hydrogen bonds in ferro-electric crystals.

595. P. G. De Gennes, *Comments Solid State Phys.*, **1**, 65 (1968). IR, Neutron Scattering: Broadening of O—H band in alcohols.

596. W. H. De Jeu, *J. Phys. Chem.*, **74**, 822 (1970). NMR: ^{13}C, acetone in H_2O and H_2SO_4/H_2O.

597. W. H. De Jeu, *Chem. Phys. Lett.*, **7**, 153 (1970). Theory: CNDO/2 calculation of blue shift in $n \rightarrow \pi^*$ transition due to hydrogen bonding in formaldehyde-water.

598. R. G. Delaplane and J. A. Ibers, *Acta Crystallogr., Sect. B*, **25**, 2423 (1969). X-Ray: α-Oxalic acid dihydrate and its deuterated analog.

599. R. G. Delaplane, J. A. Ibers, J. R. Ferraro, and J. J. Rush, *J. Chem. Phys.*, **50**, 1920 (1969). X-Ray, Neutron Diffraction: $HCoO_2$ and $DCoO_2$.

599a. J. E. Del Bene, *J. Amer. Chem. Soc.*, **95**, 5460 (1973). Theory: LCAO SCF STO-3G on ROH...NH_3.

599b. J. E. Del Bene, *J. Chem. Phys.*, **58**, 3139 (1973). Theory: LCAO SCF STO-3G on ROH...OCH_2.

599c. J. E. Del Bene, *J. Amer. Chem. Soc.*, **95**, 6517 (1973). Theory: $n \rightarrow \pi^*$ transitions, ROH...OCH_2.

600. J. E. Del Bene, *J. Chem. Phys.*, **56**, 4923 (1972). Theory: $(H_2O_2)_2$, $H_2O-H_2O_2$ dimers, SCF with minimal Slater basis expanded in terms of Gaussians.

600a. J. E. Del Bene, *J. Chem. Phys.*, **57**, 1899 (1972). Theory: Dimers of NH_2OH, H_2O, HOF.

601. J. E. Del Bene, *J. Chem. Phys.*, **55**, 4633 (1971). Theory: Dimers and trimers containing CH_3OH and H_2O; SCF calculations with Slater basis expanded in terms of Gaussians.

601a. J. E. Del Bene and F. T. Marchese, *J. Chem. Phys.*, **58**, 926 (1973). Theory: HF...HCN, bond to lone pair and to π system.

601b. J. E. Del Bene and J. A. Pople, *J. Chem. Phys.*, **58**, 3605 (1973). Theory: Analysis of recent $(H_2O)_n$ calculations.

602. J. E. Del Bene and J. A. Pople, *J. Chem. Phys.*, **55**, 2296 (1971). Theory: SCF calculations of $(HF)_n$; n = 3, . . . 16 with minimal Slater basis expanded in Gaussians.

603. J. E. Del Bene and J. A. Pople, *J. Chem. Phys.*, **52**, 4858 (1970). Theory: SCF calculations on small water polymers using a minimal Slater basis expanded in Gaussians.

604. J. E. Del Bene and J. A. Pople, *Chem. Phys. Lett.*, **4**, 426 (1969). Theory: Minimal basis SCF calculations on small water polymers.

605. P. V. De Marco, E. Farkas, D. Doddrell, B. L. Mylari, and E.
      Wenkert, *J. Amer. Chem. Soc.*, 90, 5480 (1968). NMR: Pyridine-
      induced solvent shifts.

606. L. A. Dement'eva, A. V. Iogansen, and G. A. Kurkchi, *Opt.
      Spektrosk.*, 29, 868 (1970); *Chem. Abstr.*, 74, 69700y (1971).
      IR: Caprolactam adducts, Fermi resonance.

606a. L. A. Dement'eva and G. A. Kurkchi, *Zh. Prikl. Spektrosk.*, 16,
      482 (1972). IR: Thermodynamic data, adducts of cyclohexylhydro-
      peroxide.

607. A. B. Dempster and G. Zerbi, *J. Chem. Phys.*, 54, 3600 (1971).
      IR: Crystalline CH_3OH, CH_3OD.

609. G. S. Denisov, *Dokl. Akad. Nauk SSSR*, 134, 1131 (1960). IR:
      Δv_{CO}, cyclohexanone in halo-substituted acetic acids.

610a. G. S. Denisov, G. V. Gusakova, and A. L. Smolyansky, *J. Mol.
      Struct.*, 15, 377 (1973). IR: Hydrogen bond and proton transfer
      in complexes of trioctylamine with halogenoacetic acids.

610b. G. S. Denisov, J. Starosta, and V. M. Shraiber, *Opt. Spektrosk.*,
      35, 447 (1973); *Chem. Abstr.*, 79, 145524q (1973). Far-IR:
      Pyridine adducts.

611. M. De Paz, S. Ehrenson, and L. Friedman, *J. Chem. Phys.*, 52, 3362
      (1970). Theory: CNDO/2 study of H^+ and OH^- hydrated ions.

612. D. A. Deranleau, *J. Amer. Chem. Soc.*, 91, 4044, 4050 (1969).
      K for weak complexes.

612a. L. I. Derevyanko, M. N. Tsarevskaya, and V. S. Khlevnyuk, *Zh.
      Obshch. Khim.*, 42, 2083 (1972); *Chem. Abstr.*, 78, 20814g (1973).
      IR: Adducts of haloacetic acids with amines, dioxanes.

612b. H. O. Desseyn, W. A. Jacob, and M. A. Herman, *Spectrochim. Acta*,
      28A, 1329 (1972). IR, NMR: Oxamides, thiooxamides.

613. D. F. De Tar and R. W. Novak, *J. Amer. Chem. Soc.*, 92, 1361
      (1970). IR, Electronic: Carboxylic acid-amine equilibria.

614. S. Detoni and D. Hadži, *Spectrochim. Acta*, 20, 949 (1964). IR:
      Self-association, organophosphoric and phosphinic acids.

615. S. Detoni, D. Hadži, R. Smerkolj, J. Hawranek, and L. Sobczyk,
      J. Chem. Soc. A, 2851 (1970). IR: Adducts of isothiocyanic acid,
      correlations of integrated intensity and bandwidth with Δv.

616. R. Devoto, O. M. Sorarrain, and L. M. Boggia, *Z. Phys. Chem.
      (Leipzig)*, 247, 282 (1971). Theory: Empirical potential func-
      tions.

617. M. J. S. Dewar and M. Shanshal, *J. Chem. Soc. A*, 25 (1971).
      Theory: MINDO/1 studies of hydrogen bonding.

618. M. M. Dhingra, G. Govil, C. L. Khetrapal, and K. V. Ramiah, *Proc.
      Indian Acad. Sci.*, *Sect. A*, 62, 90 (1965). NMR: Self-association,
      aminomethylpyridines.

618a. B. Dickens, E. Prince, L. W. Schroeder, and W. E. Brown, *Acta
      Crystallogr.*, *Sect. B*, 29, 2057 (1973). X-Ray: Calcium hydrogen
      phosphate.

619. P. Diehl and P. M. Henrichs, *J. Magn. Resonance*, 5, 134 (1971).
     NMR: Intramolecular, salicylaldehyde, nematic phase.

620. G. H. F. Diercksen, *Chem. Phys. Lett.*, 4, 373 (1969). Theory:
     Gaussian calculations on water dimer.

621. G. H. F. Diercksen, *Theor. Chim. Acta*, 21, 335 (1971). Theory:
     Large Gaussian basis study of water dimer.

622. G. H. F. Diercksen and W. P. Kraemer, *Theor. Chim. Acta*, 23, 387
     (1972). Theory: Gaussian SCF calculations on $H^+ \cdot H_2O$, $Li^+ \cdot H_2O$,
     $Na^+ \cdot H_2O$.

623. G. H. F. Diercksen and W. P. Kraemer, *Chem. Phys. Lett.*, 5, 570
     (1970). Theory: Gaussian SCF calculations on $(FHOH)^-$.

624. G. H. F. Diercksen and W. P. Kraemer, *Chem. Phys. Lett.*, 6, 419
     (1970). Theory: Gaussian calculations on HF dimer.

624a. G. H. F. Diercksen, W. P. Kraemer, and W. Von Niessen, *Theor.
      Chim. Acta*, 28, 67 (1972). Theory: Ammonia-water system.

624b. G. H. F. Diercksen, W. Von Niessen, and W. P. Kraemer, *Theor.
      Chim. Acta*, 31, 205 (1973). Theory: $FH_2^+ \ldots FH$.

625. A. M. Dierckx, P. Huyskens, and T. Zeegers-Huyskens, *J. Chim.
     Phys.*, 62, 336 (1965). IR: K, adducts of substituted phenols
     with C_5H_5N.

626. T. Di Paulo, C. Bourdéron, and C. Sandorfy, *Can. J. Chem.*, 50,
     3161 (1972). IR: Influence of mechanical and electrical
     anharmonicity on intensities.

627. T. Di Paulo and C. Sandorfy, *Can. J. Chem.*, 51, 1441 (1973).
     IR: Self-association, acetone oxime-d_6.

628. W. Ditter and W. A. P. Luck, *Ber. Bunsenges, Phys. Chem.*, 73, 526
     (1969). Near-IR: Intramolecular, alcohols.

628a. H. P. Dixon, H. D. B. Jenkins, and T. C. Waddington, *J. Chem.
      Phys.*, 57, 4388 (1972). Experimental hydrogen bond strength in
      HF_2^-.

629. W. B. Dixon, *J. Phys. Chem.*, 74, 1396 (1970). NMR: Self-
     association, methanol.

630. R. E. Dodd, R. E. Miller, and W. F. K. Wynne-Jones, *J. Chem. Soc.*,
     2790 (1961). IR: Intramolecular, maleate ion.

631. D. Doddrell, *Diss. Abstr. Int. B*, 30, 4969 (1970). NMR: Pyridine
     and benzene solvent shifts.

632. D. Doddrell, E. Wenkert, and P. V. De Marco, *J. Mol. Spectrosc.*,
     32, 162 (1969). NMR: Intramolecular, *o*-trifluoromethylphenol.

633. D. Dombrowski, J. Fruwert, and G. Geiseler, *Z. Phys. Chem.
     (Liepzig)*, 252, 154 (1973). Self-association, disubstituted
     hydroxylamine derivatives.

634. J. Donohue, *Acta Crystallogr., Sect. B*, 25, 2418 (1969). X-Ray:
     Hydrogen bonding in 2-mercapto-6-methylpurine monohydrate.

635. J. Donohue, *J. Mol. Biol.*, 45, 231 (1969). Review: N—H...S
     distances.

636. J. Donohue, *Arch. Biochem. and Biophys.*, <u>128</u>, 591 (1968). BIOL:
      Revised dimensions for the purine and pyrimidine bases of DNA
      and RNA.

637. J. Donohue in *Structural Chemistry and Molecular Biology*, A. Rich
      and N. Davidson, eds., W. H. Freeman, San Francisco, California,
      1968. X-Ray: Review.

638. J. Donohue, *Acta Crystallogr., Sect. B*, <u>24</u>, 1558 (1968). Review:
      Structures of carboxylic acid dimers, sixteen references.

639. J. Donohue, *J. Chem. Educ.*, <u>40</u>, 598 (1963). Review: A short
      discussion of textbook errors on the hydrogen bond.

640. J. Donohue, *Proc. Natl. Acad. Sci. U. S.*, <u>42</u>, 60 (1956). BIOL:
      Hydrogen-bonded helical configurations of polynucleotides.

641. J. Donohue and K. N. Trueblood, *J. Mol. Biol.*, <u>2</u>, 363 (1960).
      BIOL: Base pairing in DNA.

642. D. E. Dorman and J. D. Roberts, *Proc. Natl. Acad. Sci. U. S.*, <u>65</u>,
      19 (1970). BIOL, NMR: ^{13}C spectra, nucleotides.

643. C. Dorval and Th. Zeegers-Huyskens, *Spectrochim. Acta*, <u>29A</u>, 1805
      (1973). IR: K, adducts of substituted phenols with $MeC(O)NMe_2$.

644. J. Dos Santos, F. Cruege, and P. Pineau, *J. Chim. Phys.*, <u>67</u>, 826
      (1970). IR: Self-association, alcohols, pyrrole.

645. P. Douzou, *J. Chim. Phys. Physicochim. Biol.*, <u>65</u>, 167 (1968).
      BIOL: Molecular associations of guanylic acid.

646. P. Douzou in *Molecular Associations in Biology*, B. Pullman, ed.,
      Academic Press, New York, 1968, p. 447. BIOL: Discussion,
      solvent polarity and molecular associations.

647. R. S. Drago, *J. Chem. Educ.*, <u>51</u>, 300 (1974). Review, acid-base
      chemistry.

648. R. S. Drago, *Chem. Brit.*, <u>3</u>, 516 (1967). Review, enthalpies of
      donor-acceptor adducts.

649. R. S. Drago, *Record Chem. Progress*, <u>26</u>, 115 (1965). Review,
      enthalpies of donor-acceptor adducts.

650. R. S. Drago and T. D. Epley, *J. Amer. Chem. Soc.*, <u>91</u>, 2883 (1969).
      CAL: Thermodynamic data, substituted phenols with bases in C_6H_{12}.

651. R. S. Drago, M. S. Nozari, and G. C. Vogel, *J. Amer. Chem. Soc.*,
      <u>94</u>, 90 (1972). CAL: Evaluating and eliminating solvent effects
      on thermodynamic data, m-F-C_6H_4OH adducts.

652. R. S. Drago, N. O'Bryan, and G. C. Vogel, *J. Amer. Chem. Soc.*, <u>92</u>,
      3924 (1970). CAL: Thermodynamic data, adducts of 2-butanol and
      t-butanol.

653. R. S. Drago, G. C. Vogel, and T. E. Needham, *J. Amer. Chem. Soc.*,
      <u>93</u>, 6014 (1971). Four-parameter equation for predicting enthal-
      pies of adduct formation.

654. R. S. Drago and B. B. Wayland, *J. Amer. Chem. Soc.*, <u>87</u>, 3571
      (1965). Empirical relationship for predicting enthalpies.

655. R. S. Drago, B. B. Wayland, and R. L. Carlson, *J. Amer. Chem. Soc.*,
      <u>85</u>, 3125 (1963). UV: Thermodynamic data, adducts of phenol with
      Me_2SO.

657. K. Dressler and O. Schnepp, *J. Chem. Phys.*, 33, 270 (1960). UV: Self-association, vacuum UV spectra of solid and vapor for NH_3 and H_2O.

658. M. Dreyfus, B. Maigret, and A. Pullman, *Theor. Chim. Acta*, 17, 109 (1970). Theory: Gaussian calculations on cyclic formamide dimer.

659. M. Dreyfus and A. Pullman, *Compt. Rend. Acad. Sci., Ser. C*, 271, 457 (1970). Theory: Interpretation of *ab initio* results in terms of classical components of the theory of intermolecular forces.

660. M. Dreyfus and A. Pullman, *Theor. Chim. Acta*, 19, 20 (1970). Theory: Gaussian SCF study of hydrogen bond in peptide units.

661. W. C. Drinkard and D. Kivelson, *J. Phys. Chem.*, 62, 1494 (1958). NMR: Water, methanol in Me_2CO and Me_2SO.

662. E. P. Dudek and G. O. Dudek, *J. Org. Chem.*, 32, 823 (1967). NMR: Intramolecular, thiocarboxamides.

663. G. O. Dudek, *J. Org. Chem.*, 30, 548 (1965). NMR, IR: Intramolecular, keto-enol equilibria.

664. G. O. Dudek, *Spectrochim. Acta*, 19, 691 (1963). NMR: Self-association, hydroxyacetonapthones and hydroxynaphthaldehydes.

665. G. O. Dudek and E. P. Dudek, *Tetrahedron*, 23, 3245 (1967). NMR: Keto-enol equilibria, multiple hydrogen bonds.

666. W. C. Duer and G. L. Bertrand, *J. Amer. Chem. Soc.*, 92, 2587 (1970). CAL: Reliability of calorimetric methods.

666a. F. A. L. Dullien and G. H. Shroff, *J. Phys. Chem.*, 76, 2463 (1972). Kinetics: Diffusion of $(CH_3COOH)_2$.

667. H. Dunken and P. Fink, *Z. Chem.*, 2, 117 (1962). IR, Raman: Self-association, acetic acid.

668. H. Dunken and H. Fritzsche, *Spectrochim. Acta*, 20, 785 (1964). IR: Self-association, alcohols.

669. H. Dunken and H. Fritzsche, *Z. Chem.*, 2, 379 (1962). IR: Thermodynamic data, adducts of indole with bases.

670. H. Dunken and H. Fritzsche, *Z. Chem.*, 2, 345, 347 (1962). IR: Thermodynamic data, association of phenol with bases.

671. H. Dunken and H. Fritzsche, *Z. Chem.*, 1, 249 (1961). UV: Thermodynamic data, adducts of phenol with bases in CCl_4.

672. H. Dunken and H. Fritzsche, *Z. Chem.*, 1, 127 (1961). Near-IR: Thermodynamic data, adducts of phenol with bases in CCl_4.

673. J. T. R. Dunsmuir and A. P. Lane, *Spectrochim. Acta*, 28A, 45 (1972). IR: Hydrogen bonding in complex ammonium halides.

674. J. C. Duplan, J. Delmau, and M. Davidson, *Bull. Soc. Chim. Fr.*, 4081 (1968). NMR: Intramolecular, 5-hydroxymethyl-5-methyl-1,3-dioxane and 5-(2-hydroxy-2-propyl)-1,3-dioxane.

674a. J. P. Dupont, J. D'hondt, and Th. Zeegers-Huyskens, *Bull. Soc. Chim. Belg.*, 80, 369 (1971). UV: K, adducts of substituted phenols with $\overline{E}t_3N$.

674b. J. P. Dupont, D. Neerinck, and L. Lamberts, *Ann. Chim. (Paris)*, **8**, 21 (1973). ΔH, phenol adducts of amines.

675. J. R. Durig, S. F. Bush, and E. E. Mercer, *J. Chem. Phys.*, **44**, 4238 (1966). IR: Hydrazine association.

676. J. R. Durig, W. C. Harris, and D. W. Wertz, *J. Chem. Phys.*, **50**, 1449 (1969). IR, Raman: Gas and liquid spectra of hydrazine derivatives.

677. G. Durocher and C. Sandorfy, *J. Mol. Spectrosc.*, **22**, 347 (1967). Near-IR: Solvent effects on overtone frequencies.

678. G. Durocher and C. Sandorfy, *J. Mol. Spectrosc.*, **15**, 22 (1965). IR: Anharmonicity in moderately strong O—H...Y bonds.

679. D. Dutting, W. Karau, F. Melchers, and H. G. Zachau, *Biochim. Biophys. Acta*, **108**, 194 (1965). BIOL: Possible base pairing between nucleotide sequences of serine tRNA.

680. F. Duus, P. Jakobsen, and S. O. Lawesson, *Tetrahedron*, **24**, 5323 (1968). IR, NMR, Electronic: β-Thioketo-thiolesters.

681. F. Duus and S. O. Lawesson, *Arkiv för Kemi*, **29**, 127 (1968). NMR, IR: Intramolecular, monothio-β-dicarbonyl compounds.

682. L. K. Dyall, *Spectrochim. Acta*, **25A**, 1727 (1969). IR: Intramolecular, o-anilines.

683. L. K. Dyall, *Spectrochim. Acta*, **25A**, 1423 (1969). IR: $\Delta\nu_{N-H}$ for anilines in various solvents.

684. L. K. Dyall, *Aust. J. Chem.*, **20**, 93 (1967). IR: Intramolecular, aniline derivatives, solvent effects.

685. L. K. Dyall, *Spectrochim. Acta*, **17**, 291 (1961). IR: Intramolecular, intermolecular, substituted 2-nitroanilines.

686. L. K. Dyall, *Aust. J. Chem.*, **13**, 230 (1960). IR: Self-association, diazoaminobenzene.

687. L. K. Dyall and J. E. Kemp, *Spectrochim. Acta*, **22**, 483 (1966). IR: Solvent effects on group frequencies, detection of intramolecular hydrogen bonding.

688. L. K. Dyall and J. E. Kemp, *Spectrochim. Acta*, **22**, 467 (1966). IR: Solvent effects on group frequencies, detection of intramolecular hydrogen bonding.

689. T. R. Dyke, B. J. Howard, and W. Klemperer, *J. Chem. Phys.*, **56**, 2442 (1972). Microwave: $(HF)_2$, $(DF)_2$, (HF-DF).

690. J. W. Eastes, M. H. Aldridge, M. J. Kamlet, *J. Chem. Soc. B*, 922 (1969). UV: Aniline derivatives in hydroxylic solvents.

691. L. Eberson in *Chem. Carboxylic Acids and Esters*, S. Patai, ed., Interscience, New York, 1969, p. 211. Review.

692. L. Eberson and S. Forsen, *J. Phys. Chem.*, **64**, 767 (1960). NMR: Intramolecular, potassium salts of alkyl-substituted succinic acids.

693. L. Eberson and I. Wadso, *Acta Chem. Scand.*, **17**, 1552 (1963). Intramolecular, dicarboxylic acids, thermodynamic data for ionization in H_2O.

693a. C. Eckart, *Phys. Rev.*, <u>36</u>, 878 (1930). Fit of approximate to exact wavefunction.

694. Z. Eckstein, P. Gluzinski, W. Sobotka, and T. Urbanski, *J. Chem. Soc.*, 1370 (1961). IR: Spectra of aliphatic nitro-alcohols.

695. C. Edmiston, J. Doolittle, K. Murphy, K. C. Tang, and W. Willson, *J. Chem. Phys.*, <u>52</u>, 3419 (1970). Theory: CI calculations with a Gaussian basis on linear H_2He^+.

696. A. N. Egorochkin, N. S. Vyazankin, S. Ya. Khorshev, S. E. Skobeleva, V. A. Yablokov, and A. P. Tarabarina, *Dokl. Akad. Nauk SSSR*, <u>194</u>, 1326 (1970); *Chem. Abstr.*, <u>74</u>, 52617d (1971). IR: Organometallic hydroperoxides, Ph_3MOOH (M = C, Si, Ge, Sn).

697. B. E. Eichingen and M. Fixman, *Biopolymers*, <u>9</u>, 205 (1970). BIOL: Helix-coil transition in heterogeneous chains, DNA model.

698. M. Eigen, *Angew. Chem., Int. Ed., (English)*, <u>3</u>, 1 (1964). Kinetics: Review, proton transfer.

699. M. Eigen, *Angew. Chem.*, <u>75</u>, 489 (1963). Kinetics: Intramolecular, dicarboxylic acids.

700. M. Eigen and L. de Maeyer, "Investigation of Rates and Mechanisms of Reactions," in *Technique of Organic Chemistry*, Vol. VIII, Part II, S. L. Friess, E. S. Lewis, and A. Weissberger, eds., Interscience, New York, 1963, Chapter 18. Kinetics: Relaxation methods.

701. M. Eigen and D. Poerschke, *J. Mol. Biol.*, <u>53</u>, 123 (1970). BIOL: Thermodynamics of the helix-coil transition of oligoriboadenylic acids of acidic pH.

702. H. Eisenber, *Ann. N. Y. Acad. Sci.*, <u>164</u>, 25 (1969). BIOL: Sedimentation, partial specific volumes, and preferential solvation in DNA solutions.

703. D. Eisenberg and W. Kauzman, *The Structure and Properties of Water*, Oxford University Press, Oxford, 1969, 296 pp. Review.

703a. J. Ekrene and T. Gramstad, *Spectrochim. Acta*, <u>28A</u>, 2465 (1972). IR: Thermodynamic data, phenol adducts of triphenylalkylidenephosphoranes.

703b. M. A. El-Bayoumi and M. Kasha, *J. Chem. Phys.*, <u>34</u>, 2181 (1961). UV: Fluorescence spectra of acridine-carbazole complex, 77°K.

704. S. H. Eletr and C. T. O'Konski, *J. Chem. Phys.*, <u>54</u>, 4312 (1971). ^{14}N nuclear quadrupole resonance, hydrogen bonding in $NH_3 \cdot \frac{1}{2}H_2O$.

705. E. L. Eliel and H. D. Banks, *J. Amer. Chem. Soc.*, <u>94</u>, 171 (1972). IR: Intramolecular, 3-hydroxymethyl-tetrahydropyran, 5-hydroxymethyl-1,3-dioxane.

706. N. V. Elsakov and A. A. Petrov, *Opt. Spektrosk.*, <u>16</u>, 148 (1964); [English, p. 77]. NMR: Association of acetylene compounds with bases.

707. N. V. Elsakov and A. A. Petrov, *Opt. Spektrosk.*, <u>16</u>, 797 (1964); [English, p. 434]. NMR: Vinylacetylene derivatives.

708. E. L. Elson, I. E. Scheffler, and R. L. Baldwin, *J. Mol. Biol.*, <u>54</u>, 401 (1970). BIOL: Helix formation, electrostatic effects.

708a. R. V. Emanuel, *Mol. Phys.*, 19, 399 (1970). Theory: Calculation of nuclear spin-spin coupling constant in H_2.

709. M. T. Emerson, E. Grunwald, M. L. Kaplan, and R. A. Kromhout, *J. Amer. Chem. Soc.*, 82, 6307 (1960). NMR, Kinetics: Mean lifetime of hydrogen bonds in water-amine systems.

710. M. Emilia, L. Saraiva, and V. M. S. Gil, *Rev. Port. Quim.*, 13, 83 (1971); *Chem. Abstr.*, 76, 125967w (1972). NMR: Intramolecular, salicylalanilines.

711. J. W. Emsley, J. Feeney, and L. H. Sutcliffe, *High Resolution Nuclear Magnetic Resonance Spectroscopy*, Vol. 1, Pergamon, Oxford, 1965. Review.

712. J. B. F. N. Engberts, *Recl. Trav. Chim. Pays-Bas*, 87, 992 (1969). NMR: α-Diazosulfones as proton donors.

713. J. B. F. N. Engberts, K. Hovius, and G. Zuidema, *Recl. Trav. Chim. Pays-Bas*, 90, 633 (1971). IR: $\Delta\nu_{OH}$, phenol adducts of N,N-disubstituted sulfinamides and sulfonamides.

714. J. B. F. N. Engberts, Th. A. J. W. Wajer, C. Kruls, and Th. J. de Boer, *Recl. Trav. Chim. Pays-Bas*, 88, 795 (1969). IR, NMR: Phenol adducts of C-nitroso compounds, $\Delta\nu$.

715. J. B. F. N. Engberts and G. Zuidema, *Recl. Trav. Chim. Pays-Bas*, 89, 1202 (1970). IR: $\Delta\nu_{OH}$, phenol adducts of sulfinic esters and sulfinyl chlorides.

716. J. B. F. N. Engberts and G. Zuidema, *Recl. Trav. Chim. Pays-Bas*, 89, 741 (1970). IR, NMR: α-Diazosulfones and α-diazoketones as proton donors and proton acceptors.

717. R. M. Epand and H. A. Scheraga, *J. Amer. Chem. Soc.*, 89, 3888 (1967). BIOL: Enthalpy of stacking in single-stranded polyriboadenylic acid.

718. T. D. Epley and R. S. Drago, *J. Paint Techol.*, 41, 500 (1969). CAL: Review, phenol adducts in different solvents.

719. T. D. Epley and R. S. Drago, *J. Amer. Chem. Soc.*, 89, 5770 (1967). CAL: Thermodynamic data, phenol adducts of bases in CCl$_4$, $\Delta\nu$-ΔH relationship.

720. L. M. Epshtein, Z. S. Novikova, and L. D. Ashkinadze, *Khim. Primen. Fosforong. Soedin, Tr. Konf. 4th,* 1969; *Chem. Abstr.*, 78, 135130y (1973). IR: Adducts of PhC≡CH or EtOH with R$_3$P.

721. A. N. Ermakov, A. V. Karyakin, G. A. Muradova, I. N. Marov, and L. P. Kazanskii, *Zh. Neorg. Khim.*, 13, 2985 (1968); *Chem. Abstr.*, 70, 33871h (1969). IR: Hydrogen bond in $Zr(SO_4)_2 \cdot 4H_2O$, $-\Delta H =$ 8.5 kcal/bond.

722. D. F. Evans and Sister M. A. Matesich, *J. Solution Chem.*, 2, 193 (1973). Ionic association in hydrogen bonding solvents.

723. D. F. Evans, *J. Chem. Soc.*, 5575 (1963). NMR: ^{13}C-H coupling constants correlated with hydrogen bonding ability of solvents.

724. H. B. Evans and J. H. Goldstein, *Spectrochim. Acta*, 24A, 73 (1968). NMR: Intramolecular, malonic acids.

725. J. C. Evans and G. Y-S. Lo, *J. Phys. Chem.*, 73, 448 (1969).
     IR: BrHBr⁻ at 20°K.

726. J. C. Evans and G. Y-S. Lo, *J. Phys. Chem.*, 71, 3942 (1967).
     IR: Hydrogen dihalide ions, BrHBr⁻ and BrDBr⁻.

727. J. C. Evans and G. Y-S. Lo, *J. Phys. Chem.*, 71, 3697 (1967).
     ³⁵Cl nuclear quadrupole resonance of symmetrical ClHCl⁻.

728. J. C. Evans and G. Y-S. Lo, *J. Phys. Chem.*, 70, 2702 (1966).
     ³⁵Cl nuclear quadrupole resonance, unsymmetrical ClHCl⁻.

729. J. C. Evans and G. Y-S. Lo, *J. Phys. Chem.*, 70, 11, 20, 543
     (1966). IR: Hydrogen dihalide ions.

730. P. Excoffon and Y. Marechal, *Spectrochim. Acta*, 28A, 269 (1972).
     IR: Self-association, carboxylic acid dimers.

731. D. P. Eyman and R. S. Drago, *J. Amer. Chem. Soc.*, 88, 1617 (1966).
     NMR: Chemical shifts for thirty phenol-base adducts in CH₂Cl₂.

732. E. M. Eyring, L. D. Rich, L. L. McCoy, R. C. Graham, and N.
     Taylor, *Advan. Chem. Phys.*, 21, 237 (1971). Kinetics: Deproto-
     nation kinetics of intramolecularly hydrogen-bonded acids.

733. M. Falk, *Can. J. Chem.*, 49, 1137 (1971). IR: Solid and liquid
     (CH₃)₃N·10H₂O.

734. M. Falk, K. A. Hartman, Jr., and R. C. Lord, *J. Amer. Chem. Soc.*,
     85, 391 (1963). BIOL: Hydration of DNA, spectroscopic study.

735. M. Falk, K. A. Hartman, Jr., and R. C. Lord, *J. Amer. Chem. Soc.*,
     85, 387 (1963). BIOL: IR, hydration of DNA.

736. M. Falk, K. A. Hartman, Jr., and R. C. Lord, *J. Amer. Chem. Soc.*,
     84, 3843 (1962). BIOL: Hydration of DNA, gravimetric study.

737. M. Falk, A. G. Poole, and C. G. Goymour, *Can. J. Chem.*, 48, 1536
     (1970). BIOL: IR, state of water in the hydration shell of DNA.

737a. M. Falk and E. Whalley, *J. Chem. Phys.*, 34, 1554 (1961). IR:
     Self-association, methanol.

738. Y. Fang and J. R. de la Vega, *Chem. Phys. Lett.*, 6, 117 (1970).
     Theory: SCF calculations on 2H₂O → H₃O⁺ + OH⁻ using Gaussian
     expansion of a Slater basis.

739. E. L. Farguhar, M. Downing, and S. J. Gill, *Biochemistry*, 7, 1224
     (1968). BIOL: Enthalpy, self-association, pyrimidine deriva-
     tives in H₂O.

740. V. C. Farmer and R. H. Thomson, *Spectrochim. Acta*, 16, 559 (1960).
     IR: Intramolecular, *o*-nitro-N-methylaniline and *o*-ethoxycarbonyl
     N-methylaniline.

741. G. D. Fasman, C. Lindblow, and L. Grossman, *Biochemistry*, 3. 1015
     (1964). BIOL: Helical conformations of poly C, ORD.

742. E. I. Fedin, P. O. Okulevich, V. V. Korshak, E. S. Krongauz, and
     A. M. Berlin, *Dokl. Akad. Nauk SSSR*, 186, 617 (1969); *Chem. Abstr.*,
     71, 60384d (1969). NMR: Intramolecular, 4,4'-*bis*(acetoacetyl)-
     diphenyl oxide.

743. J. F. Feeney and L. H. Sutcliffe, *Proc. Chem. Soc.*, 118 (1961).
     NMR: K$_n$, self-association, EtNH₂, Et₂NH, isobutylamine.

744. J. F. Feeney and L. H. Sutcliffe, *J. Chem. Soc.*, 1123 (1962).
      NMR: K_n, self-association, $EtNH_2$, Et_2NH, isobutylamine.

745. J. F. Feeney and S. M. Walker, *J. Chem. Soc. A*, 1148 (1966).
      NMR: Self-association, 2-methoxyethanol, EtOH, MeOH, t-BuOH
      in CCl_4.

746. G. Felcher and I. Pelah, *J. Chem. Phys.*, 52, 905 (1970). Neutron
      Scattering: KH_2PO_4.

747. I. Feldman and R. P. Agarwal, *J. Amer. Chem. Soc.*, 90, 7329
      (1968). BIOL: PMR, adenosine 5'-monophosphate and adenosine
      5'-triphosphate in D_2O.

748. G. Felsenfeld, D. D. Davies, and A. Rich, *J. Amer. Chem. Soc.*,
      79, 2023 (1957). BIOL: Formation of a three-stranded polynucleo-
      tide molecule (poly A), and (poly U).

749. G. Felsenfeld and A. Rich, *Biochim. Biophys. Acta*, 26, 457 (1957).
      BIOL: Formation of 2- and 3-stranded polyribonucleotides.

749a. G. Ferraris, D. W. Jones, and J. Yerkess, *Chem. Commun.*, 1566
      (1971). Neutron Diffraction: Symmetrical hydrogen bonds in
      $Ca(H_2AsO_4)_2$.

750. J. R. Ferraro and D. F. Peppard, *J. Phys. Chem.*, 67, 2639 (1963).
      NMR: Self-association, organophosphorus acids.

751. J. A. Ferretti and L. Paolillo, *Biopolymers*, 7, 155 (1969).
      NMR: Poly-L-alanine.

752. L. F. Ferstandig, *J. Amer. Chem. Soc.*, 84, 1323 (1962). IR:
      Association of benzyl isocyanide with n-amyl alcohol.

753. L. F. Ferstandig, *J. Amer. Chem. Soc.*, 84, 3553 (1962). NMR:
      K, acetylenic proton associated with RCN, RNC.

754. H. Feuer, D. Pelle, D. M. Braunstein, and C. N. R. Rao, *Spectro-
      chim. Acta*, 25A, 1393 (1969). NMR, IR: Hydroxylamine self-
      association and hetero-association.

754a. Yu. Ya. Fialkov and O. V. Chviruk, *Zh. Obshch. Khim.*, 42, 7
      (1972); *Chem. Abstr.*, 76, 159098d (1972). IR: Binary amine
      systems.

754b. Yu. Ya. Fialkov, G. A. Puchkovskaya, and V. V. Vashchinskaya,
      Zh. Obshch. Khim., 43, 482 (1973); [English, p. 484]. IR: Effect
      of solvent on self-association of acetic acid.

755. A. Ficarra and Suei-Rong Huang, *J. Mol. Spectrosc.*, 33, 175
      (1970). NMR: Intramolecular, 1,8-diaminoperchloronaphthalene.

756. R. A. Fifer and J. Schiffer, *J. Chem. Phys.*, 52, 2664 (1970).
      IR: Force constants, H_2O.

757. J. Figueras, *J. Amer. Chem. Soc.*, 93, 3255 (1971). Electronic:
      Effect of hydrogen bonding on phenol blue spectrum.

758. R. H. Figueroa, E. Roig, and H. H. Szmant, *Spectrochim. Acta*, 22,
      1107 (1966). IR: Association of sulfoxides with phenols.

759. M. Fild, M. F. Swiniarski, and R. R. Holmes, *Inorg. Chem.*, 9, 839
      (1970). IR: Gas phase, $MeOH-Me_3N$ adduct.

760. A. Finch, P. N. Gates, K. Radcliffe, R. N. Dickson, and F. F. Bentley, *Chemical Applications of Far-Infrared Spectroscopy*, Academic Press, 1970, p. 103. Far-IR: Review.

761. J. F. Finch and A. Klug, *J. Mol. Biol.*, 46, 597 (1969). BIOL, X-Ray: Two double-helical forms of polyriboadenylic acid.

762. T. J. V. Findlay, J. S. Keniry, A. D. Kidman, and V. A. Pickles, *Trans. Faraday Soc.*, 63, 846 (1967). NMR: Thermodynamic data, CHCl₃-C₅H₅N adduct.

763. T. J. V. Findlay and A. D. Kidman, *Aust. J. Chem.*, 18, 521 (1965). IR: Thermodynamic data, adducts of alcohols with pyridine.

763a. R. B. Fiorito and R. Meister, *J. Chem. Phys.*, 56, 4605 (1972). NMR: Translational relaxation, hydrogen-bonded liquids.

764. R. A. Firestone, *J. Org. Chem.*, 36, 702 (1971). Theory: Application of Linnett Theory to hydrogen bonds.

765. W. Firshein and R. G. Gillmor, *Science*, 169, 66 (1970). BIOL: Macromolecular content and stimulation of enzymatic activity by poly A.

766. S. F. Fischer, G. L. Hofacker, and M. A. Ratner, *J. Chem. Phys.*, 52, 1934 (1970). Theory: Model, double-minimum and coupling.

767. S. F. Fischer, G. L. Hofacker, and J. R. Sabin, *Phys. Kondens. Mater.*, 8, 268 (1969); *Chem. Abstr.*, 71, 64324g (1969). Theory: Proton-phonon coupling.

768. E. Fishman and Tun Li Chen, *Spectrochim. Acta*, 25A, 1231 (1969). IR: Intramolecular, butanediols.

768a. M. C. Flanigan and J. R. de la Vega, *Chem. Phys. Lett.*, 21, 521 (1973). Theory: Proton tunneling in H₅O₂⁺.

769. A. N. Fletcher, *J. Phys. Chem.*, 74, 216 (1970). Effect of solvent on hydrogen bond equilibria.

769a. A. N. Fletcher, *J. Phys. Chem.*, 76, 2562 (1972). Near-IR: Self-association, ethanol-*d*.

770. A. N. Fletcher, *J. Phys. Chem.*, 73, 2217 (1969). Near-IR: 1:1 adduct of CCl₄ with 1-octanol.

771. A. N. Fletcher and C. A. Heller, *J. Phys. Chem.*, 72, 1839 (1968). IR, Near-IR: Self-association, alcohols.

772. A. N. Fletcher and C. A. Heller, *J. Phys. Chem.*, 71, 3742 (1967). IR: Self-association, alcohols.

773. A. E. Florin and M. Alei, Jr., *J. Chem. Phys.*, 47, 4268 (1967). NMR: H₂¹⁷O liquid and vapor.

774. R. Foglizzo and A. Novak, *J. Chem. Phys.*, 50, 5366 (1969). Far-IR, Raman: Pyridinium halides.

775. A. Foldes and C. Sandorfy, *Can. J. Chem.*, 49, 505 (1971). Theory: Model calculations on hydrogen bond vibration frequencies.

776. A. Foldes and C. Sandorfy, *J. Mol. Spectrosc.*, 20, 262 (1966). IR: Anharmonicity and strong hydrogen bonds.

777. U. Folli, D. Iarossi, and F. Taddei, *J. Chem. Soc., Perkin Trans.* 2, 848 (1973). NMR: Intramolecular, o-MeS(O)C_6H_4OH, o-MeC(O)C_6H_4OH.

778. F. K. Fong, J. P. McTague, S. K. Garg, and C. P. Smith, *J. Phys. Chem.*, 70, 3567 (1966). NMR: Self-association, 2,6-disubstituted phenols and anilines.

779. W. F. Forbes, *Can. J. Chem.*, 40, 1891 (1962). IR: Intramolecular, o-nitrobenzaldehyde.

780. W. F. Forbes, A. R. Knight, and D. L. Coffen, *Can. J. Chem.*, 38, 728 (1960). UV: Self-association, substituted benzoic acids; intramolecular, o-substituted benzoic acids.

781. T. A. Ford and M. Falk, *Can. J. Chem.*, 46, 3579 (1968). IR: Hydrogen bonding in water and ice.

782. S. Forsen, *J. Chem. Phys.*, 31, 852 (1959). NMR: Potassium hydrogen maleate in Me_2SO.

783. S. Forsen, *Acta Chem. Scand.*, 13, 1472 (1959). NMR: Self-association, ethyl mercaptan.

784. S. Forsen and B. Akermark, *Acta Chem. Scand.*, 17, 1712 (1963). NMR: Intramolecular, 3-nitro-salicylaldehyde.

785. S. Forsen, W. E. Frankle, P. Laszlo, and J. Lubochinsky, *J. Magn. Resonance*, 1, 327 (1969). NMR: Intramolecular, dimedone-aldehyde adducts.

786. S. Forsen and M. Nilsson, *Acta Chem. Scand.*, 14, 1333 (1960). NMR: Intramolecular, methyl and ethyl diacetoacetates.

787. S. Forsen, M. Nilsson, and C. A. Wachtmeister, *Acta Chem. Scand.*, 16, 583 (1962). NMR: Intramolecular, usnic acid.

788. A. Foster, A. Haines, and M. Stacey, *Tetrahedron*, 16, 177 (1961). IR: Intramolecular, acyclic alcohols.

789. R. Foster and C. A. Fyfe in *Progress in Nuclear Magnetic Resonance Spectroscopy*, J. W. Emsley, J. Feeney, and L. H. Sutcliffe, eds., Vol. 4, Pergamon Press, 1969, p. 1. NMR: Equilibrium constant expressions.

790. R. Foster and C. A. Fyfe, *Trans. Faraday Soc.*, 61, 1626 (1965). NMR: Determination of equilibrium constants for 1:1 adducts.

791. I. F. Franchuk and L. I. Kalinina, *Zh. Prikl. Spektrosk.*, 15, 896 (1971). IR, NMR: Ph_3CO_2H, π-hydrogen bond.

791a. I. F. Franchuk and L. I. Kalinina, *Zh. Fiz. Khim.*, 46, 1673 (1972); [English, 961]. NMR: Kinetics, proton exchange between 1,1'-dihydroperoxycyclohexyl peroxide and methanol.

792. A. Francina, A. Lamotte, and J. C. Merlin, *Compt. Rend. Acad. Sci., Ser. C*, 267, 763 (1968); *Chem. Abstr.*, 70, 14976e (1969). NMR: $[H_3PO_3][H_2PO_3]$ joined by hydrogen bond.

793. E. U. Franck, *J. Solution Chem.*, 2, 339 (1973). Raman, IR: High-pressure, HD in H_2O.

794. H. S. Frank, *Federation Proc.*, 24, S1 (1965). BIOL: Structure of water.

795. H. S. Frank, *Proc. Roy. Soc.*, A247, 481 (1958). Hydrogen bond in water and ice.

796. H. S. Frank and W.-Y. Wen, *Discussions Faraday Soc.*, 24, 133 (1957). Cooperative nature of hydrogen bond in water.

797. R. E. Franklin and R. G. Gosling, *Acta Crystallogr.*, 6, 673 (1953). BIOL, X-Ray: DNA, A and B forms.

798. A. Fratiello and D. C. Douglass, *J. Mol. Spectrosc.*, 11, 465 (1963). NMR: Chemical shifts for dioxane-water and pyridine-water mixtures.

799. A. Fratiello and J. P. Luongo, *J. Amer. Chem. Soc.*, 85, 3072 (1963). IR: Dioxane and pyridine in aqueous mixtures.

799a. A. Fratiello, R. E. Schuster, G. A. Vidulich, J. Bragin, and D. Lui, *J. Amer. Chem. Soc.*, 95, 633 (1973). NMR: 1,1,1,3,3,3-$(CF_3)_2CHOH-N(C_2H_5)_3$ adduct at $-125°C$.

799b. L. Fredin and B. Nelander, *J. Mol. Struct.*, 16, 217 (1973). Theory: CNDO/2 on $H_2O...Cl_2$.

800. H. H. Freedman, *J. Amer. Chem. Soc.*, 83, 2900 (1961). IR: Intramolecular, phenolic hydroxyl to azomethine nitrogen.

801. T. C. French and G. G. Hammes, *J. Amer. Chem. Soc.*, 87, 4669 (1965). Kinetics: D_2O solvent isotope effects for isomerization of ribonuclease.

802. M. N. Frey, M. S. Lehmann, T. F. Koetzle, and W. C. Hamilton, *Acta Crystallogr.*, Sect. B, 29, 876 (1973). Neutron Diffraction: L-serine•H_2O, DL-serine.

803. M. Freymann and R. Freymann, *Compt. Rend. Acad. Sci.*, 248, 677 (1959). NMR: Pyrrole-pyridine, pyrrole-oxygenated bases.

804. I. Fric, W. Guschlbauer, and A. Holy, *FEBS Lett.*, 9, 261 (1970). BIOL: Optical properties of chemically synthesized polynucleotides.

805. H. B. Friedrich and W. B. Person, *J. Chem. Phys.*, 44, 2161 (1966). Theory: Charge-transfer theory for hydrogen bonds.

806. H. Fritzsche, *Spectrochim. Acta*, 21, 799 (1965). IR: Asymmetry of bonded OH peak in phenol-carbonyl systems.

807. H. Fritzsche, *Ber. Bunsenges, Physik. Chem.*, 68, 459 (1964). IR: Phenol adducts of bases in CCl_4.

808. H. Fritzsche, *Z. Naturforsch.*, 19A, 1132 (1964). IR: $\Delta\nu_{OH}$ for phenol adducts, correlation of $\overline{\Delta\nu}_{OH}$ and ionization potential.

809. H. Fritzsche, *Spectrochim. Acta*, 17, 352 (1961). IR: Dioxane-water mixtures.

810. H. Fritzsche, *Z. Chem.*, 13, 35 (1973). IR: Phenol-ketone adducts.

811. A. J. Fry and R. G. Reed, *J. Amer. Chem. Soc.*, 93, 553 (1971). Electrochemistry: Anomalous proton-donor effects in dimethylformamide.

812. E. Fuchs and P. Hanawalt, *J. Mol. Biol.*, 52, 301 (1970): BIOL: Isolation and characterization of DNA replication complex from *E. coli*.

813. E. Fujimoto, Y. Takeoka, and K. Kozima, *Bull. Chem. Soc. Jap.*,
      43, 991 (1970). IR: Intramolecular, *trans*-2-halocyclohexanols.

814. H. Fujita, A. Imamura, and C. Nagata, *J. Theor. Biol.*, 28, 143
      (1970). BIOL: Tautomerism of anticodon base and code degeneracy.

815. H. Fujita, A. Imamura, and C. Nagata, *Bull. Chem. Soc. Jap.*, 42,
      1467 (1969). BIOL: Electronic structures of base components of
      nucleic acids and their tautomers calculated by CNDO/2 method.

816. H. Fujita, A. Imamura, and C. Nagata, *Bull. Chem. Soc. Jap.*, 41,
      2017 (1968). BIOL: ASMO-SCF-CI calculations on the nucleic
      acid bases.

817. Y. Fujiwara and S. Fujiwara, *Bull. Chem. Soc. Jap.*, 36, 574
      (1963). NMR: Thermodynamic data, water-acetaldehyde adduct.

818. T. Fukuroi, Y. Fujiwara, S. Fujiwara, and K. Fujii, *Analyt. Chem.*,
      40, 879 (1968). NMR: Intramolecular, 2,4-pentanediol.

819. K. Fukushima and B. J. Zwolinski, *J. Chem. Phys.*, 50, 737 (1969).
      IR: Normal-coordinate treatment of acetic acid monomer and dimer.

820. W. Fuller, *Advan. Sci.*, 26, 267 (1970). BIOL: Review, structure
      and function of nucleic acids and proteins.

821. W. Fuller, S. Arnott, and J. Creek, *Biochem. J.*, 114, 26P (1969).
      BIOL: Molecular model for tRNA.

822. S. Furberg and L. H. Jensen, *Acta Crystallogr., Sect. B*, 26, 1260
      (1970). BIOL, X-Ray: Thiocytosine.

823. V. P. Gaidaenko, I. M. Ginzburg, and D. V. Ioffe, *Opt. Spektrosk.*,
      27, 620 (1969); *Chem. Abstr.*, 72, 37300k (1970). IR: Intramo-
      lecular, amino alcohols.

824. G. E. Gajnos and F. E. Karasz, *J. Phys. Chem.*, 76, 3464 (1972).
      BIOL: Helix-coil transition of poly (γ-benzyl L-glutamate).

825. G. A. Galkin, A. V. Kiselev, and V. I. Lygin, *Russ. J. Phys.
      Chem.*, 41, 20 (1967). IR: Frequency shifts of hydroxyl groups
      as an indication of base adsorption on silica.

826. A. N. Garg and P. S. Goel, *J. Inorg. Nucl. Chem.*, 31, 697 (1969).
      Mössbauer: $H_4Fe(CN)_6$.

827. A. W. Garrison, L. H. Keith, and A. L. Alford, *Spectrochim. Acta*,
      25A, 77 (1969). NMR: Self-association, 2,4-dichlorophenyl
      methyl isopropylphosphoramidate.

828. E. A. Gastilovich *et al.*, *Opt. Spektrosk.*, supplement 2, 178
      (1963); [English, 90 (1966)]. PhC≡CH adducts of amines.

828a. E. A. Gastilovich and D. N. Shigorin, *Usp. Khim.*, 42, 1353 (1973).
      IR: Review, monosubstituted acetylene interactions.

829. E. A. Gastilovich, D. N. Shigorin, E. P. Gracheva, I. A. Chekula,
      and M. F. Shostakovskii, *Opt. Spektrosk.*, 10, 595 (1961);
      [English, 312]. IR: Adducts of acetylenes.

830. E. A. Gastilovich, K. V. Zhukova, D. N. Shigorin, O. G. Yarosh,
      and I. S. Aknurina, *Opt. Spektrosk.*, 29, 41 (1970); *Chem. Abstr.*,
      73, 135538j (1970). IR: Spectra of acetylene derivatives dis-
      solved in CCl_4, C_6H_6, deuteriopyridine, Et_3N.

831. L. Gattlin and J. C. Davis, Jr., *J. Amer. Chem. Soc.*, $\underline{84}$, 4464
      (1962). BIOL, NMR: Comparison of ribose and deoxyribose nucleo-
      sides.

832. J. Gaultier and C. Hauw, *Acta Crystallogr.*, *Sect. B*, $\underline{25}$, 546
      (1969). IR, X-Ray: Bifurcated hydrogen bonds in naphthoquinones.

833. H. A. Gebbie, W. J. Burroughs, J. Chamberlain, J. E. Harries, and
      R. G. Jones, *Nature*, $\underline{221}$, 143 (1969). IR: Strength of water
      dimer bond = 5.2 ± 1.5 kcal/mole.

834. E. P. Geiduschek, E. N. Brody, and D. L. Wilson in *Molecular
      Associations in Biology*, B. Pullman, ed., Academic Press, New
      York, 1968, p. 163. BIOL: Discussion, RNA transcription.

835. E. P. Geiduschek and T. T. Herskovits, *Arch. Biochem. Biophys.*,
      $\underline{95}$, 114 (1961). BIOL: Nonaqueous solutions of DNA, reversible
      and irreversible denaturation in methanol.

836. G. Geiseler and E. Stockel, *Spectrochim. Acta*, $\underline{17}$, 1185 (1961).
      IR: Self-association, alcohols.

837. H. Geisenfelder and H. Zimmermann, *Ber. Bunsenges, Physik. Chem.*,
      $\underline{67}$, 480 (1963). X-Ray: Liquid formic acid.

838. J. Gelas and R. Rambaud, *Bull. Soc. Chim. Fr.*, 1300 (1969).
      IR: Intramolecular, 2-hydroxyalkyl-1,3-dioxolanes, 2-methyl-
      and 2-*tert*-4-hydroxyalkyl-1,3-dioxolanes.

838a. L. G. Gelenskaya and A. V. Iogansen, *Zh. Prikl. Spektrosk.*, $\underline{16}$,
      108 (1972); *Chem. Abstr.*, $\underline{77}$, 4715z (1972). IR: Thermodynamic
      data, adducts of phenol with hydrocarbons.

839. J. Gendell, W. R. Miller, Jr., and G. K. Fraenkel, *J. Amer. Chem.
      Soc.*, $\underline{91}$, 4369 (1969). EPR: Intramolecular, hydroxysemiquinone
      radicals.

840. W. O. George, J. H. S. Green, and D. Pailthorpe, *J. Mol. Struct.*,
      $\underline{10}$, 297 (1971). IR: 2-Hydroxycarboxylic acids.

841. W. O. George and A. J. Porter, *J. Mol. Struct.*, $\underline{17}$, 152 (1973).
      IR: Monoalkylesters of fumaric and maleic acids.

842. R. Gerdil, *Acta Crystallogr.*, $\underline{14}$, 333 (1961). BIOL, X-Ray:
      Crystals of thymine monohydrate.

843. F. Gerson, *Helv. Chim. Acta*, $\underline{44}$, 471 (1961). Theory: Sixth
      power oscillator as basis for symmetric double-minimum potential
      in hydrogen bonds.

844. S. Ghersetti, *Boll. Sci. Fac. Chim. Ind. Bologna*, $\underline{27}$, 35 (1969);
      Chem. Abstr., $\underline{71}$, 34741y (1969). IR: Thermodynamic data, phenol
      adducts in various solvents.

845. S. Ghersetti, *Boll. Sci. Fac. Chim. Ind. Bologna*, $\underline{27}$, 17 (1969);
      Chem. Abstr., $\underline{71}$, 34740x (1969). IR: Solvent effects on Δv_{OH}
      for phenol adducts.

845a. S. Ghersetti, S. Giorgianni, P. L. Capucci, and G. Spunta,
      Spectrochim. Acta, $\underline{29A}$, 1207 (1973). Far-IR: Phenol adducts of
      substituted C_5H_5NO.

845b. S. Ghersetti, S. Giorgianni, A. Mangini, and G. Spunta, *Spectrosc.
      Lett.*, $\underline{5}$, 111 (1972). Far-IR: Phenol-pyridine adduct.

846. S. Ghersetti and A. Lusa, *Boll. Sci. Fac. Chim. Ind. Bologna*, <u>27</u>, 3 (1969); *Chem. Abstr.*, <u>71</u>, 34727y (1971). IR: Thermodynamic data for phenol-diphenyl sulfone in C_2Cl_4.

847. S. Ghersetti and A. Lusa, *Spectrochim. Acta*, <u>21</u>, 1067 (1965). IR: Thermodynamic data, adducts of substituted phenols with Ph_2SO.

847a. S. Ghersetti, G. Maccagnani, A. Mangini, and F. Montanari, *J. Heterocycl. Chem.*, <u>6</u>, 859 (1969). IR, UV: Thermodynamic data, CH_3OH adducts of substituted C_5H_5NO in C_2Cl_4.

848. B. Ghosh and S. Basu, *J. Chim. Phys. Physicochim. Biol.*, <u>65</u>, 1587 (1968). Fluorescence: α- and β-naphthol OH association with π-bases.

849. B. Ghosh and S. Basu, *Trans. Faraday Soc.*, <u>61</u>, 2097 (1965). UV: Naphthols, O—H...π bonding.

850. A. Gierer and G. S. Schramm, *Nature*, <u>177</u>, 702 (1956). BIOL: Genetic role for virus RNA.

851. C. Giessner-Prettre, *Compt. Rend. Acad. Sci.*, <u>252</u>, 3238 (1961). NMR: Amines in aromatic solvents.

852. C. Giessner-Prettre and B. Pullman, *Compt. Rend. Acad. Sci., Ser. D*, <u>270</u>, 866 (1970). BIOL: Formation of complementary purine-pyrimidine pairs in nonaqueous solution.

853. C. Giessner-Prettre and B. Pullman, *Compt. Rend. Acad. Sci.*, <u>261</u>, 2521 (1965). BIOL: Quantum mechanical calculations, ring currents in purines and pyrimidines.

854. E. Giglio, *Nature*, <u>222</u>, 339 (1969). Theory: Calculation of crystal structures using empirical potential functions for van der Waals and hydrogen bonds.

854a. P. A. Giguere, *J. Phys. Chem.*, <u>76</u>, 3675 (1972). IR: HDO, questions assignments of Ref. [277a].

855. C. H. Giles, R. B. McKay, and W. Good, *J. Chem. Soc.*, 5434 (1961). Refractive Index: Method of continuous variations applied to detection of hydrogen bonds.

856. S. J. Gill, M. Downing, and G. F. Sheats, *Biochemistry*, <u>6</u>, 272 (1967). BIOL: Enthalpy of self-association of purine derivatives in H_2O.

856a. S. J. Gill and L. Noll, *J. Phys. Chem.*, <u>76</u>, 3065 (1972). BIOL: Self-association, diketopiperazine in H_2O.

857. S. G. W. Ginn and J. L. Wood, *J. Chem. Phys.*, <u>46</u>, 2735 (1967). Far-IR: Self-association, acetic acid vapor, proton tunneling.

858. S. G. W. Ginn and J. L. Wood, *Proc. Chem. Soc., (London)*, 370 (1964). Far-IR: Phenol-amine complexes.

859. S. G. W. Ginn and J. L. Wood, *Spectrochim. Acta*, A<u>23</u>, 611 (1967). Far-IR: Phenol-amine complexes, phenol dimer and trimer.

860. S. G. W. Ginn and J. L. Wood, *Chem. Commun.*, 628 (1965). Far-IR: Deuterium band stretching vibration.

861. S. G. W. Ginn and J. L. Wood, *Nature*, <u>200</u>, 467 (1963). Far-IR: Gas-phase spectra of $MeOH-NMe_3$.

862. I. M. Ginzburg, *Opt. Spektrosk.*, 17, 57 (1964); [English, 28].
      IR: Thermodynamic data, *p*-chlorophenol association with ethers.

863. I. M. Ginzburg, *Opt. Spektrosk.*, supplement 2, 237 (1963);
      [English, 123 (1966)]. IR: Ether adducts of CF_3COOH.

864. I. M. Ginzburg and M. A. Abramovich, *Opt. Spektrosk.*, supplement
      2, 230 (1963); [English, 119 (1966)]. IR: Ether adducts of
      CH_3OD and CCl_3COOH.

865. I. M. Ginzburg, N. N. Bessonova, and E. S. Korbelainen, *Opt.
      Spektrosk.*, 27, 771 (1969); *Chem. Abstr.*, 72, 49214y (1970).
      IR: Effect of hydrogen bonding on ν_{CO} of diacetyl.

866. I. M. Ginzburg, D. V. Ioffe, and N. N. Bessonova, *Opt. Spektrosk.*,
      30, 234 (1971); *Chem. Abstr.*, 74, 132590e (1971). IR: Intramo-
      lecular, N-methyl amides of substituted acids.

867. I. M. Ginzburg and V. S. Khlevnyuk, *Opt. Spektrosk.*, 26, 183
      (1969); *Chem. Abstr.*, 71, 26194g (1969). IR: Effect of solvents
      on ν_{CO} Fermi doublet in cyclopentanone.

868. I. M. Ginzburg and L. A. Loginova, *Opt. Spektrosk.*, 20, 241
      (1966); [English, 130]. Self-association, thioacetic and thio-
      benzoic acids.

869. I. M. Ginzburg and L. A. Loginova, *Dokl. Akad. Nauk SSSR*, 156,
      1382 (1964). IR: Intramolecular, thiosalicylic and salicylic
      acid.

869a. I. M. Ginzburg and B. P. Tarasov, *Zh. Obshch. Khim.*, 42, 2740
      (1972); *Chem. Abstr.*, 78, 135496x (1973). IR: Self-association,
      haloacetic acids, methyl haloacetates.

870. A. M. Giuliani, *J. Chem. Soc., Dalton Trans.*, 497 (1972). NMR:
      Intramolecular, N—H...X in $Zn[EtNHC(S)NHEt]_2X_2$.

871. A. T. Gladishev and Ya. K. Syrkin, *Compt. Rend. Acad. Sci. URSS*,
      20, 145 (1938). Vapor pressure studies of $HCl—OMe_2$ adduct.

872. P. K. Glasoe and L. Eberson, *J. Phys. Chem.*, 68, 1560 (1964).
      Deuterium isotope effect in the ionization of succinic acids.

873. P. K. Glasoe, S. Hallock, M. Hove, and J. M. Duke, *Spectrochim.
      Acta*, 27A, 2309 (1971). IR: Self-association, $(C_6H_5COOH)_2$,
      $(C_6H_5COOD)_2$ in CCl_4.

874. P. K. Glasoe and J. R. Hutchison, *J. Phys. Chem.*, 68, 1562 (1964).
      Deuterium isotope effect in the ionization of substituted malonic
      acids.

875. D. Glaubiger, D. A. Lloyd, and I. Tinoco, Jr., *Biopolymers*, 6,
      409 (1968). BIOL: Model for base pair interaction.

876. D. N. Glew and N. S. Rath, *Can. J. Chem.*, 49, 837 (1971). IR:
      Correlations, HOH, DOH, CH_3OH ν_{OH} frequencies and band shapes
      with solvent basicity.

877. P. Gloux and B. Lamotte, *Mol. Phys.*, 24, 23 (1972). ESR and
      ENDOR study of γ-induced radicals, imidazole, triazole.

878. Yu. M. Glubokov, V. V. Yastrebov, and S. S. Korovin, *Zh. Neorg.
      Khim.*, 14, 1082 (1969); *Chem. Abstr.*, 71, 17414b (1969). NMR:
      ^1H chemical shifts in $Bu_3PO_4—HCl$ and $Bu_3PO_4—HCl—H_2O$ systems.

879. A. Goel, A. S. N. Murthy, and C. N. R. Rao, *J. Chem. Soc. A*, 190 (1971). Theory: CNDO/2 calculations on water dimer.

880. A. Goel, A. S. N. Murthy, and C. N. R. Rao, *J. Chem. Soc. D*, 423 (1970). Theory: CNDO/2 calculations on polywater.

881. A. Goel and C. N. R. Rao, *Trans. Faraday Soc.*, 67, 2828 (1971). IR, Theory: Octyne adducts, thermodynamic data; CNDO/2 calculations on hydrogen bonds formed by C_2H_2, HCN, HF.

882. H. J. Gold, *J. Amer. Chem. Soc.*, 93, 6387 (1971). BIOL, Theory: CNDO/2 study of the double hydrogen bond between glucose and 2-pyridone.

882a. M. A. Goldman and M. T. Emerson, *J. Phys. Chem.*, 77, 2295 (1973). NMR: Self-association, acetic acid.

882b. I. P. Gol'dshtein, E. N. Guryanova, and T. I. Perepelkova, *Zh. Obshch. Khim.*, 42, 2091 (1972); *Chem. Abstr.*, 78, 9114h (1973). ΔH, dipole moments, adducts of carboxylic acids with amines.

883. M. Goldstein, C. B. Mullins, and H. A. Willis, *J. Chem. Soc. B*, 321 (1970). NMR: K for phenylacetylene adducts.

884. R. Goldstein and S. S. Penner, *J. Quant. Spectrosc. Radiat. Transfer*, 4, 441 (1964). IR: Self-association, H_2O.

885. L. Golic and J. C. Speakman, *J. Chem. Soc.*, 2530 (1965). X-Ray: Acid salts of trifluoroacetates.

885a. A. Gomez, L. Lamberts, and P. Huyskens, *Bull. Soc. Chim. Fr.*, 1734 (1972). CAL: Self-association, substituted anilines.

885b. A. Gomez, J. Mullens, and P. Huyskens, *J. Phys. Chem.*, 76, 4011 (1972). Aniline-water interactions.

886. C. Gonzales, M. Pieber, and J. Toha, *Z. Naturforsch, B*, 23, 397 (1968). BIOL: Quantitative evaluation of the wobble hypothesis for a polynucleotide system.

886a. B. Z. Gorbunov and Yu. I. Naberukhim, *J. Mol. Struct.*, 14, 113 (1972). IR: Intensity, OD band of HDO mixtures.

887. J. E. Gordon, *J. Amer. Chem. Soc.*, 94, 650 (1972). NMR: Dependence of acidity and basicity of H_2O on hydrogen-bonded structure.

888. J. E. Gordon, *J. Phys. Chem.*, 67, 19 (1963). K for $(ArCOO)_2H^-$.

889. J. E. Gordon, *J. Org. Chem.*, 26, 738 (1961). IR: Correlation of association constants with frequency shifts.

889a. M. S. Gordon and R. D. Koob, *J. Amer. Chem. Soc.*, 95, 5863 (1973). Theory: INDO, intramolecular, acetylacetone, trifluoroacetylacetone.

889b. M. S. Gordon and D. E. Tallman, *Chem. Phys. Lett.*, 17, 385 (1972). Theory: INDO on HCOOH, $(H_2O)_2$, pyruvic acid.

890. W. Gordy, *J. Chem. Phys.*, 9, 215 (1941). IR: $\Delta\nu_{OD}$ versus pK_a for D_2O adducts, $\Delta\nu_{HCl}$ versus pK_a for HCl adducts.

891. W. Gordy and S. C. Stanford, *J. Chem. Phys.*, 9, 204 (1941). IR: $\Delta\nu_{OD}$ versus pK_a for CH_3OD adducts.

892. W. Gordy and S. C. Stanford, *J. Chem. Phys.*, 8, 170 (1940). IR: $\Delta\nu_{OD}$ versus pK_a for CH_3OD adducts.

893. M. V. Gorellik *et al.*, *Khim. Geterotsikl. Soedin.*, 7, 238 (1971); *Chem. Abstr.*, 75, 34977n (1971). IR: Intramolecular, α-naphthol derivatives with heterocyclic N in *peri* position.

893a. S. Goren, *Phys. Lett.*, 42A, 185 (1972). Theory: INDO, $(NH_3)_x$.

893b. T. M. Gorrie, E. M. Engler, R. C. Bingham, and P. von R. Schleyer, *Tetrahedron Lett.*, 3039 (1972). IR: Intramolecular, dihedral angle, noradamantane-3,4-diol and 15 related 1,2-diols.

894. A. Goudot, *Compt. Rend. Acad. Sci.*, 256, 1776 (1963). Theory: Energy levels of electrons in hydrogen bonds of DNA and RNA.

895. S. R. Gough and A. H. Price, *J. Phys. Chem.*, 73, 459 (1969). Dielectric properties of pyridine complexes with dichloro- and trichloroacetic acids.

896. G. Govil, A. D. H. Clague, and H. J. Bernstein, *J. Chem. Phys.*, 49, 2821 (1968). NMR: Gas phase, Me_2O—HCl adduct.

897. G. Govil and C. L. Khetrapal, *Current Sci. (India)*, 34, 116 (1965). NMR: Intramolecular, 5-hydroxy-chromones.

897a. G. Govil and A. Saran, *J. Chem. Soc. A*, 3624 (1971). BIOL, Theory: Effect of hydrogen bond on peptide chain length.

898. I. P. Gragerov, *Dokl. Akad. Nauk SSSR*, 193, 818 (1970); *Chem. Abstr.*, 74, 91349n (1971). Theory: Strength of HX acids.

899. I. P. Gragerov and V. K. Pogorelyi, *Dokl. Akad. Nauk SSSR*, 185, 1052 (1969); *Chem. Abstr.*, 71, 29783j (1969). NMR: Kinetics, proton exchange of thiophenols with methanol.

900. L. L. Graham and C. Y. Chang, *J. Phys. Chem.*, 75, 784 (1971). NMR: Self-association, N-monosubstituted amides in dioxane.

901. L. L. Graham and C. Y. Chang, *J. Phys. Chem.*, 75, 776 (1971). NMR: Self-association, N-monosubstituted amides in CCl_4.

902. R. Grahn, *Arkiv Fysik*, 21, 13 (1962). Theory: Approximate LCAO MO treatment of the hydrated proton $H_9O_4^+$.

903. T. Gramstad, *Spectrochim. Acta*, 26A, 741 (1970). Near-IR: Phenol complex of carbethoxymethylenetriphenylphosphorane.

904. T. Gramstad, *Spectrochim. Acta*, 20, 729 (1964). Near-IR: Thermodynamic data, adduct of phenol with tetrahydropyran.

905. T. Gramstad, *Spectrochim. Acta*, 19, 1698 (1963). Near-IR: Thermodynamic data, adducts of phenol and pentachlorophenol with 2,6-dimethyl-4-pyrone.

906. T. Gramstad, *Spectrochim. Acta*, 19, 1391 (1963). Near-IR: Thermodynamic data, adducts of phenol with $(C_2H_5O)_2P(O)OCH_2CH_2N(CH_3)_2$.

907. T. Gramstad, *Spectrochim. Acta*, 19, 829 (1963). IR: Thermodynamic data, adducts of phenol with sulfoxides, amine oxides, Δν-ΔH relationship.

908. T. Gramstad, *Spectrochim. Acta*, 19, 497 (1963). Near-IR: Thermodynamic data, adducts of phenol and pentachlorophenol with bases, Δν-ΔH relationship.

909. T. Gramstad, *Acta Chem. Scand.*, 16, 807 (1961). Near-IR: Thermodynamic data, adducts of phenol, pentachlorophenol, and α-naphthol with amines.

910. T. Gramstad, *Acta Chem. Scand.*, 15, 1337 (1961). Near-IR: Thermodynamic data, adducts of pentachlorophenol, α-naphthol, and methanol with organophosphorus compounds.

911. T. Gramstad and E. D. Becker, *J. Mol. Struct.*, 5, 253 (1970). NMR: Self-association and hetero-association, phenol in various solvents.

912. T. Gramstad and W. J. Fuglevik, *Spectrochim. Acta*, 21, 503 (1965). Near-IR: Thermodynamic data, indole adducts of phosphoryl compounds.

913. T. Gramstad and W. J. Fuglevik, *Spectrochim. Acta*, 21, 343 (1965). Near-IR: Thermodynamic data, adducts of phenol with lactones.

914. T. Gramstad and W. J. Fuglevik, *Acta Chem. Scand.*, 16, 1369 (1962). Near-IR: Thermodynamic data, adducts of phenol, pentachlorophenol with amides, $\Delta\nu$-ΔH relationship.

914a. T. Gramstad and O. Mundheim, *Spectrochim. Acta*, 28A, 1405 (1972). NMR: Thermodynamic data, chloroform adducts with phosphoryl donors.

915. T. Gramstad and J. Sandström, *Spectrochim. Acta*, 25A, 31 (1969). IR: Thermodynamic data, adducts of phenol with thioamides and nitriles, $\Delta\nu$-ΔH relationship.

916. T. Gramstad and H. J. Storesund, *Spectrochim. Acta*, 26A, 426 (1970). IR: Correlation of IR intensities with $\Delta\nu_{OH}$.

917. T. Gramstad and G. Van Binst, *Spectrochim. Acta*, 22, 1681 (1966). Near-IR: Thermodynamic data, adducts of pentafluorophenol with triphenylphosphine oxide.

917a. T. Gramstad and O. Vikane, *Spectrochim. Acta*, 28A, 2131 (1972). NMR: Thermodynamic data, chloroform adducts of amides, sulfoxides.

918. I. Gränacher, *Helv. Phys. Acta*, 34, 272 (1961). NMR: Correlation of chemical shifts with frequency shift.

919. N. Granboulon and R. Franklin, *J. Virol.*, 2, 129 (1968). BIOL: Electron microscopic comparison of the ultrastructure of single-stranded RNA and double-stranded replicative form and replicative intermediate.

920. V. A. Granzhan, S. V. Semenenko, and P. M. Zaitsev, *Zh. Prikl. Spektrosk.*, 12, 922 (1970); *Chem. Abstr.*, 73, 55235t (1970). IR: Intramolecular, nitrophenols.

921. V. A. Granzhan, S. V. Semenenko, and P. M. Zaitsev, *Zh. Prikl. Spektrosk.*, 9, 407 (1968); *Chem. Abstr.*, 70, 57011d (1969). IR: Intramolecular, *o*-nitrophenol derivatives.

923. W. B. Gratzer, *Eur. J. Biochem.*, 10, 184 (1969). BIOL: Association of nucleic acid bases in aqueous solution.

924. W. B. Gratzer, M. Haynes, and R. A. Garrett, *Biochemistry*, 9, 4410 (1970). BIOL, X-Ray: Complexes of DNA, RNA, with poly-L-lysine.

925. W. B. Gratzer and C. W. F. McClare, *J. Amer. Chem. Soc.*, 89, 4224 (1967). BIOL, UV: Hypochromic effect from pairing of purine and pyrimidine bases.

926. E. Grech and M. Szafran, *Rocz. Chem.*, <u>46</u>, 2365 (1972). IR: (BHB)$^+$ClO$_4$$^-$ B=quinoline, lepidine, quinaldine.

927. D. W. Green, F. S. Mathews, and A. Rich, *J. Biol. Chem.*, <u>237</u>, 3573 (1962). BIOL, X-Ray: 1-methyluracil dimer.

928. E. A. Green, R. Shiono, and R. D. Rosenstein, *Chem. Commun.*, 53 (1971). BIOL, X-Ray: Nucleoside β-uridine.

929. G. Green and H. R. Mahler, *Biopolymers*, <u>6</u>, 1509 (1968). BIOL: Absorption, CD, and ORD spectra for DNA in H$_2$O and ethylene glycol.

930. J. H. S. Green, W. Kynaston, and H. A. Gebbie, *Spectrochim. Acta*, <u>19</u>, 807 (1963). Far-IR: Self-association, phenol and *m*-cresol.

931. R. D. Green, *Can. J. Chem.*, <u>47</u>, 2407 (1969). NMR: Formamide-bromide ion association.

932. R. D. Green and J. S. Martin, *J. Amer. Chem. Soc.*, <u>90</u>, 3659 (1968). NMR, IR: Association of trihalomethanes with halide ions.

933. R. D. Green, J. S. Martin, W. B. Cassie, and J. B. Hyne, *Can. J. Chem.*, <u>47</u>, 1639 (1969). NMR: Association of alcohols with halide ions.

933a. C. Gref, B. Sabourault, and J. Bourdais, *Tetrahedron Lett.*, 1957 (1972). NMR, IR: Intramolecular, 2-oxo-3-indoline-carboxamides.

934. M. J. Gregory and M. J. R. Loadman, *J. Chem. Soc. B*, 1862 (1971). IR: Self-association, succinamic acids.

935. E. Greinacher, W. Lüttke, and R. Mecke, *Z. Electrochem.*, <u>59</u>, 23 (1955). IR: Anharmonicity of overtone bands for H$_2$O—solvent spectra.

936. C. E. Griffin, E. J. Fendler, and B. D. Martin, *Spectrochim. Acta*, <u>25A</u>, 710 (1969). NMR: Self-association of 1-(β-hydroxyethoxy)-2,4-dinitrobenzene in CH$_3$CN.

936a. P. J. F. Griffiths and G. D. Morgan, *Spectrochim. Acta*, <u>28A</u>, 1899 (1972). IR: Self-association, heterocyclic thioamides.

937. V. S. Griffiths and G. Socrates, *J. Mol. Spectrosc.*, <u>21</u>, 302 (1966). NMR: K$_n$ for self-association of substituted phenols.

937a. Z. I. Grigorovich, Yu. I. Malov, and V. Ya. Rosolovskii, *Izv. Akad. Nauk SSSR, Ser. Khim.*, 265 (1973); *Chem. Abstr.*, <u>78</u>, 152159g (1973). NMR: H(ClO$_4$)$_2$$^-$ in HClO$_4$ and CHCl$_3$.

937b. E. P. Grimsrud and P. Kebarle, *J. Amer. Chem. Soc.*, <u>95</u>, 7939 (1973). Gas Phase: Ion cyclotron resonance, thermodynamic data, H$^+$(CH$_3$OH)$_n$, H$^+$(CH$_3$OCH$_3$)$_n$.

938. R. Grinter, *J. Mol. Spectrosc.*, <u>17</u>, 240 (1965). Theory: Simple Hückel theory calculations on intramolecular bonds.

939. J. P. Grolier and A. Viallard, *J. Chim. Phys. Physicochim. Biol.*, <u>69</u>, 203 (1972). Thermodynamic study of ester-alcohol molecular interactions.

940. D. Grünberger, A. Holý, and F. Sorm, *Biochim. Biophys. Acta*, <u>134</u>, 484 (1967). BIOL: Coding properties of some tri-ribonucleoside diphosphates containing inosine.

941. M. Grunberg-Manago, B. F. C. Clark, M. Revel, P. S. Rudland, and
     J. Dondon, *J. Mol. Biol.*, $\underline{40}$, 33 (1969). BIOL: Stability of
     ribosomal complexes with tRNA and synthetic mRNA.

942. J. Grundy and L. J. Morris, *Spectrochim. Acta*, $\underline{20}$, 695 (1964).
     IR: Intramolecular, O—H...π, methyl esters of unsaturated long-
     chain acids.

943. E. Grunwald and W. C. Coburn, *J. Amer. Chem. Soc.*, $\underline{80}$, 1322
     (1958). IR: K, adducts of ethanol.

944. E. Grunwald and Dodd-Wing Fong, *J. Phys. Chem.*, $\underline{73}$, 650 (1969).
     Self-association of $Al(H_2O)_5OH^{2+}$ in water.

945. E. Grunwald and M. S. Puar, *J. Phys. Chem.*, $\underline{71}$, 1842 (1967).
     NMR: Proton exchange of phenol in aqueous acid.

946. E. Grunwald and E. K. Ralph, *Accounts Chem. Res.*, $\underline{4}$, 107 (1971).
     Kinetics: Hydrogen-bonded solvation complexes of amines in H_2O.

947. H. U. Güdel, *J. Chem. Phys.*, $\underline{56}$, 4984 (1972). Theory: Gaussian
     calculations on models for the hydrogen bonds in $H_3Co(CN)_6$ and
     $D_3Co(CN)_6$.

948. H. U. Güdel, A. Ludi, and P. Fischer, *J. Chem. Phys.*, $\underline{56}$, 674
     (1972). Neutron Diffraction: N—D—N bond in $D_3Co(CN)_6$.

948a. M. Guerin and M. Gomel, *J. Chim. Phys. Physicochim. Biol.*, $\underline{70}$,
      953, 960 (1973). Solvent effects on K for phenol-pyridine and
      pyrrole-THF adducts.

948b. C. Guidotti, U. Lamanna, and M. Maestro, *Theor. Chim. Acta*, $\underline{26}$,
      147 (1972). Theory: LCAO SCF on $(H_2O)_5$.

949. R. M. Guidry and R. S. Drago, Submitted for publication, *J. Chem.
     Educ.* Review, equilibrium constant calculations.

949a. R. M. Guidry and R. S. Drago, *J. Amer. Chem. Soc.*, $\underline{95}$, 759 (1973).
      Allowance for intramolecular bonding in E and C parameter evalua-
      tions.

949b. W. Guild, Jr., *J. Chem. Educ.*, $\underline{49}$, 171 (1972). Review of the
      theories attributing sugar sweetness to inter- and intramolecular
      hydrogen bonding.

950. J. Guilemé, M. Chabanel, and B. Wojtkowiak, *Spectrochim. Acta*,
     $\underline{27A}$, 2355 (1971). IR: Self-association, α-ethylenic and α-
     acetylenic carboxylic acids.

951. J. Guillermet and A. Novak, *J. Chim. Phys. Physicochim. Biol.*, $\underline{66}$,
     68 (1969). IR: $MgX_2 \cdot nH_2O$ and deuterated derivatives.

954. A. Guradta, R. E. Klinck, and J. B. Stothers, *Can. J. Chem.*, $\underline{45}$,
     213 (1967). NMR: Self-association, aromatic aldehydes.

955. D. Gurka and R. W. Taft, *J. Amer. Chem. Soc.*, $\underline{91}$, 4794 (1969).
     NMR: Equilibrium constants for adducts of *p*-fluorophenol with
     sixty-two bases in CCl_4.

956. D. Gurka, R. W. Taft, L. Joris, and P. von R. Schleyer, *J. Amer.
     Chem. Soc.*, $\underline{89}$, 5957 (1967). NMR: *p*-FC_6H_4OH adducts.

957. G. V. Gusakova, G. S. Denisov, A. L. Smolyanskii, and V. M.
     Shraiber, *Dokl. Akad. Nauk SSSR*, $\underline{193}$, 1065 (1970). IR: Thermo-
     dynamic data, adducts of acetic acid derivatives with pyridine.

958. G. V. Gusakova, E. V. Ryl'tsev, and A. L. Smolyanskii, *Opt.*
      Spektrosk., 34, 461 (1973); [English, 262]. IR: Thermodynamic
      data, Me₃PO adducts of carboxylic acids.

959. W. Guschlbauer, A. Ruet, and P. Fromageot, *Compt. Rend. Acad.*
      Sci., Ser. D, 265, 287 (1967). BIOL: Spectrophotometric pKs
      for adenosine, guanosine, uridine, uracil, cytidine and pseudo-
      uridine.

960. G. D. Guthrie and R. L. Sinsheimer, *J. Mol. Biol.*, 2, 297 (1960).
      BIOL: Genetic role of DNA.

961. V. Gutmann, D. E. Hagen, and K. Utvary, *Monatsh.*, 91, 869 (1960).
      IR: ν_{NH} and ν_{PO} shifts in phenylphosphonic diamides.

962. C. Haas and D. F. Hornig, *J. Chem. Phys.*, 32, 1763 (1960). IR:
      Inter- and intramolecular potentials and the spectrum of ice.

963. J. F. Habener, B. S. Bynum, and J. Shack, *J. Mol. Biol.*, 49, 157
      (1970). BIOL: Destabilized secondary structure of newly repli-
      cated HeLa DNA.

964. V. Hach, R. F. Raimondo, D. M. Cartlidge, and E. C. McDonald,
      Tetrahedron Lett., 3175 (1970). IR: Intramolecular, thujanols.

965. D. C. Haddix and P. C. Lauterbur, *Nat. Bur. Stand. (U.S.), Spec.*
      Publ. 1967, No. 301, pp. 403-407. NMR: Trichloroacetic acid,
      single crystal.

966. S. A. Hady, I. Nahringbauer, and I. Olovsson, *Acta Chem. Scand.*,
      23, 2764 (1969). X-Ray: Crystal structure of hydrazinium
      acetate.

967. D. Hadži, *J. Chem. Soc. A*, 418 (1970). IR: Adducts of tri- and
      dihaloacetic acids with Me₃SeO, Me₃AsO, Me₃NO.

968. D. Hadži, *Pure Appl. Chem.*, 11, 435 (1965). IR: Strong hydrogen
      bonds, double-potential minima, review.

968a. D. Hadži, *Chimia*, 26, 7 (1972). IR: Review, strong hydrogen
      bonds.

969. D. Hadži, *J. Chem. Soc.*, 5128 (1962). IR: Adducts of HCl, HBr,
      and Cl₃CCOOH with bases.

970. D. Hadži, *J. Chem. Phys.*, 34, 1445 (1962). Far-IR: Double-
      minimum potential, strong hydrogen bonds, phosphates.

971. D. Hadži, ed., *Hydrogen Bonding*, Pergamon Press, 1959. Review.

971a. D. Hadži, B. Drel, and A. Novak, *Spectrochim. Acta*, 29A, 1745
      (1973). IR, Raman: Single minimum in NaH(CH₃COO)₂, KH(CF₃COO)₂,
      KH(CH₃COO)₂.

972. D. Hadži, C. Klofutar, and S. Oblak, *J. Chem. Soc. A*, 905 (1968).
      IR: Δν for adducts of acids with oxygen bases.

973. D. Hadži and N. Kobilarov, *J. Chem. Soc. A*, 439 (1966). IR:
      Adducts of chloroacetic acids with bases.

974. D. Hadži, A. Novak, and J. E. Gordon, *J. Phys. Chem.*, 67, 1118
      (1963). IR: Adducts of phenols with phenoxides and other oxygen
      bases.

975. D. Hadži, I. Petrov, and M. Zitko, *Proc. Intern. Meeting Mol. Spectrosc.*, *4th*, Bologna, A2, 1959, 794 (1962). IR: Self-association of alcohols-OD and phenols-OD.

975a. D. Hadži, M. Obradović, B. Orel, and T. Solmajer, *J. Mol. Struct.*, 14, 439 (1972). IR: Symmetrical bond, RbH(CCl$_3$COO)$_2$.

976. D. Hadži and L. Premru, *Bull. Sci. Fac. Chim. Ind. Bologna*, 18, 148 (1960). IR: Self-association of anthranilic acid.

976a. D. Hadži and J. Rajnvajn, *J. Chem. Soc. Faraday*, *1*, 69, 151 (1973). IR: Thermodynamic data, chloroacetic acid adducts.

977. D. Hadži, H. Ratajczak, and L. Sobczyk, *J. Chem. Soc. A*, 48 (1967). Dielectric studies of adducts of oxygen bases with acids.

977a. A. T. Hagler, H. A. Scheraga, and G. Nemethy, *J. Phys. Chem.*, 76, 3229 (1972). Theory: Structure of liquid water.

978. A. Hall and J. L. Wood, *Spectrochim. Acta*, 24A, 1109 (1968). Far-IR: Phenol-pyridine complex.

978a. A. Hall and J. L. Wood, *Spectrochim. Acta*, 28A, 2331 (1972). Far-IR: Phenol adducts of pyridine derivatives.

979. A. Hall and J. L. Wood, *Spectrochim. Acta*, 23A, 1257, 2657 (1967). IR: Structure of ν_{OH}.

980. H. E. Hallam, *J. Mol. Struct.*, 3, 43 (1969). IR: Review of frequency shifts in vapor phase and solution.

981. H. E. Hallam in *Infrared Spectroscopy and Molecular Structure*, M. M. Davis, ed., Elsevier Publishing Co., New York, 1963, Chapter 12. IR: Review of solvent effects on frequency shifts.

982. H. E. Hallam and T. C. Ray, *Trans. Faraday Soc.*, 58, 1299 (1962). IR: Solvent effects on C—X frequencies.

983. H. E. Hallam and T. C. Ray, *Nature*, 189, 915 (1961). IR: Solvent effects on ν_{NH} of pyrrole.

984. J. D. Halliday, H. D. W. Hill, and R. E. Richards, *Chem. Commun.*, 219 (1969). NMR: Cs resonance shift in H$_2$^{16}O, D$_2$O, and H$_2$^{18}O, isotope effect.

984a. H. Hamano, *Bull. Chem. Soc. Jap.*, 30, 741 (1957). Theory: Approximate SCF on HF$_2$$^-$.

985. A. N. Hambly and B. V. O'Grady, *Aust. J. Chem.*, 17, 860 (1964). IR: Intra- and intermolecular bonding, aniline derivatives.

986. A. N. Hambly and B. V. O'Grady, *Aust. J. Chem.*, 16, 459 (1963). IR: Intramolecular, primary aromatic amines with o-carbonyl groups.

987. A. N. Hambly and B. V. O'Grady, *Aust. J. Chem.*, 15, 626 (1962). IR: Intramolecular, amines and o-aniline derivatives.

988. L. D. Hamilton, *Nature*, 218, 633 (1968). BIOL: Discussion, DNA models.

989. W. C. Hamilton and J. A. Ibers, *Hydrogen Bonding in Solids*, W. A. Benjamin, New York, 1968. Review.

990. R. M. Hamlin, Jr., R. C. Lord, and A. Rich, *Science*, 148, 1734
      (1965). IR: Association of 1-cyclohexyluracil with 9-ethyl-
      adenine in $CHCl_3$.

991. R. M. Hammaker, R. M. Clegg, L. K. Patterson, P. E. Rider, and
      S. L. Rock, *J. Phys. Chem.*, 72, 1837 (1968). IR, Near-IR:
      Self-association, alcohols.

992. R. M. Hammaker, L. K. Patterson, and K. C. Lin, *J. Phys. Chem.*,
      72, 4346 (1968). NMR: Di-*tert*-butylcarbinol dimer.

993. G. G. Hammes, *Accounts Chem. Res.*, 1, 321 (1968). Kinetics:
      Relaxation spectrometry.

994. G. G. Hammes, *Advan. Protein Chem.*, 23, 1 (1968). Kinetics:
      Relaxation methods, biological systems.

995. G. G. Hammes and P. J. Lillford, *J. Amer. Chem. Soc.*, 92, 7578
      (1970). Kinetics: Self-association, 2-pyridone in Me_2SO.

996. G. G. Hammes and A. C. Park, *J. Amer. Chem. Soc.*, 90, 4151
      (1968). Kinetics: Self-association of 1-cyclohexyluracil,
      9-ethyladenine.

997. G. G. Hammes and A. C. Park, *J. Amer. Chem. Soc.*, 91, 956 (1969).
      Kinetics: Self-association, 2-pyridone, 2-thiopyridine,
      mephobarbital, thiopental.

998. G. G. Hammes and H. O. Spivey, *J. Amer. Chem. Soc.*, 88, 1621
      (1966). Kinetics: Self-association of 2-pyridone.

999. P. R. Hammond, *J. Chem. Soc.*, 479 (1964). Error analysis for
      Benesi-Hildebrand plots.

1000. P. Handler, ed., *Biology and the Future of Man*, Oxford Univer-
      sity Press, New York, 1970, Chapter 2. BIOL: Survey, molecular
      biology.

1001. D. Hankins, J. W. Moskowitz, and F. H. Stillinger, *J. Chem. Phys.*,
      53, 4544 (1971). Theory: SCF Gaussian study of pairs and
      triplets of H_2O molecules.

1002. D. Hankins, J. W. Moskowitz, and F. H. Stillinger, *Chem. Phys.
      Lett.*, 4, 527 (1970). Theory: Non-additivity of hydrogen bond
      energy in water trimers.

1003. S. Hanlon, *Biochem. Biophys. Res. Commun.*, 23, 861 (1966).
      BIOL: Role of London dispersion forces in the maintenance of
      the DNA helix.

1004. M. W. Hanna and A. L. Ashbaugh, *J. Phys. Chem.*, 68, 811 (1964).
      NMR: 1:1 π complexes.

1005. E. S. Hanrahan, *Spectrochim. Acta*, 22, 1243 (1966). IR:
      Malonic acids.

1005a. E. S. Hanrahan and B. D. Bruce, *Spectrochim. Acta*, 23A, 2497
      (1967). IR: Self-association, *p*-substituted benzoic acids.

1006. J. A. Happe, *J. Phys. Chem.*, 65, 72 (1961). NMR: K_x for self-
      association of pyrrole; K for pyrrole-pyridine.

1007. R. Haque and D. R. Buhler, *J. Amer. Chem. Soc.*, 94, 1824 (1972).
      NMR: Adducts of hexachlorophene with amides and polypeptides.

1008.   N. Harada, Y. Mori, and I. Tanaka, *Mol. Photochem.*, 2, 153
        (1970). IR, NMR: Intramolecular, nitrone photoisomerization.

1008a. A. H. Hardin and K. B. Harvey, *Spectrochim. Acta*, 29A, 1139
        (1973). IR: Ice I.

1009.   S. A. Harrell and D. H. McDaniel, *J. Amer. Chem. Soc.*, 86, 4497
        (1964). ΔH for HF + F⁻ ⟶ HF₂⁻

1010.   J. E. Harries, W. J. Burroughs, and H. A. Gebbie, *J. Quant.
        Spectrosc. Radiat. Transfer*, 9, 799 (1969). Millimeter wave-
        length spectroscopic observations of water dimer.

1012.   H. Hartig and W. W. Brandt, *J. Phys. Chem.*, 69, 335 (1965).
        IR, Electronic: Association of secondary amines with tetra-
        hydrofuran.

1013.   K. A. Hartman, Jr., *J. Phys. Chem.*, 70, 270 (1966). NMR:
        Effect of temperature and electrolytes in ¹H spectrum of H₂O.

1014.   K. A. Hartman and G. J. Thomas, Jr., *Science*, 170, 740 (1970).
        BIOL, IR: Secondary structure of rRNA.

1015.   K. A. Hartman, Jr. and A. Rich, *J. Amer. Chem. Soc.*, 87, 2033
        (1965). BIOL: Tautomeric form of helical polyribocytidylic
        acid.

1016.   A. E. V. Haschemeyer and H. M. Sobell, *Acta Crystallogr.*, 18,
        525 (1965). BIOL, X-Ray: 1:1 complex of adenosine and 5-
        bromouridine.

1017.   A. E. V. Haschemeyer and H. M. Sobell, *Acta Crystallogr.*, 19,
        125 (1965). BIOL, X-Ray: 1:1 complex of deoxyguanosine and
        5-bromodeoxycytidine.

1018.   A. E. V. Haschemeyer and H. M. Sobell, *Nature*, 202, 969 (1964).
        BIOL, X-Ray, UV: Deoxyguanosine-5-bromodeoxycytidine crystals.

1019.   A. E. V. Haschemeyer and H. M. Sobell, *Proc. Natl. Acad. Sci.
        U. S.*, 50, 872 (1963). BIOL, X-Ray: Adenosine adduct of
        5-bromouridine.

1020.   M. Hasegawa, *J. Phys. Soc. Jap.*, 28, 266 (1970). Theory:
        CNDO/2 calculations on water dimer.

1021.   M. Hasegawa, K. Daiyasu, and S. Yomosa, *J. Phys. Soc. Jap.*, 28,
        1304 (1970). BIOL, Theory: VB calculations on hydrogen bonds
        in guanine-cytosine base pair.

1022.   M. Hasegawa, K. Daiyasu, and S. Yomosa, *J. Phys. Soc. Jap.*, 27,
        999 (1969). Theory: VB calculation using a 4-electron model
        for O—H...O.

1023.   J. L. Haslam and E. M. Eyring, *J. Phys. Chem.*, 71, 4470 (1967).
        Kinetics: Intramolecular, temperature-jump relaxation technique.

1024.   J. L. Haslam, E. M. Eyring, W. W. Epstein, G. A. Christiansen,
        and M. H. Miles, *J. Amer. Chem. Soc.*, 87, 1 (1965). Kinetics:
        Intramolecular, *cis*-cyclopropanepolycarboxylic acids.

1025.   J. V. Hatton and R. E. Richards, *Mol. Phys.*, 5, 153 (1962).
        NMR: Polar solutes associated with aromatic solvents.

1026.   J. V. Hatton and R. E. Richards, *Mol. Phys.*, 5, 139 (1962).
        NMR: Polar solutes associated with aromatic solvents.

1027.   J. V. Hatton and R. E. Richards, *Trans. Faraday Soc.*, 57, 28
        (1961). NMR: Acetylenic protons associated with bases.

1028.   J. V. Hatton and R. E. Richards, *Mol. Phys.*, 3, 253 (1960).
        NMR: N-substituted amides associated with aromatic solvents.

1029.   J. V. Hatton and W. G. Schneider, *Can. J. Chem.*, 40, 1285
        (1962). NMR: Polar solutes in toluene.

1030.   M. Haurie and A. Novak, *Spectrochim. Acta*, 21, 1217 (1965).
        IR: Crystals of CH_3COOH, CH_3COOD, CD_3COOH.

1031.   M. Haurie and A. Novak, *J. Chim. Phys.*, 62, 137, 146 (1965).
        IR, Raman: Self-association, vapor phase, CH_3COOH, CH_3COOD,
        CD_3COOH, CD_3COOD.

1031a. J. P. Hawranek, J. Oszust, and L. Sobczyk, *J. Phys. Chem.*, 76,
        2112 (1972). IR: Dipole moments and spectra of 2,6-dichloro-
        4-nitrophenol adducts.

1032.   J. P. Hawranek and L. Sobczyk, *Acta Phys. Pol.*, A39, 651 (1971).
        IR, Dipole Moment: Polarity of Ph_2NH, 2,4,6-trimethylphenol.

1033.   J. P. Hawranek and L. Sobczyk, *Acta Phys. Pol.*, A39, 639 (1971).
        IR: K, complexes of Ph_2NH and $2,4,6-Me_3C_6H_2OH$ with various
        bases.

1034.   R. W. Hay and P. P. Williams, *J. Chem. Soc.*, 2270 (1964). NMR:
        Intramolecular, *o*-hydroxy-aldehydes, hydroxy-ketones, hydroxy-
        esters, β-diketones, and β-keto-esters.

1035.   S. Hayashi, H. Hara, and N. Kimura, *Bull. Inst. Chem. Res.*,
        Kyoto Univ., 46, 213 (1968). IR: Self-association, spectra
        of $Me(CH_2)_nCOOH$ and $Me(CH_2)_nCOOD$ where n = 1-16.

1036.   H. W. Hayer and K. S. Birdi, *Biopolymers*, 6, 1507 (1968). BIOL:
        Detection and enthalpy of vaporization of bound water in DNA
        by DTA.

1037.   F. N. Hayes, E. H. Lilly, R. L. Ratliff, D. A. Smith, and D. L.
        Williams, *Biopolymers*, 9, 1105 (1970). BIOL: Thermal transi-
        tions, poly deoxyribonucleotides.

1038.   W. P. Hayes and C. J. Timmons, *Spectrochim. Acta*, 21, 529 (1965).
        UV: Solvent effects on $n \rightarrow \pi^*$ bands of ketones.

1039.   M. Haynes, R. A. Garrett, and W. B. Gratzer, *Biochemistry*, 9,
        4410 (1970). BIOL: Structure of nucleic acid-poly base
        complexes.

1040.   J. B. Hays, M. E. Magar, and B. H. Zimm, *Biopolymers*, 8, 531
        (1969). BIOL: Persistence length of DNA, worm-like chain model.

1041.   R. Heess and H. Kriegsmann, *Spectrochim. Acta*, A, 24, 2121
        (1968). IR: Intensity changes of organic carbonyl compounds
        dissolved in $CHCl_3$.

1042.   G. K. Helmkamp and N. S. Kondo, *Biochim. Biophys. Acta*, 145, 27
        (1967). BIOL: NMR, preferred orientation in purine stacking.

1043.   G. K. Helmkamp and N. S. Kondo, *Biochim. Biophys. Acta*, 157, 242
        (1968). BIOL: NMR, vapor pressure osmometry, association
        behavior of various alkyl-substituted purines in H_2O.

1044.   G. K. Helmkamp and P. O. P. Ts'o, *J. Amer. Chem. Soc.*, $\underline{83}$, 138
        (1961). BIOL: Secondary structure of nucleic acids in organic
        solvents.

1045.   J. Heidberg, J. A. Weil, G. A. Janusonis, and J. K. Anderson,
        J. Chem. Phys., $\underline{41}$, 1033 (1964). NMR: Intramolecular, nitro
        groups *ortho* to amino group in nitroaromatic amines.

1045a. W. Heitler and F. London, *Z. Physik*, $\underline{44}$, 455 (1927). The
        original VB calculation on H_2.

1046.   C. H. Henrickson, A. D. Wilks, D. J. Pietrzyk, and D. P. Eyman,
        Chemical Instrumentation, $\underline{1}$, 145 (1968). CAL: Calorimeter.

1046a. D. B. Henson and C. A. Swenson, *J. Phys. Chem.*, $\underline{77}$, 2401 (1973).
        IR: Thermodynamic data, H_2O-amide adducts.

1047.   N. F. Hepfinger and P. A. Clarke, *J. Org. Chem.*, $\underline{34}$, 2572 (1969).
        NMR: Association of nitromethane with methanol.

1048.   D. W. Herlocker, R. S. Drago, and V. I. Meek, *Inorg. Chem.*, $\underline{5}$,
        2009 (1966). IR: Δv_{OH}, adducts of phenol with pyridine-N-
        oxides.

1049.   A. D. Hershey and M. Chase, *J. Gen. Physiol.*, $\underline{36}$, 39 (1952).
        BIOL: Genetic role of DNA.

1050.   W. Hertl, *J. Phys. Chem.*, $\underline{77}$, 1473 (1973). IR: Surface amino
        groups on silica.

1051.   W. Hertl and M. L. Hair, *J. Phys. Chem.*, $\underline{72}$, 4676 (1968). IR:
        Hydrogen bonding between adsorbed gases and surface hydroxyl
        groups on silica.

1051a. G. Herzberg, *J. Mol. Spectrosc.*, $\underline{33}$, 147 (1970). Revision of
        bond energy of H_2.

1052.   F. K. Hess and K. H. Pook, *Can. J. Chem.*, $\underline{47}$, 1151 (1969).
        NMR: Intramolecular, acylamino biphenyls.

1052a. T. Hidaka, *J. Phys. Soc. Jap.*, $\underline{33}$, 635 (1972). Theory:
        Delocalization of hydrogen bond in KH_2PO_4.

1053.   R. J. Highet, P. F. Highet, and J. C. N. Ma, *Tetrahedron Lett.*,
        1049 (1966). NMR: Intramolecular, dihydrotazettine methine
        alcohol.

1053a. S. Higuchi, E. Kuno, S. Tanaka, and H. Kamada, *Bull. Chem. Soc.
        Jap.*, $\underline{45}$, 3011 (1972). IR: Solvent effect on aniline inten-
        sities.

1054.   J. H. Hildebrand, *J. Phys. Chem.*, $\underline{72}$, 1841 (1968). BIOL:
        Criticism of the term "hydrophobic bond."

1055.   J. C. Hindman, *J. Chem. Phys.*, $\underline{44}$, 4582 (1966). NMR: Self-
        association, H_2O.

1055a. J. Hine, *J. Amer. Chem. Soc.*, $\underline{94}$, 5766 (1972). Kinetics:
        Hydrogen-bonded intermediates, proton exchange reactions in
        hydroxylic solvents.

1056.   J. F. Hinton and E. S. Amis, *Z. Phys. Chem. (Frankfurt am Main)*,
        $\underline{60}$, 159 (1968). NMR: N-methylacetamide mixtures with water.

1056a. J. F. Hinton and K. H. Ladner, *Spectrochim. Acta*, 28A, 1731 (1972). NMR: Aqueous N-methylformamide.

1057. H. J. Hinz, W. Haar, and T. Ackermann, *Biopolymers*, 9, 923 (1970). BIOL: Thermodynamics of the helix-random coil transition, complexes of poly (I + C) and poly I.

1058. E. Hirano and K. Kojima, *Bull. Chem. Soc. Jap.*, 39, 1216 (1966). IR: Thermodynamic data, MeOH—NEt$_3$, gas phase, solvents.

1059. K. Hirano, *Bull. Chem. Soc. Jap.*, 41, 731 (1968). BIOL: Raman spectra of DNA in H$_2$O.

1060. K. Hirota, *Z. Physik. Chem. (Frankfurt)*, 35, 222 (1962). UV: Fluorescence and absorption spectra of 3-hydroxy-2-naphthoic acid derivatives.

1061. K. Hirota and Y. Nakai, *Bull. Chem. Soc. Jap.*, 32, 769 (1959). Far-IR: Formic acid.

1062. M. Hirota, M. Tabei, and T. Tezuka, *Tetrahedron*, 27, 301 (1971). NMR: Intramolecular, *p*-chloro derivative of 2,6-diformylphenols.

1062a. M. Hirota and G. Hirano, *Bull. Chem. Soc. Jap.*, 45, 1448 (1972). Intramolecular, *o*-methoxy-phenoxyacetic and phenylthioacetic acids.

1063. J. O. Hirschfelder in *Molecular Biophysics*, B. Pullman and M. Weissbluth, eds., Academic Press, New York, 1965, p. 325. BIOL: Intermolecular forces, discussion.

1064. T. Hisano and M. Ichikawa, *Chem. Pharm. Bull.*, 20, 163 (1972). IR: Thermodynamic data, adducts of phenols with quinazolinones.

1065. G. Hitte, E. Smissman, and R. West, *J. Amer. Chem. Soc.*, 82, 1207 (1960). IR: Intramolecular, 3-piperidinols.

1066. M. Hoeke and A. L. Koevoet, *Rec. Trav. Chim.*, 82, 17 (1963). IR: K, adducts of 1-butanol with ketones.

1067. G. L. Hofacker, *Hydrogen Bonding, Papers Symp., Ljubljana*, 375 (1957). Theory: MO treatment of the hydrogen bond.

1067a. R. Hoffmann, *J. Chem. Phys.*, 39, 1397 (1963). The Extended Hückel method.

1067b. T. E. Hogen-Esch, *J. Amer. Chem. Soc.*, 95, 639 (1973). UV: Hydrogen-bonded alkali carbanions in protic media.

1068. D. N. Holcomb and I. Tinoco, Jr., *Biopolymers*, 3, 121 (1965). BIOL: Conformation of polyriboadenylic acid.

1069. J. R. Holden and C. Dickinson, *J. Phys. Chem.*, 73, 1199 (1969). X-Ray: Survey of structures of *o*-substituted anilines.

1070. R. W. Holley, J. Apgar, G. A. Everett, J. T. Madison, M. Marguisee, S. H. Merrill, J. R. Penswick, and A. Zamir, *Science*, 147, 1462 (1965). BIOL: Complete nucleotide sequence of an alanine tRNA.

1071. J. R. Holmes, D. Kivelson, and W. C. Drinkard, *J. Amer. Chem. Soc.*, 84, 4677 (1962). NMR, IR: H$_2$O in various organic solvents.

1072. W. B. Holzapfel, *J. Chem. Phys.*, 56, 712 (1972). Theory: Study of hydrogen bonds in ice VII using empirical potential functions to represent atom-atom interactions.

1073. J. Homer and M. C. Cooke, *J. Chem. Soc. A*, 1984 (1969).
Mechanism of complex formation.

1074. J. Homer and M. C. Cooke, *J. Chem. Soc. A*, 777 (1969). NMR:
Mechanism of complex formation of aliphatic molecules with
benzene.

1075. J. Homer and M. C. Cooke, *J. Chem. Soc. A*, 773 (1969). NMR:
K, complexes of chloroethylenes with C_6H_6.

1076. J. Homer, E. J. Hartland, and C. J. Jackson, *J. Chem. Soc. A*,
931 (1970). NMR: Calculation of equilibrium constants.

1077. J. Homer and P. J. Huck, *J. Chem. Soc. A*, 277 (1968). NMR:
K, adducts of nitroform with π-bases.

1078. R. B. Homer and C. D. Johnson in *The Chemistry of Amides*,
J. Zabicky, ed., Interscience, 1970, p. 223. Review, hydrogen
bonding of amides.

1079. K. Hoogsten, *Molecular Associations in Biology*, B. Pullman, ed.,
Academic Press, New York, 1968, p. 21. BIOL: Review, hydrogen
bonding between purines and pyrimidines.

1080. K. Hoogsten, *Acta Crystallogr.*, 16, 907 (1963). BIOL, X-Ray:
1-Methylthymine and 9-methyladenine 1:1 complexes, R = 0.081.

1081. H. P. Hopkins, Jr. and F. C. Marler, III, *J. Phys. Chem.*, 74,
4164 (1970). IR: Intramolecular, *o*-trifluoromethylphenol.

1082. R. F. W. Hopmann, *Ber. Bunsenges, Phys. Chem.*, 77, 52 (1973).
BIOL, Kinetics: $(Uracil)_2$, uracil-adenine.

1083. M. Horák, J. Moravec, and J. Pliva, *Spectrochim. Acta*, 21, 919
(1965). IR: Collision complexes of phenols.

1084. M. Horák and J. Pliva, *Spectrochim. Acta*, 21, 911 (1965). IR:
Theory of solvent shifts.

1084a. M. Horák, V. Sara, and J. Moravec, *Collect. Czech. Chem. Commun.*,
37, 1990 (1972). IR: Thermodynamic data, phenol adducts of
quinuclidine and quinolizidine.

1085. M. Horák, J. Smolikova, and J. Pitha, *Collect. Czech. Chem.
Commun.*, 26, 2891 (1961). IR: Intramolecular, *o*-nitrophenols.

1086. D. F. Hornig, *J. Chem. Phys.*, 40, 3119 (1964). IR: Self-
association, H_2O.

1087. H. Hosoya, J. Tanaka, and S. Nagakura, *J. Mol. Spectrosc.*, 8,
257 (1962). UV: Self-association, benzoic acid and methyl
benzoate.

1088. B. B. Howard, *J. Chem. Phys.*, 39, 2524 (1963). Theory: $(HF)_2$,
calculation of electrostatic contribution to hydrogen bond
energy using an approximate method and Slater basis orbitals.

1089. B. B. Howard, C. F. Jumper, and M. T. Emerson, *J. Mol. Spectrosc.*,
10, 117 (1963). NMR: Thermodynamic data, adducts of $CHCl_3$ with
bases in C_6H_{12}, CCl_4.

1090. F. B. Howard, J. Frazier, M. N. Lipsett, and H. T. Miles,
Biochem. Biophys. Res. Commun., 17, 93 (1964). BIOL: IR, two-
and three-stranded helix formation between poly C and guanosine
mononucleotides and oligonucleotides.

1091. F. B. Howard, J. Frazier, M. F. Singer, and H. T. Miles, *J. Mol. Biol.*, 16, 415 (1966). BIOL: IR, ORD, helix formation between polyribonucleotides and purines, purine nucleosides and nucleotides, base-pairing specificity.

1093. J. R. Hoyland and L. B. Kier, *Theor. Chim. Acta*, 15, 1 (1969). Theory: CNDO/2 calculations on several small hydrogen-bonded systems.

1094. F. Hruska, G. Kotowycz, and T. Schaefer, *Can. J. Chem.*, 43, 3188 (1965). NMR: Hydrogen bond shifts of RCl in Me_2SO.

1095. F. E. Hruska and S. S. Danyluk, *Biochim. Biophys. Acta*, 157, 238 (1968). BIOL: PMR, thermal effects on base stacking in oligonucleotides in H_2O.

1096. F. E. Hruska, A. A. Grey, and I. C. P. Smith, *J. Amer. Chem. Soc.*, 92, 4088 (1970). BIOL: NMR, molecular conformation of pseudouridine in H_2O.

1096a. S. J. Hu, E. Goldberg, and S. I. Miller, *Org. Magn. Resonance*, 4, 683 (1972). NMR: Thermodynamic data, thiol-base adducts.

1097. S. J. Hu and S. I. Miller, *Org. Magn. Resonance*, 5, 197 (1973). NMR: Self-association, alkanethiols, thermodynamic data.

1099. T. N. Huckerby, *Annual Rev. NMR Spectrosc.*, 3, E. F. Mooney, ed., Academic Press, New York, 1970. NMR: Review, short discussion of hydrogen bonding.

1100. R. A. Hudson, R. M. Scott, and S. N. Vinogradov, *Spectrochim. Acta*, 26A, 337 (1970). Electronic: *p*-Nitrophenol adducts with bases.

1100a. R. A. Hudson, R. M. Scott, and S. N. Vinogradov, *J. Phys. Chem.*, 76, 1989 (1972). UV: 3,4-Dinitrophenol adducts.

1101. C. M. Huggins and D. R. Carpenter, *J. Phys. Chem.*, 63, 238 (1959). NMR: $SiHCl_3$ versus $CHCl_3$.

1101a. C. M. Huggins and G. C. Pimentel, *J. Phys. Chem.*, 60, 1615 (1956). IR: Correlation of $\Delta \nu$ with $\nu_{\frac{1}{2}}$ and B.

1102. C. M. Huggins, G. C. Pimentel, and J. N. Shoolery, *J. Phys. Chem.*, 60, 1311 (1956). NMR: Limiting slope method for measuring oligimer constants.

1103. M. L. Huggins, *Angew. Chem., Int. Ed. Engl.*, 10, 147 (1971). Review: Some of the early history of the theory of the hydrogen bond.

1104. M. Huke and K. Goerlitzer, *Arch. Pharm. (Weinheim, Ger.)*, 302, 423 (1969). NMR: Intramolecular, 3-coumaranones.

1105. I. M. Hunsberger, *J. Amer. Chem. Soc.*, 72, 5626 (1959). NMR: Self-association, 2-aceto-3-hydroxynaphthaldehydes.

1106. W. T. Huntress, Jr., *J. Phys. Chem.*, 73, 103 (1969). NMR: Nuclear spin relaxation times of $CDCl_3$ and $CHCl_3$ in C_6H_6.

1107. P. V. Huong and M. Couzi, *J. Chim. Phys.*, 67, 1994 (1970). IR: Association of HF and DF with organic bases.

1108. P. V. Huong and M. Couzi, *J. Chim. Phys.*, 66, 1309 (1969). IR: Gas phase, HF, DF.

1108a. P. V. Huong and M. Couzi, *Abstracts, International Conference on Hydrogen Bonding*, Ottawa, Canada, p. 12, 1972. IR: Multi-molecular hydrogen-bonded complexes of hydrogen halides, band-width.

1109. P. V. Huong and A. Graja, *Chem. Phys. Lett.*, 13, 162 (1972). IR: Far-IR: Phenol-methyl pyridine oxide adduct.

1110. P. V. Huong and J. C. Lassegues, *Spectrochim. Acta*, 26A, 269 (1970). IR: Effect of solvents on ν_{OH} of phenol and $\overline{\nu}_{NH}$ of pyrrole.

1111. P. V. Huong and G. Turrell, *J. Mol. Spectrosc.*, 25, 185 (1968). IR: $\Delta\nu_{C-H}$, adducts of 1-heptyne.

1112. W. J. Hurley, I. D. Kuntz, Jr., and G. E. Leroi, *J. Amer. Chem. Soc.*, 88, 3199 (1966). Far-IR: Self-association, cresols and chlorophenols.

1112a. P. Huyskens, *Ind. Chim. Belg.*, 37, 15 (1972). CAL: Alcohol-amine adducts of variable stoichiometry.

1113. P. Huyskens, Th. Zeegers-Huyskens, and A. M. Dierckx, *Ann. Soc. Sci.*, Bruxelles, Ser. I, 78, 175 (1964); *Chem. Abstr.*, 63, 12339b (1965). IR, NMR: Self-association, butanols, solvent effects.

1114. D. A. Ibbitson and L. F. Moore, *J. Chem. Soc. B*, 76, 80 (1967). Near-IR: Self-association, alcohols, dielectric constant studies.

1115. D. A. Ibbitson and L. F. Moore, *Chem. Commun.*, 339 (1965). IR: Self-association, ethanol in CCl_4.

1116. J. A. Ibers, *Ann. Rev. Phys. Chem.*, 16, 384 (1965). Review: Hydrogen bonding in solids.

1117. J. A. Ibers, *J. Chem. Phys.*, 41, 25 (1964). IR: Theory, KHF_2, KDF_2.

1117a. J. A. Ibers, *J. Chem. Phys.*, 40, 402 (1964). Neutron Diffraction: Symmetrical hydrogen bond in KHF_2.

1118. J. A. Ibers, *J. Phys. (Paris)*, 25, 474 (1964). Neutron Diffraction: Symmetrical hydrogen bond in HF_2^-.

1119. K. Igarashi, F. Watari, and K. Aida, *Spectrochim. Acta*, 25A, 1743 (1969). IR: Adducts of phenol with alkyl thiocyanates.

1120. Y. I'Haya and T. Shibuya, *Bull. Chem. Soc. Jap.*, 38, 1144 (1965). IR: Self-association, benzoic acid.

1121. Y. Ikegami, T. Ikenoue, and S. Seto, *Kogyo Kagaku Zasshi*, 68, 1415 (1965); *Chem. Abstr.*, 64, 190a (1966). NMR: Intramolecular, tropolone.

1122. M. Ikehara, S. Uesugi, and M. Lasumoto, *J. Amer. Chem. Soc.*, 92, 4735 (1970). BIOL: Highly stacked dinucleoside monophosphates derived from adenine 8-cyclonucleosides.

1123. K. U. Ingold, *Can. J. Chem.*, 40, 111 (1962). IR: *o*-Alkyl phenols in vapor phase.

1124. K. U. Ingold and D. R. Taylor, *Can. J. Chem.*, 39, 471 (1961). IR: Effect of *o*-alkyl groups on ν_{OH} of phenols.

1125. R. B. Inman and R. L. Baldwin, *J. Mol. Biol.*, 5, 172 (1962).
       BIOL: Helix-random coil transitions in synthetic DNAs of
       alternating sequence.

1126. R. B. Inman and R. L. Baldwin, *J. Mol. Biol.*, 5, 185 (1962).
       BIOL: Hybrid double helices from two alternating DNA copolymers.

1127. D. Inners and G. Felsenfeld, *J. Mol. Biol.*, 50, 373 (1970).
       BIOL: Conformation of polyribouridylic acid in H_2O using light
       scattering, viscometry, sedimentation velocity.

1128. R. G. Inskeep, F. E. Dickson, and J. M. Kelliher, *J. Mol.
       Spectrosc.*, 4, 477 (1960). IR: Gas phase, enthalpy for
       methanol-ether adduct.

1129. R. G. Inskeep, F. E. Dickson, and H. M. Olson, *J. Mol. Spectrosc.*,
       5, 284 (1960). IR: Self-association, CH_3OD vapor.

1130. W. H. Inskeep, D. L. Jones, W. T. Silfvast, and E. M. Eyring,
       Proc. Natl. Acad. Sci. U. S., 59, 1027 (1967). Kinetics:
       Intramolecular, azodyes in H_2O.

1131. A. V. Iogansen, *Teor. Eksp. Khim.*, 7, 312 (1971); *Chem. Abstr.*,
       75, 101824a (1971). Effect of medium on hydrogen bonding.

1132. A. V. Iogansen, *Teor. Eksp. Khim.*, 7, 302 (1971); *Chem. Abstr.*,
       75, 101848m (1971). Effect of medium on hydrogen bonding.

1133. A. V. Iogansen, *Opt. Spektrosk.*, supplement 3, 228 (1967);
       [English, 113 (1968)]. IR: Fermi resonance and ν_{AH} band
       structure.

1134. A. V. Iogansen *et al.*, *Zh. Prikl. Spektrosk.*, 15, 1046 (1971).
       IR: Adducts of (3-aminopropyl)dibutylborane, Fermi resonance.

1135. A. V. Iogansen, *Dokl. Akad. Nauk SSSR*, 184, 1350 (1969); *Chem.
       Abstr.*, 71, 17266e (1969). IR: Model for intensities.

1136. A. V. Iogansen, *Dokl. Akad. Nauk SSSR*, 164, 610 (1965). IR:
       Correlation of ΔH with integrated band intensity.

1137. A. V. Iogansen, G. A. Kurkchi, and B. V. Rassadin, *Zh. Prikl.
       Spektrosk.*, 11, 1054 (1969); *Chem. Abstr.*, 72, 94862u (1970).
       IR: Intensity of ν_{NH} bands in hydrogen bonds.

1138. A. V. Iogansen, G. A. Kurkchi, and O. V. Levina, *Zh. Fiz. Khim.*,
       43, 2915 (1969); *Chem. Abstr.*, 72, 59957g (1970). Gas Chroma-
       tography: Enthalpies for adducts of HCl and HBr with various
       bases.

1139. A. V. Iogansen and B. V. Rassadin, *Zh. Prikl. Spektrosk.*, 11,
       828 (1969); *Chem. Abstr.*, 72, 84489p (1970). IR: Adducts of
       phenol and correlation $-\Delta H = 5.0$ $(\Delta\Gamma^{\frac{1}{2}})$.

1140. A. V. Iogansen and B. V. Rassadin, *Zh. Prikl. Spektrosk.*, 6, 492
       (1967); [English, 323]. IR: Spectral parameters for adducts of
       phenol with ethers.

1141. A. V. Iogansen and M. Sh. Rozenberg, *Dokl. Akad. Nauk SSSR*, 197,
       117 (1971); *Chem. Abstr.*, 75, 12529r (1971). IR: Relationship
       of ΔH and $(\nu_{OH})^2$.

1142. M. Ito, *J. Mol. Spectrosc.*, 4, 125 (1960). UV: Self-
       association, phenols.

1143. M. Ito, K. Inuzuka, and S. Iminishi, *J. Amer. Chem. Soc.*, 82,
       1317 (1960). UV: Effect of hydrogen bonding on the $n \rightarrow \pi^*$
       band of ketones.

1144. M. Ito, M. Suzuki, and T. Yokoyama, *J. Chem. Phys.*, 50, 2949
       (1969). Raman: Crystal spectra of HCl, DCl, HBr, DBr.

1145. R. Itoh, *J. Phys. Soc. Jap.*, 22, 698 (1967). Theory: Proton
       transfer in hydrogen bonds.

1146. K. Itoh and T. Shimanouchi, *Spectrochim. Acta*, 25A, 290 (1969).
       Far-IR: Acetanilide derivatives and N-methylacetamide.

1147. K. Itoh and T. Shimanouchi, *Biopolymers*, 5, 921 (1967).
       Far-IR: Self-association of N-methylacetamide.

1148. K. Itoh and S. Takehiko, *J. Mol. Spectrosc.*, 42, 86 (1972).
       IR, Raman: Crystalline formamide.

1149. R. Itoh, *J. Phys. Soc. Jap.*, 22, 698 (1967). Theory: Solution
       of the Schrödinger equation for the behavior of a proton in an
       empirical double-minimum potential.

1149a. S. S. Ivanchev, Yu. N. Anisimov, and B. A. Khomenko, *Zh. Prikl.
       Spektrosk.*, 16, 508 (1972). IR: Thermodynamic data, cumene
       hydroperoxide adducts with aromatic bases.

1150. V. T. Ivanov, I. A. Laine, N. D. Abdulaev, L. B. Senyavina,
       E. M. Popov, Yu. A. Ovchinnikov, and M. M. Shemyakin, *Biochem.
       Biophys. Res. Commun.*, 34, 803 (1969). IR, NMR: Intramolecular,
       valinomycin.

1151. T. M. Ivanova, L. I. Denisova, and V. A. Batyakina, *Isv. Akad.
       Nauk SSSR, Ser. Khim.*, 435 (1971); *Chem. Abstr.*, 75, 27606n
       (1971). IR: Intramolecular, aromatic sulfoxides, sulfides.

1152. H. Iwamura, *Tetrahedron Lett.*, 2227 (1970). IR: Intramolecular,
       O—H...π bonding in β-phenylethanols.

1152a. H. Iwamura, *Biochim. Biophys. Acta*, 308, 333 (1973). NMR:
       Adduct of 1-propyl-4-thiouracil with 9-propyladenine.

1153. H. Iwamura, N. J. Leonard, and J. Eisinger, *Proc. Natl. Acad.
       Sci., U. S.*, 1025 (1970). BIOL: Synthetic spectroscopic
       models, stacking interactions in tRNA.

1153a. S. Iwata and K. Morokuma, *J. Amer. Chem. Soc.*, 95, 7563 (1973).
       Theory: Energy partitioning in HOH...OCH$_2$.

1153b. S. Iwata and K. Morokuma, *Chem. Phys. Lett.*, 19, 94 (1973).
       Theory: Excited states of H$_2$O...CH$_2$O and H$_2$O$_2$...CH$_2$O.

1154. L. M. Jackman and N. S. Bowman, *J. Amer. Chem. Soc.*, 88, 5565
       (1966). NMR: Self-association, *erythro*- and *threo*-2-butanol-
       3-d, 2-butanol in CCl$_4$.

1155. J. Jadzyn and J. Malecki, *Acta Phys. Pol. A*, 41, 599 (1972).
       Dipole moments of phenol adducts.

1156. R. J. Jakobsen and J. W. Brasch, *Spectrochim. Acta*, 21, 1753
       (1965). Far-IR: Self-association, phenols.

1157. R. J. Jakobsen, J. W. Brasch, and Y. Mikawa, *J. Mol. Struct.*, 1,
       309 (1967). IR: Solid alcohols.

1157a. R. J. Jakobsen and J. E. Katon, *Develop. Appl. Spectrosc.*, 10, 107 (1972). IR: Review, high pressure studies of $ClCH_2COOH$ crystals.

1157b. R. J. Jakobsen and J. E. Katon, *Spectrochim. Acta*, 29A, 1953 (1973). Far-IR: Self-association, halogenated carboxylic acids.

1158. R. J. Jakobsen, Y. Mikawa, and J. W. Brasch, *Spectrochim. Acta*, 25A, 839 (1969). Far-IR: Self-association, n-alkyl acids.

1159. R. J. Jakobsen, Y. Mikawa, and J. W. Brasch, *Applied Spectrosc.*, 22, 641 (1968). Far-IR: Review.

1160. R. J. Jakobsen, Y. Mikawa, and J. W. Brasch, *Applied Spectrosc.*, 24, 333 (1970). IR: Effect of pressure on ν_{OH} of alcohols.

1161. R. J. Jakobsen, Y. Mikawa, and J. W. Brasch, *Spectrochim. Acta*, 23A, 2199 (1967). Far-IR: Self-association of formic acid.

1162. W. Jakubetz and P. Schuster, *Tetrahedron*, 27, 101 (1971). Theory: CNDO calculations on dimer between HF and benzene and pyridine.

1163. K. C. James and P. R. Noyce, *Spectrochim. Acta*, 27A, 691 (1971). IR: Association of testosterone propionate with $CHCl_3$.

1163a. R. Janoschek, *Theor. Chim. Acta*, 29, 57 (1973). Theory: $HCl...Cl^-$, IR, Raman.

1164. R. Janoschek, E. G. Weidemann, H. Pfeiffer, and G. Zundel, *J. Amer. Chem. Soc.*, 94, 2387 (1972). Theory: Gaussian SCF calculations on double-minimum potential in $H_5O_2^+$.

1164a. R. Janoschek, E. G. Weidemann, and G. Zundel, *J. Chem. Soc. Faraday 2*, 69, 505 (1973). IR: Theory, double-minimum potential, $\overline{H_5O_2}^+$.

1165. C. D. Jardetzky, *J. Amer. Chem. Soc.*, 82, 229 (1960). BIOL, PMR: Ribose conformations, purines, pyrimidines, ribose nucleosides and nucleotides in H_2O.

1166. C. D. Jardetzky, *J. Amer. Chem. Soc.*, 84, 62 (1962). BIOL, PMR: Ribose conformation of mononucleotides in H_2O.

1167. C. D. Jardetzky and O. Jardetzky, *J. Amer. Chem. Soc.*, 82, 222 (1960). BIOL, PMR: Structure of purines, pyrimidines, ribose nucleosides and nucleotides.

1168. O. Jardetzky, *Biopolymers, Symp.*, 1, 501 (1964). BIOL, PMR: Studies on nucleotide interactions.

1169. T. Jasinski, Z. Kokot, and W. Grabowska, *Rocz. Chem.*, 42, 1925 (1968); *Chem. Abstr.*, 70, 100230p (1969). UV: K, reactions of benzoic acid derivatives with N,N'-diphenylguanidine in $(MeO)_4Si$.

1170. T. Jasinski, E. Kwiatkowski, and K. Kozubek, *Rocz. Chem.*, 42, 1561 (1968); *Chem. Abstr.*, 70, 118670s (1969). UV: Thermodynamic data, phenol adducts with amides in CCl_4.

1171. W. Jasiobedzki, *Rocz. Chem.*, 43, 43 (1969); *Chem. Abstr.*, 71, 2858a (1969). IR, UV: Intramolecular, haloallenic alcohols and unsaturated ketols.

1172. S. R. Jaskunas, C. R. Cantor, and I. Tinoco, Jr., *Biochemistry*, 7, 3164 (1968). BIOL: ORD, association of complementary oligonucleotides in H_2O.

453

1172a. M. Jaszuński and A. J. Sadlej, *Chem. Phys. Lett.*, 15, 41 (1972).
       Theory: ^{1}H NMR shift in $(H_2O)_2$.

1172b. M. Jaszuński and A. J. Sadlej, *Theor. Chim. Acta*, 30, 257 (1973).
       Theory: ^{1}H NMR shift, $(H_2O)_2$.

1173. G. A. Jeffrey and E. J. Fasiska, *Carbohyd. Res.*, 21, 187 (1972).
       BIOL, X-Ray: Intramolecular, potassium D-gluconate monohydrate.

1174. G. A. Jeffrey and Y. Kinoshita, *Acta Crystallogr.*, 16, 20 (1963).
       BIOL, X-Ray: Cytosine monohydrate.

1175. H. Jehle in *Molecular Biophysics*, B. Pullman and M. Weissbluth,
       eds., Academic Press, New York, 1965, p. 359. BIOL: Proposed
       mechanisms for replication and transcription of DNA.

1176. H. Jehle and W. C. Parke in *Structural Chemistry and Molecular
       Biology*, A. Rich and N. Davidson, eds., W. F. Freeman, San
       Francisco, 1968, p. 399. BIOL: Discussion, replication and
       transcription of DNA.

1177. C. A. Jennings, Ph.D. Thesis, Southern Illinois University,
       1970, p. 132. NMR: Self-association, effect on spin-spin
       coupling for alcohols.

1177a. C. A. Jennings and D. W. Slocum, *Tetrahedron Lett.*, 3547 (1972).
       NMR: Intramolecular, J_{HCCH} in 2-haloethanols.

1178. U. Jentschura and E. Lippert, *Z. Phys. Chem.*, 75, 88 (1971).
       UV: Prediction of charge-transfer band in vacuum ultraviolet.

1179. U. Jentschura, H. Prigge, and E. Lippert, *Chem. Soc. (London)*,
       Spec. Publ., 20, 189 (1966). NMR: Association chemical shifts
       for alcohols, phenols, carboxylic acids.

1179a. G. J. Jiang and G. R. Anderson, *J. Phys. Chem.*, 77, 1764 (1973).
       Theory: Semi-empirical potential function for HF_2^-.

1180. M. D. Joesten, Ph.D. Thesis, University of Illinois, 1962.
       UV, IR: Thermodynamic data for phenol adducts, frequency shift
       correlations.

1181. M. D. Joesten and R. S. Drago, *J. Amer. Chem. Soc.*, 84, 2696
       (1962). UV: Thermodynamic data, phenol adducts with amides in
       CCl_4.

1182. M. D. Joesten and R. S. Drago, *J. Amer. Chem. Soc.*, 84, 3817
       (1962). UV: Thermodynamic data, adducts of phenol with bases
       in CCl_4, $\Delta\nu$-ΔH relationship.

1182a. A. Johansson, P. Kollman, and S. Rothenberg, *Theor. Chim. Acta*,
       26, 97 (1972). Theory: $(HCN)_2$, $(HCN)_3$.

1182b. A. Johansson and P. A. Kollman, *J. Amer. Chem. Soc.*, 94, 6196
       (1972). Theory: Amide-water.

1182c. A. Johansson, P. Kollman, and S. Rothenberg, *Chem. Phys. Lett.*,
       16, 123 (1972). Theory: HF-HCN dimer.

1182d. A. Johansson, P. Kollman, and S. Rothenberg, *Chem. Phys. Lett.*,
       18, 276 (1973). Theory: LCAO SCF on 1,3-propanediol.

1182e. A. Johansson, P. Kollman, and S. Rothenberg, *Theor. Chim. Acta*,
       29, 167 (1973). Theory: LCAO SCF on $(HCN)_2$, $(H_2O)_2$, $(HF)_2$.

1183. S. L. Johnson and K. A. Rumon, *J. Phys. Chem.*, 69, 74 (1965). IR: Solid 1:1 pyridine-benzoic acid adducts, single- and double-minimum potentials.

1184. M. D. Johnston, Jr., F. P. Gasparro, and I. D. Kuntz, Jr., *J. Amer. Chem. Soc.*, 91, 5715 (1969). NMR: Comparison of dipolar, hydrogen-bonding, and charge-transfer effects, thermodynamic data, adducts of $CHCl_3$, CH_2Cl_2, $CHCl_2CN$ with bases.

1185. F. Jona and G. Shirane, *Ferroelectric Crystals*, Pergamon, New York, 1969, 402 pp. Review.

1186. A. J. Jones, D. M. Grant, M. W. Winkley, and R. K. Robins, *J. Phys. Chem.*, 74, 2684 (1970). BIOL: ^{13}C, 23 pyrimidine nucleosides.

1187. A. J. Jones, D. M. Grant, M. W. Winkley, and R. K. Robins, *J. Amer. Chem. Soc.*, 92, 4079 (1970). BIOL: ^{13}C NMR, pyrimidine and purine nucleosides.

1188. D. Jones, Ph.D. Thesis, University of Wales, 1965. IR: Frequency shifts for hydrogen-bonded systems in vapor phase.

1189. D. A. K. Jones and J. G. Watkinson, *J. Chem. Soc.*, 2371 (1964). IR: Intramolecular, ΔH for o-halophenols.

1190. D. A. K. Jones and J. G. Watkinson, *J. Chem. Soc.*, 2366 (1964). IR: Thermodynamic data, adducts of phenols with organic halides.

1191. D. A. K. Jones and J. G. Watkinson, *Chem. Ind. (London)*, 661 (1960). IR: Intramolecular, ΔH for o-halophenols.

1191a. F. M. Jones, III, D. Eustace, and E. Grunwald, *J. Amer. Chem. Soc.*, 94, 8941 (1972). NMR: Kinetics of exchange, acetic acid-aniline systems.

1191b. R. G. Jones and J. R. Dyer, *J. Amer. Chem. Soc.*, 95, 2465 (1973). NMR: ^1H, $H(CF_3COO)_2^-$.

1192. R. L. Jones, *Spectrochim. Acta*, 22, 1555 (1966). IR: Self-association, secondary amides.

1193. R. L. Jones, *Spectrochim. Acta*, 20, 1879 (1964). IR: Steric inhibition of hydrogen bonding in secondary amides.

1194. W. J. Jones, R. M. Seel, and N. Sheppard, *Spectrochim. Acta*, 25A, 385 (1969). IR: Gas phase, $(HCN)_2$ and $NH_3—HCN$.

1195. P. G. Jönsson, *Acta Crystallogr., Sect. B*, 27, 893 (1971). Neutron Diffraction: Acetic acid.

1195a. P. G. Jönsson, *Acta Chem. Scand.*, 26, 1599 (1972). X-Ray: 1:1 adduct, acetic acid with phosphoric acid.

1196. P. G. Jönsson and W. C. Hamilton, *J. Chem. Phys.*, 56, 4433 (1972). Neutron Diffraction: Self-association, $(CCl_3COOH)_2$.

1197. N. Joop and H. Zimmermann, *Z. Physik. Chem. (Frankfurt)*, 42, 61 (1964). NMR: Self-association, crystalline imidazole.

1198. C. F. Jordan, A. P. Stefani, and E. M. Williams, *J. Miss. Acad. Sci.*, 13, 45 (1967); *Chem. Abstr.*, 70, 52640y (1969). IR: Self-association of disubstituted acetamidines.

1199. D. O. Jordan, *The Chemistry of Nucleic Acids*, Butterworth, London, 1960. BIOL: Review, nucleic acid compositions.

1200. F. Jordan and B. Pullman, *Theor. Chim. Acta*, 9, 242 (1968). BIOL: Preferred conformations of nucleosides.

1200a. F. Jordan and H. D. Sostman, *J. Amer. Chem. Soc.*, 94, 7898 (1972). Theory: CNDO/2 and MINDO, protonated adenine tautomers.

1201. L. Joris, J. Mitsky, and R. W. Taft, *J. Amer. Chem. Soc.*, 94, 3438 (1972). NMR: ^{19}F, K, adducts of p-F-C$_6$H$_4$OH in various solvents.

1202. L. Joris, P. von R. Schleyer, and R. Gleiter, *J. Amer. Chem. Soc.*, 90, 327 (1968). IR: Intramolecular, double bonds, triple bonds, cyclopropane rings.

1203. L. Joris, P. von R. Schleyer, and E. Osawa, *Tetrehedron*, 24, 4759 (1968). IR: Conformational heterogeneity in saturated alcohols.

1204. M. L. Josien, *J. Chim. Phys.*, 61, 245 (1964). IR: Solvent effects on ν_{AH}.

1205. M. L. Josien, *Pure and Appl. Chem.*, 4, 33 (1962). IR: Solvent effects on ν_{AH}.

1206. M. L. Josien, P. Dizabo, and P. Saumagne, *Bull. Soc. Chim. Fr.*, 423, 684 (1957). IR: Dimerization of thiophenol.

1207. M. L. Josien and N. Fuson, *J. Chem. Phys.*, 22, 1169, 1264 (1954). IR: Solvent effects on group frequencies.

1208. M. L. Josien and J. Lascombe, *Colloq. Spectrosc. Int.*, *Ottawa*, *13th*, 1967, pp. 40-57. IR: Effect of solvents on IR spectra of proton donors.

1209. M. L. Josien and P. Pineau, *Compt. Rend. Acad. Sci.*, 250, 2559 (1960). IR: Self-association of dichloroethanol.

1210. R. Jost, P. Rimmelin, and J. M. Sommer, *J. Chem. Soc. D*, 1243 (1971). NMR: Intramolecular, protonated haloacetones.

1211. T. H. Jukes, *Curr. Top. Microbiol. Immunol.*, 49, 178 (1969). BIOL: Review, mechanism of tRNA amino acid specificity and anticodon relationship to mRNA.

1212. T. H. Jukes and L. Gatlin, *Prog. Nucl. Acid Res. Mol. Biol.*, 11, 303 (1971). BIOL: Review and discussion, coding mechanism.

1213. A. Julg and M. Bonnet, *Theor. Chim. Acta*, 1, 6 (1962). Theory: Semiempirical calculation of the effect of hydrogen bond formation on the energy levels of phenol.

1214. S. Julia, D. Varech, Th. Burer, and Hs. H. Gunthard, *Helv. Chim. Acta*, 43, 1623 (1960). IR: Intramolecular, 1,3-diols.

1215. C. F. Jumper, M. T. Emerson, and B. B. Howard, *J. Chem. Phys.*, 35, 1911 (1961). NMR: K$_2$ for self-association of CHCl$_3$ in C$_6$H$_{12}$.

1215a. A. S. Kabankin, G. M. Zhidomirov, and A. L. Buchachenko, *Zh. Strukt. Khim.*, 13, 423 (1972); *Chem. Abstr.*, 77, 94813g (1972). ESR: Nitroxide radicals with alcohols.

456

1216. R. E. Kagarise, *Spectrochim. Acta*, 19, 629 (1963). IR: K, 1:1 adducts of acetone with $CDCl_3$, $CDCl_2Br$, $CDBr_3$.

1216a. R. E. Kagarise and K. B. Whetsel, *Spectrochim. Acta*, 17, 869 (1961). IR: Determination of frequency shifts by differential spectroscopy.

1217. T. Kagiya, Y. Sumida, and T. Inoue, *Bull. Chem. Soc. Jap.*, 41, 767, 773, 779 (1968). IR: $\Delta\nu_{OD}$, CH_3OD in various solvents, $\Delta\nu_{CO}$, acetophenone in various solvents.

1218. T. Kagiya, Y. Sumida, and T. Tachi, *Bull. Chem. Soc. Jap.*, 43, 3716 (1970). IR: $\Delta\nu_{OH}$, alcohols in various solvents.

1219. T. Kagiya, Y. Sumida, T. Watanabe, and T. Toshihiro, *Bull. Chem. Soc. Jap.*, 44, 923 (1971). IR: Adducts of Me_3SiOH.

1220. R. Kaiser, *Can. J. Chem.*, 41, 430 (1963). NMR: Chemical shifts for $CHCl_3$ with dioxane.

1222. N. R. Kallenbach, *J. Mol. Biol.*, 37, 445 (1968). BIOL: Intra-molecular complementary base-pair interactions.

1223. K. Kalnins and B. G. Belen'ki, *Izv. Akad. Nauk SSSR, Ser. Khim.*, 1552 (1968). IR: Spectra of amino acid salts of octadecanesul-fonic acid.

1224. B. Kamb, *Science*, 172, 231 (1971). "Anomalous water."

1224a. U. Kambayashi, *Bull. Chem. Soc. Jap.*, 36, 1173 (1963). IR: Self-association, dihalomethanes.

1225. H. Kamei, *Bull. Chem. Soc. Jap.*, 41, 2269 (1968). NMR: Formamide, pure; in acetone, dioxane, H_2O solutions.

1225a. M. Kamiya, *Chem. Pharm. Bull.*, 17, 1854 (1969). NMR: Intra-molecular, o-haloanilines.

1226. M. J. Kamlet, E. G. Kayser, J. W. Eastes, and W. H. Gilligan, *J. Amer. Chem. Soc.*, 95, 5210 (1973). Electronic: Effect of hydrogen bonding on spectra of N,N-diethyl-4-nitroaniline.

1227. M. J. Kamlet, R. R. Minesinger, E. G. Kayser, M. H. Aldridge, and J. W. Eastes, *J. Org. Chem.*, 36, 3852 (1971). Electronic: Hydrogen bonding of hydroxylic solvents with N-(4-nitrophenyl) aziridine.

1227a. M. J. Kamlet, R. R. Minesinger, and W. H. Gilligan, *J. Amer. Chem. Soc.*, 94, 4744 (1972). UV: Correlation of bathochromic shifts of 4-nitroaniline, 4-nitrophenol with pK_{HB} and Δ.

1227b. I. Kampschulte-Scheuing and G. Zundel, *J. Phys. Chem.*, 74, 2363 (1970). IR: Tunneling of protons in hydrogen bonds, continuous absorbance, p-toluenesulfonic acid.

1228. Y. Kanda, R. Shimada, and Y. Takenoshita, *Spectrochim. Acta*, 19, 1249 (1963). UV: Phosphorescent spectra of benzoic acid in various solvents.

1229. C. R. Kanekar, G. Govil, C. L. Khetrapal, and M. M. Dhingra, *Proc. Indian Acad. Sci., Sect. A*, 65, 265 (1967). NMR: Self-association, aromatic thiols.

1229a. S. Kang, *J. Mol. Struct.*, 17, 127 (1973). Theory: Cytidine.

1230. G. S. Karentnikov and L. Ya. Golishnikova, *Zh. Fiz. Khim.*, 42, 1885 (1968). IR: H_2O—MeCN system.

1231. J. D. Karkas and E. Chargaff, *Proc. Natl. Acad. Sci. U. S.*, 56, 664 (1966). BIOL: Integrity of the DNA template.

1232. J. D. Karkas, R. Rudner, and E. Chargaff, *Proc. Natl. Acad. Sci. U. S.*, 60, 915 (1968). BIOL: Template functions in the enzymic formation of polyribonucleotides.

1233. J. Karle and L. Brockway, *J. Amer. Chem. Soc.*, 66, 574 (1944). Structure of acetic and formic acid dimers.

1233a. R. Karlsson and D. Losman, *J. Chem. Soc., Chem. Commun.*, 626 (1972). X-Ray: Symmetrical bond in the hemihydrochloride of coccinelline.

1234. V. B. Kartha, R. N. Jones, and R. E. Robertson, *Proc. Indian Acad. Sci.*, A58, 216 (1963). IR: Thermodynamic data, adducts of p-chlorophenol with alkyl sulfonyl compounds.

1235. A. V. Karyakin and G. A. Muradova, *Zh. Fiz. Khim.*, 45, 1054 (1971); *Chem. Abstr.*, 75, 53268y (1971). IR: ΔH, water adducts of organic bases.

1236. A. V. Karyakin, G. A. Muradova, and L. Ya. Golishnikova, *Kolebatel'nye Specktry. Neorg. Khim.*, 267 (1971); *Chem. Abstr.*, 75, 69089b (1971). IR: Bond energies for H_2O-anion hydrogen bonds.

1237. T. R. Kasturi, B. N. Mylari, A. Balasubramanian, and C. N. R. Rao, *Can. J. Chem.*, 40, 2272 (1962). UV: 1,2-dicyanoesters.

1238. Y. Kasugai, Y. Arata, and S. Fujiwara, *Bull. Chem. Soc. Jap.*, 44, 2557 (1971). NMR: ^{1}H and ^{14}N, 1:1 H_2O—C_5H_5N adduct.

1239. J. E. Katon and R. L. Kleinlein, *Spectrochim. Acta*, 29A, 791 (1973). IR, Raman: Self-association, bromoacetic acid.

1239a. B. Katz, A. Ron, and O. Schnepp, *J. Chem. Phys.*, 47, 5303 (1967). Far-IR: Matrix studies, $(HCl)_2$.

1240. L. Katz, *J. Mol. Biol.*, 44, 279 (1969). BIOL, PMR: Association of adenine and uracil derivatives in $CHCl_3$.

1241. L. Katz and S. Penman, *Biochem. Biophys. Res. Commun.*, 27, 456 (1967). BIOL: Influence of coprecipitants on the solvent denaturation of double-stranded RNA.

1242. L. Katz and S. Penman, *J. Mol. Biol.*, 15, 220 (1966). NMR: Purine-pyrimidine chemical shifts in Me_2SO.

1243. L. Katz, K. Tomita, and A. Rich, *Acta Crystallogr.*, 21, 754 (1966). BIOL: X-Ray, 1:1 complex of 9-ethyladenine with 1-methyl-5-bromouracil.

1244. L. Katz, K. Tomita, and A. Rich, *J. Mol. Biol.*, 13, 340 (1965). BIOL: X-Ray, 9-ethyladenine and 1-methyl-5-bromouracil 1:1 complex.

1244a. M. V. Kaulgud and K. J. Patil, *Acustica*, 26, 292 (1972). Sound propagation, collision factors, inter- and intramolecular hydrogen bonds.

1245. R. Kavcic and B. Plesmiar, *J. Org. Chem.*, 35, 2033 (1970).
       IR: Correlation of frequency shifts with rates of epoxidation
       by *p*-nitroperoxybenzoic acid.

1246. H. Kaye, *J. Amer. Chem. Soc.*, 92, 5777 (1970). BIOL, CD:
       Interaction of poly-9-vinyladenine with poly U.

1247. W. I. Kaye and R. Poulson, *Nature*, 193, 675 (1962). UV: Self-
       association, methanol.

1248. D. R. Kearns, D. Patel, R. G. Shulman, and T. Yamane, *J. Mol.
       Biol.*, 61, 265 (1971). BIOL, NMR: Base pairing in tRNAs.

1249. E. R. Kearns, *J. Phys. Chem.*, 65, 314 (1961). CAL: Thermody-
       namic data, adducts of $CHCl_3$ with acetone.

1250. P. Kebarle, S. K. Searles, A. Zolla, J. Scarborough, and M.
       Arshadi, *J. Amer. Chem. Soc.*, 89, 6393 (1967). Ion cyclotron
       resonance, $H^+(H_2O)_n$ dissociation energies.

1251. Z. Kecki, *Postepy Fiz.*, 20, 277 (1969); *Chem. Abstr.*, 72, 84699g
       (1970). NMR: Proton exchange in methanol.

1251a. Z. Kecki and D. Cieslak, *Acta Phys. Pol. A*, 44, 95 (1973).
       NMR: Self-association, alcohols in organic solvents.

1252. W. E. Keder and L. L. Burger, *J. Phys. Chem.*, 69, 3075 (1965).
       NMR: Self-association, tri-*n*-octylammonium salts in CCl_4.

1253. D. E. Kennell, *Prog. Nucl. Acid Res. Mol. Biol.*, 11, 259 (1971).
       BIOL: Review, nucleic acid hybridization.

1254. R. C. Kerber and A. Porter, *J. Amer. Chem. Soc.*, 91, 366 (1969).
       NMR, UV: Association of 1-nitroindene and 9-nitrofluorene
       salts with proton donors.

1255. A. V. Kerimbekov and N. A. Kerimbekova, *Zh. Fiz. Khim.*, 42, 2996
       (1968); *Chem. Abstr.*, 70, 77156h (1969). IR: *p*-Phenylenediamine-
       chloranil adduct.

1256. A. S. Kertes, H. Gutmann, O. Levy, and G. Markovits, *Israel J.
       Chem.*, 6, 463 (1968). IR: Long-chain normal aliphatic amine
       salts.

1257. N. R. Kestner and O. Sinanoğlu, *J. Chem. Phys.*, 38, 1730 (1963).
       BIOL: Effect of solvent on base-pair energies.

1257a. L. F. Keyser and G. W. Robinson, *J. Chem. Phys.*, 45, 1694 (1966).
       IR: Matrix studies, HCl, DCl polymers.

1258. A. K. Khairetdinova and I. S. Perelygin, *Opt. Spektrosk.*, 26, 62
       (1969); *Chem. Abstr.*, 70, 86841e (1969). IR: Adducts of iso-
       propyl alcohol derivatives with acetonitrile in CCl_4.

1259. L. V. Khazhakyan and I. A. Gyulbaryan, *Arm. Khim. Zh.*, 21, 387
       (1968); *Chem. Abstr.*, 70, 14905f (1969). IR: Intramolecular,
       intermolecular, alkoxy isophthalic acids.

1260. Yu. A. Khon, *Zh. Fiz. Khim.*, 45, 960 (1971). Correlation of ΔH
       with pK_a.

1261. N. N. Khovratovich and N. A. Borisevich, *Opt. Spektrosk.*,
       supplement 2, 248 (1963); [English, 130 (1966)]. IR: Intra-
       and intermolecular, phthalimide derivatives.

1262. N. N. Khovratovich and N. A. Borisevich, *Opt. Spektrosk.*,
       supplement 3, 123 (1967); [English, 62 (1968)]. IR: Intramo-
       lecular, ν_{NH} intensities, phthalimides.

1263. E. Kiehlmann, B. C. Menon, and N. McGillivray, *Can. J. Chem.*,
       51, 3177 (1973). NMR: Intramolecular, 1,1,1-trifluoro-2-
       hydroxy-4-alkanones.

1264. A. P. Kilimov *et al.*, *Zh. Obshch. Khim.*, 37, 768 (1967);
       [English, 722]. IR: Aminothiols.

1265. Jong Taik Kim, *Daehan Hwahak Hwoejee*, 14, 147 (1970); *Chem.
       Abstr.*, 74, 26378g (1971). IR: Association of substituted
       ureas with $CHCl_3$ and *p*-dioxane.

1266. S. H. Kim, P. Schofield, and A. Rich, *Cold Spring Harbor Symp.
       Quant. Biol.*, 34, 153 (1969). BIOL, X-Ray: tRNA crystals.

1267. S. H. Kim and A. Rich, *Science*, 166, 1621 (1969). BIOL:
       Crystalline tRNA, three-dimensional Patterson function at 12Å
       resolution.

1268. S. H. Kim and A. Rich, *J. Mol. Biol.*, 42, 87 (1969). BIOL,
       X-Ray: 1-Methylcytosine and 5-fluorouracil complexes.

1269. S. H. Kim and A. Rich, *Science*, 158, 1046 (1967). BIOL, X-Ray:
       5-Fluorouracil and 9-ethylhypoxanthine complexes.

1270. L. Kimtys, H. Jonaitis, and P. Kaikaris, *Liet. Fiz. Rimkinys*, 9,
       577 (1969); *Chem. Abstr.*, 72, 116592k (1970). NMR: Self-
       association, carboxylic acids.

1271. F. S. Kinoyan and A. V. Mushegyan, *Arm. Khim. Zh.*, 23, 399
       (1970); *Chem. Abstr.*, 73, 125271c (1970). IR: Self-association,
       tertiary alkylethynylcarbinols in CCl_4.

1271a. H. H. Kirchner and W. Richter, *Z. Phys. Chem. (Frankfurt am
       Main)*, 81, 274 (1972). IR: Intramolecular, $PhCH_2CH_2OH$.

1272. A. Kirck, Per G. Jönsson, and I. Olovsson, *Inorg. Chem.*, 8, 2775
       (1969). X-Ray: 1:1 adduct of acetic acid with fluorosulfuric
       acid.

1273. J. G. Kirkwood in W. West and R. T. Edwards, *J. Chem. Phys.*, 5,
       14 (1937). Model for solvent effects on frequency shifts based
       on dielectric constant.

1274. H. Kiriyama and R. Kiriyama, *Mem. Inst. Sci. Ind. Res.*, *Osaka
       University*, 27, 77 (1970). NMR, IR: $M(HCO_3)_2 \cdot 2H_2O$ where M =
       Mg, Zn.

1275. T. L. Kiseler and T. A. Avdonina, *Mol. Biol.*, 3, 113 (1969).
       BIOL: Specific complexes between t- and mRNAs in a ribosome-
       free system.

1276. S. Kishida and K. Nakamota, *J. Chem. Phys.*, 41, 1558 (1964).
       IR: Normal coordinate analysis of monomeric and dimeric formic
       and acetic acids.

1276a. H. Kistenmacher, H. Popkie, and E. Clementi, *J. Chem. Phys.*, 58,
       5627 (1973). Theory: LCAO SCF on $H_2O...Cl^-$, $H_2O...F^-$.

1277. T. Kitao and C. H. Jarboe, *J. Org. Chem.*, 32, 407 (1967). IR:
       Thermodynamic data, adducts of methanol with pyridine derivatives

1278. C. Kittel, *Amer. J. Phys.*, 36, 610 (1968). BIOL, X-Ray: Structure of DNA.

1279. A. Kivinen and J. Murto, *Suomen Kemistilehti B*, 42, 190 (1969). Near-IR: Thermodynamic data for adducts of 2,2,2-trifluoro-ethanol and 1,1,1,3,3,3-hexafluoro-2-propanol with 1,4-dioxane in CCl₄.

1280. A. Kivinen, J. Murto, and L. Kilpi, *Suomen Kemistilehti B*, 40, 301 (1967). IR: Thermodynamic data for adducts of fluoroalcohols with tetrahydrofuran.

1281. A. Kivinen and J. Murto, *Suomen Kemistilehti B*, 40, 6 (1967). Near-IR: Self-association, 2,2,2-trifluoro-2-propanol, 1,1,1,3,3,3-hexafluoro-2-propanol.

1281a. T. Kjallman and I. Olovsson, *Acta Crystallogr.*, *Sect. B*, 28, 1692 (1972). X-Ray: $(H_5O_2)_2SO_4$ and $(D_5O_2)_2SO_4$ at -190°C.

1282. D. C. Kleinfelter, *J. Amer. Chem. Soc.*, 89, 1734 (1967). NMR: Intramolecular, π-cloud, norbornanols.

1283. J. Klicnar, F. Kristek, V. Bekarek, and M. Vecera, *Collect. Czech. Chem. Commun.*, 34, 553 (1969). IR: Intramolecular, *o*-hydroxy derivatives of Schiff bases.

1284. I. M. Klotz, *Federation Proc.*, 24, S24 (1965). BIOL: Role of water structure in macromolecules.

1285. I. M. Klotz and S. B. Farnham, *Biochemistry*, 7, 3879 (1968). Stability of N—H...O=C amide bond in apolar solvents.

1286. E. Kluk and H. Kluk, *Acta Phys. Pol. A*, 43, 271 (1973). NMR: Self-association, aliphatic alcohols.

1286a. E. Knoezinger, *Beckman Rep.*, 25 (1972); *Chem. Abstr.*, 77, 81496f (1972). IR, UV: Dimeric formic acid.

1287. R. G. Knubovets, M. L. Afans'ev, and S. P. Gabuda, *Spectrosc. Lett.*, 2, 121 (1969). NMR: ¹⁹F, ¹H, apatite crystal.

1287a. T. Kobayashi et al., *Bull. Soc. Chem. Jap.*, 45, 1494 (1972). IR: Intramolecular, *o*-substituted benzenethiols.

1287b. T. Kobayashi and M. Hirota, *Chem. Lett.*, 975 (1972). NMR: Intramolecular, *o*-fluorophenol and *o*-fluorobenzenethiol.

1287c. H. J. Koehler, *Z. Chem.*, 12, 65 (1972). Theory: CNDO, $(HCOOH)_2$.

1287d. F. Kohler, E. Liebermann, G. Miksch, and C. Kainz, *J. Phys. Chem.*, 76, 2764 (1973). Thermodynamic data, acetic acid-triethylamine.

1288. J. P. Kokko, J. H. Goldstein, and L. Mandell, *J. Amer. Chem. Soc.*, 83, 2909 (1961). BIOL, PMR: Tautomerism and substituent effects in some pyrimidines and related nucleosides.

1289. A. Kolbe, *Tetrahedron Lett.*, 1049 (1969). IR: Hydrogen-deuterium isotope effects of hydrogen bonds between alcohols and tertiary amines.

1289a. J. Koller and A. Ažman, *Croat. Chem. Acta*, 44, 283 (1972). Theory: Localized orbitals in HF, HF_2^-, HCO_2H, $(HCO_2H)_2$, H_2O, $(H_2O)_2$.

1290. J. Koller, B. Borstnik, and A. Ažman, *Croat. Chem. Acta*, 41, 175 (1969). Theory: CNDO/2 and INDO calculations on FHF⁻.

1290a. J. Koller, S. Kaiser, and A. Ažman, *J. Mol. Struct.*, 13, 305 (1972). Theory: CNDO, $H_2O...HOH$.

1291. P. A. Kollman, *J. Amer. Chem. Soc.*, 94, 1837 (1972). Theory: Discussion and analysis of recent work (mostly theoretical) on hydrogen bonds in small polyatomics.

1292. P. A. Kollman and L. C. Allen, *Chem. Revs.*, 72, 283 (1972). Review of theory including recent *ab initio* work.

1293. P. A. Kollman and L. C. Allen, *J. Amer. Chem. Soc.*, 93, 4991 (1971). Theory: *Ab initio* Gaussian basis calculations on first-row dimers.

1294. P. A. Kollman and L. C. Allen, *J. Amer. Chem. Soc.*, 92, 753 (1970). Theory: CNDO studies of several first-row dimers.

1295. P. A. Kollman and L. C. Allen, *J. Amer. Chem. Soc.*, 92, 6101 (1970). Theory: Gaussian SCF calculations on HF_2^- and $H_5O_2^+$.

1296. P. A. Kollman and L. C. Allen, *J. Chem. Phys.*, 52, 5085 (1970). Theory: Gaussian SCF calculations on HF dimer and H_2O—HF.

1297. P. A. Kollman and L. C. Allen, *Theor. Chim. Acta*, 18, 399 (1970). Theory: $(HF)_2$ and $(H_2O)_2$, partition hydrogen bond energy into electrostatic, delocalization and correlation contributions.

1298. P. A. Kollman and L. C. Allen, *J. Chem. Phys.*, 51, 3286 (1969). Theory: Near Hartree-Fock limit, Gaussian basis calculations on water dimers.

1299. P. A. Kollman and A. D. Buckingham, *Mol. Phys.*, 21, 567 (1971). Theory: Comments on structure of water dimer.

1300. P. A. Kollman, J. F. Liebman, and L. C. Allen, *J. Amer. Chem. Soc.*, 92, 1142 (1970). Theory: Gaussian SCF studies on Li analogue of hydrogen bond.

1300a. W. Kołos and C. C. J. Roothaan, *Revs. Mod. Phys.*, 32, 219 (1960). Theory: Accurate wavefunction for H_2.

1300b. W. Kołos and L. Wolniewicz, *J. Chem. Phys.*, 49, 404 (1968). Accurate wavefunctions for H_2.

1301. I. M. Kolthoff and M. K. Chantooni, Jr., *J. Amer. Chem. Soc.*, 93, 3843 (1971). IR: Intramolecular, *o*-hydroxybenzoic acids with anions.

1302. I. M. Kolthoff and M. K. Chantooni, Jr., *J. Amer. Chem. Soc.*, 92, 7025 (1970). Dissociation constants of substituted benzoic acids.

1303. I. M. Kolthoff and M. K. Chantooni, Jr., *J. Amer. Chem. Soc.*, 91, 4621 (1969). Hydrogen bonding of substituted phenols with Cl^-.

1304. I. M. Kolthoff and M. K. Chantooni, Jr., *J. Amer. Chem. Soc.*, 91, 25 (1969). Association of sulfate and bisulfate ions with *p*-bromophenol, H_2O, and MeOH in acetonitrile.

1305. I. M. Kolthoff and M. K. Chantooni, Jr., *J. Phys. Chem.*, 70, 856 (1966). $K_{HA_2}^-$ for substituted benzoic acids.

1306. I. M. Kolthoff and M. K. Chantooni, Jr., *J. Phys. Chem.*, 66,
       1675 (1962). K for $H_2SO_4 \cdot HSO_4^-$.

1306a. A. I. Kol'tsov and G. M. Kheifets, *Usp. Khim.*, 40, 1646 (1971);
       [English, 773]. NMR: Intramolecular, keto-enol tautomerism.

1306b. A. I. Kol'tsov and G. M. Kheifets, *Usp. Khim.*, 41, 877 (1972);
       [English, 452]. NMR: Intramolecular, review of tautomerism.

1307. J. M. Konarski, *Bull. Acad. Polon. Sci., Ser. Sci., Math. Astron.
       Phys.*, 13, 813 (1965). Theory: Energy bands in polypeptides.

1308. I. I. Kondilenko, V. E. Pogorelov, and Khuong Hue, *Opt. Spek-
       trosk.*, 26, 956 (1969); *Chem. Abstr.*, 71, 75967v (1969). Raman:
       Polarized line width of acetone in various solvents.

1309. N. S. Kondo, H. M. Holmes, L. M. Stempel, and P. O. P. Ts'o,
       Biochemistry, 9, 3479 (1970). BIOL: Conformation and interac-
       tion of dinucleoside mono- and diphosphates.

1310. R. Konig and G. Malewski, *Spectrochim. Acta*, 24A, 219 (1968).
       IR: Self-association, secondary sulfonamides.

1311. J. H. Konnert, J. W. Gibson, I. L. Karle, M. N. Khattak, and
       S. Y. Wang, *Nature*, 227, 953 (1970). BIOL, X-Ray: Dimers of
       uracil and 6-methyl-uracil.

1312. J. H. Konnert, I. L. Karle, and J. Karle, *Acta Crystallogr.,
       Sect. B*, 26, 770 (1970). BIOL, X-Ray: Structure of dihydro-
       thymidine.

1313. R. Konopka, B. Pedzisz, and M. Jurewicz, *Acta Phys. Pol. A*, 40,
       751 (1971). IR: Influence of steric effects on hydrogen bond
       in alcohols.

1314. E. V. Konovalov, Yu. P. Egorov, R. V. Belinskaya, V. N. Boiko,
       and L. M. Yagupol'skii, *Zh. Prikl. Spektrosk.*, 14, 484 (1971);
       Chem. Abstr., 75, 42673a (1971). IR: Intramolecular, *o*-
       substituted phenols containing —CF_3 group.

1315. E. V. Konovalov, Yu. P. Egorov, L. M. Yagupol'skii, and R. V.
       Belinskaya, *Teor. Eksp. Khim.*, 5, 134 (1969); *Chem. Abstr.*, 71,
       12345v (1969). IR: Intramolecular, *o*-$F_3CC_6H_4CH_2OH$.

1316. G. J. Korinek and W. J. Schneider, *Can. J. Chem.*, 35, 1157
       (1957). NMR: $CHCl_3$ as proton donor.

1317. V. S. Korobkov, *Opt. Spektrosk.*, 17, 938 (1964); [English, 509].
       IR: Bandwidth in hydrogen-bonded systems.

1318. A. V. Korshunov, V. F. Shabanov, and V. E. Volkov, *Izv. Vyssh.
       Ucheb. Zaved., Fiz.*, 12, 143 (1969); *Chem. Abstr.*, 72, 7715g
       (1970). Raman: Temperature studies, low-frequency spectra of
       pyrocatechol and 2,4,6-trichloroaniline crystals.

1319. A. V. Kotov, V. A. Zarinskii, and V. M. Bokina, *Izv. Akad. Nauk
       SSSR, Ser. Khim.*, 1319 (1969); *Chem. Abstr.*, 71, 74600h (1969).
       IR: CH_3COOH—H_2SO_4 mixtures.

1320. S. Kovac, M. Dandarova, and A. Piklerova, *Tetrahedron*, 27, 2831
       (1971). IR: Self-association, 2-hydroxy-8-acetyl-naphthalene,
       propyl coumarate, methyl coumarate.

1321. S. Kovac and G. Eglinton, *Tetrahedron*, <u>25</u>, 3599 (1969). NMR, IR: Self-association of ethyl coumarate, *m*-hydroxyacetophenone, and *bis*(1-hydroxy-4-methylphenyl) methane.

1322. S. Kovac, E. Solcaniova, and J. Baxa, *Tetrahedron*, <u>27</u>, 2823 (1971). IR, NMR: Intramolecular, alkyl substituted *bis*(hydroxyphenyl) alkanes.

1322a. S. Kovac, E. Solcaniova, E. Beska, and P. Rapos, *J. Chem. Soc., Perkin Trans.*, *2*, 105 (1973). IR: Self-association, 3,4-disubstituted 5-hydroxyfuran-2(5H)-ones.

1323. T. V. Kozlova, N. A. Kozlov, and V. V. Zharkov, *Zh. Fiz. Khim.*, <u>45</u>, 2110 (1971); *Chem. Abstr.*, <u>75</u>, 135280k (1971). IR: Self-association, substituted ureas.

1324. V. Kh. Kozlovskii, *Fiz. Tverd. Tela*, <u>5</u>, 3294 (1963); *Chem. Abstr.*, <u>60</u>, 3608d (1964). Theory: Hydrogen bond in ferroelectrics.

1325. E. Kozlowska, *Compt. Rend. Acad. Sci.*, <u>267B</u>, 696 (1968). IR: Crystalline HNO_3, DNO_3.

1326. M. H. Krackov, C. M. Lee, and H. G. Mautner, *J. Amer. Chem. Soc.*, <u>87</u>, 892 (1965). Self-association of 2-pyridone, 2-pyridthione, and 2-pyridselenone.

1326a. W. P. Kraemer and G. H. F. Diercksen, *Theor. Chim. Acta*, <u>27</u>, 265 (1972). Theory: LCAO SCF on $2H_2O \cdot F^-$.

1327. W. P. Kraemer and G. H. F. Diercksen, *Theor. Chim. Acta*, <u>23</u>, 398 (1972). Theory: Gaussian SCF calculations on $OH^- \cdot H_2O$.

1328. W. P. Kraemer and G. H. F. Diercksen, *Theor. Chim. Acta*, <u>23</u>, 393 (1972). Theory: Gaussian SCF calculations on $Li^+ \cdot 2H_2O$.

1329. W. P. Kraemer and G. H. F. Diercksen, *Chem. Phys. Lett.*, <u>5</u>, 463 (1970). Theory: Gaussian calculations on $H_3O^+ \cdot H_2O$.

1330. H. Krakauer and J. M. Sturtevant, *Biopolymers*, <u>6</u>, 491 (1968). BIOL: Heats of the helix-coil transitions of poly A—poly U complexes.

1331. E. Krakower and L. R. Reeves, *Trans. Faraday Soc.*, <u>59</u>, 2528 (1963). NMR: Kinetics, proton transfer between methanol and *o*-chlorophenol.

1331a. G. H. Krause and H. Hoyer, *Z. Naturforsch. B*, <u>27</u>, 663 (1972). IR: Intramolecular, 3-nitrosalicylaldehydes, 2-hydroxy-3-nitroacetophenones.

1331b. D. N. Kravtsov *et al.*, *Izv. Akad. Nauk SSSR, Ser. Khim.*, 1710 (1972); *Chem. Abstr.*, <u>77</u>, 163880r (1972). NQR: Intramolecular, *o*-halophenols.

1332. R. W. Kreilick and S. I. Weissman, *J. Amer. Chem. Soc.*, <u>84</u>, 306 (1962). Kinetics: Proton transfer between diamagnetic RH and paramagnetic R• by ESR, NMR.

1332a. G. C. Kresheck, D. Kierleber, and R. J. Albers, *J. Amer. Chem. Soc.*, <u>94</u>, 8889 (1972). NMR: Self-association, N-deutero, N-methylacetamide.

1333.   C. B. Kretschmer and R. Wiebe, *J. Amer. Chem. Soc.*, 76, 2579
        (1954). P-V-T relationships of alcohol vapors.

1333a. S. Krimm, K. Kuroiwa, and T. Rebane, *Conformation of Biopolymers*,
        G. N. Ramachandran, ed., Academic Press, New York, 1967, Vol. 2,
        p. 439. IR: Polyglycine II.

1334.   V. G. Krishna and L. Goodman, *J. Chem. Phys.*, 33, 381 (1960).
        UV: Fluorescence and absorption spectra of pyrazine in various
        solvents.

1335.   R. S. Krishnan and R. S. Katiyar, *Bull. Chem. Soc. Jap.*, 42,
        2098 (1969). Raman: β-Alanine.

1336.   J. Kroon and J. A. Kanters, *Acta Crystallogr.*, *Sect. B*, 28, 714
        (1972). X-Ray: Potassium hydrogen *meso*-tartrate.

1337.   K. Kroon, J. A. Kanters, and A. F. Peerdeman, *Nature (London)*,
        Phys. Sci., 229A, 120 (1971). X-Ray: Position of hydrogen atom
        in A-type hydrogen bond, potassium hydrogen *meso*-tartrate.

1338.   J. Kroon, J. A. Kanters, A. F. Peerdeman, and A. Vos, *Nature
        (London)*, *Phys. Sci.*, 232A, 107 (1971). X-Ray: Position of
        hydrogen atom in A-type hydrogen bonds.

1339.   P. J. Krueger, *Tetrahedron*, 26, 4753 (1970). IR: Intramolecu-
        lar, *o*-aminophenols and *o*-aminothiophenols.

1340.   P. J. Krueger, *Can. J. Chem.*, 45, 2143 (1967). IR: Intramolecu-
        lar, ethyleneamines, diamines.

1341.   P. J. Krueger, *Can. J. Chem.*, 45, 2135 (1967). IR: Intramolecu-
        lar, *o*-phenylenediamines.

1342.   P. J. Krueger, *Can. J. Chem.*, 42, 201 (1964). Theory: Intra-
        molecular, *o*-substituted anilines.

1343.   P. J. Krueger, *Can. J. Chem.*, 41, 363 (1963). IR: Intramolecu-
        lar, *o*-substituted anilines.

1344.   P. J. Krueger, *Can. J. Chem.*, 40, 2300 (1962). IR: Intramolecu-
        lar, *o*-substituted anilines.

1344a. P. J. Krueger and B. F. Hawkins, *Can. J. Chem.*, 51, 3250 (1973).
        IR: Intramolecular, OH...π in *p*-X-benzyl alcohols.

1345.   P. J. Krueger and H. D. Mettee, *Tetrahedron Lett.*, 1587 (1966).
        IR: Intramolecular, 2-phenylethanol.

1346.   P. J. Krueger and H. D. Mettee, *Can. J. Chem.*, 43, 2970 (1965).
        IR: Intramolecular, ethanolamine derivatives.

1347.   P. J. Krueger and H. D. Mettee, *Can. J. Chem.*, 43, 2888 (1965).
        IR: Intramolecular, cyanoalcohols.

1348.   P. J. Krueger and H. D. Mettee, *J. Mol. Spectrosc.*, 18, 131
        (1965). IR: Intramolecular, ethylene glycol.

1349.   P. J. Krueger and H. D. Mettee, *Can. J. Chem.*, 42, 326, 340
        (1964). IR: Intramolecular, mono-, di-, and trihaloethanols.

1350.   P. J. Krueger and H. D. Mettee, *Can. J. Chem.*, 42, 288 (1964).
        IR: K, methanol adducts with bases.

1351.   P. J. Krueger and J. Skolik, *Tetrahedron*, 23, 1799 (1967). NMR:
        Intramolecular, monocations of sparteine- and α-isosparteine-N-
        oxides.

1352. P. J. Krueger and D. W. Smith, *Develop. Appl. Spectrosc.*, 4, 197 (1965). IR: Intramolecular, N—H...X.

1352a. M. Krumpolc, V. Bazant, and V. Chvalovsky, *Collect. Czech. Chem. Commun.*, 38, 711 (1973). IR: $\Delta\nu$, $Me_3Ge(CH_2)_nOH$ and $Me_3C(CH_2)_nOH$ adducts with phenol and THF.

1352b. K. Krynicki and J. G. Powles, *J. Magn. Resonance*, 6, 539 (1972). NMR: Relaxation times, liquid HBr in DBr, liquid HCl in DCl.

1353. T. Kubota, *J. Amer. Chem. Soc.*, 88, 211 (1966). UV: Thermodynamic data, adducts of phenol and α-naphthol with trimethylamine oxide.

1354. T. Kubota, Y. Oishi, K. Nishikida, and H. Miyazaki, *Bull. Chem. Soc. Jap.*, 43, 1622 (1970). ESR: Effect of EtOH, PhOH, H_2O on ESR spectra of anion radicals of aromatic amine 1-oxides.

1355. T. Kubota, M. Yamakawa, M. Takasuka, K. Iwatani, H. Akazawa, and I. Tanaka, *J. Phys. Chem.*, 71, 3597 (1967). UV: Phenol adduct with nitrile-N-oxides.

1356. T. Kubota, M. Takasuka, and Y. Matsui, *Shionogi Kenkyusho Nempo*, 16, 63 (1966); *Chem. Abstr.*, 66, 23977a (1967). IR: Intramolecular, correlation of ν_{OH} with H—X bond length in OH—X.

1357. E. Küchler and J. Derkosch, *Z. Naturforsch.*, 21b, 209 (1966). IR: Association of isopropylidene-trityladenosine and isopropylidene-trityluridine in CCl_4.

1358. L. P. Kuhn, *J. Amer. Chem. Soc.*, 74, 2492 (1952). IR: Intramolecular, correlation of $\Delta\nu$ with O—H...O distance for 1,2-diols.

1359. L. P. Kuhn and R. E. Bowman, *Spectrochim. Acta*, A23, 189 (1967). IR: Intermolecular, adducts of phenol with olefins.

1360. L. P. Kuhn and R. E. Bowman, *Spectrochim. Acta*, 17, 650 (1961). IR: Intramolecular, catechol, pyrogallol in CCl_4 and $CHCl_3$.

1361. L. P. Kuhn and G. G. Kleinspehn, *J. Org. Chem.*, 28, 721 (1963). IR: Intramolecular, 2,2'-dipyrrylmethanes and 2,2'-dipyrrylmethenes.

1362. L. P. Kuhn, P. von R. Schleyer, W. F. Baitinger, and L. Eberson, *J. Amer. Chem. Soc.*, 86, 650 (1964). IR: Intramolecular, 1,2-diols and 1,4-diols.

1363. L. P. Kuhn and R. A. Wires, *J. Amer. Chem. Soc.*, 86, 2161 (1964). IR: Intramolecular, monomethyl ethers of diols, thermodynamic data.

1364. L. P. Kuhn, R. A. Wires, W. Ruoff, and H. Kwart, *J. Amer. Chem. Soc.*, 91, 4790 (1969). IR: Intramolecular, 2-hydroxyalkylpyridines, 2-hydroxyalkylpiperidines.

1365. M. Kuhn, W. Luttke, and R. Mecke, *Z. Anal. Chem.*, 57, 680 (1963). IR: Intramolecular, β-nitroethanol.

1366. N. Kulevsky and P. M. Froehlich, *J. Amer. Chem. Soc.*, 89, 4839 (1967). IR: Self-association, thiolactams.

1366a. N. Kulevsky and L. Lewis, *J. Phys. Chem.*, 76, 3502 (1972). Near-IR: Thermodynamic data, phenol-diazine N-oxide adducts.

1367.   A. M. Kuliev, M. A. Salimov, F. N. Mamedov, and N. Yu. Ibragimov,
        Dokl. Akad. Nauk SSSR, 184, 1141 (1969). IR: Intramolecular,
        Mannich bases.

1367a. A. M. Kuliev, B. R. Gasanov, E. A. Agaeva, and S. B. Bilalov,
        Zh. Prikl. Spektrosk., 19, 506 (1973); *Chem. Abstr.*, 79, 125661d
        (1973). IR: Intramolecular, β-hydroxyamines, β-hydroxysulfides.

1368.   A. M. Kuliev and B. Yu. Sultanov, *Dokl. Akad. Nauk Azerb. SSR*,
        28, 29 (1972); *Chem. Abstr.*, 78, 123758w (1973). IR: Intra-
        molecular, thiophenols.

1369.   V. Kumar and A. S. N. Murthy, *Indian J. Pure Appl. Phys.*, 8,
        234 (1970). Theory: Hydrogen bond resonance energy in DNA
        base pairs.

1370.   I. D. Kuntz, Jr., F. P. Gasparro, M. D. Johnston, Jr., and R. P.
        Taylor, *J. Amer. Chem. Soc.*, 90, 4778 (1968). Benesi-Hildebrand
        equation, concentration scales.

1371.   R. Kuopio and A. Kivinen, *Suomen Kemistilehti A*, 44, 161 (1971).
        IR: Review.

1372.   V. A. Kuprievich, *Int. J. Quantum Chem.*, 1, 561 (1967). BIOL:
        π-Electron structures of A, G, C, T, and U in their ground,
        ionized, singlet, and triplet excited states.

1373.   G. A. Kurkchi and A. V. Iogansen, *Opt. Spektrosk.*, supplement 3,
        128 (1967); [English, 65 (1968)]. IR: Bands of 1-alkynes in
        isooctane, CH_3CN, $(CH_3)_2CO$.

1374.   G. A. Kurkchi, S. F. Shakhova, and L. E. Sergeeva, *Zh. Fiz.
        Khim.*, 46, 2302 (1972); [English, 1313]. GLC: ΔH, diacetylene-
        N-methyl pyrrolidinone adduct, 2.5 kcal/mole^{-1}/≡CH bond.

1375.   L. N. Kurkovskaya, A. M. Kuliev, B. Yagshiev, N. N. Shapet'ko,
        F. N. Mamedov, and A. G. Bairamova, *Dokl. Akad. Nauk SSSR*, 197,
        842 (1971); *Chem. Abstr.*, 75, 82224m (1971). NMR: Intramolecu-
        lar, aminomethyl (alkyl) phenols, thiobisphenols and bisphenols.

1376.   L. N. Kurkovskaya, N. N. Shapet'ko, and N. N. Magdesieva, *Teor.
        Eksp. Khim.*, 8, 688 (1972); *Chem. Abstr.*, 78, 22265c (1973).
        NMR: Intramolecular, selenophene β-diketones.

1377.   L. N. Kurkovskaya, N. N. Shapet'ko, Yu. S. Adreichikov, and R.
        F. Saraeva, *Zh. Strukt. Khim.*, 13, 1026 (1972); *Chem. Abstr.*,
        78, 57183b (1973). NMR: Intramolecular, carbalkoxyl deriva-
        tives of β-dicarbonyl compounds.

1378.   L. N. Kurkovskaya *et al.*, *Teor. Eksp. Khim.*, 9, 261 (1973);
        Chem. Abstr., 79, 25342g (1973). NMR: Intramolecular, β-
        naphthylamineazobenzene derivatives.

1379.   K. Kurosaki, *Nippon Kagaku Zasshi*, 83, 655 (1962). IR: Spectra
        of $CHCl_3 — (C_2H_5)_2O$ mixtures in CCl_4.

1380.   L. A. Kutulya, Yu. N. Surov, N. S. Pivnenko, S. V. Tsukerman,
        and V. F. Lavrushin, *Zh. Obshch. Khim.*, 41, 895 (1971); *Chem.
        Abstr.*, 75, 69096b (1971). IR, NMR: Trifluoroacetic acid
        adducts of chalcones, RCH=CHCOR'.

1381.   A. Kvivk, P. G. Jonsson, and R. Liminger, *Acta Chem. Scand.*, 26,
        1087 (1972). Neutron Diffraction: Hydrazinium *bis*(dihydrogen
        phosphate).

1382.   J. S. Kwiatkowski, *Theor. Chim. Acta*, <u>10</u>, 47 (1968). BIOL:
        Pariser-Parr-Pople calculations, purine and some of its 6-
        substituted derivatives.

1383.   Y. Kyogoku, *Kagaku No Ryoiki*, <u>22</u>, 364 (1968); *Chem. Abstr.*, <u>69</u>,
        97074n (1968). BIOL: IR, associations between adenine and
        thymine analogs and between guanine and cytosine analogs in
        CDCl$_3$.

1384.   Y. Kyogoku, R. C. Lord, and A. Rich, *Biochim. Biophys. Acta*,
        <u>179</u>, 10 (1969). BIOL: IR, hydrogen bonding specificity of
        hypoxanthine and other nucleic acid derivatives in CHCl$_3$.

1385.   Y. Kyogoku, R. C. Lord, and A. Rich, *Nature*, <u>218</u>, 69 (1968).
        IR: Self-association, mephobarbital.

1386.   Y. Kyogoku, R. C. Lord, and A. Rich, *J. Amer. Chem. Soc.*, <u>89</u>,
        496 (1967). IR: Association constants for dimerization of
        9-ethyladenine and 1-cyclohexyluracil, and their 1:1 complex.

1387.   Y. Kyogoku, R. C. Lord, and A. Rich, *Proc. Natl. Acad. Sci.
        U. S.*, <u>57</u>, 250 (1967). BIOL: IR, association constants,
        purine, pyrimidine analogs.

1388.   Y. Kyogoku, R. C. Lord, and A. Rich, *Science*, <u>154</u>, 518 (1966).
        BIOL: Hydrogen bonding specificity of nucleic acid purines and
        pyrimidines in solution.

1389.   Y. Kyogoku, M. Tsuboi, and T. Shimanouchi, *Nature*, <u>189</u>, 120
        (1961). BIOL: An IR criterion of the base pairing in nucleic
        acid structure.

1390.   I. A. Kyuntsel and Yu. I. Rozenberg, *Opt. Spektrosk.*, <u>34</u>, 597
        (1973); [English, 341]. Quadrupole relaxation, crystalline
        CHCl$_3$ adducts.

1391.   L. L. Labana and H. M. Sobell, *Proc. Natl. Acad. Sci. U. S.*, <u>57</u>,
        459 (1967). BIOL: X-Ray, 2:1 complex of 1-methyl-5-iodouracil
        and 9-ethyl-2,6-diamino purine.

1392.   M. Labanauskas, P. G. Connors, J. D. Young, R. M. Bock, J. W.
        Anderegg, and W. W. Beeman, *Science*, <u>166</u>, 1530 (1969). BIOL:
        Structural studies on tRNA, crystallographic analysis.

1393.   R. H. Laby and G. F. Walker, *J. Phys. Chem.*, <u>74</u>, 2369 (1970).
        IR, X-Ray: Alkylammonium-vermiculite complexes.

1394.   L. N. Labzovskii, *Teor. Eksp. Khim.*, <u>5</u>, 260 (1969). BIOL:
        Electron correlation, structure and properties of DNA.

1395.   J. Ladik in *Electronic Aspects of Biochemistry*, B. Pullman, ed.,
        Academic Press, 1964, p. 203. BIOL: Quantum mechanical calcu-
        lations on DNA.

1396.   J. Ladik and K. Appel, *Theor. Chim. Acta*, <u>4</u>, 132 (1966). BIOL:
        Pariser-Parr-Pople calculations on DNA constituents.

1397.   J. Ladik and G. Biczó, *Acta Chim.*, <u>63</u>, 53 (1970). BIOL:
        Electronic structure, monosubstituted purines.

1398.   J. Ladik and G. Biczó, *Acta Chim.*, <u>62</u>, 401 (1969). BIOL:
        π-Electron densities, monosubstituted pyrimidines.

1399. J. Ladik, D. K. Rai, and K. Appel, *J. Mol. Spectrosc.*, 27, 72
       (1968). BIOL: Semiempirical SCF-LCAO calculations, homopoly-
       nucleotides.

1400. J. Ladik and K. Sundaram, *J. Mol. Spectrosc.*, 29, 146 (1969).
       Theory: Pariser-Parr-Pople calculations, DNA base pairs.

1401. J. H. Lady and K. B. Whetsel, *J. Phys. Chem.*, 71, 1421 (1967).
       Near-IR: Thermodynamic data, adducts of aniline, N-methylani-
       line with bases.

1402. J. H. Lady and K. B. Whetsel, *Spectrochim. Acta*, 21, 1669
       (1965). IR, Near-IR: Intramolecular, *o*-anilines.

1403. J. H. Lady and K. B. Whetsel, *J. Phys. Chem.*, 68, 1001 (1964).
       Near-IR: Self-association, aniline.

1404. R. F. Lake and H. W. Thompson, *Proc. Roy. Soc. (London)*, A291,
       469 (1966). Far-IR: Self-association, alcohol.

1405. A. V. Lakshminarayanan and V. Sasisekharan, *Biophys. Acta*, 204,
       49 (1970). BIOL: Calculated conformations of a monomer unit
       of a polynucleotide chain, puckering of the ribose moiety.

1406. A. V. Lakshminarayanan and V. Sasisekharan, *Biopolymers*, 8, 489
       (1969). BIOL: Stereochemistry of nucleic acids and polynucleo-
       tides, conformational energy of a ribose-phosphate unit.

1407. A. V. Lakshminarayanan and V. Sasisekharan, *Biopolymers*, 8, 475
       (1969). BIOL: Stereochemistry of nucleic acids and polynucleo-
       tides, conformational energy of base-sugar units.

1408. L. Lamberts, *J. Chim. Phys.*, 62, 1404 (1965). CAL: ΔH, EtOH-
       acetone adduct.

1408a. L. Lamberts, *Z. Physik. Chem. Neue Folge*, 73, 159 (1970).
       CAL: Comparison of pure base and dilution methods.

1408b. L. Lamberts, *Ind. Chim. Belg.*, 36, 347 (1971). CAL: Review.

1409. B. Lamm, *Acta Chem. Scand.*, 19, 2316 (1965). NMR: Intramolecu-
       lar, N-alkylanilines.

1411. R. Langridge, D. A. Marvin, W. E. Seeds, H. R. Wilson, C. W.
       Hooper, M. H. F. Wilkins, and L. D. Hamilton, *J. Mol. Biol.*, 2,
       38 (1960). BIOL: Molecular configuration of DNA.

1412. R. Langridge and A. Rich, *Acta Crystallogr.*, 13, 1052 (1960).
       BIOL: Polyribonucleotide chains can combine together to form
       biological complexes.

1413. R. Langridge, H. R. Wilson, C. W. Hooper, M. H. F. Wilkins, and
       L. D. Hamilton, *J. Mol. Biol.*, 2, 19 (1960). BIOL: X-Ray
       study of lithium salt of DNA.

1414. G. F. Lanthier, W. A. Graham, and A. G. William, *Inorg. Chem.*,
       8, 172 (1969). Self-association of Me_2BOH.

1414a. L. A. LaPlanche, H. B. Thompson, and M. T. Rogers, *J. Phys.
       Chem.*, 69, 1482 (1965). NMR: Self-association, N-monosubsti-
       tuted amides.

1415. C. Laruelle, *Tetrahedron Lett.*, 2235 (1970). NMR: Intramolecu-
       lar, dicoumarols and thiocoumarols.

1416.   J. Lascombe, M. Haurie, and M. L. Josien, *J. Chim. Phys.*, 59,
        1233 (1962). IR: Self-association, monocarboxylic acids.

1416a. B. N. Laskorin, V. V. Yakshin, and B. N. Sharapov, *Dokl. Akad.*
        Nauk SSSR, 211, 350 (1973); *Chem. Abstr.*, 79, 114901f (1973).
        IR: Self-association, dialkyl phosphoramidates.

1417.   J. C. Lassegues, J. C. Cornut, P. V. Huong, and Y. Grenie,
        Spectrochim. Acta, A27, 73 (1971). IR: Solid HX-ether adducts.

1417a. J. C. Lassegues and P. V. Huong, *Chem. Phys. Lett.*, 17, 444
        (1972). IR: Gas phase, $(CH_3)_2O$—HCl.

1418.   J. C. Lassegues and P. V. Huong, *Method. Phys. Anal.*, 5, 69
        (1969). IR: K for adducts of phenol and pyrrole in fourteen
        solvents.

1418a. P. Laszlo in *Progress in Magnetic Nuclear Resonance Spectros-*
        copy, Vol. 3, J. W. Emsley, J. Feeney, and L. H. Sutcliffe,
        eds., Pergamon, Oxford, 1967, pp. 279-310. NMR: Solvent
        effects.

1418b. W. M. Latimer and W. H. Rodebush, *J. Amer. Chem. Soc.*, 42, 1419
        (1920). First definitive paper on hydrogen bonding.

1419.   J. Lauranson and P. Pineau, *J. Chim. Phys.*, 65, 1937 (1968).
        IR: K, 1:1 adducts of diphenylamine and p-bromoaniline with
        bases in CCl_4.

1420.   J. Lauranson, P. Pineau, and J. Lascombe, *J. Chim. Phys.*, 63,
        635 (1966). IR: K, adducts of aromatic amines with pyridine.

1421.   C. Laurence and B. Wojtkowiak, *Compt. Rend. Acad. Sci.*, 264C,
        1216 (1967). IR: $\Delta\nu_{AH}$ versus σ correlation.

1422.   P. C. Lauterbur, *Ann. N. Y. Acad. Sci.*, 70, 841 (1958). NMR:
        Intramolecular, ^{13}C chemical shifts of o-substituted methyl
        benzoates.

1423.   A. Lautie and A. Novak, *J. Chem. Phys.*, 56, 2479 (1972).
        Raman: ν_{NH}, pyrrole.

1423a. A. Lautie and A. Novak, *Can. J. Spectrosc.*, 17, 113 (1972).
        IR, Raman: Crystallized pyrroles at -180°C.

1424.   W. S. Layne, H. H. Jaffé, and H. Zimmer, *J. Amer. Chem. Soc.*,
        85, 1816 (1963). UV: Hydrogen bonds between sulfuric acid and
        N-nitrosoamines.

1425.   W. S. Layne, H. H. Jaffé, and H. Zimmer, *J. Amer. Chem. Soc.*,
        85, 435 (1963). UV: Adducts of N-nitrosoamines with tri-
        chloroacetic acid in cyclohexane.

1425a. G. R. Leader, *Anal. Chem.*, 45, 1700 (1973). NMR: ^{19}F.
        $-C(CF_3)_2OH$ probe.

1425b. S. Leavell and R. F. Curl, Jr., *J. Mol. Spectrosc.*, 45, 428
        (1973). Microwave: Intramolecular, 2-nitrophenol.

1426.   R. Lebedev, *Izv. Vyssh. Ucheb. Zaved., Fiz.*, 15, 148, 159
        (1972); *Chem. Abstr.*, 77, 94906q, 95035s (1972). IR: Self-
        association, tetrahydro-1,3-thiazines, rhodanines, azolidones.

1427. G. I. Lebedeva, *Latv. PSR Zinat. Akad. Vestis*, 26 (1970); *Chem. Abstr.*, 73, 13683g (1970). IR: Aminoindenones in solid phase and in Me_2CO, $HCONMe_2$.

1428. G. I. Lebedeva and J. Freimanis, *Latv. PSR Zinat. Akad. Vestis*, *Kim. Ser.*, 711 (1970); *Chem. Abstr.*, 74, 93010u (1971). IR: Aminoindenones.

1429. G. I. Lebedeva, J. Freimanis, and R. Gaile, *Zh. Fiz. Khim.*, 44, 1450 (1970); *Chem. Abstr.*, 73, 65557p (1970). IR: Cyclic β-amino-vinyl ketones in EtOH, dimethylformamide, and dichloro-ethane.

1430. M. B. Ledger and P. Suppan, *Spectrochim. Acta*, 23A, 641 (1967). UV: Effect of intra- and intermolecular hydrogen bonding on $n \to \pi^*$ band of carbonyl group.

1431. G. C. Y. Lee, J. H. Prestegard, and S. I. Chan, *J. Amer. Chem. Soc.*, 94, 951 (1972). BIOL: PMR, cytosine tautomerism.

1431a. M. K. Lee, M. S. Thesis, University of Wisconsin, 1963. Near-IR: Thermodynamic data, phenol adducts.

1432. W. A. Lees and A. Buraway, *Tetrahedron*, 19, 419 (1963). UV: Effect of hydrogen bonding on electronic spectra.

1433. L. LeGall, A. LeNarvor, J. Lauransan, and P. Saumagne, *Compt. Rend. Acad. Sci.*, *Ser. C*, 268, 1285 (1969). IR: Thermodynamic data, adducts of maleimide and succinimide.

1434. G. W. Lehman and J. P. McTague, *J. Chem. Phys.*, 49, 3170 (1968). BIOL: Melting of DNA, base-pair sequences.

1435. M. S. Lehmann and F. K. Larsen, *Acta Chem. Scand.*, 25, 3859 (1971). Neutron Diffraction: Hydrogen bonds in $KH_3(SeO_3)_2$.

1436. S. S. Lehrer and G. D. Fasman, *J. Amer. Chem. Soc.*, 87, 4687 (1965). UV: Fluorescence spectra of phenol derivatives.

1437. L. A. Leites, N. A. Ogorodnikova, and L. I. Zakharkin, *J. Organometal. Chem.*, 15, 287 (1968). IR: Adducts of decahalo-o-carboranes with Me_2SO and dimethylformamide.

1437a. L. A. Leites, L. E. Vinogradova, N. A. Ogorodnikova, and L. I. Zakharkin, *Zh. Prikl. Spektrosk.*, 16, 488 (1972); *Chem. Abstr.*, 77, 26972h (1972). IR: Thermodynamic data, adducts of B-decachlorocarboranes.

1438. B. Lemanceau, C. Lussan, and N. Souty, *J. Chim. Phys.*, 59, 148 (1962). NMR: *t*-Butanol in various solvents.

1439. B. Lemanceau, C. Lussan, N. Souty, and J. Biais, *J. Chim. Phys.*, 61, 195 (1964). NMR: Self-association, alcohols and amines.

1440. R. U. Lemieux, *Can. J. Chem.*, 39, 116 (1960). BIOL: PMR, configuration and conformation of thymidine in aqueous solution.

1441. A. T. Lemley, J. H. Roberts, K. R. Plowman, and J. J. Lagowski, *J. Phys. Chem.*, 77, 2185 (1973). Raman: Liquid ammonia.

1442. M. Leng, F. Pochon, and A. M. Michelson, *Biochim. Biophys. Acta*, 169, 338 (1968). BIOL: Fluorescence of a variety of modified mono- and dinucleotides at 20°C in H_2O.

1443. P. Lengyel, *Cold Spring Harbor Symp. Quant. Biol.*, <u>34</u>, 828
      (1969). BIOL: Review, translation of m-RNA.

1443a. J. Lennard-Jones and J. A. Pople, *Proc. Roy. Soc.*, <u>A205</u>, 155
       (1951). Theory: Electrostatic calculations of hydrogen bond
       in $(H_2O)_2$.

1443b. H. LeNours, P. Dorval, and P. Saumagne, *Compt. Rend. Acad. Sci.*,
       Ser. C, <u>276</u>, 1703 (1973). IR: Me_2SO adducts, thermodynamic
       data.

1443c. B. R. Lentz and H. A. Scheraga, *J. Chem. Phys.*, <u>58</u>, 5296 (1973).
       Theory: LCAO SCF on $(H_2O)_4$, $(H_2O)_3$.

1444. N. J. Leonard, H. Iwamura, and J. Eisinger, *Proc. Natl. Acad.
      Sci. U. S.*, <u>64</u>, 352 (1969). BIOL: Stacking interactions in
      tRNA, the anticodon-adjacent base.

1445. J. P. Le Rolland and R. Freymann, *Compt. Rend. Acad. Sci., Ser.
      C*, <u>276</u>, 931 (1973). IR, X-Ray: Pyrimidine-purine association.

1445a. J. P. Le Rolland and R. Freymann, *Compt. Rend. Acad. Sci., Ser.
       D*, <u>276</u>, 727 (1973). IR: Self-association, pyrimidine bases.

1446. W. A. Lester and M. Krauss, *J. Chem. Phys.*, <u>52</u>, 4775 (1970).
      Theory: Gaussian calculations of interaction potential, includ-
      ing angular dependence, between Li and HF.

1447. T. M. Letcher, *Spectrum*, <u>8</u>, 372 (1970); *Chem. Abstr.*, <u>73</u>,
      126745x (1970). Review: Influence of hydrogen bonding on
      intermolecular forces in liquids.

1448. Yu. N. Levchuk, V. G. Koval, and I. F. Tsymbal, *Zh. Prikl.
      Spektrosk.*, <u>14</u>, 735 (1971); *Chem. Abstr.*, <u>75</u>, 69004v (1971).
      IR: Intramolecular, steroid ketones.

1449. J. L. Leviel and Y. Marechal, *J. Chem. Phys.*, <u>54</u>, 1104 (1971).
      IR: Anharmonicity in carboxylic acid dimers.

1451. G. C. Levy, *J. Magn. Resonance*, <u>6</u>, 453 (1972). NMR: ^{13}C spin-
      lattice relaxation, self-association of phenol, aniline.

1452. G. C. Levy and S. Winstein, *J. Amer. Chem. Soc.*, <u>90</u>, 3574 (1968).
      NMR: Intramolecular, protonated β-phenyl ketones.

1452a. G. N. Lewis, *Valence and the Structure of Atoms and Molecules*,
       Chemical Catalog Co., New York, 1923, Chapter 12.

1453. V. P. Lezina, V. F. Bystrov, B. E. Zaitsev, N. A. Andronova,
      L. D. Smirnov, and K. M. Dyumaev, *Teor. Eksp. Khim.*, <u>5</u>, 247
      (1969); *Chem. Abstr.*, <u>71</u>, 38169r (1969). NMR: Intramolecular,
      3-hydroxypyridines.

1454. N. C. Li, *American Chemical Society, Div. Fuel Chem., Preprints*,
      <u>11</u>, 241 (1967). NMR: Thermodynamic data for proton donors
      with bases.

1455. W. Libus and W. Moska, *Bull. Acad. Pol. Sci., Ser. Sci. Chim.*,
      <u>17</u>, 675 (1969); *Chem. Abstr.*, <u>73</u>, 3327e (1970). UV: Thermody-
      namic data, adducts of p-nitrophenol with pyridine derivatives.

1456. W. I. Libus and W. Moska, *Bull. Acad. Pol. Sci., Ser. Sci. Chim.*,
      <u>17</u>, 669 (1969); *Chem. Abstr.*, <u>72</u>, 137058f (1970). Electronic:
      Association of p-nitrophenol with pyridine.

1456a. W. Libus and W. Moska, *Bull. Acad. Pol. Sci., Ser. Sci. Chim.*,
       20, 897 (1972). UV: *p*-Nitrophenol adducts.

1457. R. L. Lichter and J. D. Roberts, *J. Phys. Chem.*, 74, 912 (1970).
       NMR: ^{13}C, $CHCl_3$—solvents.

1457a. G. Lichtfus, F. Lemaire, and Th. Zeegers-Huyskens, *Spectrochim.*
       Acta, 28A, 2069 (1972). IR: K, adducts of substituted phenols
       with substituted anilines.

1458. G. Lichtfus and Th. Zeegers-Huyskens, *J. Mol. Struct.*, 9, 343
       (1971). Far-IR: Adducts of C_5H_5N or $C_6H_5NH_2$ with phenols.

1458a. G. Lichtfus and Th. Zeegers-Huyskens, *Spectrochim. Acta*, 28A,
       2081 (1972). IR: ν_{NH_2}, adducts of substituted phenols with
       substituted anilines.

1459. U. Liddel and E. D. Becker, *Spectrochim. Acta*, 10, 70 (1957).
       IR: Estimating fraction of OH protons in ethanol which are not
       involved in hydrogen bonding.

1460. U. Liddel and N. F. Ramsey, *J. Chem. Phys.*, 19, 1608 (1951).
       NMR: Temperature dependence of OH proton resonance of ethanol.

1461. E. J. Lien, J. T. Chou, and G. A. Guadauskas, *Spectrosc. Lett.*,
       5, 293 (1972). NMR: Intramolecular, ^1H chemical shifts for
       benzimidazolinones, comparison with cyclic ureas.

1462. M. N. Lilly, *Leicester Chem. Rev.*, 1, 27 (1962). Review:
       Physical properties of hydrogen bond.

1463. Y. Y. Lim and R. S. Drago, *J. Amer. Chem. Soc.*, 93, 891 (1971).
       CAL, IR, ESR: Adducts of the free radical 2,2,6,6-tetramethyl-
       piperidine-N-oxyl with various alcohols.

1464. M. L. Lin and R. M. Scott, *J. Phys. Chem.*, 76, 587 (1972).
       UV: Thermodynamic data, *p*-Cl-phenol adducts.

1465. S. H. Lin in *Physical Chemistry, An Advanced Treatise*, H. Eyring,
       D. Henderson, and W. Jost, eds., Academic Press, New York, 1970,
       Vol. V, Chapter 8. Review of theory, does not include the newer
       SCF calculations.

1466. T. F. Lin, S. D. Christian, and H. E. Affsprung, *J. Phys. Chem.*,
       69, 2980 (1965). NMR: Self-association, H_2O in 1,2-dichloro-
       ethane.

1467. Tien-Sung Lin and E. Fishman, *Spectrochim. Acta*, A23, 491 (1967).
       IR: Intramolecular, ΔH for *o*-halophenols and deuterated deriva-
       tives in vapor phase.

1468. W. Lin and S. Tsay, *J. Phys. Chem.*, 74, 1037 (1970). NMR:
       K for 1:1 complexes of $CHCl_3$.

1469. J. J. Lindberg and C. Majani, *Suomen Kemistilehti*, 38B, 21
       (1965). IR: Phenol adducts of Me_2SO.

1470. J. J. Lindberg and C. Majani, *Suomen Kemistilehti*, 37B, 21
       (1964). Raman, IR: Association of acetic acid with Me_2SO.

1471. J. J. Lindberg and C. Majani, *Acta Chem. Scand.*, 17, 1477 (1963).
       Raman: Me_2SO—H_2O mixtures.

1471a. R. Lindemann and G. Zundel, *J. Chem. Soc., Faraday 2*, <u>68</u>, 979
(1972). IR: Strong hydrogen bonds, proton transfer, adducts
of carboxylic acids with heterocyclic nitrogen bases.

1471b. P. Lindner and J. R. Sabin, *Int. J. Quantum Chem.*, <u>7</u>, 301 (1973).
Theory: CNDO/2, short A-type hydrogen bond.

1472. R. H. Linnell, *J. Chem. Phys.*, <u>34</u>, 698 (1961). UV: Pyridazine-
water adducts.

1473. R. H. Linnell, M. Aldo, and F. Raab, *J. Chem. Phys.*, <u>36</u>, 1401
(1962). NMR: Self-association, pyrrolidine.

1474. J. W. Linnett, *The Electronic Structures of Molecules*, Methuen
and Co., Ltd., London, 1964. Theory.

1474a. B. G. Liorber, M. P. Sokolov, Z. M. Khammatova, and A. I.
Razumov, *Zh. Obshch. Khim.*, <u>43</u>, 438 (1973); [English, p. 435].
IR: Intramolecular, phosphorylated glycols.

1475. T. E. Lipatova and Yu. N. Nizel'skii, *Teor. Eksp. Khim.*, <u>4</u>, 662
(1968); *Chem. Abstr.*, <u>70</u>, 14911e (1969). UV: Association of
methanol with copper bisacetylacetonate.

1476. A. A. Lipovskii and T. A. Dem'yanova, *Zh. Prikl. Spektrosk.*, <u>15</u>,
550 (1971). IR: Association of alkylammonium ions with alco-
hols and phenols.

1477. A. A. Lipovskii and T. A. Dem'yanova, *Zh. Prikl. Spektrosk.*, <u>9</u>,
239 (1968); *Chem. Abstr.*, <u>70</u>, 15587r (1969). IR: Thermodynamic
data for association of anions with aliphatic alcohols.

1478. E. Lippert and D. Oechssler, *Z. Physik. Chem. (Frankfurt)*, <u>29</u>,
403 (1961). UV: Self-association, acetic acid and halogenated
acetic acids.

1479. E. Lippert, D. Oechssler, and H. Feldbauer, *Z. Physik. Chem.
(Frankfurt)*, <u>29</u>, 397 (1961). NMR: Trifluoroacetic acid and
perfluoromethylcyclohexane.

1480. E. Lippert and W. Schroler, *Ber. Bunsenges, Phys. Chem.*, <u>73</u>,
1027 (1969). IR: Adducts of phenol with 1,1,1-tri-chloroethane.

1481. E. R. Lippincott and R. Schroeder, *J. Chem. Phys.*, <u>23</u>, 1099
(1955). One-dimensional model of the hydrogen bond.

1482. E. R. Lippincott and A. Srinivasa Rao, *J. Chem. Phys.*, <u>41</u>, 3006
(1964). NMR: Dependence of line broadening in hydrogen-bonded
solids on the hydrogen bond length.

1483. E. R. Lippincott, C. E. Weir, and A. V. Valkenburg, Jr., *J. Chem.
Phys.*, <u>32</u>, 612 (1960). IR: Self-association, H_2O.

1484. M. N. Lipsett, L. A. Heppel, and D. F. Bradley, *J. Biol. Chem.*,
<u>236</u>, 857 (1961). BIOL: Complex formation between oligonucleo-
tides and polymers, stabilities of small lengths of hydrogen-
bonded helices.

1485. M. N. Lipsett, L. A. Heppel, and D. F. Bradley, *Biochim. Biophys.
Acta*, <u>41</u>, 173 (1960). BIOL: Complex formation between poly U
and adenine riboöligonucleotides as short as the dinucleotide.

1486. W. Liptay, *Z. Elektrochem.*, <u>65</u>, 375 (1961). Spectroscopic
method for determining equilibrium constants.

1487. B. I. Lirova, A. L. Smolyanskii, and A. A. Tager, *Zh. Prikl. Spektrosk.*, 15, 491 (1971); *Chem. Abstr.*, 76, 46594u (1972). IR: Poly (vinyl alcohol), effect of Me_2SO.

1488. W. M. Litchman, M. Alei, Jr., and A. E. Florin, *J. Amer. Chem. Soc.*, 91, 6574 (1969). NMR: ^{15}N chemical shifts of NH_3 in various solvents.

1489. W. M. Litchman, M. Alei, Jr., and A. E. Florin, *J. Chem. Phys.*, 50, 1031 (1969). NMR: ^{15}N chemical shifts of NH_3 and ND_3 with temperature.

1491. A. B. Littlewood and F. W. Willmott, *Anal. Chem.*, 38, 1031 (1966). Gas-Liquid Chromatography: Application of GLC to solute-solvent interactions.

1492. A. B. Littlewood and F. W. Willmott, *Trans. Faraday Soc.*, 62, 3287 (1966). NMR: Self-association of 1-dodecanol in squalene.

1492a. C. H. Lochmueller, J. M. Harris, and R. W. Souter, *J. Chromatogr.*, 71, 405 (1973). NMR: Chiral ureide-amide systems.

1494. G. H. Loew and S. Chang, *Theor. Chim. Acta*, 27, 273 (1972). Theory: Extended Hückel on $(HCN)_4$.

1495. J. W. Longsworth, *Photochem. Photobiol.*, 8, 589 (1968). BIOL: Instrumentation and techniques for measuring fluorescence and phosphorescence of biological materials.

1496. M. C. S. Lopes and H. W. Thompson, *Spectrochim. Acta*, 24A, 1367 (1968). IR: Thermodynamic data, phenol and *t*-butanol adducts of nitriles.

1497. R. C. Lord and T. J. Porro, *Z. Elektrochem.*, 64, 672 (1960). IR: Dimerization of caprolactam, K, ΔH, ΔS.

1498. R. C. Lord and G. J. Thomas, Jr., *Develop. Appl. Spectrosc.*, 6, 179 (1967). BIOL, IR: Base-pair interactions in nonaqueous solvents.

1499. V. Lorenzelli, *Ann. Chim. (Rome)*, 53, 1018 (1963). Far-IR: Spectra of liquid RCOOH (R = H, CH_3, C_2H_5, C_3H_7).

1500. V. Lorenzelli and A. Alemagna, *Compt. Rend. Acad. Sci.*, 257, 2977 (1963). Far-IR: Self-association of pyrrole.

1501. V. Lorenzelli and A. Alemagna, *Compt. Rend. Acad. Sci.*, 256, 3626 (1963). Near-IR; Carboxylic acids.

1502. H. Loth and D. Beer, *Arch. Pharm. (Weinheim)*, 304, 65 (1971); *Chem. Abstr.*, 74, 87192p (1971). NMR, IR: Adduct of 1-methylbarbital with dimethylacetamide.

1503. J. E. Lowder, *J. Quant. Spectrosc. Radiat. Transfer*, 11, 153 (1971). IR: Self-association, H_2O, increase in integrated intensities.

1504. J. E. Lowder, L. A. Kennedy, K. G. P. Sulzmann, and S. S. Penner, *J. Quant. Spectrosc. Radiat. Transfer*, 10, 17 (1970). IR: Self-association, H_2S in gas phase.

1505. P.-O. Löwdin in *Molecular Associations in Biology*, B. Pullman, ed., Academic Press, 1968, p. 539. BIOL: Discussion, theoretical approach to molecular associations in biology.

1506. P.-O. Löwdin, *Adv. Quant. Chem.*, 2, 213 (1965). BIOL: Review
       and discussion, quantum theory of the DNA molecule.

1507. P.-O. Löwdin in *Electronic Aspects of Biochemistry*, B. Pullman,
       ed., Academic Press, New York, 1964, p. 167. BIOL: DNA
       replication, incorporation errors and proton transfer.

1508. P.-O. Löwdin, *Biopolymers, Symp. No. 1*, 161 (1964). Theory:
       Proton tunneling in DNA.

1509. P.-O. Löwdin, *Biopolymers, Symp. No. 1*, 293 (1964). BIOL:
       Some aspects of quantum biology.

1510. P.-O. Löwdin, *Rev. Mod. Phys.*, 35, 724 (1963). Theory:
       Proton tunneling in DNA.

1511. A. Lowenstein and Y. Margalit, *J. Phys. Chem.*, 69, 4152 (1965).
       NMR: Thermodynamic data, MeOH adducts of CH_3CN, CH_3NC.

1512. J. Lower and J. Ferguson, *J. Org. Chem.*, 30, 3000 (1965). NMR,
       IR: Intramolecular, benzoylacetones.

1513. B. Lubas, T. Wilezok, and O. K. Daszkiewicz, *Biopolymers*, 5,
       967 (1967). BIOL: Thermal transition of DNA measured by NMR
       spin-echo technique.

1514. G. Luck and C. Zimmer, *Biochem. Biophys. Acta*, 169, 466 (1968).
       BIOL: ORD, DNA from various sources at neutral and acidic pH.

1515. G. Luck, C. Zimmer, and G. Snatzke, *Biochem. Biophys. Acta*, 169,
       548 (1968). BIOL: CD of protonated DNA.

1516. W. A. P. Luck, *Med. Welt*, 87 (1970). Review: Hydrogen bonding
       in H_2O.

1517. W. A. P. Luck, *Naturwiss Rundsch.*, 21 (6), 236 (1968). BIOL:
       Hydrogen bonding discussed as it may relate to biochemical
       systems.

1518. W. A. P. Luck, *Naturwissenschaften*, 54, 601 (1967). Review,
       water.

1519. W. A. P. Luck, *Ber. Bunsenges, Phys. Chem.*, 69, 626 (1965).
       Near-IR: Self-association, H_2O.

1520. W. A. P. Luck, *Z. Elektrochem.*, 68, 895 (1964). IR: Self-
       association, H_2O.

1521. W. A. P. Luck, *Z. Elektrochem.*, 67, 186 (1963). Near-IR:
       Self-association, H_2O.

1522. W. A. P. Luck and W. Ditter, *Ber. Bunsenges, Phys. Chem.*, 75,
       163 (1971). IR: Intramolecular, unsaturated alcohols.

1523. W. A. P. Luck and W. Ditter, *Ber. Bunsenges, Phys. Chem.*, 72,
       365 (1968). Near-IR: Self-association, CH_3OH and C_2H_5OH near
       critical point.

1524. F. I. Luknitskii and B. A. Vovsi, *Zh. Org. Khim.*, 4, 2055 (1968);
       Chem. Abstr., 70, 28231v (1969). NMR: Association of β-
       trichloromethyl-β-propiolactone with aldoximes.

1525. H. Lumbroso, *J. Chim. Phys.*, 61, 132 (1964). Dipole moments of
       pyrrole-base adducts.

1526. L. Lunazzi and F. Taddei, *Spectrochim. Acta*, 24A, 1479 (1968).
       NMR: Association shifts of CHCl₃ with bases.

1527. J. O. Lundgren, *Acta Crystallogr.*, *Sect. B*, 28, 1684 (1972).
       X-Ray: Picrylsulfonic acid tetrahydrate.

1527a. J. O. Lundgren and J. M. Williams, *J. Chem. Phys.*, 58, 788
       (1973). Neutron Diffraction: *p*-Toluene sulfonic acid monohy-
       drate.

1528. S. Lunell and G. Sperber, *J. Chem. Phys.*, 46, 2119 (1967).
       Theory: Motion of protons in hydrogen bonds of DNA base pairs,
       a semiempirical study.

1529. C. Lussan, *J. Chim. Phys.*, 60, 1100 (1963). NMR: Self-
       association and 1:1 adducts of alcohols.

1530. C. Lussan, B. Lemanceau, and N. Souty, *Compt. Rend. Acad. Sci.*,
       254, 1980 (1962). NMR: *t*-Butanol adducts of acetone and
       dioxane.

1530a. P. Lutgen, M. P. Van Damme, and Th. Zeegers-Huyskens, *Bull. Soc.
       Chim. Belges*, 75, 824 (1966). IR: K, adducts of substituted
       phenols with triethylamine.

1531. A. E. Lutskii, *Teor. Eksp. Khim.*, *Akad. Nauk Ukr. SSR*, 1, 815
       (1965); *Chem. Abstr.*, 64, 8937b (1966). Theory: Relation
       between overlap integral and hydrogen bond formation.

1531a. A. E. Lutskii, *Zh. Strukt. Khim.*, 13, 534 (1972). IR: Badger-
       Bauer correlation.

1532. A. E. Lutskii, *Zh. Obshch. Khim.*, 32, 4099 (1962). Theory:
       On the nature of forces in the hydrogen bond.

1532a. A. E. Lutskii, L. P. Batrakova, L. A. Fedotova, and I. P.
       Kovalev, *Zh. Prikl. Spektrosk.*, 17, 138 (1972). IR: Δν for
       diphenylamine adducts.

1532b. A. E. Lutskii, L. P. Batrakova, and L. A. Fedotova, *Zh. Obshch.
       Khim.*, 42, 1820 (1972); *Chem. Abstr.*, 78, 15130b (1973). IR:
       Phenol adducts of Group VI donors.

1533. A. E. Lutskii and E. I. Goncharova, *Opt. Spektrosk.*, supplement
       3, 198 (1967); [English, 98 (1968)]. IR: Self-association,
       aniline, cyclohexylamine, *n*-hexylamine.

1533a. A. E. Lutskii and E. I. Goncharova, *Ukr. Khim. Zh.*, 38, 1223
       (1972); *Chem. Abstr.*, 78, 96971b (1973). IR: (C₂H₅)₂S adducts.

1534. A. E. Lutskii, V. A. Granzhan, and S. V. Semenenko, *Zh. Strukt.
       Khim.*, 10, 56 (1969); *Chem. Abstr.*, 70, 101309h (1969). IR:
       Intramolecular, 2,6-disubstituted phenols.

1535. A. E. Lutskii, V. A. Granzhan, Ya. A. Shuster, and P. M. Zaitsev,
       Zh. Prikl. Spektrosk., 11, 913 (1969); *Chem. Abstr.*, 72, 66042h
       (1970). IR: Intramolecular, unsaturated alcohols.

1536. A. E. Lutskii and V. N. Konelskaya, *Zh. Obshch. Khim.*, 30, 3773,
       3782 (1960). UV: Intramolecular, N-methyl and N-phenyl substi-
       tuted nitrobenzenes, nitroacetanilides and nitro-N-acetyldi-
       phenylamines.

1537. A. E. Lutskii, I. I. Men'shova, L. A. Fedotova, and M. G.
       Voronkov, *Zh. Obshch. Khim.*, 39, 879 (1966); *Chem. Abstr.*, 71,
       43916m (1969). UV: Picric acid adducts with amines.

1538. A. E. Lutskii and I. S. Romodanov, *Spektrosk. At. Mol.*, 366
       (1969); *Chem. Abstr.*, 74, 17818y (1971). NMR: Intramolecular,
       proton hyperfine splitting of OH.

1539. A. E. Lutskii and S. V. Semenenko, *Zh. Fiz. Khim.*, 43, 356
       (1969); *Chem. Abstr.*, 70, 101318k (1969). IR: Intramolecular,
       intermolecular *o*- and *p*-substituted phenols and anilines associ-
       ated with dimethylformamide.

1540. A. E. Lutskii and G. I. Sheremeteva, *Opt. Spektrosk.*, supplement
       3, 20 (1967); [English, 11 (1968)]. IR: Association of X—H
       with π electrons (X = O, S, N).

1540a. A. E. Lutskii, Ya. A. Shuster, V. A. Granzhan, and P. M. Zaitsev,
       Zh. Prikl. Spektrosk., 16, 673 (1972); *Chem. Abstr.*, 77, 40884n
       (1972). IR: Correlation of $\Delta\nu$ with ionization potential for
       alcohol adducts.

1540b. A. E. Lutskii, Ya. A. Shuster, V. A. Granzhan, and P. M. Zaitsev,
       Zh. Prikl. Spektrosk., 16, 870 (1972); *Chem. Abstr.*, 77, 66434r
       (1972). IR: Thermodynamic data, adducts of R_3COH with
       $HC(O)N(CH_3)_2$.

1541. A. E. Lutskii and Z. A. Tret'yak, *Ukr. Khim. Zh.*, 34, 1095
       (1968); *Chem. Abstr.*, 70, 67439p (1969). UV: Spectra of *o*-
       and *p*-nitrophenol in Me_2CO, dioxane, Et_2O, and Et_3N.

1542. A. E. Lutskii and S. N. Vragova, *Zh. Prikl. Spektrosk.*, 15, 935
       (1971); *Chem. Abstr.*, 76, 65938a (1972). IR: Phenol adducts
       of C_3H_7Cl, CH_3NO_2, and CH_3CN.

1543. Z. Luz and G. Yagil, *J. Phys. Chem.*, 70, 554 (1966). NMR: ^{17}O
       NMR spectra for aqueous solutions of electrolytes.

1544. V. Luzzati, D. Luzzati, and F. Masson, *J. Mol. Biol.*, 5, 375
       (1962). BIOL: Structure of DNA in solution by small-angle
       X-ray scattering techniques.

1545. R. E. Lyle, D. H. McMahon, W. E. Krueger, and C. K. Spicer,
       J. Org. Chem., 31, 4164 (1966). IR, NMR: Intramolecular, 3-
       piperidinol derivatives.

1546. B. M. Lynch, C. M. Chen, and Y. Y. Wigfield, *Can. J. Chem.*, 46,
       1141 (1968). NMR: Intramolecular, *o*-nitro-acetanilides.

1547. B. M. Lynch, B. C. Macdonald, and J. G. K. Webb, *Tetrahedron*,
       24, 3895 (1968). NMR: Correlation of amino proton shifts with
       σ for anilines in $(CH_3)_2SO$.

1548. B. J. McCarthy and R. M. Church, *Ann. Rev. Biochem.*, 39, 131
       (1970). BIOL: Review, molecular hybridization reactions,
       DNA/DNA and DNA/RNA.

1549. A. L. McClellan, *J. Chem. Educ.*, 44, 547 (1967). Review:
       Significance of hydrogen bonds in biological structures.

1550. A. L. McClellan and S. W. Nicksic, *J. Phys. Chem.*, 69, 446
       (1965). NMR: Self-association and intermolecular association
       of halomethanes and haloethanes.

1551. A. L. McClellan, S. W. Nicksic, and J. C. Guffy, *J. Mol. Spec-
       trosc.*, 11, 340 (1963). NMR: CHCl₃—Me₂SO adduct.

1552. B. McConnell and P. H. Von Hippel, *J. Mol. Biol.*, 50, 317 (1970).
       BIOL: Rate of exchange of hydrogen-bonded hydrogen atoms in DNA.

1553. B. McConnell and P. H. Von Hippel, *J. Mol. Biol.*, 50, 297 (1970).
       BIOL: Hydrogen exchange as a probe of the dynamic structure of
       DNA, general acid-base catalysis.

1554. L. L. McCoy, *J. Amer. Chem. Soc.*, 89, 1673 (1967). Geometry of
       intramolecular hydrogen bonding in 1,2-dicarboxylic acids.

1556. D. H. McDaniel and W. G. Evans, *Inorg. Chem.*, 5, 2180 (1966).
       Experimental: Hydrogen bond in HSHSH⁻ is greater than 7 kcal/
       mole and perhaps as large as 14 kcal/mole.

1557. D. H. McDaniel and R. E. Valleé, *Inorg. Chem.*, 2, 996 (1963).
       Thermodynamic data for ClHCl⁻, BrHBr⁻, IHI⁻.

1558. N. T. McDevitt, R. E. Witkowski, and W. G. Fateley, *Nat. Bur.
       Stand. (U. S.), Spec. Publ. 1967*, No. 301, 339-41. Far-IR:
       Methods, lattice parameter, high-pressure diamond cell, tempera-
       ture studies.

1559. C. C. McDonald and W. D. Phillips, *Magn. Resonance Biol. Syst.,
       Proc. Int. Conf., 2nd, Stockholm*, 3 (1966). BIOL: NMR, poly A
       in neutral aqueous solution at room temperature probably exists
       in a single-stacked, flexible rod-like conformation.

1560. M. P. McDonald and L. D. R. Wilford, *Spectrochim. Acta*, 29A,
       1407 (1973). IR: Pinacol hydrates.

1560a. T. R. R. McDonald, *Acta Crystallogr.*, 13, 113 (1960). X-Ray:
       NH₄HF₂.

1561. A. O. McDougall and F. A. Long, *J. Phys. Chem.*, 66, 429 (1962).
       Relative hydrogen bonding of deuterium.

1561a. W. McFarlane and D. S. Rycroft, *Mol. Phys.*, 24, 893 (1972).
       NMR: Effect of deuterium substitution on P—H coupling con-
       stants in (MeO)₂P(O)H.

1561b. F. R. McFeely and G. A. Somorjai, *J. Phys. Chem.*, 76, 914
       (1973). Vaporization kinetics of hydrogen-bonded liquids.

1562. S. McGavin, *Bull. Math. Biophys.*, 31, 797 (1969). BIOL: A
       four-strand structure for DNA is indicated as stereochemically
       possible.

1562a. B. L. McGaw and J. A. Ibers, *J. Chem. Phys.*, 39, 2677 (1963).
       Neutron Diffraction: NaHF₂ and NaDF₂.

1563. D. E. McGreer and M. M. Mocek, *J. Chem. Educ.*, 40, 358 (1963).
       NMR: Self-association, alcohols.

1564. R. F. McGuire, F. A. Momany, and H. A. Scheraga, *J. Phys. Chem.*,
       76, 375 (1972). Theory: Hydrogen bond potential functions from
       CNDO/2 calculations for use in polypeptide structure calcula-
       tions.

1565. R. T. McIver, Jr. and J. R. Eyler, *J. Amer. Chem. Soc.*, 93, 6334
       (1971). Gas-phase acidities of H₂S and HCN.

1566. P. C. McKinney and G. M. Barrow, *J. Chem. Phys.*, 31, 294 (1959). Theory: Approximate MO treatment.

1567. R. D. McLachlan and R. A. Nyquist, *Spectrochim. Acta*, 20, 1397 (1964). IR: Intramolecular, α-substituted secondary amides.

1568. C. S. McLaughlin, J. Dondon, M. Grunberg-Manago, A. M. Michelson, and G. Sanders, *J. Mol. Biol.*, 32, 521 (1968). BIOL: Thermal stability of the mRNA-amino acyl tRNA-ribosome complex.

1568a. A. D. McLean and M. Yoshimine, *Suppl. to IBM J. Research*, (1967). Theory: SCF calculations on HF_2^-.

1569. J. C. McManus, Y. Harano, and M. J. D. Low, *Can. J. Chem.*, 47, 2545 (1969). IR: Adsorbed acetone on silica.

1570. E. G. McRae, *J. Phys. Chem.*, 61, 562 (1957). Equation relating absorption frequency to intrinsic solvent effects.

1571. B. Macchia, F. Macchia, and L. Monti, *Gazz. Chim. Ital.*, 100, 35 (1970). IR: Intramolecular, alkyl and aryl substituted cyclohexanols.

1571a. A. L. Macdonald, J. C. Speakman, and D. Hadži, *J. Chem. Soc. Perkin Trans.* 2, 68, 825 (1972). Neutron Diffraction: $KH(CF_3CO)_2$ and $KD(CF_3CO)_2$, short symmetrical hydrogen bonds.

1572. G. E. Maciel, P. D. Ellis, and D. C. Hofer, *J. Phys. Chem.*, 71, 2160 (1967). NMR: Deuterium isotope effect.

1573. G. E. Maciel and R. V. James, *J. Amer. Chem. Soc.*, 86, 3893 (1964). NMR: ^{13}C chemical shifts for phenol in hydrogen bonding solvents.

1574. G. E. Maciel and R. V. James, *Inorg. Chem.*, 3, 1650 (1964). NMR: ^{31}P chemical shifts of $(C_6H_5)_3PO$ in hydrogen bonding solvents.

1575. G. E. Maciel and J. J. Natterstad, *J. Chem. Phys.*, 42, 2752 (1965). NMR: ^{13}CO chemical shifts of ketones in hydrogen bonding solvents.

1576. G. E. Maciel and G. C. Ruben, *J. Amer. Chem. Soc.*, 85, 3903 (1963). NMR: Solvent effects, ^{13}C chemical shift, carbonyl group of acetone.

1577. G. E. Maciel and G. B. Savitsky, *J. Phys. Chem.*, 68, 437 (1964). NMR: Intramolecular, ^{13}C chemical shifts.

1578. G. E. Maciel and D. D. Traficante, *J. Amer. Chem. Soc.*, 88, 220 (1966). NMR: ^{13}C chemical shifts, C=O of CH_3COOH.

1580. R. M. MacQueen, J. A. Eddy, and P. J. Lena, *Nature*, 220, 1112 (1968). Far-IR: Water vapor dimer lines at 49.50 and 21.21 cm^{-1} in atmosphere.

1581. M. E. Magar and R. F. Steiner, *Biochim. Biophys. Acta*, 224, 80 (1970). BIOL: Enthalpy of association of nucleosides by vapor pressure osmometry.

1582. M. D. Magee and S. Walker, *J. Chem. Phys.*, 55, 3068 (1971). Dielectric: Intramolecular, *o*-substituted phenols.

1582a. M. D. Magee and S. Walker, *J. Chem. Soc. Faraday 2*, <u>69</u>, 161
       (1973). Relaxation times, 2-nitrophenol, 8-hydroxyquinoline
       in different solvents.

1583. M. D. Magee and S. Walker, *J. Chem. Phys.*, <u>50</u>, 1019 (1969).
       Dielectric rotational lifetimes for chloroform adducts.

1584. L. B. Magnusson, *Mol. Phys.*, <u>21</u>, 571 (1971). IR: Structure of
       $(H_2O)_2$.

1585. L. B. Magnusson, *J. Phys. Chem.*, <u>74</u>, 4221 (1970). IR: Spectrum
       of water dimer in CCl_4, cyclic structure.

1586. M. M. Maguire and R. West, *Spectrochim. Acta*, <u>17</u>, 369 (1961).
       Near-IR: Self-association, phenol, p-cresol, and p-chlorophenol.

1586a. T. O. Maier and R. S. Drago, *Inorg. Chem.*, <u>11</u>, 1861 (1972).
       Evaluation of spectrophotometric and calorimetric methods for
       use with multiple equlibria.

1587. W. Maier, *J. Chim. Phys.*, <u>61</u>, 239 (1964). Ultrasonic Absorp-
       tion: Self-association of benzoic acid.

1588. W. Maier, *Z. Elektrochem.*, <u>64</u>, 132 (1960). Kinetics: Self-
       association, benzoic acid.

1589. V. Maitra, S. N. Cohen, and J. Hurwitz, *Cold Spring Harbor Symp.
       Quant. Biol.*, <u>31</u>, 113 (1966). BIOL: Specificity of initiation
       and synthesis of RNA and DNA templates.

1590. L. I. Maklakov and V. N. Nikitin, *Opt. Spektrosk.*, <u>18</u>, 509
       (1964); [English, 286]. IR: Crystalline formic acid.

1590a. Z. B. Maksimovic, A. Miksa-Spiric, and S. V. Ribnikar, *J. Inorg.
       Nucl. Chem.*, <u>35</u>, 1239 (1973). NMR: Thermodynamic data, $CHCl_3$
       adducts with ketones.

1591. E. Malawer and C. Marzzacco, *J. Mol. Spectrosc.*, <u>46</u>, 341 (1973).
       Electronic: Benzophenone in $EtOH-H_2O$ mixtures.

1591a. J. Malecki and J. Jadzyn, *Magn. Resonance Relat. Phenomena,
       Proc. Congr. AMPERE, 16th 1970*, I. Ursu, ed., p. 525; *Chem.
       Abstr.*, <u>78</u>, 21450x (1973). Solvent effects.

1592. Yu. I. Malyshenko, *Radiotekk. Elektron.*, <u>14</u>, 522 (1969).
       Radiowave: Absorption coefficients of H_2O at 1.35 mm.

1592a. L. P. Mamchur, G. I. Elagin, and I. D. Zaikin, *Zh. Fiz. Khim.*,
       <u>47</u>, 1861 (1973); *Chem. Abstr.*, <u>79</u>, 114982h (1973). IR: Self-
       association, ethers of glycerol alkoxyhydrins.

1593. F. D. Mamedov, V. M. Tatevskii, and M. A. Salimov, *Zh. Prikl.
       Spektrosk.*, <u>10</u>, 620 (1969); *Chem. Abstr.*, <u>71</u>, 25206a (1971).
       UV: Association of phenol derivatives with Et_3N.

1594. V. I. Mamonov, G. G. Dvoryantseva, I. P. Boiko, E. B. Sysoeva,
       and B. V. Unkovskii, *Zh. Org. Khim.*, <u>5</u>, 1299 (1969); *Chem.
       Abstr.*, <u>71</u>, 101134t (1969). IR: Intramolecular, 1,3-dimethyl-
       and 1,2,5-trimethyl-4-(alkoxycarbonyl)-4-piperidinols.

1595. L. Manojlovic-Muir, *Nature*, <u>224</u>, 686 (1969). Neutron Diffrac-
       tion: O—H...S bond in $BaS_2O_3 \cdot H_2O$.

1596. R. P. Marchi and H. Eyring, *J. Phys. Chem.*, <u>68</u>, 221 (1964).
       Theory: Structure of liquid water.

1596a. M. M. Marciacq-Rousselot, *Ann. Chim. (Paris)*, 6, 367 (1971). NMR: Intra- and intermolecular, aliphatic thiols.

1597. S. H. Marcus and S. I. Miller, *J. Phys. Chem.*, 68, 331 (1964). NMR: Self-association constants of thiols.

1598. Y. Marechal, *Chem. Phys. Lett.*, 13, 237 (1972). IR, Theory: Anharmonicity of ν_{AH} in hydrogen bonds.

1599. Y. Marechal, *Commis. Energ. At. (Fr.) Rapp.*, 1969, CEA-R-3683, 111 pp.; *Chem. Abstr.*, 71, 26205m (1969). IR: Theory of IR spectra of hydrogen-bonded systems.

1600. Y. Marechal and A. Witkowski, *J. Chem. Phys.*, 48, 3697 (1968). IR, Theory: Coupling of ν_{OH} in carboxylic acid dimers, theory, band broadening.

1601. Y. Marechal and A. Witkowski, *Theor. Chim. Acta*, 9, 116 (1967). Theory: Coupling of ν_{AH} with hydrogen bond vibrations.

1602. Y. Marechal and W. Witkowski, *J. Chim. Phys. Physicochim. Biol.*, 65, 1279 (1968). Theory: Vibrational spectra of hydrogen-bonded crystals.

1602a. E. M. Marek *et al.*, *Zh. Prikl. Spektrosk.*, 19, 130 (1973); *Chem. Abstr.*, 79, 65304c (1973). IR: $\Delta\nu$, phenol adducts of poly-fluorinated ketones and esters.

1602b. M. G. Marenchic and J. M. Sturtevant, *J. Phys. Chem.*, 77, 544 (1973). CAL: Self-association, purine bases in H_2O.

1603. R. A. Marino, *J. Chem. Phys.*, 57, 4560 (1972). NQR: Hexamethyl-enetetramine triphenol, ^{14}N.

1604. R. E. Marsh in *Structural Chemistry and Molecular Biology*, A. Rich and N. Davidson, eds., W. F. Freeman, San Francisco, 1968, p. 484. BIOL: Discussion, hydrogen bonding in purine and pyrimidine bases.

1604a. K. M. Marstokk and H. Mollendal, *J. Mol. Struct.*, 16, 259 (1973). Microwave: Intramolecular, glycolaldehydes.

1604b. D. Martin and K. Oehler, *J. Prakt. Chem.*, 314, 93 (1972). IR: Thermodynamic data, phenol adducts with π donors.

1605. J. Martin, *Compt. Rend. Acad. Sci.*, 256, 3651 (1963). NMR: Self-association of 3-ethylpentan-3-ol.

1606. M. Martin, *J. Chim. Phys.*, 59, 736 (1962). NMR: Self-association, substituted phenols and alcohols.

1607. M. Martin and M. Quilbeuf, *Compt. Rend. Acad. Sci.*, 252D, 4151 (1961). NMR: Self-association, phenols and alcohols.

1608. D. E. Martire and P. Riedl, *J. Phys. Chem.*, 72, 3478 (1968). GLC: ΔH, alcohols associated with di-*n*-octyl ether and di-*n*-octyl ketone.

1609. G. Marx and M. Suliman, *Z. Chem.*, 6, 353 (1966). Theory: Tunneling frequency in hydrogen bonds.

1609a. M. P. Marzocchi, C. W. Fryer, and M. Bambagiotti, *Spectrochim. Acta*, 21, 155 (1965). Far-IR: Hexamethylenetetramine•HX.

1610. C. Marzzacco, *J. Amer. Chem. Soc.*, 95, 1774 (1973). UV:
       Pyrazine in EtOH—H$_2$O at 77°K, phosphorescence, absorption
       spectra.

1610a. A. A. Mashkovskii and S. E. Odinokov, *Dokl. Akad. Nauk SSSR*,
       204, 1165 (1972); *Chem. Abstr.*, 77, 94912p (1972). IR: Strong
       hydrogen bonds and proton transfer, RCOOH-base.

1610b. F. N. Masri and J. L. Wood, *J. Mol. Struct.*, 14, 201 (1972).
       IR: [C$_5$H$_5$NHN(CH$_3$)$_3$]$^+$.

1610c. F. N. Masri and J. L. Wood, *J. Mol. Struct.*, 14, 217 (1972).
       IR: [(CH$_3$)$_3$NHN(CH$_3$)$_3$]$^+$.

1611. W. Masschelein, *Spectrochim. Acta*, 18, 1557 (1962). IR: Self-
       association, cyclohexanol.

1612. J. Massoulie, *J. Chim. Phys. Physicochim. Biol.*, 65, 66 (1968).
       BIOL: Helical combinations of poly A and poly U in acid medium.

1613. N. Mataga, *Bull. Chem. Soc. Jap.*, 36, 654 (1963). UV: Fluores-
       cence and absorption spectra of naphthylamines and aminobenzoic
       acids.

1614. N. Mataga and Y. Kaifu, *Mol. Phys.*, 7, 137 (1963). UV: Evidence
       for charge-transfer in β-naphthol-triethylamine adduct from
       fluorescence measurements.

1615. N. Mataga, Y. Kawasaki, and Y. Torihashi, *Theor. Chim. Acta*, 2,
       168 (1964). UV: Mechanism of proton-transfer reactions.

1615a. R. L. Matcha, *J. Amer. Chem. Soc.*, 95, 7505 (1973). Relativis-
       tic effects in chemical bond.

1616. J. L. Mateos, R. Cetina, and O. Chao, *Chem. Commun.*, 519 (1965).
       NMR: K, adducts of dimethylamine, *m*-chloroaniline, piperidine,
       and pyrrolidine with pyridine.

1617. J. L. Mateos, R. Cetina, and O. Chao, *Bol. Inst. Quim. Univ. Nal.
       Auton. Mex.*, 17, 189 (1965). NMR: K$_2$, self-association of
       amines.

1618. F. S. Mathews and A. Rich, *J. Mol. Biol.*, 8, 89 (1964). BIOL,
       X-Ray: 9-Ethyladenine and 1-methyluracil 1:1 complex.

1619. F. S. Mathews and A. Rich, *Nature*, 201, 179 (1964). BIOL, X-Ray:
       Crystal and molecular structure of 1-methylcytosine.

1620. A. Mathis, *J. Chim. Phys. Physicochim. Biol.*, 65, 46 (1968).
       BIOL: DNA behavior in ethylene glycol in the 0°-130°C range,
       central X-Ray diffusion.

1620a. R. Mathis *et al.*, *Spectrochim. Acta*, 29A, 63, 79 (1973). IR:
       Self-association, 1,3,2-diazaphosphorinanes.

1621. R. Mathis-Noel, M. T. Maurette, Ch. Godechot, and A. Lattes,
       Bull. Soc. Chim. Fr., 3047 (1970). IR, NMR: Inter- and intra-
       molecular aminoalcohols.

1622. R. Mathis-Noel, F. Flouquet, and A. Secches, *Ann. Chim. (Paris)*,
       4, 195 (1969). IR: N-chlorobenzamide in various solvents.

1623. R. Mathur, E. D. Becker, R. B. Bradley, and N. C. Li, *J. Phys.
       Chem.*, 67, 2190 (1963). NMR: Thermodynamic data, benzenethiol
       adducts with bases in CCl$_4$.

1624.   R. Mathur, S. M. Wang, and N. C. Li, *J. Phys. Chem.*, 68, 2140
(1964). NMR: Thermodynamic data, benzenethiol adducts with
N-methylacetamide in CCl_4.

1624a. R. Mathur-De Vre, *Spectrochim. Acta*, 28A, 1451 (1972). NMR:
Self-association, axial and equatorial alcohols.

1625.   E. I. Matrosov and M. I. Kabachnik, *Spectrochim. Acta*, 28A, 313
(1972). IR: Intramolecular, phosphinyl-α-hydroxyalkyl group.

1625a. E. I. Matrosov, M. I. Kabachnik, and S. T. Ioffe, *Zh. Obshch.
Khim.*, 42, 2625 (1972); [English, 2617]. IR: Intramolecular,
formylmethyl phosphonic acid ester.

1626.   E. I. Matrosov and M. I. Kabachnik, *Spectrochim. Acta*, 28A, 191
(1972). IR: Intramolecular, *cis*- and *trans*-enols.

1626a. T. Matsui, L. G. Hepler, and D. V. Fenby, *J. Phys. Chem.*, 77,
2397 (1973). CAL: Adducts of $CHCl_3$ with Et_3N, Me_2SO, Me_2CO.

1627.   T. Matsuo, *J. Phys. Chem.*, 72, 1819 (1968). NMR: Solvent
effects.

1628.   A. Matsuyama and H. Baba, *Bull. Chem. Soc. Jap.*, 44, 854 (1971).
Electronic: Proton transfer in photo-excited p-hydroxybenzo-
phenone.

1628a. A. Matsuyama and A. Imamura, *Bull. Chem. Soc. Jap.*, 45, 2196
(1972). Theory: CNDO, phenol-ammonia, p-nitrophenol-ammonia.

1628b. I. V. Matyash, D. A. Zhogolev, and A. M. Kalinichenko, *Magn.
Resonance Relat. Phenomena, Proc. Congr. AMPERE, 16th 1970*,
p. 393; *Chem. Abstr.*, 78, 22269g (1973). NMR: Wide-line spectra, H_2O.

1629.   M. Maurin, *Bull. Soc. Chim. Fr.*, 1497 (1962). Review:
Hydrogen bond in ionic crystals.

1630.   G. Mavel, *J. Chim. Phys.*, 61, 182 (1964). NMR: Chemical shifts
for $CHCl_3$ adducts.

1632.   F. Mazza, H. M. Sobell, and G. Kartha, *J. Mol. Biol.*, 43, 407
(1969). X-Ray: 9-Ethyl-2-aminopurine adduct of 1-methyl-5-
bromouracil.

1633.   K. C. Medhi, S. B. Banerjee, and G. S. Kastha, *Indian J. Phys.*,
36, 457 (1962). IR: Association of toluidines in various solvents.

1634.   K. C. Medhi and G. S. Kastha, *Indian J. Phys.*, 38, 483 (1964).
IR: Aminopyridines.

1635.   D. W. Meek and C. S. Springer, Jr., *Inorg. Chem.*, 5, 445 (1966).
NMR: Self-association of tri(2-aminoethyl) and tri(2-N-
methylaminoethyl) borates.

1635a. E. Meeussen and P. Huyskens, *J. Chim. Phys.*, 63, 845 (1966).
Self-association, n-butanol.

1636.   M. K. Meilahn and M. E. Munk, *J. Org. Chem.*, 34, 1440 (1969).
IR: Intramolecular, 2-amino-1,2-diphenylethanol.

1637.   T. G. Meister, *Opt. Spektrosk.*, 30, 220 (1971). Effect of
hydrogen bonds on electronic band shifts.

1638.   A. Mellier and P. Crouigneau, *Compt. Rend. Acad. Sci.*, *Ser. B*, 276, 921 (1973). IR: Polarized spectra, benzophenone-diphenyl-amine adduct.

1638a. P. Merlet, S. D. Peyerimhoff, and R. J. Buenker, *J. Amer. Chem. Soc.*, 94, 8301 (1972). Theory: Gaussian calculation of $(H_3N—H...NH_3)^+$.

1638b. A. Meunier, B. Levy, and G. Berthier, *Theor. Chim. Acta*, 29, 49 (1973). Theory: Correlation in $H_2O...NH_3$.

1639.   J. R. Merrill, *J. Phys. Chem.*, 65, 2023 (1961). NMR: Intramo-lecular, substituted 2-hydroxybenzophenones.

1639a. S. Meshitsuka, H. Takahashi, K. Higasi, and B. Schrader, *Bull. Chem. Soc. Jap.*, 45, 1664 (1972). Far-IR: Benzoic acid cry-stals.

1640.   R. J. Mesley, *Spectrochim. Acta*, 26A, 1427 (1970). IR: 5,5-disubstituted barbituric acids and salts.

1641.   R. P. Messmer, *Science*, 168, 479 (1970). Theory: Polywater.

1642.   H. D. Mettee, *J. Phys. Chem.*, 77, 1762 (1973). IR: Self-association, $(HCN)_2$.

1642a. W. Meyer, W. Jakobetz, and P. Schuster, *Chem. Phys. Lett.*, 21, 97 (1973). Theory: Correlation in $H_5O_2^+$.

1643.   W. C. Meyer and J. T. K. Woo, *J. Phys. Chem.*, 73, 2989 (1969). NMR: Association of methyl N-vinylcarbamate and Me_2SO.

1644.   A. J. Michell, *Aust. J. Chem.*, 23, 833 (1970). IR: Low temper-ature spectra of carbohydrates.

1645.   A. M. Michelson in *Molecular Associations in Biology*, B. Pullman, ed., Academic Press, New York, 1968, p. 93. BIOL: Review, oligonucleotide interactions.

1646.   A. M. Michelson and C. Monny, *Biochim. Biophys. Acta*, 149, 107 (1967). BIOL: Oligonucleotides and their association with polynucleotides.

1647.   A. M. Michelson, C. Monny, and A. M. Kapuler, *Biochim. Biophys. Acta*, 217, 7 (1970). BIOL: ORD, CD, poly(8-bromoguanylic acid).

1648.   A. M. Michelson, T. L. V. Ulbricht, R. T. Emerson, and R. J. Swan, *Nature*, 209, 873 (1966). BIOL: ORD, oligoadenylic acids, factors stabilizing helical polynucleotides.

1649.   W. J. Middleton and R. V. Lindsey, *J. Amer. Chem. Soc.*, 86, 4948 (1964). NMR: Fluoroalcohols and diols in bases.

1649a. R. Mierzecki, *Rocz. Chem.*, 46, 1375 (1972). IR, Raman: Chloroform adducts.

1650.   Y. Mikawa, J. W. Brasch, and R. J. Jakobsen, *J. Mol. Struct.*, 3, 103 (1969). IR, Far-IR: Polarized spectra of propanoic acid crystals.

1651.   Y. Mikawa, J. W. Brasch, and R. J. Jakobsen, *J. Mol. Spectrosc.*, 24, 314 (1967). Far-IR: Crystalline formic acid.

1652.   Y. Mikawa and R. J. Jakobsen, *J. Mol. Spectrosc.*, 33, 178 (1970). IR, Raman: Crystalline formic acid.

1653.   Y. Mikawa, R. J. Jakobsen, and J. W. Brasch, *J. Chem. Phys.*, $\underline{45}$,
        4750 (1966). IR: Formic acid crystals.

1654.   D. W. Miles, S. J. Hahn, R. K. Robins, M. J. Robins, and H.
        Eyring, *J. Phys. Chem.*, $\underline{72}$, 1483 (1968). BIOL: ORD, CD,
        adenine nucleosides substituted in the carbohydrate ring.

1655.   D. W. Miles, M. J. Robins, R. K. Robins, and H. Eyring, *Proc.
        Natl. Acad. Sci. U. S.*, $\underline{62}$, 22 (1969). BIOL: CD curves of
        seventeen purine nucleoside derivatives.

1656.   D. W. Miles, M. J. Robins, R. K. Robins, M. W. Winkley, and H.
        Eyring, *J. Amer. Chem. Soc.*, $\underline{91}$, 831 (1969). BIOL: CD data
        for twelve cytosine nucleoside derivatives.

1657.   D. W. Miles, M. J. Robins, R. K. Robins, M. W. Winkley, and H.
        Eyring, *J. Amer. Chem. Soc.*, $\underline{91}$, 824 (1969). BIOL: CD data
        for sixteen uracil nucleoside derivatives.

1658.   H. T. Miles in *Methods in Enzymology, Vol. XII, Nucleic Acids,
        Part B*, L. Grossman and K. Moldave, eds., Academic Press, New
        York, 1968, p. 256. BIOL: IR spectroscopy, measurement of
        nucleoside binding to polynucleotides.

1659.   H. T. Miles, *Nature*, $\underline{195}$, 459 (1962). BIOL: IR, base pairing
        in nucleic acids.

1660.   H. T. Miles, *Proc. Natl. Acad. Sci. U. S.*, $\underline{47}$, 791 (1961).
        BIOL: IR, tautomeric forms in a polynucleotide helix and their
        bearing on the structure of DNA.

1661.   M. H. Miles, E. M. Eyring, W. W. Epstein, and M. T. Anderson,
        J. Phys. Chem., $\underline{70}$, 3490 (1966). Kinetics: Proton transfer,
        malonic acids.

1662.   M. H. Miles, E. M. Eyring, W. W. Epstein, and R. E. Ostlund,
        J. Phys. Chem., $\underline{69}$, 467 (1965). Kinetics: Intramolecular,
        2,2-disubstituted acids.

1664.   D. B. Millar and M. MacKenzie, *Biochim. Biophys. Acta*, $\underline{204}$, 82
        (1970). BIOL: Helix-coil transition of poly U studied by
        various optical and hydrodynamic methods.

1665.   D. J. Millen and O. A. Samsonov, *J. Chem. Soc.*, 3085 (1965).
        IR: Gas-phase spectra of nitric acid-ether systems.

1666.   D. J. Millen and O. A. Samsonov, *Chem. Ind.*, 1694 (1963). IR:
        Gas-phase spectra of nitric acid-ether systems.

1667.   D. J. Millen and J. Zabicky, *J. Chem. Soc.*, 3080 (1965). IR:
        Gas-phase spectra, MeOH adducts with NH_3, methylamines, and
        aziridine.

1668.   D. J. Millen and J. Zabicky, *Nature*, $\underline{196}$, 889 (1962). IR:
        Gas-phase spectra of $CH_3OH-N(CH_3)_3$ adduct.

1669.   C. Miller, R. Miller, and W. Rogers, *J. Amer. Chem. Soc.*, $\underline{80}$,
        1562 (1958). IR: Intramolecular, α-hydroxy phosphoryl com-
        pounds.

1670.   Jeffrey H. Miller and H. M. Sobell, *J. Mol. Biol.*, $\underline{24}$, 345
        (1967). BIOL: IR, preferential association between alkylated
        adenine analogs and several uracil derivatives.

1670a. John H. Miller and H. P. Kelly, *Phys. Rev.*, *A*, 4, 480 (1971).
       Perturbation calculation on water monomer.

1670b. P. J. Miller, R. A. Butler, and E. R. Lippincott, *J. Chem. Phys.*,
       57, 5451 (1972).   IR, Raman: Symmetric O—H—O in $CsH(CF_3CO_2)_2$
       and $KH(CF_3CO_2)_2$.

1670c. K. Minakata and M. Iwasaki, *J. Chem. Phys.*, 57, 4758 (1972).
       ESR: Proton tunneling, carboxylic acids.

1671. V. I. Minkin, V. A. Bren, E. N. Malysheva, Yu. A. Zhdanov, and
       I. D. Sadekov, *Reakts. Sposobnost Org. Soedin.*, 6, 34 (1969);
       Chem. Abstr., 71, 101163b (1971).   Effect of intramolecular
       hydrogen bonding on pK_a of *o*-hydroxybenzaldehyde anils.

1671a. R. R. Minesinger, E. G. Kayser, and M. J. Kamlet, *J. Org. Chem.*,
       36, 1342 (1971).   UV: Bathochromic shifts, 4-nitroaniline.

1672. A. P. Minton, *Trans. Faraday Soc.*, 67, 1226 (1971).   Theory:
       Extended Hückel study of systems of two, three, four, and eight
       water molecules.

1673. J. Mirek, I. Holak, and T. Holak, *Rocz. Chem.*, 44, 2355 (1970);
       Chem. Abstr., 75, 19373u (1971).   NMR, IR: Intramolecular,
       anilides of 2,4-*bis*(dimethylamino) benzoic acid.

1674. E. J. Mitchell, Jr., Ph.D. Thesis, University of Pittsburgh,
       1972.   CAL: Thermodynamic data, *p*-F-phenol adducts.

1675. S. S. Mitra, *J. Chem. Phys.*, 36, 3286. (1962).   IR: Enthalpies
       for phenol-nitriles.

1676. J. Mitsky, L. Joris, and R. W. Taft, *J. Amer. Chem. Soc.*, 94,
       3442 (1972).   NMR, IR: Use of ^{19}F NMR and IR to obtain equi-
       librium constants for adducts of 5-fluoroindole.

1677. T. Miyata, H. Suzuki, and S. Yomosa, *J. Phys. Soc. Jap.*, 25,
       1428 (1968).   BIOL: π-Electronic structures, DNA bases, normal
       tautomeric and ionized forms.

1678. T. Miyazawa and K. S. Pitzer, *J. Chem. Phys.*, 30, 1076 (1959).
       Far-IR: Vapor phase and solid nitrogen matrix studies of formic
       acid.

1678a. W. Moffitt, *Proc. Roy. Soc.*, A210, 245 (1951).   Theory:
       The atoms-in-molecules method.

1679. S. C. Mohr, W. D. Wilk, and G. M. Barrow, *J. Amer. Chem. Soc.*,
       87, 3048 (1965).   IR: Association of H_2O with various bases.

1679a. C. Moliton and M. Gerbier, *Compt. Rend. Acad. Sci.*, 274B, 1363
       (1972).   IR: Solid CH_3CN—$CHCl_3$ and C_2H_5CN—$CHCl_3$ at -180°C.

1680. F. Moll and H. Thoma, *Arch. Pharm. (Weinheim)*, 301, 872 (1968).
       IR: Intramolecular, hydroxyethylpyrroles.

1681. F. Moll, *Arch. Pharm.*, 299, 429 (1966).   IR: Intramolecular,
       acetylene alcohols.

1681a. K. D. Moller and W. G. Rothschild, *Far-Infrared Spectroscopy*,
       Wiley-Interscience, 1971, Chapter 6.   Review.

1682. F. A. Momany, R. F. McGuire, J. F. Yan, and H. A. Scheraga,
       J. Phys. Chem., 74, 2424 (1970).   Theory: CNDO/2 calculations
       on conformations of model polypeptides.

1683.   W. B. Moniz, C. F. Poranski, Jr., and T. N. Hall, *J. Amer. Chem. Soc.*, 88, 190 (1966). NMR: Effect of hydrogen bonding on J_{HCOH} in primary alcohols.

1683a. C. E. Moore, *Ionization Potentials and Ionization Limits Derived from the Analyses of Optical Spectra*, National Bureau of Standards NSRDS-NBS 34, Washington, 1970.

1684.   L. S. Moore and J. P. O'Connell, *J. Chem. Phys.*, 55, 2606 (1971). Theory: Comparison of empirical potential function results with MO calculations in water dimers.

1684a. M. A. Moore and H. C. W. L. Williams, *J. Phys. C*, 5, 3168, 3185, 3222 (1972). Theory: Hydrogen-bonded ferroelectrics.

1685.   D. J. Morantz and M. S. Waite, *Spectrochim. Acta*, 27A, 1133 (1971). Near-IR: Intramolecular, alkane diols.

1686.   A. R. Morgan and R. D. Wells, *J. Mol. Biol.*, 37, 63 (1968). BIOL: Three-stranded nucleic acid complexes formed from double-stranded DNA and single-stranded RNA polymers.

1686a. K. J. Morgan, *J. Chem. Soc.*, 2343 (1961). IR: Self-association, perfluoroalkyl benzimidazoles.

1687.   R. S. Morgan, *Biosystems*, 5, 95 (1973). Models for ribose-ribose hydrogen bonds between ribopolynucleotide chains.

1687a. N. Mori, M. Aihara, Y. Asabe, and Y. Tsuzuki, *Bull. Chem. Soc. Jap.*, 45, 1786 (1972). IR: Intramolecular, β-alkylaminoalkanoates.

1688.   N. Mori, Y. Asabe, J. Tatsumi, and Y. Tsuzuki, *Bull. Chem. Soc. Jap.*, 43, 3227 (1970). IR: Intramolecular, N-alkylaminoalkanonitriles.

1689.   N. Mori, Y. Asano, T. Irie, and Y. Tsuzuki, *Bull. Chem. Soc. Jap.*, 42, 482 (1969). IR: Intramolecular, salicylic, mandelic, and α-hydroxyisobutyric acids in CCl_4.

1690.   N. Mori, Y. Asano, and Y. Tsuzuki, *Bull. Chem. Soc. Jap.*, 42, 488 (1969). IR: Intramolecular, correlation of ν_{OH} and ν_{CO} of methyl salicylates with Hammett σ.

1691.   N. Mori, S. Kaido, K. Suzuki, M. Nakamura, and Y. Tsuzuki, *Bull. Chem. Soc. Jap.*, 44, 1858 (1971). IR: Intramolecular, SH of mercapto compounds.

1692.   N. Mori, E. Nakamura, and Y. Tsuzuki, *Bull. Chem. Soc. Jap.*, 40, 2191 (1967). IR: Intramolecular, ω-N,N-diethylaminoalcohols.

1693.   N. Mori, S. Ōmura, N. Kobayashi, and Y. Tsuzuki, *Bull. Chem. Soc. Jap.*, 38, 2149 (1965). IR: Intramolecular, aliphatic hydroxycarboxylates.

1694.   N. Mori, S. Ōmura, and Y. Tsuzuki, *Bull. Chem. Soc. Jap.*, 38, 1631 (1965). IR: Intramolecular, phenylalkanediols.

1695.   N. Mori, S. Ōmura, H. Yamakawa, and Y. Tsuzuki, *Bull. Chem. Soc. Jap.*, 38, 1627 (1965). IR: Intramolecular, cyanoalcohols.

1696.   N. Mori, S. Ōmura, O. Yamamoto, T. Suzuki, and Y. Tswaki, *Bull. Chem. Soc. Jap.*, 36, 1401 (1963). NMR: Intramolecular, α- and β-hydroxycarboxylates.

1697.   N. Mori, M. Yoshifuji, Y. Asabe, and Y. Tsuzuki, *Bull. Chem. Soc. Jap.*, 44, 1137 (1971). IR: Intramolecular, 1-tetralols.

1698.   T. Mori, M. Hasegawa, and S. Yomosa, *J. Phys. Soc. Jap.*, 28, 188 (1970). BIOL: π-Electronic contribution to the conformation of DNA.

1699.   R. M. Moriarty, *J. Org. Chem.*, 28, 1296 (1963). NMR: N-methylamides-pyridine.

1700.   I. Morishima, K. Endo, and T. Yonezawa, *J. Amer. Chem. Soc.*, 93, 2048 (1971). Theory: NMR contact shifts caused by hydrogen bonding, nitroxide radical.

1701.   I. Morishima, K. Endo, T. Yonezawa, *Chem. Phys. Lett.*, 9, 203 (1971). NMR: ^{1}H and ^{13}C shifts for adducts of CHCl$_3$, CH$_2$Cl$_2$, C$_6$H$_5$C≡CH with di-*t*-butyl nitroxide radical.

1702.   I. Morishima, K. Endo, and T. Yonezawa, *Chem. Phys. Lett.*, 9, 143 (1971). Theory: INDO studies of nitroxide radical-methanol, correlation of computed charge density on hydroxyl proton with NMR contact shifts.

1703.   I. Morishima, K. Toyoda, K. Yoshikawa, and T. Yonezawa, *J. Amer. Chem. Soc.*, 95, 8627 (1973). NMR: ^{1}H, ^{13}C, α,γ-*bis*(diphenylene)-β-phenylallyl radical, contact shift studies, π hydrogen bonds.

1703a. I. Morishima, K. Endo, and T. Yonezawa, *J. Chem. Phys.*, 58, 3146 (1973). NMR, Theory: Thermodynamic data, adducts of di-*t*-butyl nitroxide, INDO calculations on dimers with (CH$_3$)$_2$NO.

1704.   H. Morita and S. Nagakura, *J. Mol. Spectrosc.*, 41, 54 (1972). Near-IR: Self-association, carboxylic acids.

1704a. H. Morita and S. Nagakura, *J. Mol. Spectrosc.*, 42, 536 (1972). IR: Self-association, formic and acetic acids.

1705.   H. Morita and S. Nagakura, *Theor. Chim. Acta*, 11, 279 (1968). BIOL: UV spectra, electronic structure, cytosine, isocytosine and their anions and cations.

1705a. H. Morita and S. Nagakura, *Theor. Chim. Acta*, 27, 325 (1972). Theory: CNDO/2 calculations for (H$_2$O)$_2$, (CH$_3$OH)$_2$, (HCOOH)$_2$, H maleate.

1706.   A. G. Moritz, *Spectrochim. Acta*, 20, 1642 (1964). IR: Intramolecular, *o*-nitroaniline.

1707.   A. G. Moritz, *Spectrochim. Acta*, 18, 671 (1962). IR: Intramolecular, *o*-anilines.

1708.   A. G. Moritz, *Spectrochim. Acta*, 17, 365 (1961). IR: Self-association of aniline and 1:1 adduct with CHCl$_3$.

1709.   A. G. Moritz, *Spectrochim. Acta*, 16, 1176 (1960). IR: Intramolecular, secondary aromatic amides with *o*-nitro substituents.

1710.   K. Morokuma, *J. Chem. Phys.*, 55, 1236 (1971). Theory: Minimal Slater basis SCF calculations on hydrogen bond between ⟩C=O and water.

1711.   K. Morokuma, *Chem. Phys. Lett.*, 4, 358 (1969). Theory: Symmetric versus antisymmetric hydrogen bonds in water, CNDO/2 calculations.

1712.   K. Morokuma, H. Kato, T. Yonezawa, and K. Fukui, *Bull. Chem.*
        Soc. Jap., 38, 1263 (1965). Theory: Extended Hückel calcula-
        tions on formic acid monomer and dimer, and intramolecular
        hydrogen bond in acetylacetone, salicylaldehyde, benzyl alcohol
        and allyl alcohol.

1713.   K. Morokuma and L. Pedersen, *J. Chem. Phys.*, 48, 3275 (1968).
        Theory: Gaussian basis calculations on water dimer.

1714.   M. Morokuma and J. R. Winick, *J. Chem. Phys.*, 52, 1301 (1970).
        Theory: Minimal Slater basis calculations on water dimer.

1715.   J. D. Morrison, R. M. Salinger, and F. L. Pilar, *J. Org. Chem.*,
        34, 1497 (1969). IR: Adducts of phenol with ketones.

1716.   T. H. Morton and J. L. Beauchamp, *J. Amer. Chem. Soc.*, 94, 3671
        (1972). Ion Cyclotron Resonance: Intramolecular, gas-phase,
        strong hydrogen bonds, protonated dimethoxylalkanes.

1716a. J. W. Moskowitz and M. C. Harrison, *J. Chem. Phys.*, 43, 3550
        (1965). SCF calculation on water monomer.

1717.   I. Motoyama and C. H. Jarboe, *J. Phys. Chem.*, 71, 2723 (1967).
        IR: Thermodynamic data, alcohol adducts with ethers.

1718.   S. P. Moulik, A. K. Chatterjee, and K. K. Sengupta, *Spectrochim.*
        Acta, 29A, 365 (1973). Hydrogen-bonded ion pair, p-nitrophenol
        with ethylenediamine.

1719.   K. B. Moxon, R. H. Still, and F. J. Swinbourne, *J. Chin. Chem.*
        Soc., 16, 100 (1969). NMR: Chemical shifts of benzoic acid in
        various solvents.

1720.   M. Mrnka and J. Celeda, *Chem. Prum.*, 19, 304 (1969); *Chem.*
        Abstr., 71, 85106c (1969). IR: Association of HCl with
        $(C_8H_{17})_3N$ and H_2O.

1721.   M. Mrnka and J. Celeda, *Chem. Prum.*, 18, 572 (1968); *Chem.*
        Abstr., 70, 81484f (1969). Extraction of HNO_3 by tri-n-
        octylamine.

1722.   D. K. Mukherjee and S. B. Banerjee, *Indian J. Phys.*, 42, 325
        (1968). IR: Guaiacol in tetrahydrofuran, CCl_4, $CHCl_3$, C_6H_6,
        and Et_2O.

1723.   S. Mukherjee, S. R. Palit, and S. K. De, *J. Phys. Chem.*, 74,
        1389 (1970). Electronic: Association of ethylene trithiocar-
        bonate with thiophenols.

1723a. S. Mukherjee, S. K. De, and S. R. Palit, *Indian J. Chem.*, 11,
        574, 577 (1973). Electronic: Adducts of α-naphthylamine,
        β-naphthol, ΔH.

1724.   E. Muller and J. B. Hyne, *Can. J. Chem.*, 46, 3587 (1968). NMR,
        IR: Sulfanes (H_2S_x).

1724a. J. P. Muller, G. Vercruysse, and Th. Zeegers-Huyskens, *J. Chim.*
        Phys. Physicochim. Biol., 69, 143 (1972). IR: Substituted
        phenol-tetramethylurea adducts, thermodynamic data.

1725.   N. Muller, *J. Chem. Phys.*, 43, 2555 (1965). NMR: Structural
        models for H_2O and chemical shift data.

1726. N. Muller and O. R. Hughes, *J. Phys. Chem.*, 70, 3975 (1966).
       NMR: Self-association, benzoic acid.

1727. N. Muller and P. I. Rose, *J. Phys. Chem.*, 69, 2564 (1965).
       NMR: K, acetic acid complexes in acetic anhydride, acetone,
       1,4-dioxane.

1728. N. Muller and P. I. Rose, *J. Amer. Chem. Soc.*, 85, 2173 (1963).
       NMR: Acetic acid in acetic anhydride.

1729. N. Muller and R. C. Reiter, *J. Chem. Phys.*, 42, 3265 (1965).
       NMR: Temperature dependence of proton chemical shifts involved
       in hydrogen bonding.

1730. N. Muller and P. Simon, *J. Phys. Chem.*, 71, 568 (1967). NMR:
       1:1 and 1:2 complex of H_2O with dioxane in CCl_4.

1731. R. S. Mulliken, *J. Chim. Phys.*, 20, 20 (1964). Theory: Charge-
       transfer theory as applied to hydrogen bonding.

1732. R. S. Mulliken, *J. Amer. Chem. Soc.*, 74, 811 (1952). Theory:
       Charge-transfer theory.

1733. R. S. Mulliken and W. B. Person, *Molecular Complexes*, Wiley-
       Interscience, 1969. Review: Charge-transfer complexes.

1734. W. S. Muney and J. F. Coetzee, *J. Phys. Chem.*, 66, 89 (1962).
       Conductivity: Ion-pair hydrogen bonding.

1735. S. Murahashi, B. Ryntani, and K. Hatada, *Bull. Chem. Soc. Jap.*,
       32, 1001 (1959). IR: $\Delta\nu_{CH}$ for acetylene adducts.

1736. Y. Murakami and J. Sunamoto, *J. Chem. Soc., Perkin Trans.* 2,
       1231 (1973). NMR: Pyridyl alcohols and phenols in Me_2SO.

1737. R. A. Murphy, Ph.D. Thesis, University of Texas, Austin, Texas,
       1967. NMR: K_2, self-association of piperidine in cyclohexane.

1738. R. A. Murphy and J. C. Davis, Jr., *J. Phys. Chem.*, 71, 3361
       (1967). NMR: Interference of ^{13}C satellite with Et_2NH self-
       association studies.

1739. R. A. Murphy and J. C. Davis, Jr., *J. Phys. Chem.*, 72, 3111
       (1968). NMR: Self-association of aliphatic secondary amines.

1739a. W. F. Murphy and H. J. Bernstein, *J. Phys. Chem.*, 76, 1147
       (1972). Raman: Assignments, H_2O.

1740. J. N. Murrell, *Chem. Brit.*, 5, 107 (1969). Theory: Review of
       the quantum mechanics of the hydrogen bond.

1741. J. N. Murrell and V. M. S. Gil, *Trans. Faraday Soc.*, 61, 402
       (1965). NMR: Pyridine derivative in CCl_4 and C_6H_6.

1742. J. N. Murrell, N. Randić, and D. R. Williams, *Proc. Roy. Soc.*,
       A284, 566 (1965). Theory: A double perturbation expansion in
       powers of interaction potential and overlap. This method is
       applied in later papers to the hydrogen bond.

1742a. J. N. Murrell and G. Shaw, *J. Chem. Phys.*, 46, 1768 (1967).
       Double perturbation theory.

1743. A. S. N. Murthy, *Indian J. Chem.*, 3, 143 (1965). IR: Effect
       of concentration on ν_{NH} of aniline.

1744. A. S. N. Murthy, Ph.D. Thesis, Indian Institute of Science,
       Bangalore, India, 1964. NMR: MeOH and t-BuOH adducts with
       bases.

1745. A. S. N. Murthy, A. Balasubramanian, T. R. Kasturi, and C. N. R.
       Rao, $Can.$ $J.$ $Chem.$, $\underline{40}$, 2267 (1962). UV: Solvent effects on
       keto-enol equilibria.

1746. A. S. N. Murthy, S. N. Bhat, and C. N. R. Rao, $J.$ $Chem.$ $Soc.$ A,
       1251 (1970). Theory: Extended Hückel and CNDO/2 calculations
       on several organic dimers.

1747. A. S. N. Murthy, R. E. Davis, and C. N. R. Rao, $Theor.$ $Chim.$
       $Acta$, $\underline{13}$, 81 (1969). Theory: Extended Hückel and CNDO/2 cal-
       culations on methanol and formic acid dimers and trimers.

1748. A. S. N. Murthy and C. N. R. Rao, $J.$ $Mol.$ $Struct.$, $\underline{6}$, 253
       (1970). Theory: Review.

1749. A. S. N. Murthy and C. N. R. Rao, $Bhagavantam$ $Vol.$ 1969, 1;
       $Chem.$ $Abstr.$, $\underline{73}$, 91361q (1970). Theory: Semiempirical.

1750. A. S. N. Murthy and C. N. R. Rao, $Chem.$ $Phys.$ $Lett.$, $\underline{2}$, 123
       (1968). Theory: Water dimer, extended Hückel, and CNDO/2
       calculations.

1751. A. S. N. Murthy and C. N. R. Rao, $Appl.$ $Spectrosc.$ $Rev.$, $\underline{2}$, 69
       (1968). Review of hydrogen bond.

1752. A. S. N. Murthy, C. N. R. Rao, B. D. N. Rao, and P. Venkate-
       swarlu, $Trans.$ $Faraday$ $Soc.$, $\underline{58}$, 855 (1962). NMR: Self-
       association, thiolbenzoic acid.

1753. A. S. N. Murthy, K. G. Rao, and C. N. R. Rao, $J.$ $Amer.$ $Chem.$
       $Soc.$, $\underline{92}$, 3544 (1970). Theory: Extended Hückel and CNDO/2
       calculations on hydrogen bond in formamide and N-methylacetamide
       dimers.

1754. J. Murto and A. Kivinen, $Suomen$ $Kemistilehti$, $\underline{B40}$, 14 (1967).
       IR: Intramolecular, fluoroalcohols.

1755. A. N. Murty and R. F. Curl, Jr., $J.$ $Chem.$ $Phys.$, $\underline{46}$, 4176 (1967).
       Microwave: Intramolecular, allyl alcohol.

1756. T. S. S. R. Murty, $Can.$ $J.$ $Chem.$, $\underline{48}$, 184 (1970). IR: Self-
       association, alcohols.

1757. T. S. S. R. Murty and K. S. Pitzer, $J.$ $Phys.$ $Chem.$, $\underline{73}$, 1426
       (1969). IR: Self-association of trifluoroacetic acid.

1758. R. S. Musa and M. Eisner, $J.$ $Chem.$ $Phys.$, $\underline{30}$, 227 (1959).
       Kinetics: Self-association of t-butanol, ultrasonic absorption
       measurements.

1759. H. Musso and K. H. Bantel, $Chem.$ $Ber.$, $\underline{102}$, 686 (1969). IR,
       NMR: Intramolecular, t-butyl substituted o-nitrophenols and
       o-aminophenols.

1760. H. Musso and S. von Grunelius, $Chem.$ $Ber.$, $\underline{92}$, 3101 (1959).
       IR: Intramolecular, hydroxybiphenyl derivatives.

1761. H. Musso and I. Seeger, $Chem.$ $Ber.$, $\underline{93}$, 796 (1960). IR: Intra-
       molecular, $peri$-hydroxy-naphthaquinones.

1762. Y. Nagai and O. Simamura, *Bull. Chem. Soc. Jap.*, 35, 132 (1962).
       IR: Self-association, acetic and chloroacetic acids.

1763. S. Nagakura, *J. Chim. Phys.*, 61, 217 (1964). UV: Evidence for
       charge-transfer in hydrogen maleate anion.

1763a. S. Nagakura, *J. Amer. Chem. Soc.*, 80, 520 (1958). Equation for
       determining K from spectroscopic data.

1764. S. Nagakura, K. Kaya, and H. Tsubomura, *J. Mol. Spectrosc.*, 13,
       1 (1964). Vacuum UV: Self-association, formic acid, acetic
       acid, ethyl acetate, charge-transfer.

1764a. G. Nagarajan, *Z. Naturforsch.*, 27A, 221 (1972). IR: Adduct of
       o-cresol with γ-butyrolactone.

1765. Y. Nakai and K. Hirota, *Nippon Kagaku Zasshi*, 81, 881 (1960);
       Chem. Abstr., 54, 22040 (1960). Far-IR: Self-association of
       gaseous formic and acetic acids.

1766. N. Nakagawa and S. Fujiwara, *Bull. Chem. Soc. Jap.*, 33, 1634
       (1960). NMR: Phenylacetylene association with bases.

1767. K. Nakamoto and S. Kishida, *J. Chem. Phys.*, 41, 1554 (1964).
       IR: Normal coordinate analysis of monomeric and dimeric formic
       and acetic acids.

1768. K. Nakamoto, M. Margoshes, and R. E. Rundle, *J. Amer. Chem. Soc.*,
       77, 6480 (1955). IR: Correlation of $\Delta\nu_{OH}$ with O...O distance.

1769. K. Nakamoto, Y. A. Sarma, and G. T. Behnke, *J. Chem. Phys.*, 42,
       1662 (1965). IR: Normal coordinate analysis of the acid maleate
       ion.

1770. K. Nakamoto, Y. A. Sarma, and H. Ogoshi, *J. Chem. Phys.*, 42,
       1177 (1965). IR: Normal coordinate analysis of the acid carbo-
       nate ion.

1771. N. I. Nakano and S. J. Igarashi, *Biochemistry*, 9, 577 (1970).
       BIOL: Interactions of mixed purines, pyrimidines.

1772. M. Nakano, N. I. Nakano, and T. Higuchi, *J. Phys. Chem.*, 71,
       3954 (1967). NMR: Calculation of K, phenol adducts.

1773. H. Naora, *J. Theor. Biol.*, 19, 183 (1968). BIOL: Mechanism
       for simultaneous read-out of genetic information from both
       strands of DNA.

1775. H. A. Nash and D. F. Bradley, *J. Chem. Phys.*, 45, 1380 (1966).
       Theory: Semiempirical calculation of configuration of hydrogen-
       bonded nucleotide base pairs.

1776. H. A. Nash and D. F. Bradley, *Biopolymers*, 3, 261 (1965). BIOL:
       Interaction between trinucleotides, electrostatic contribution,
       theoretical calculations.

1777. J. Nasielski and E. Van der Donckt, *Spectrochim. Acta*, 19, 1989
       (1963). UV: Adducts of carboxylic acids with pyridine, quino-
       line, acridine.

1777a. A. Neckel, P. Kuzmany, and G. Vinek, *Z. Naturforsch.*, A26, 569
       (1971). Experimental hydrogen bond strength, HF_2^-.

1778. D. Neerinck and L. Lamberts, *Bull. Soc. Chim. Belges*, 75, 484
       (1966). CAL: Thermodynamic data, adducts of phenol with sub-
       stituted pyridines.

1779.   D. Neerinck and L. Lamberts, *Bull. Soc. Chim. Belges*, 75, 473
        (1966). CAL: Thermodynamic data, adducts of pyridine with
        alcohols and substituted phenols.

1779a. D. Neerinck, A. van Audenhaege, and L. Lamberts, *Ann. Chim.*, 43
        (1969). CAL: Adducts of alcohols.

1780.   J. A. Nelder and R. Mead, *Computer J.*, 8, 308 (1965). Computer
        program for minimum-seeking method.

1781.   J. Nelson, *Spectrochim. Acta*, 26A, 235 (1970). IR: Adducts of
        hydrazoic acid with bases, thermodynamic data.

1782.   J. Nelson, *Spectrochim. Acta*, 26A, 109 (1970). IR: Adducts of
        isocyanic acid with bases, thermodynamic data.

1783.   R. C. Nelson, R. W. Hemwall, and G. D. Edwards, *J. Paint Tech-
        nol.*, 42, 636 (1970). IR: Use of $\Delta\nu_{AH}$ to predict miscibility.

1783a. R. D. Nelson, Jr., D. R. Lide, and A. A. Maryott, *Selected
        Values of Electric Dipole Moments for Molecules in the Gas
        Phase*, National Bureau of Standards, NSRDS-NBS 10, Washington,
        1967.

1784.   J. Nematollahi, F. Shihab, and J. Sprowls, *J. Pharm. Sci.*, 60,
        622 (1971). Solubility of alkyl benzoates.

1785.   G. Nemethy, *Federation Proc.*, 24, 538 (1965). BIOL: Comparison
        of models for water and aqueous solutions.

1785a. G. Nemethy and A. Ray, *J. Phys. Chem.*, 77, 64 (1973). UV:
        Phenol-solvent mixtures.

1786.   G. Nemethy and H. A. Scheraga, *J. Chem. Phys.*, 36, 3382 (1962).
        IR: Self-association, H_2O.

1787.   G. Nemethy and H. A. Scheraga, *J. Phys. Chem.*, 66, 1773 (1962).
        BIOL: Structure of water and hydrophobic bonding in proteins,
        thermodynamics.

1788.   G. Nemethy, H. A. Scheraga, and W. Kauzmann, *J. Phys. Chem.*, 72,
        1842 (1968). BIOL: Comments on hydrophobic bonding.

1789.   R. C. Neuman, Jr., W. R. Woolfenden, and V. Jonas, *J. Phys.
        Chem.*, 73, 3177 (1969). NMR: Effect of hydrogen bonding on
        barrier to rotation about amide bonds.

1789a. D. Neumann and J. W. Moskowitz, *J. Chem. Phys.*, 55, 1720 (1971).
        Theory: SCF calculation, H_2O.

1790.   R. A. Newmark and C. R. Cantor, *J. Amer. Chem. Soc.*, 90, 5010
        (1968). BIOL: NMR, interactions of guanosine and cytidine in
        Me_2SO.

1791.   M. D. Newton and S. Ehrenson, *J. Amer. Chem. Soc.*, 93, 4971
        (1971). Theory: Gaussian SCF calculations on $H_3O^+(H_2O)_n$ and
        $OH^-(H_2O)_n$, n = 0, ...4.

1792.   Soon Ng, *Spectrochim. Acta*, 28A, 321 (1972). NMR: $CHCl_3$-
        aniline.

1792a. Soon Ng, *J. Magn. Resonance*, 7, 370 (1972). NMR: Vicinal H—H
        and H—F coupling constants in CH_3CHCl_2 and CF_3CHCl_2.

1793.   J. W. Nibler and G. C. Pimentel, *J. Chem. Phys.*, 47, 710 (1967).
IR: 20°K spectra of ClHCl⁻, ClDCl⁻, ClHBr⁻, ClHI⁻, ClDI⁻.

1794.   R. S. Niedzielski, R. S. Drago, and R. L. Middaugh, *J. Amer.*
Chem. Soc., 86, 1694 (1964).   IR: $\Delta\nu_{OH}$, phenol adducts of sulfur compounds.

1794a. R. K. Nigam and B. S. Mahl, *Indian J. Chem.*, 10, 1167 (1972).
NMR: CH_2Cl_2 complexes with π-bases.

1795.   G. A. Nikiforov and V. V. Ershov, *Usp. Khim.*, 39, 1369 (1970).
Review: Sterically hindered phenols, ninety-eight references.

1796.   V. N. Nikitin and B. Z. Volchek, *Vipokomolekul. Soedin.*, 2, 1015
(1960); *Chem. Abstr.*, 55, 10069a (1961). IR: Polyamides.

1796a. S. D. Nikonovich and A. V. Rukosueva, *Zh. Strukt. Khim.*, 13, 939
(1972); [English, p. 875]. IR: Intramolecular, enamines.

1797.   M. Nilsson, *Svensk. Kem. Tidskr.*, 73, 458 (1961). NMR: Intramolecular, enols.

1798.   J. Ninio, *Biochimie*, 53, 485 (1971). BIOL: Nucleic acid repre-
sentations, topology.

1799.   T. Nishikawa and K. Someno, *Bull. Chem. Soc. Jap.*, 44, 851
(1971). ESR: Phenyl nitric oxide radicals.

1800.   K. Nishimoto, *Bull. Chem. Soc. Jap.*, 40, 2493 (1967). BIOL:
MO calculations, uracil, lumazine, alloxazine, their tautomeric
isomers.

1802.   S. Nishimura, C. H. Ke, and N. C. Li, *J. Phys. Chem.*, 72, 1297
(1968). NMR: Thermodynamic data, adducts of $CHCl_3$ with bases.

1803.   S. Nishimura and N. C. Li, *J. Phys. Chem.*, 72, 2908 (1968).
IR: Thermodynamic data, adducts of aniline and N-methylaniline
with tri-*n*-butylphosphate and tri-*n*-octylphosphine oxide.

1804.   T. Nishiwaki, *Tetrahedron*, 23, 2657 (1967). NMR: Self-
association, 2,6-dimethyl-4-hydroxypyrimidine, 2-ethylthio-4-
hydroxypyridine.

1805.   P. N. Noble and R. N. Kortzeborn, *J. Chem. Phys.*, 52, 5375
(1970). Theory: Gaussian SCF calculations on HF_2^- and HF_2.

1806.   D. C. Nonhebel, *Tetrahedron*, 24, 1869 (1968). NMR: Intramolecular, β-diketones, *o*-hydroxyaldehydes, *o*-hydroxyketones.

1807.   D. C. Nonhebel, *J. Chem. Soc. C*, 676 (1968). IR, NMR, Electronic: Intramolecular, triacylmethanes.

1807a. R. Nouwen and P. Huyskens, *J. Mol. Struct.*, 16, 459 (1973).
Dipole moments, phenol-pyridine adducts.

1808.   A. Novak and A. Lautie, *Nature*, 216, 1202 (1967). IR: Purine,
N—H...N.

1808a. A. Novak, J. Belloc, E. Foglizzo, and F. Romain, *J. Chim. Phys.*
Physicochim. Biol., 69, 1609 (1972). IR, Raman: Crystals of
$NaH(CH_3CO_2)_2$ and deuterated derivatives.

1809.   A. Novak and E. Whalley, *Spectrochim. Acta*, 16, 521 (1960).
IR: Intramolecular, fluoral, chloral, bromal hydrates.

1810. L. Novakovic, *J. Phys. Chem. Solids*, 27, 1469 (1966). Theory: Calculated frequency spectrum in hydrogen-bonded ferroelectrics.

1811. M. S. Nozari and R. S. Drago, *J. Amer. Chem. Soc.*, 92, 7086 (1970). IR, CAL: Association of pyrrole with bases, thermodynamic data.

1811a. M. S. Nozari and R. S. Drago, *J. Amer. Chem. Soc.*, 94, 6877 (1972). CAL: Solvent effects, thermodynamic data for *m*-fluorophenol adducts.

1812. R. H. Nuttall, D. W. A. Sharp, and T. C. Waddington, *J. Chem. Soc.*, 4965 (1960). IR: Alkyl and aryl ammonium salts.

1813. F. Nuzzo, A. Brega, and A. Falaschi, *Proc. Natl. Acad. Sci. U. S.*, 65, 1017 (1970). BIOL: DNA replication in mammalian cells, size of newly synthesized helices.

1814. R. A. Nyquist, *Spectrochim. Acta*, 25A, 47 (1969). IR: Intramolecular, $(RO)_2P(S)(SH)$ and $(PhO)_2P(S)(SH)$.

1815. R. A. Nyquist, *Spectrochim. Acta*, 19, 1655 (1963). IR: Intramolecular, O—H deformation, *o*-phenols.

1816. J. Oakes, *J. Chem. Soc. Faraday Trans. 2*, 69, 1311 (1973). NMR: Chemical shifts, relaxation times, alcohol/water mixtures.

1817. E. J. O'Brien, *Acta Crystallogr.*, 23, 92 (1967). BIOL, X-Ray: 1:1 complexes of 9-ethylguanine.

1818. J. P. O'Connell and J. M. Prausnitz, *Ind. Eng. Chem. Fundam.*, 8, 453 (1969). Strength of hydrogen bond in water vapor estimated as 6 kcal/mole.

1819. A. Ocvirk, A. Ažman, and D. Hadži, *Theor. Chim. Acta*, 10, 187 (1968). Theory: CNDO/2 calculations on formic acid dimer and $(CF_3COO...H...OOCCF_3)^-$.

1819a. S. E. Odinokov and A. V. Iogansen, *Spectrochim. Acta*, 28A, 2343 (1972). IR: Fermi resonance, carboxylic acid adducts.

1820. S. E. Odinokov, A. V. Iogansen, and A. K. Dzizenko, *Zh. Prikl. Spektrosk.*, 14, 418 (1971); *Chem. Abstr.*, 75, 12716z (1971). IR: Association of carboxylic acids with bases.

1821. S. E. Odinokov, O. B. Maximov, and A. K. Dzizenko, *Spectrochim. Acta*, 25A, 131 (1969). IR: Adducts of benzoic and salicylic acids with Me₂SO, dioxane, Me₂CO, and sulfolane.

1822. K. Ogawa, S. Matsvoka, and K. Senda, *J. Phys. Soc. Jap.*, 22, 662 (1967). NMR: Temperature dependence of salicylaldehyde signal.

1823. H. Ogoshi and K. Nakamoto, *J. Chem. Phys.*, 45, 3113 (1966). IR: Intramolecular, normal coordinate analysis of acetylacetone, hexafluoroacetylacetone.

1824. H. Ogoshi and Z. Yoshida, *J. Chem. Soc. D*, 176 (1970). IR: Intramolecular, O—D deformation, β-dicarbonyls.

1824a. K. Ogura and H. Sobue, *Polym. J.*, 3, 153 (1972). IR: Styrene-methacrylic acid co-polymers.

1825. A. Ohno, D. J. Grosse, and R. E. Davis, *Tetrahedron Lett.*, 959 (1968). IR: $\Delta\nu_{CO}$, cyclopropanone in $CHCl_3$.

1826. N. Oi and J. F. Coetzee, *J. Amer. Chem. Soc.*, 91, 2478 (1969).
       IR: Solvent effects on rotational isomers of 1,2-dichloroethane
       and 1,1,2,2-tetrachloroethane.

1827. N. Oi and J. F. Coetzee, *J. Amer. Chem. Soc.*, 91, 2473 (1969).
       IR: Effect of solvents on C—Cl stretching frequencies.

1828. R. Okazaki, T. Okazaki, K. Sakabe, K. Sugimoto, and A. Sugino,
       Proc. Natl. Acad. Sci. U. S., 59, 598 (1968). BIOL: Discon-
       tinuous mechanism for DNA replication.

1829. M. Ōki, K. Akashi, G. Yamamoto, and H. Iwamura, *Bull. Chem. Soc.
       Jap.*, 44, 1683 (1971). NMR: Intramolecular, internal rotation
       of biphenyls.

1830. M. Ōki and M. Hirota, *Nippon Kagaku Zasshi*, 86, 115 (1965).
       IR: Intramolecular, review of carboxylic acids.

1831. M. Ōki and M. Hirota, *Bull. Chem. Soc. Jap.*, 37, 209, 213 (1964).
       IR, UV: Intramolecular, *o*-methoxybenzoic, *o*-aryloxybenzoic, and
       methoxynaphthoic acids.

1832. M. Ōki and M. Hirota, *Bull. Chem. Soc. Jap.*, 36, 290 (1963).
       IR: Intramolecular, α-alkoxy- and α-aryloxyacetic acids.

1833. M. Ōki and M. Hirota, *Bull. Chem. Soc. Jap.*, 35, 1048 (1962).
       IR: Intramolecular, *o*-aryloxybenzoic acids.

1834. M. Ōki and M. Hirota, *Spectrochim. Acta*, 17, 583 (1961). IR:
       Intramolecular, α-aryloxy fatty acids.

1835. M. Ōki and M. Hirota, *Bull. Chem. Soc. Jap.*, 34, 374, 378 (1961).
       IR: Intramolecular, alkoxy fatty acids.

1836. M. Ōki and M. Hirota, *Nippon Kagaku Zasshi*, 81, 855 (1960).
       IR: Intramolecular, α-keto- and α-alkoxycarboxylic acids.

1837. M. Ōki, M. Hirota, and S. Hirofuji, *Spectrochim. Acta*, 22, 1537
       (1966). IR: Intramolecular, methoxynaphthoic acids and methyl
       hydroxynaphthoates.

1838. M. Ōki and H. Iwamura, *J. Amer. Chem. Soc.*, 89, 576 (1967).
       UV: Intramolecular, O—H...π bonding in 2-hydroxybiphenyl.

1839. M. Ōki and H. Iwamura, *Bull. Chem. Soc. Jap.*, 36, 1 (1963).
       NMR: Intramolecular, O—H...π in aryldimethylcarbinols.

1840. M. Ōki and H. Iwamura, *Bull. Chem. Soc. Jap.*, 35, 1744 (1962).
       IR: Intramolecular, fluoroalcohols.

1841. M. Ōki and H. Iwamura, *Bull. Chem. Soc. Jap.*, 35, 1552 (1962).
       NMR: Intramolecular, π-cloud, aryldimethylcarbinols.

1842. M. Ōki and H. Iwamura, *Bull. Chem. Soc. Jap.*, 35, 283 (1962).
       IR: Intramolecular, carboxylic O—H...π.

1843. M. Ōki and H. Iwamura, *Bull. Chem. Soc. Jap.*, 33, 1600 (1960).
       IR: Intramolecular, O—H...π in allyldimethyl- and benzyldi-
       methyl carbinols.

1844. M. Ōki and H. Iwamura, *Bull. Chem. Soc. Jap.*, 33, 427, 717
       (1960). IR: Intramolecular, ΔH, 2-allylphenol, 2-benzylphenol,
       2-isopropenylphenol, 2-hydroxybiphenyl, benzyldimethyl carbinol.

1845. M. Ōki and H. Iwamura, *Bull. Chem. Soc. Jap.*, 33, 681 (1960).
      IR: Intramolecular, O—H...π, 2-(ω-alkenyl) phenols and 2-(ω-
      phenylalkyl) phenols.

1846. M. Ōki and H. Iwamura, *Bull. Chem. Soc. Jap.*, 32, 950 (1959).
      IR: Rotational isomers of alcohols.

1847. M. Ōki, H. Iwamura, and T. Nishida, *Bull. Chem. Soc. Jap.*, 41,
      656 (1968). NMR: Effect of hydrogen bonding on the rates of
      inversion of 1-hydroxy-5,7-dihydrodibenz[c,e]oxepin.

1847a. M. Ōki and T. Yoshida, *Bull. Chem. Soc. Jap.*, 44, 1336 (1971).
      IR: Intramolecular, cyanoalcohols.

1848. G. A. Olah, S. J. Kuhn, and S. H. Flood, *J. Amer. Chem. Soc.*,
      83, 4581 (1961). IR: $\Delta\nu_{XH}$ for proton donors in halobenzenes.

1848a. Zh. P. Ol'Khovskaya, *Ukr. Fiz. Zh.*, 16, 860 (1971); *Chem. Abstr.*,
      75, 53187w (1971). Theory: Proton density in hydrogen bond of
      nucleic acids.

1849. G. Olofsson and I. Wirbrant, *Acta Chem. Scand.*, 25, 1408 (1971).
      CAL: Self-association, $(CH_3COOH)_2$ in 1,2-dichloroethane.

1850. T. Olsen, *Acta Chem. Scand.*, 24, 3081 (1970). NMR: K, $CHCl_3$—
      $(Me_2N)_3PO$.

1851. P. L. Olympia, Jr., *Chem. Phys. Lett.*, 5, 593 (1970). Theory:
      Calculations of stretching frequency and deuteron quadrupole
      coupling constants in C—H...O and C—D...O bonds.

1852. P. L. Olympia, Jr. and B. M. Fung, *J. Chem. Phys.*, 51, 2976
      (1969). Theory: Deuteron quadrupole coupling constants in
      hydrogen bonded systems.

1853. T. T. Omarov and O. V. Agashkin, *Isv. Akad. Nauk Kaz. SSR, Ser.
      Khim.*, 18, 38 (1968); *Chem. Abstr.*, 70, 15548d (1969). IR:
      α- and β-decahydronaphthols, thermodynamic data for association
      with Bu_2O.

1854. S. Onari, *J. Phys. Soc. Jap.*, 26, 214 (1969). BIOL: Vacuum
      UV absorption spectra of poly A, poly G, poly C, and poly U.

1855. E. Ōsawa, K. Kitamura, and Z. Yoshida, *J. Amer. Chem. Soc.*, 89,
      3814 (1967). IR: ΔH, methanol-diphenylcyclopropanone adduct.

1856. E. Ōsawa and Z. Yoshida, *Spectrochim. Acta*, 23A, 2029 (1967).
      IR: Effect of solvent on ν_{OH} of phenyl-ether and phenol-π-base
      complexes.

1857. H. Oshima, *Bull. Chem. Soc. Jap.*, 34, 846 (1961). UV: Solvent
      effects on $n \rightarrow \pi^*$ transitions of aliphatic aldehydes.

1860. R. J. Ouellette, *Can. J. Chem.*, 43, 707 (1965). NMR: Correla-
      tion of chemical shift with σ⁻ for adducts of phenol derivatives
      with Me_2SO.

1861. R. J. Ouellette, *J. Amer. Chem. Soc.*, 86, 4378 (1964). NMR:
      Adducts of cyclohexanol derivatives with pyridine and dimethyl-
      sulfoxide.

1862. R. J. Ouellette, K. Liptak, and G. E. Booth, *J. Org. Chem.*, 32,
      2394 (1967). NMR: Intramolecular, conformational analysis,
      isomeric alcohols containing bicyclo[2.2.1]heptane skeleton.

1863. R. J. Ouellette, D. L. Marks, and D. Miller, *J. Amer. Chem. Soc.*, 89, 913 (1967). NMR: Intramolecular, π-cloud, arylcarbinols.

1864. K. Ozeki, N. Sakabe, and J. Tanaka, *Acta Crystallogr.*, *Sect. B*, 25, 1038 (1969). BIOL, X-Ray: Crystal structure of thymine.

1865. R. P. Ozerov, *Vodorodnaya Svyaz.*, *Akad. Nauk SSSR*, *Inst. Khim. Fiz.*, *Sb. Statei*, 92 (1964). Neutron Diffraction: Review.

1866. R. Paetzold, H. D. Schumann, and A. Simar, *Z. Anorg. Allgem. Chem.*, 305, 88 (1960). IR, Raman: Self-association, selenic and methyl-selenic acids.

1867. G. Pala, *Nature*, 204, 1190 (1964). NMR: Intramolecular, 2-methyl-4-propionylphenol.

1868. S. R. Palit, S. Mukherjee, and S. K. De, *J. Phys. Chem.*, 75, 2404 (1971). Electronic: N—H...π, α-naphthylamine in π-bases.

1869. S. R. Palit, S. Mukherjee, A. K. Ghosh, and S. K. De, *J. Indian Chem. Soc.*, 47, 1053 (1970). UV: ΔH, phenol adducts of ethylene trithiocarbonate.

1870. M. H. Palmer and N. M. Scollick, *J. Chem. Soc. B*, 1353 (1968). NMR: Salicylic and 2-hydroxyisophthalic acids and esters.

1870a. T. S. Pang and Soon Ng, *Spectrochim. Acta*, 29A, 207 (1973). NMR: Adducts of halogenated hydrocarbons.

1871. G. Pannetier, M. Kern, L. Abello, and G. Djega-Mariadassou, *Compt. Rend. Acad. Sci.*, *Ser. C*, 264, 1016 (1967). IR: Self-association, cyclohexylamine, N-methyl-cyclohexylamine.

1872. L. Paolillo and E. D. Becker, *J. Magn. Resonance*, 2, 168 (1970). NMR: [15]N chemical shifts.

1873. L. Paoloni, *J. Chem. Phys.*, 30, 1045 (1959). Theory: Early incomplete calculations on molecular fragments.

1873a. R. Pariser and R. G. Parr, *J. Phys. Chem.*, 21, 466, 767 (1953). The Pariser-Parr-Pople method.

1874. A. J. Parker, *Quart. Rev. (London)*, 16, 163 (1962). Review: Solvation of anions.

1875. B. R. Parker and G. P. Khare, *J. Mol. Spectrosc.*, 41, 195 (1972). IR: DNA, correlation of theoretical and observed frequencies.

1876. F. S. Parker and K. R. Bhaskar, *Biochemistry*, 7, 1286 (1968). IR: Thermodynamic data, adducts of cholesterol with triacetin, tributyrin, and trilaurin.

1877. A. Parmigioni, A. Perotti, and V. Riganti, *Gazz. Chim. Ital.*, 91, 1148 (1961). NMR: Self-association, acetic acid.

1878. G. S. Parry, *Acta Crystallogr.*, 7, 313 (1954). BIOL, X-Ray: Uracil crystals.

1879. W. Partenheimer, T. D. Epley, and R. S. Drago, *J. Amer. Chem. Soc.*, 90, 3886 (1968). CAL: Solvent effects on enthalpies of association for *m*-trifluoromethyl- and *t*-butyl-phenol with N,N-dimethylacetamide.

1879a. D. J. Patel, *Biochemistry*, 12, 667 (1973). NMR: [1]H, ^{13}C, antamanide conformations.

1880. W. G. Paterson and D. M. Cameron, *Can. J. Chem.*, 41, 198 (1963). NMR: $CHCl_3$, $CHBr_3$ as proton donors.

1881. W. G. Paterson and N. R. Tipman, *Can. J. Chem.*, 40, 2122 (1962). NMR: Self-association, *p*-substituted phenols.

1882. L. K. Patterson and R. M. Hammaker, *Spectrochim. Acta*, A23, 2333 (1967). NMR: Self-association, di-*t*-butylcarbinol.

1883. R. C. Paul, P. Singh, and S. L. Chadha, *Indian J. Chem.*, 6, 673 (1968). IR: Adducts of Me_2SO with amides.

1884. L. Pauling, *J. Amer. Chem. Soc.*, 58, 94 (1936). IR: Intramolecular, *o*-chlorophenol.

1885. L. Pauling, *The Nature of the Chemical Bond*, 3rd ed., Cornell University Press, Ithaca, New York, 1960.

1886. L. Pauling and R. B. Corey, *Arch. Biochem. Biophys.*, 65, 164 (1956). BIOL, X-Ray: Purine, pyrimidine base pairs.

1887. B. M. Pava and F. E. Stafford, *J. Phys. Chem.*, 72, 4628 (1968). IR: Self-association, gaseous $H_2C_2O_4$, $D_2C_2O_4$.

1887a. Z. Pawlak, *Rocz. Chem.*, 46, 249 (1972). Adducts of chloride and carboxylate ions with carboxylic acids.

1888. Z. Pawlak, *Rocz. Chem.*, 46, 2069 (1972). K, $RCOO...H...OOCR^-$ in CH_3NO_2.

1888a. Z. Pawlak, *Rocz. Chem.*, 47, 641 (1973). IR, NMR: K, AHA_1^-, $A(HA_1)_2^-$, acetic acid with acetate, benzoate in Me_2CO.

1889. B. D. Pearson, *Tetrahedron*, 12, 32 (1961). UV: Intramolecular, nitronaphthylamines.

1889a. B. D. Pearson, *Proc. Chem. Soc.*, 78 (1962). UV: Arylamines.

1890. V. I. Pechenaya and V. I. Danilov, *Chem. Phys. Lett.*, 11, 539 (1971). Theory: Semiempirical calculation of hydrogen potential in 7-azaindole dimer.

1891. L. Pedersen, *Chem. Phys. Lett.*, 4, 280 (1969). Theory: INDO calculations on possible contributing structures of polywater.

1892. J. H. Peet, *Educ. Chem.*, 7, 199 (1970). Review: Experiments to illustrate hydrogen bonding.

1892a. A. Pellegrini, D. R. Ferro, and G. Zerbi, *Mol. Phys.*, 26, 577 (1973). IR: Disordered crystals, MeOH.

1893. V. V. Pen'kovskii, *Zh. Fiz. Khim.*, 42, 2994 (1968); *Chem. Abstr.*, 70, 77155g (1969). IR: Interpretation of spectra for *p*-phenylenediamine-chloranil adduct.

1894. R. E. Penn and R. F. Curl, Jr., *J. Chem. Phys.*, 55, 651 (1971). Microwave: Intramolecular, 2-aminoethanol.

1895. Yu. A. Pentin and O. S. Anisimova, *Opt. Spektrosk.*, 26, 68 (1969); *Chem. Abstr.*, 70, 82581x (1969). IR: Self-association, piperidine.

1896. C. Perchard, A. M. Bellocq, and A. Novak, *J. Chim. Phys.*, 62, 1344 (1965). Far-IR: Imidazole and deuterated derivatives.

1897. C. Perchard and A. Novak, *J. Chem. Phys.*, 48, 3079 (1968). Far-IR: Self-association, imidazole.

1898. J. P. Perchard and M. L. Josien, *J. Chim. Phys. Physicochim. Biol.*, 65, 1856 (1968). IR: Self-association, twelve isotopic species of ethanol in liquid and solid states.

1899. I. S. Perelygin, *Opt. Spektrosk.*, 12, 353 (1962). IR: Effect of anions on ν_{OH} of ethanol in acetonitrile.

1899a. I. S. Perelygin and A. M. Afanas'eva, *Zh. Prikl. Spektrosk.*, 19, 500 (1973). IR: Thermodynamic, data, adducts of acetonitrile with halo-substituted acetic acids.

1899b. I. S. Perelygin and T. F. Akhunov, *Zh. Prikl. Spektrosk.*, 18, 696 (1973); *Chem. Abstr.*, 79, 52657v (1973). IR: Thermodynamic data, chloro-substituted phenol adducts with Me_2CO, $HC(O)NMe_2$.

1900. I. S. Perelygin and T. F. Akhunov, *Opt. Spektrosk.*, 30, 679 (1971); *Chem. Abstr.*, 75, 27627v (1971). IR: Adducts of chloro-substituted phenols with CH_3CN, correlations.

1901. I. S. Perelygin and T. F. Akhunov, *Opt. Spektrosk.*, 29, 516 (1970); *Chem. Abstr.*, 74, 7977k (1971). IR: Intramolecular, substituted phenols.

1901a. I. S. Perelygin and T. F. Akhunov, *Opt. Spektrosk.*, 33, 246 (1972); *Chem. Abstr.*, 77, 145817w (1972). IR: Self-association, chloro-phenol derivatives.

1901b. I. S. Perelygin and T. F. Akhunov, *Opt. Spektrosk.*, 34, 891 (1973); *Chem. Abstr.*, 79, 31098y (1973). IR: Effect of temperature on ν_{OH} of chloro-phenols.

1902. I. S. Perelygin and A. K. Khairetdinova, *Dokl. Akad. Nauk SSSR*, 184, 1109 (1969); *Chem. Abstr.*, 70, 105795m (1969). IR: Self-association, RR'R''COH where R = Me, $ClCH_2$, $CHCl_2$, CCl_3.

1903. I. S. Perelygin and M. A. Klimchuk, *Opt. Spektrosk.*, 29, 37 (1970); *Chem. Abstr.*, 73, 125251w (1970). IR: Adducts of chloro-substituted alcohols with HCl.

1904. I. S. Perelygin and A. B. Shaikhova, *Opt. Spektrosk.*, 31, 205 (1971); [English, 110]. IR: Adducts of H_2O with acetonitrile.

1905. I. S. Perelygin and A. B. Shaikhova, *Teor. Eksp. Khim.*, 7, 278 (1971); *Chem. Abstr.*, 75, 69002t (1971). IR: Effect of $LiClO_4$ on ν_{OH} of CH_3OH and ν_{OD} of HOD in basic solvents.

1906. H. H. Perkampus, F. Miligy, and A. Kerim, *Spectrochim. Acta*, 24A, 2071 (1968). IR: Adducts of alcohols with pyridine derivatives, thermodynamic data.

1907. V. I. Permogorov, V. G. Debabov, I. A. Slodkova, and B. A. Rebentish, *Biochim. Biophys. Acta*, 199, 556 (1970). BIOL: CD, UV, DNA conformation.

1908. A. Perotti and M. Cola in *Nuclear Magnetic Resonance in Chemistry*, B. Pesce, ed., Academic Press; New York, 1965, p. 249. NMR: Self-association of pyrazole in CCl_4.

1909. A. Perotti and M. Cola, *Gazz. Chim. Ital.*, 91, 1153 (1961). NMR: Self-association, acetic acid.

1910.   J. W. Perram, *J. Chem. Phys.*, 49, 4245 (1968). IR: Inaccura-
        cies in fitting Gaussian components to experimental curves.

1910a. J. W. Perram, *Advances Mol. Relaxation Processes*, 3, 51 (1972).
        Review, structural models of water.

1911.   J. W. Perram and S. Levine, *Mol. Phys.*, 21, 701 (1971). Theory:
        Study of hydrogen bond network in liquid water.

1912.   N. C. Perrins and J. P. Simons, *Trans. Faraday Soc.*, 65, 390
        (1969). UV: Effect of hydrogen bonding on the photochemical
        addition to the benzene ring.

1913.   W. B. Person, *J. Amer. Chem. Soc.*, 87, 167 (1965). Reliability
        of formation constants of weak complexes.

1914.   A. Peterlin and M. Pintar, *J. Chem. Phys.*, 34, 1730 (1961).
        NMR: Temperature dependence of NMR absorption in solid acetic
        acid and its chloro- and acid-chloride derivatives.

1914a. V. S. Petrosyan, E. Ya. Davydov, and A. O. Reutov, *Izv. Akad.
        Nauk SSSR, Ser. Khim.*, 1730 (1973); *Chem. Abstr.*, 79, 141296u
        (1973). NMR: Self-association, HCl in organic solvents.

1915.   A. A. Petrov and N. V. Elsakov, *Zh. Obshch. Khim.*, 33, 319
        (1963); [English, 314]. NMR: Effect of solvents on NMR spec-
        tra of diacetylene and vinylacetylene.

1916.   A. A. Petrov, N. V. Elsakov, and V. B. Lebedev, *Opt. Spektrosk.*,
        16, 1013 (1964); [English, 547]. NMR: Association of acetylene
        compounds with solvents.

1917.   A. A. Petrov, N. V. Elsakov, and V. B. Lebedev, *Opt. Spektrosk.*,
        17, 679 (1964); [English, 367]. NMR: Adducts of acetylenic
        compounds.

1918.   A. A. Petrov, V. B. Lebedev, and I. G. Savich, *Zh. Obshch. Khim.*,
        32, 658 (1962); [English, 655]. IR, NMR: Acetylenic aldehydes
        and ketones.

1919.   A. A. Petrov and T. V. Yakovleva, *Opt. Spektrosk. (USSR)*
        [English Transl.], 7, 479 (1959). IR: Self-association of
        tertiary acetylenic amines.

1920.   A. K. Petrov, *Izv. Sib. Otd. Akad. Nauk SSSR, Ser. Khim. Nauk*,
        121 (1969); *Chem. Abstr.*, 71, 101088f (1969). IR: Intramolecu-
        lar, *o*-fluorophenols.

1921.   A. K. Petrov and A. V. Sechkarev, *Opt. Spektrosk.*, supplement 3,
        250 (1967); [English, 126 (1968)]. IR: Self-association,
        pentafluorobenzoic acid.

1922.   K. I. Petrov, N. M. Sinitsyn, and M. V. Rubstov, *Zh. Neorg.
        Khim.*, 13, 2614 (1968); *Chem. Abstr.*, 70, 14582y (1969). IR:
        Intramolecular, alkylammonium salts of nitrosopentahalides.

1923.   S. M. Petrov and V. S. Pilyugin, *Zh. Prikl. Spektrosk.*, 15, 555
        (1971); *Chem. Abstr.*, 76, 13296v (1972). IR: Thermodynamic
        data, adducts of substituted phenols with $HC(O)NMe_2$.

1923a. S. M. Petrov and V. S. Pilyugin, *Zh. Prikl. Spektrosk.*, 16, 552
        (1972); *Chem. Abstr.*, 76, 160348s (1972). IR: Thermodynamic
        data, adducts of halophenols with $(CH_3)_2SO$ and $CH_3C(O)N(CH_3)_2$.

1924. S. M. Petrov and V. S. Pilyugin, *Zh. Fiz. Khim.*, 46, 163 (1972); *Chem. Abstr.*, 76, 106000j (1972). IR: Adducts of phenols with Me$_2$SO.

1924a. S. M. Petrov, V. S. Pilyugin, F. A. Fatkullina, and Z. A. Eredzhepova, *Zh. Obshch. Khim.*, 42, 685 (1972); *Chem. Abstr.*, 77, 81560x (1972). UV: Adducts of halophenols with ketones.

1925. S. D. Peyerimhoff and R. J. Buenker, *J. Chem. Phys.*, 50, 1846 (1969). Theory: Gaussian calculations on HCOOH suggest the possibility of a hydrogen bridge even in the monomer.

1926. M. Pezolet and R. Savoie, *Can. J. Chem.*, 47, 3041 (1969). Raman: Liquid and solid HCN, DCN.

1927. J. Phillips, *J. Chim. Phys.*, 62, 951 (1965). Theory: Vibrational levels and coefficients of thermal expansion for hydrogen-bonded substances calculated by numerical integration of the Schrödinger equation with empirically chosen potential function.

1928. J. Phillips and S. Bratož, *Compt. Rend. Acad. Sci.*, 254, 1937 (1962). Theory: Vibrational wavefunctions for the hydrogen bond.

1929. P. Pichat, M. V. Mathieu, and B. Imelik, *J. Chim. Phys. Physico-chim. Biol.*, 66, 845 (1969). IR: Adsorption of NH$_3$ and ND$_3$ on silica.

1929a. L. Piela, *Chem. Phys. Lett.*, 19, 134 (1973). Theory: LCAO SCF on H$_2$0...Cl$^-$.

1929b. L. Piela, *Chem. Phys. Lett.*, 15, 199 (1972). Theory: LCAO SCF Slater basis calculations, H$_2$O—NH$_3$ dimer.

1930. J. L. Pierre, *Bull. Soc. Chim. Fr.*, 3116 (1970). NMR: Intramolecular, substituted alcohols.

1931. J. L. Pierre and R. Perraud, *Tetrahedron Lett.*, 1935 (1973). NMR: Intramolecular, cyclopropylmethanol, contradicts IR evidence.

1932. R. J. Piffath and S. Saas, *Appl. Spectrosc.*, 26, 92 (1972). IR: Intramolecular, methyl esters of phenylglycolic acids.

1933. H. Pikman and S. Pinchas, *J. Inorg. Nucl. Chem.*, 32, 2441 (1970). NMR: C—H...O hydrogen bond in aqueous solutions of acetonitrile.

1934. G. C. Pimentel, *J. Amer. Chem. Soc.*, 79, 3323 (1957). UV: Role of Franck-Condon Principle in hydrogen bonding and electronic transitions.

1934a. G. C. Pimentel, *J. Chem. Phys.*, 19, 446 (1951). Theory: MO calculations, HF$_2^-$.

1935. G. C. Pimentel, M. O. Bulanin, and M. Van Thiel, *J. Chem. Phys.*, 36, 500 (1962). IR: Self-association, matrix spectra of NH$_3$, 20°K.

1936. G. C. Pimentel, S. W. Charles, and K. Rosengren, *J. Chem. Phys.*, 44, 3029 (1966). IR: Self-association, matrix spectra of hydrazoic acid.

1937. G. C. Pimentel and A. L. McClellan, *Ann. Revs. Phys. Chem.*, 22, 347 (1971). A review.

1938. G. C. Pimentel and A. L. McClellan, *The Hydrogen Bond*, W. H.
       Freeman, San Francisco, 1960. Review.

1939. G. C. Pimentel and C. H. Sederholm, *J. Chem. Phys.*, 24, 639
       (1956). IR: Correlation of $\Delta\nu_{OH}$ with bond distances.

1940. S. Pinchas, *J. Chem. Phys.*, 51, 2285 (1969). IR: Effect of
       hydrogen bonding on stretching frequencies of $[P^{18}O_4]^{3-}$.

1941. S. Pinchas, *J. Phys. Chem.*, 67, 1862 (1963). UV: Intramolecu-
       lar, *o*-nitrobenzaldehyde.

1942. Yu. E. Pinchukov, *Teor. Eksp. Khim.*, 6, 390 (1970); *Chem. Abstr.*,
       73, 113159j (1970). Theory: Mechanism of transfer of proton
       vibrational energy in hydrogen-bonded systems.

1943. Yu. E. Pinchukov, *Zh. Strukt. Khim.*, 11, 415 (1970); *Chem.
       Abstr.*, 73, 81835h (1970). Intermolecular proton transfer and
       the dielectric properties of ice crystals.

1944. P. Pineau and M. L. Josien, *Proc. Intern. Meeting Mol. Spec-
       trosc., 4th Bologna, 1959*, 2, 924 (1962). IR: Adducts of
       butanol with ketones.

1945. O. Piovesana and C. Furlani, *J. Inorg. Nucl. Chem.*, 32, 879
       (1970). IR: Intramolecular, Co(II) complexes of thioureas and
       halogens.

1946. A. C. Pipkin, *Biopolymers*, 4, 3 (1966). BIOL: Kinetics, syn-
       thesis and conformational changes in biological macromolecules.

1947. J. Pitha, *Biochemistry*, 9, 3678 (1970). BIOL: IR, hydrogen
       bonding of sugar moiety of twelve nucleosides with attached
       bases in CCl_4.

1948. J. Pitha, *J. Org. Chem.*, 35, 2411 (1970). IR: Self-association
       in axial β-hydroxycyclohexanones.

1949. J. Pitha, R. N. Jones, and P. Pithova, *Can. J. Chem.*, 44, 1045
       (1966). BIOL, IR: Hydrogen bonding between pairs of deriva-
       tives of the bases of nucleic acids and their analogues in $CHCl_3$.

1950. L. P. Pivovarevich *et al.*, *Reakts. Sposobnost Org. Soedin.*, 10,
       119 (1973). IR: $\Delta\nu$, ΔH, phenol adducts of acetophenones,
       acetylthiophenes, 2-acetylfuran, 2-acetylselenophane.

1951. A. Planckaert and C. Sandorfy, *Can. J. Chem.*, 50, 296 (1972).
       IR: Self-association, fluoroalcohols.

1952. Th. Plesser and H. Stiller, *Solid State Commun.*, 7, 323 (1969).
       Neutron Scattering: Experimental proof for proton tunneling
       in KH_2PO_4.

1953. V. G. Plotnikov and D. N. Shigorin, *Zh. Fiz. Khim.*, 41, 1750
       (1967); *Chem. Abstr.*, 67, 120270t (1967). Theory: Role of
       hydrogen atom excited orbitals in hydrogen bond formation.

1954. V. G. Plotnikov and D. N. Shigorin, *Zh. Fiz. Khim.*, 39, 2608
       (1965); *Chem. Abstr.*, 64, 10569e (1966). Theory: Approximate
       calculations showing participation of π-electrons in hydrogen
       bond formation.

1955. G. R. Plourde, *Dissertation Abstr.*, 22, 1400 (1962). IR: Self-
       association, phenol, phenol-d.

1956.   F. Pochon and A. M. Michelson, *Biochim. Biophys. Acta*, 217, 225
        (1970). BIOL: Substitution of poly G at C-8 by diazonium
        salts, hydrogen-bonded secondary structure.

1957.   F. Pochon and A. M. Michelson, *Proc. Natl. Acad. Sci. U. S.*, 53,
        1425 (1965). BIOL: Interaction between poly G and poly C,
        spectrophotometric titrations, thermal dissociation.

1958.   V. K. Pogorelyi, *Dokl. Akad. Nauk SSSR*, 204, 110 (1972); *Chem.
        Abstr.*, 77, 39775w (1972). NMR: Self-association, thiobenzoic
        acids.

1958a. V. K. Pogorelyi and I. I. Kukhtenko, *Teor. Eksp. Khim.*, 8, 684
        (1972); *Chem. Abstr.*, 78, 20801a (1973). NMR: Adducts of
        thiophenol with anisole, dimethylaniline, thioanisole.

1958b. V. K. Pogorelyi and I. I. Kukhtenko, *Teor. Eksp. Khim.*, 9, 400
        (1973); *Chem. Abstr.*, 79, 77849x (1973). NMR: Thiophenol-
        acetophenone adducts.

1959.   V. K. Pogorelyi and I. P. Gragerov, *Dokl. Akad. Nauk SSSR*, 186,
        610 (1969); *Chem. Abstr.*, 71, 60544f (1969). NMR: Association
        of thiophenols with acetone.

1960.   F. M. Pohl, *Naturwissenschaften*, 54, 616 (1967). BIOL: Model
        of DNA structure.

1961.   V. D. Pokhodenko and V. S. Kuts, *Teor. Eksp. Khim.*, 4, 387
        (1968); [English, 247]. NMR: Effect of p-substituent on di-o-
        t-butyl phenol association with t-butanol.

1961a. J. Pola, Z. Papouskova, and V. Chvalovsky, *Collect. Czech. Chem.
        Commun.*, 38, 1522 (1973). IR: Intramolecular, phenylsilyl-
        alkanols.

1962.   D. Poland and H. A. Scheraga, *Physiol. Chem. Phys.*, 1, 389
        (1969). Equilibrium unwinding in finite chains of DNA.

1963.   D. Poland and H. A. Scheraga, *Biochemistry*, 6, 3791 (1967). Theory:
        Empirical potential functions for hydrogen bonds in polypeptides.

1964.   D. Poland, J. N. Vournakis, and H. A. Scheraga, *Biopolymers*, 4,
        223 (1966). BIOL: ORD, cooperative interactions in single-
        strand oligomers of adenylic acid.

1965.   M. Pollak and R. Rein, *J. Theor. Biol.*, 19, 333 (1968). BIOL:
        An electrical mechanism for strand separation of DNA.

1966.   M. Pollak and R. Rein, *J. Theor. Biol.*, 19, 241 (1968). BIOL:
        Examination of the energetics of Crick's wobble hypothesis.

1967.   M. Pollak and R. Rein, *J. Theor. Biol.*, 11, 490 (1966). Theory:
        DNA base pairs.

1968.   V. I. Poltev and B. I. Sukhorukov, *Biofizika*, 12, 763 (1967).
        Theory: Conformational state of polynucleotides.

1969.   V. I. Poltev and B. I. Sukhorukov, *Biofizika*, 13, 941 (1968);
        Chem. Abstr., 70, 34315s (1969). Theory: Overlap of electron
        clouds between bases in the DNA double helix.

1969a. H. Popkie, H. Kistenmacher, and E. Clementi, *J. Chem. Phys.*, 59,
        1325 (1973). Theory: Best calculation on $(H_2O)_2$ to date.

1970. J. A. Pople, *Proc. Roy. Soc.*, A239, 541, 550 (1957). NMR:
       Theory for chemical shifts in hydrides.

1970a. J. A. Pople, *Proc. Roy. Soc.*, A205, 163 (1951). Theory:
       Electrostatic model of hydrogen bond in $(H_2O)_2$.

1970b. J. A. Pople, *Trans. Faraday Soc.*, 49, 1375 (1953). The
       Pariser-Parr-Pople Method.

1970c. J. A. Pople and D. L. Beveridge, *Approximate Molecular Orbital
       Theories*, McGraw-Hill, New York, 1970.

1971. J. A. Pople, W. G. Schneider, and H. J. Bernstein, *High-
       Resolution Nuclear Magnetic Resonance*, McGraw-Hill, New York,
       1959, p. 400.

1971a. J. A. Pople, D. P. Santry, and G. A. Segal, *J. Chem. Phys.*, 43,
       S129 (1965). The CNDO method.

1972. A. I. Popov in *The Chemistry of Nonaqueous Solvents*, J. J.
       Lagowski, ed., Academic Press, New York, 1970, Vol. 3, p. 339.
       Review: Carboxylic acids as nonaqueous solvents.

1973. A. I. Popov in *The Chemistry of Nonaqueous Solvents*, J. J.
       Lagowski, ed., Academic Press, New York, 1970, Vol. 3, p. 241.
       Review: Anhydrous acetic acid solvent system.

1974. E. M. Popov, M. I. Kabachnik, and L. S. Mayants, *Russian Chem.
       Rev.*, 30, 362 (1961). IR: Intramolecular, $(C_2H_5O)_2P(S)SH$.

1975. A. L. Porte, H. S. Gutowsky, and I. M. Hunsberger, *J. Amer.
       Chem. Soc.*, 82, 5057 (1960). NMR: Intramolecular, *ortho*
       derivatives of phenol, α-naphthol, 9-phenanthrol.

1976. D. M. Porter and W. S. Brey, Jr., *J. Phys. Chem.*, 72, 650
       (1968). NMR: Pyrrole—Me_2SO, thermodynamic data.

1977. D. M. Porter and W. S. Brey, Jr., *J. Phys. Chem.*, 71, 3779
       (1967). NMR: Succinimide—Me_2SO, thermodynamic data.

1978. P. S. Portoghese and D. A. Williams, *J. Med. Chem.*, 12, 839
       (1969). NMR, IR: Intramolecular, β-methadol.

1979. A. Pospíšil, *Chem. Prum.*, 18, 500 (1968); *Chem. Abstr.*, 70,
       28265j (1969). IR: Intramolecular bonding in 2-(2'-
       hydroxyphenyl)benzotriazole derivatives.

1980. P. H. Pouwels, C. M. Knijnenburg, J. van Rotterdam, J. A. Cohen,
       and H. S. Jansz, *J. Mol. Biol.*, 32, 169 (1968). BIOL: Struc-
       ture of the replicative form of bacteriophage φX174, alkali-
       denatured double-stranded φX174 DNA.

1980a. J. P. Povolyreva, R. R. Shagidullin, and Z. G. Isaeva, *Izv. Akad.
       Nauk SSSR, Ser. Khim.*, 561 (1972); *Chem. Abstr.*, 77, 62147b
       (1972). IR: Intramolecular, unsaturated alcohols.

1981. D. L. Powell and R. West, *Spectrochim. Acta*, 20, 983 (1964).
       Near-IR: Thermodynamic data, adducts of phenol with methylethyl
       ketone.

1981a. D. L. Powell, Ph.D. Thesis, University of Wisconsin, 1962.
       Near-IR: Thermodynamic data, phenol adducts.

1982. V. M. Pozdnyakov, A. N. Savitskaya, I. B. Klimenko, L. A. Vol'f, and A. I. Meos, *Vysokomol. Soedin.*, *Ser. B*, <u>11</u>, 649 (1969); *Chem. Abstr.*, <u>72</u>, 3899x (1970). IR: Poly (vinyl alcohol) — poly (vinylpyrrolidone) and poly (vinyl alcohol) — poly (vinylcaprolactam) mixtures.

1982a. K. K. Prasad and R. V. Venkataratum, *Aust. J. Chem.*, <u>26</u>, 1263 (1973). NMR: Intramolecular, *o*-amidobenzanilides and acetanilides.

1983. J. M. Prausnitz, J. A. Hunter, W. S. Gilliam, and L. Leiserson, *U. S. Office Saline Water*, *Res. Develop. Progr. Rep. No. 306* (1968); *Chem. Abstr.*, <u>71</u>, 33584n (1969). Experimental: Estimates hydrogen bond energy as 6 kcal/mole in open-chain water dimer.

1984. J. H. Prestegard and S. I. Chan, *J. Amer. Chem. Soc.*, <u>91</u>, 2843 (1969). BIOL, PMR: Study of effect of electrolytes on conformation of uracil nucleotides and nucleosides in H_2O.

1985. M. P. Printz, *Biochemistry*, <u>9</u>, 3077 (1970). BIOL: H^3—H exchange of polyriboadenylic acid as a function of pH, Na^+ concentration.

1986. W. H. Pritchard, *Aspects Adhes.*, <u>6</u>, 11 (1969); *Chem. Abstr.*, <u>79</u>, 19455e (1973). Review, hydrogen bonding in polymer adhesion.

1987. P. L. Privalov, K. A. Kafiani, D. R. Monaselidze, G. M. Mrevlishvili, and V. A. Magaldadze, *Nukleinovye Kisloty*, *Tr. Konf.*, *2nd*, 15 (1965); *Chem. Abstr.*, <u>68</u>, 75188e (1968). BIOL: Denaturation of DNA and the state of the surrounding water.

1988. P. L. Privalov, O. B. Ptitsyn, and T. M. Birshtein, *Biopolymers*, <u>8</u>, 559 (1969). BIOL: Free energy of stabilization of native DNA structure, microcalorimeter measurements.

1989. A. W. Pross and F. Van Zeggren, *Spectrochim. Acta*, <u>16</u>, 563 (1960). IR: Self-association, acetic and butyric acids in the vapor state.

1990. E. A. Pshenichnov and N. D. Sokolov, *Int. J. Quant. Chem.*, <u>1</u>, 855 (1967). Theory: Effect of hydrogen bond formation on rate of proton transfer reactions.

1991. E. A. Pshenichnov and N. D. Sokolov, *Opt. Spektrosk.*, <u>11</u>, 16 (1961). Theory: Double-minimum potential of the hydrogen bond.

1992. E. A. Pshenichnov and N. D. Sokolov, *Dokl. Akad. Nauk SSSR*, <u>137</u>, 352 (1961). Theory: Double minimum hydrogen bond potential.

1993. M. S. Puar, B. T. Keeler, and A. I. Cohen, *J. Org. Chem.*, <u>31</u>, 219 (1971). NMR, IR: Intramolecular, β-amino α-, β-unsaturated esters.

1994. G. Pukanic, N. C. Li, W. S. Brey, Jr., and G. B. Savitsky, *J. Phys. Chem.*, <u>70</u>, 2899 (1966). NMR: Association of β-diketones with organophosphorus esters.

1995. A. D. E. Pullin, *Spectrochim. Acta*, <u>16</u>, 12 (1960). IR: Solvent effects on stretching frequencies.

1996. J. A. Pullin and R. L. Werner, *Spectrochim. Acta*, <u>21</u>, 1257 (1965). IR: Adducts of indole, pyrrole, N-methylacetamide, diphenylamine, N-methylaniline.

1997. J. A. Pullin and R. L. Werner, *Nature*, <u>206</u>, 393 (1965). IR:
       Correlation, Δν versus ΔG.

1998. A. Pullman, *Ann. N. Y. Acad. Sci.*, <u>158</u>, 65 (1969). BIOL: EHT,
       IEHT and CNDO/2 calculations for A, G, U, and C with simultan-
       eous treatment of all valence electrons.

1999. A. Pullman, *Int. J. Quant. Chem.*, *Symp. No. 2*, 1968, p. 187.
       BIOL: EHT, IEHT and CNDO/2 calculations on A, G, U, and C.

2000. A. Pullman in *Modern Quantum Chemistry*, O. Sinanoğlu, ed.,
       Academic Press, New York, 1965, Vol. 3, p. 283. Theory: Review
       of hydrogen bond in protein structure.

2001. A. Pullman in *Electronic Aspects of Biochemistry*, B. Pullman,
       ed., Academic Press, New York, 1964, p. 135. BIOL: Theoreti-
       cal, molecular aspects of mutations.

2002. A. Pullman, *Biochim. Biophys. Acta*, <u>87</u>, 365 (1964). Theory:
       Tautomeric equilibria in purines and pyrimidines.

2003. A. Pullman, *Compt. Rend. Acad. Sci.*, <u>256</u>, 5435 (1963). Theory:
       Hückel calculations, DNA base pairs.

2004. A. Pullman and H. Berthod, *Theor. Chim. Acta*, <u>10</u>, 461 (1968).
       Theory: CNDO/2 calculations on formamide dimers.

2005. A. Pullman, M. Dreyfus, and B. Mely, *Theor. Chim. Acta*, <u>17</u>, 85
       (1970). BIOL: Electron distributions in adenine, thymine and
       cytosine, probability density curves, nonempirical calculations.

2007. A. Pullman and B. Pullman, *Adv. Quantum Chem.*, <u>4</u>, 267 (1968).
       BIOL: Review, quantum mechanical calculations on purine and
       pyrimidine bases.

2008. B. Pullman in *Molecular Orbital Studies in Chemical Pharmacology*,
       L. B. Kier, ed., Springer-Verlag, New York, 1970, p. 1. BIOL:
       Review and discussion, theoretical calculations on nucleic acid
       constituents.

2009. B. Pullman in *Mol. Assn. Biol.*, *Proc. Int. Symp.*, *1967*, B. Pull-
       man, ed., Academic Press, New York, 1968, p. 1. BIOL: Discus-
       sion, molecular associations via hydrogen bonding.

2010. B. Pullman in *Molecular Biophysics*, B. Pullman and M. Weissbluth,
       eds., Academic Press, New York, 1965, p. 117. BIOL: Review,
       applications of quantum chemistry to nucleic acid constituents.

2011. B. Pullman, H. Berthod, F. B. Bergmann, Z. Neiman, H. Weiler-
       Feilchenfeld, and E. D. Bergman, *Tetrahedron*, <u>26</u>, 1483 (1970).
       BIOL: Electronic structure of purine tautomers.

2012. B. Pullman, H. Berthod, and J. Caillet, *Theor. Chim. Acta*, <u>10</u>,
       43 (1968). BIOL, X-Ray: Purine.

2014. B. Pullman, H. Berthod, and J. Sanglet, *Compt. Rend. Acad. Sci.*,
       Ser. D, <u>226</u>, 1063 (1968). BIOL: CNDO calculations, tautomeric
       forms, purine, adenine, guanine, xanthine.

2015. B. Pullman, P. Claverie, and J. Caillet, *Proc. Natl. Acad. Sci.*
       U. S., <u>57</u>, 1663 (1967). BIOL: Interaction energies in hydrogen-
       bonded purine-pyrimidine coplanar triplets, theoretical calcula-
       tions.

2016. B. Pullman, P. Claverie, and J. Caillet, *J. Mol. Biol.*, 22, 373
       (1967). BIOL: van der Waals-London in-plane interaction ener-
       gies for all possible base pairs involving A, T, C and G.

2017. B. Pullman, P. Claverie, and J. Caillet, *Proc. Natl. Acad. Sci.
       U. S.*, 55, 904 (1966). Theory: Configuration of hydrogen-
       bonded purine-pyrimidine pairs.

2018. B. Pullman and A. Pullman, *Quantum Biochemistry*, Wiley, New
       York, 1963. BIOL: Treatise, application of quantum mechanics
       to problems in biochemistry.

2019. B. Pullman and A. Pullman, *Biochim. Biophys. Acta*, 36, 343
       (1959). BIOL: First theoretical calculations on purine-
       pyrimidine base pairs, HMO method.

2020. B. Pullman and A. Pullman, *Prog. Nucl. Acid Res. Mol. Biol.*, 9,
       327 (1969). BIOL: Quantum-mechanical investigations of the
       electronic structure of nucleic acids.

2021. P. G. Puranik and V. Kumar, *Proc. Indian Acad. Sci.*, A58, 29,
       327 (1963). Theory: Charge-transfer theory of hydrogen bonding.

2021a. P. G. Puranik and B. S. Prakash, *Curr. Sci.*, 41, 502 (1972).
       Raman: Dioxane-water, HCl-dioxane.

2022. P. G. Puranik and K. Venkataramiah, *Proc. Indian Acad. Sci.*,
       A54, 146 (1961). IR: Association of toluidines in various sol-
       vents.

2023. J. M. Purcell, H. Susi, and J. R. Cavanaugh, *Can. J. Chem.*, 47,
       3655 (1969). NMR: Association of δ-valerolactam, thermodynamic
       data.

2024. K. F. Purcell and R. S. Drago, *J. Amer. Chem. Soc.*, 89, 2874
       (1967). Theoretical aspects of Δν-ΔH relationships.

2025. K. F. Purcell, J. A. Stikeleather, and S. D. Brunk, *J. Mol.
       Spectrosc.*, 32, 202 (1969). NMR, IR: Correlation of chemical
       shifts with frequency shifts.

2026. K. F. Purcell, J. A. Stikeleather, and S. D. Brunk, *J. Amer.
       Chem. Soc.*, 91, 4019 (1969). Calorimetric method, thermodynamic
       data for 1,1,1,3,3,3-hexafluoro-2-propanol as reference acid.

2027. K. F. Purcell and S. T. Wilson, *J. Mol. Spectrosc.*, 24, 468
       (1967). NMR, IR: $\Delta\nu_{OH}$ and chemical shifts for 2,2,2-trifluoro-
       ethanol.

2028. N. A. Puttnam, *J. Chem. Soc.*, 5100 (1960). IR: *Cis-* and *trans-*
       orientations for *o*-phenols.

2029. N. A. Puttnam, *J. Chem. Soc.*, 486 (1960). IR: Effect of alkyl
       groups on the self-association of *o*-phenols.

2030. R. Queignec and B. Wojtkowiak, *Bull. Soc. Chim. Fr.*, 860 (1970).
       IR, Gas Chromatography: Association of RC≡CH with N,N-
       diethylacetamide.

2031. I. D. Raacke, *Biochem. Biophys. Res. Commun.*, 31, 528 (1968).
       BIOL: Cloverleaf conformation for 5s RNAs.

2032. S. W. Rabideau, E. D. Finch, and A. B. Denison, *J. Chem. Phys.*,
       49, 4660 (1968). NMR: Proton and deuteron NMR of ice polymorphs.

2033. C. P. Rader, *J. Amer. Chem. Soc.*, 88, 1713 (1966). NMR: Effect of hydrogen bonding on J_{HCOH} of cyclohexanols in Me_2SO.

2034. I. D. Rae, *Can. J. Chem.*, 46, 2589 (1968). NMR: Intramolecular, *o*-substituted acetanilides.

2035. I. D. Rae, *Aust. J. Chem.*, 20, 1173 (1967). NMR: Intramolecular, *o*-nitroanilines.

2036. I. D. Rae, *Aust. J. Chem.*, 19, 409 (1966). NMR: Spin-spin coupling, N-methylanilines.

2037. I. D. Rae, *Chem. Commun.*, 3, 519 (1966). NMR: Self-association, *o*-nitroanilines.

2037a. R. O. Ragsdale, *U. S. Nat. Tech. Inform. Serv.*, AD Rep., *1971*, *No. 735279*; *Chem. Abstr.*, 77, 4717b (1972). Near-IR: Thermodynamic data, phenol adducts of amine oxides.

2037b. A. Rahman and F. H. Stillinger, *J. Amer. Chem. Soc.*, 95, 7943 (1973). Hydrogen bond in H_2O.

2038. D. K. Rai and J. Ladik, *J. Mol. Spectrosc.*, 27, 79 (1968). Theory: Double-well potential of the hydrogen bond in the guanine-cytosine base pair.

2039. S. Rajagopalan and G. S. Verma, *Nuovo Cimento B*, 67, 13 (1970). Ultrasonic: Mixtures of dioxane and glycerol.

2040. G. N. Ramachandran and R. Chandrasekaran, *Biopolymers*, 10, 935 (1971). Theory: Conformation of dipeptides from empirical potential functions.

2040a. G. N. Ramachandran, R. Chandrasekaran, and R. Chidambaram, *Proc. Indian Acad. Sci.*, *Sect. A*, 74, 270, 284 (1971). Theory: Empirical potential function for peptide N—H...O=C hydrogen bond.

2041. G. N. Ramachandran, C. M. Venkatachalam, and S. Krimm, *Biophys. J.*, 6, 849 (1966). Theory: Configuration of polypeptides based on semiempirical potential functions for hydrogen bond and for van der Waals forces.

2042. K. Ramaswamy, R. Pickai, and S. G. Gnanadesian, *J. Mol. Spectrosc.*, 23, 416 (1967). IR: Solvent shifts, methyl phenols.

2043. K. V. Ramiah and C. A. I. Chary, *Curr. Sci.*, 39, 276 (1970). IR: $\Delta\nu_{OH}$, association of MeOH and *p*-chlorophenol with thioamides.

2044. B. D. N. Rao, P. Venkateswarlu, A. S. N. Murthy, and C. N. R. Rao, *Can. J. Chem.*, 40, 963 (1962). NMR: Self-association, phenol, aniline, thiophenol.

2045. B. D. N. Rao, P. Venkateswarlu, A. S. N. Murthy, and C. N. R. Rao, *Can. J. Chem.*, 40, 387 (1962). NMR: Self-association, ethanol and 2,2,2-trifluoroethanol.

2045a. C. N. R. Rao in *Water: Comprehensive Treatise*, F. Franks, ed., Plenum Press, New York, 1972, Vol. 1, pp. 93-114. Theory: Water structure.

2046. C. N. R. Rao, A. Goel, and A. S. N. Murthy, *Indian J. Chem.*, 9, 56 (1971). Theory: CNDO/2 calculations on water dimers and polymers.

2047. C. N. R. Rao, A. Goel, and A. S. N. Murthy, *J. Chem. Soc. A*,
       190 (1971). Theory: CNDO/2 calculations on water dimer.

2048. C. N. R. Rao, A. Goel, K. G. Rao, and A. S. N. Murthy, *J. Phys.
       Chem.*, 75, 1744 (1971). Theory: Extended Hückel and CNDO/2
       calculations on formamide dimer and formaldehyde-water.

2049. C. N. R. Rao and A. S. N. Murthy, *Theor. Chim. Acta*, 22, 392
       (1971). Theory: CNDO/2 calculations on hydrogen bonds in
       excited states of $>$C$=$O—water and $>$C$=$O—methanol.

2050. C. N. R. Rao and A. S. N. Murthy, *Develop. Appl. Spectrosc.*, 7B,
       54 (1968). Review of spectroscopic studies of hydrogen bonding.

2050a. C. N. R. Rao and A. S. N. Murthy in *Spectroscopy in Inorganic
       Chemistry*, Vol. 1, C. N. R. Rao and J. R. Ferraro, eds., Aca-
       demic Press, New York, 1970, p. 107. Review.

2051. C. N. R. Rao and A. S. N. Murthy, *J. Sci. Ind. Res. (India)*,
       B20, 290 (1961). UV: Self-association, phenols.

2052. S. T. Rao and M. Sundaralingam, *J. Amer. Chem. Soc.*, 92, 4963
       (1970). BIOL, X-Ray: Structure of 3'-*o*-acetyladenosine.

2053. B. V. Rassadin and A. V. Iogansen, *Zh. Prikl. Spektrosk.*, 10,
       524 (1969); *Chem. Abstr.*, 71, 8027a (1969). IR: Intensity of
       ν_{OH} for adducts of phenol with phosphine oxides.

2054. B. V. Rassadin and A. V. Iogansen, *Zh. Prikl. Spektrosk.*, 11,
       184 (1969). Reply to H. Fritzche in *Spectrochim. Acta*, 21, 799
       (1965) on band broadening.

2055. B. V. Rassadin and A. V. Iogansen, *Zh. Prikl. Spektrosk.*, 6, 801
       (1967); [English, 542]. IR: Phenol adducts of carbonyl and
       phosphoryl bases.

2055a. B. V. Rassadin, A. V. Iogansen, V. A. Marchenko, F. S. Yakushin,
       and A. I. Shatenshtein, *Zh. Prikl. Spektrosk.*, 17, 1037 (1972);
       Chem. Abstr., 78, 110043j (1973). IR: Thermodynamic data,
       adducts of Me$_3$COH, phenol.

2056. J. Rassing, *Acta Chem. Scand.*, 25, 1418 (1971). Kinetics:
       Ultrasonic, self-association of amides.

2056a. J. Rassing, *Advan. Mol. Relaxation Processes*, 4, 55 (1972).
       Kinetics: Review, relaxation spectrometry.

2056b. J. Rassing, *J. Chem. Phys.*, 56, 5225 (1972). Kinetics: Ultra-
       sonic, acetic acid/acetone mixtures.

2057. J. Rassing and F. Garland, *Acta Chem. Scand.*, 24, 2419 (1970).
       Kinetics: Ultrasonic absorption (10-300 MHz) of N-methylaceta-
       mide.

2058. J. Rassing and B. N. Jensen, *Acta Chem. Scand.*, 25, 3663 (1971).
       Kinetics: Self-association, butanol.

2059. J. Rassing and B. N. Jensen, *Acta Chem. Scand.*, 24, 855 (1970).
       Kinetics: Self-association, ultrasonic relaxtion, benzyl
       alcohol.

2060. J. Rassing, O. Osterberg, and T. A. Bak, *Acta Chem. Scand.*, 21,
       1443 (1967). Kinetics: Self-association, ultrasonics, benzoic
       acid derivatives.

2060a. M. Raszka and N. O. Kaplan, *Proc. Nat. Acad. Sci.*, *U. S.*, 69, 2025 (1972). NMR: Association of mononucleotides in aqueous solution.

2060b. H. Ratajczak, *J. Mol. Struct.*, 17, 431 (1973). Theory: CNDO/2, charge-transfer theory.

2061. H. Ratajczak, *J. Mol. Struct.*, 3, 27 (1969). IR: Polarized IR transmission spectra of KH_2PO_4 and $NH_4H_2PO_4$.

2061a. H. Ratajczak, *J. Mol. Struct.*, 11, 267 (1972). IR: Polarized spectrum, KH_2AsO_4.

2061b. H. Ratajczak, *J. Phys. Chem.*, 76, 3000, 3991 (1972). Theory: Charge-transfer theory and dipole moment enhancement of adducts.

2061c. H. Ratajczak and W. J. Orville-Thomas, *J. Chem. Phys.*, 58, 911 (1973). Theory: Charge-transfer theory, enhancement of dipole moment in hydrogen-bonded adducts.

2062. H. Ratajczak and W. J. Orville-Thomas, *J. Mol. Struct.*, 1, 449 (1967). IR: Frequencies and bond length in O—H...O systems.

2062a. S. S. Rathi, M. K. Machive, K. Gopalakrishanan, and V. V. S. Murthi, *Curr. Sci.*, 41, 326 (1972). Effect of hydrogen bonding on excited states of coumarins.

2062b. K. Ravindranath and K. V. Ramiah, *Curr. Sci.*, 42, 706 (1973). IR: Amides.

2063. W. T. Raynes and M. A. Raza, *Mol. Phys.*, 17, 157 (1969). NMR: Liquid $(MeO)_4Si$.

2064. D. G. Rea, *J. Chem. Phys.*, 33, 1857 (1960). Raman: $CDCl_3$ association with bases, effect on Raman intensity.

2066. B. S. Reddy and M. A. Viswamitra, *Z. Kristallogr.*, *Kristallogeometric*, *Kristallophys.*, *Kristallchem.*, 131, 237 (1970). BIOL, X-Ray: Morphology and space group of deoxycytidylic acid-5'-monohydrate.

2067. R. L. Redington and K. C. Lin, *J. Chem. Phys.*, 54, 4111 (1971). IR: Self-association, ν_{OH}, matrix-isolated CF_3COOH, CF_3COOD, CH_3COOH, vapor-phase CF_3COOH below 500 cm^{-1}.

2068. R. L. Redington and K. C. Lin, *Spectrochim. Acta*, 27A, 2445 (1971). IR: Spectra of matrix-isolated CF_3COOH, CF_3COOD, $(CF_3COOH)_2$, $(CF_3COOD)_2$.

2069. I. H. Reece and R. L. Werner, *Spectrochim. Acta*, 24A, 1271 (1968). IR: Association of alcohols with various bases.

2070. L. W. Reeves, *Can. J. Chem.*, 39, 1711 (1961). NMR: Association chemical shifts for adducts of substituted acetic acids.

2071. L. W. Reeves, *Can. J. Chem.*, 38, 748 (1960). NMR: Intramolecular orthohydroxy azo compounds.

2072. L. W. Reeves, *Trans. Faraday Soc.*, 55, 1684 (1959). NMR: Carboxylic acids in various solvents.

2073. L. W. Reeves, E. A. Allan, and K. O. Stromme, *Can. J. Chem.*, 38, 1249 (1960). NMR: Intramolecular, phenols and naphthols.

2074. L. W. Reeves and W. G. Schneider, *Trans. Faraday Soc.*, 54, 314 (1958). NMR: Self-association, acetic acid.

2075. L. W. Reeves and W. G. Schneider, *Can. J. Chem.*, 35, 251 (1957).
NMR: ^{1}H chemical shift for CHCl$_3$—benzene complex.

2076. L. W. Reeves and C. P. Yue, *Can. J. Chem.*, 48, 3307 (1970).
NMR: ^{1}H relaxation time, H$_2$O/(CD$_3$)$_2$CO mixtures.

2077. S. Refn, *Spectrochim. Acta*, 17, 40 (1961). IR: Self-association, pyrazolones.

2077a. U. V. Reichert and H. Hartmann, *Z. Naturforsch.*, 27A, 983 (1972).
Microwave: HF at -70°C.

2078. C. Reid, *J. Chem. Phys.*, 30, 182 (1959). Theory: Semiempirical
treatment of the hydrogen bond.

2079. G. N. Reike, Jr. and R. E. Marsh, *Acta Crystallogr.*, 20, 703
(1966). BIOL, X-Ray: 5-Ethyl-6-methyluracil crystals.

2080. R. Rein, G. A. Clarke, and F. E. Harris, *J. Mol. Struct.*, 2,
103 (1968). Theory: Extended Hückel calculations on water
dimers.

2081. R. Rein, P. Claverie, and M. Pollak, *Int. J. Quant. Chem.*, 2,
129 (1968). BIOL: Calculation of London-van der Waals interactions between purine and pyrimidine bases.

2083. R. Rein, N. S. Goel, N. Fukuda, M. Pollak, and P. Claverie,
Ann. N. Y. Acad. Sci., 153, 805 (1969). BIOL: Critical analysis of approximations.

2084. R. Rein and F. E. Harris, *J. Chem. Phys.*, 45, 1797 (1966).
Theory: Pariser-Parr-Pople calculations on hydrogen bond in
guanine-cytosine base pair.

2085. R. Rein and F. E. Harris, *J. Chem. Phys.*, 43, 4415 (1965).
Theory: Semiempirical treatment of hydrogen bonds in guanine-
cytosine base pair.

2086. R. Rein and F. E. Harris, *J. Chem. Phys.*, 41, 3393 (1964).
Theory: Semiempirical treatment of the hydrogen bond in the
guanine-cytosine base pair.

2087. R. Rein and F. E. Harris, *J. Chem. Phys.*, 42, 2177 (1965).
Theory: Potential curves for proton motion in guanine-cytosine
base pair.

2088. R. Rein and F. E. Harris, *Science*, 146, 649 (1964). Theory:
Proton tunneling in the hydrogen bond of the guanine-cytosine
base pair.

2088a. R. Rein and J. Ladik, *J. Chem. Phys.*, 40, 2466 (1964). Theory:
Guanine-cytosine base pair, mutagenic effect of radiation.

2089. R. Rein and M. Pollak, *J. Chem. Phys.*, 47, 2039 (1967). Theory:
Guanine-cytosine, semiempirical perturbation method.

2090. R. Rein, M. S. Rendell, and J. P. Harlos in *Molecular Orbital
Studies in Chemical Pharmacology*, L. B. Kier, ed., Springer-
Verlag, New York, 1970, p. 191. BIOL: Molecular recognition,
evolution of nucleic acids, quantum chemical considerations.

2091. R. Rein and S. Svetina, *Int. J. Quant. Chem., Symp.*, 171 (1967).
Theory: Vibrational levels of proton in hydrogen bond of
guanine-cytosine base pairs.

2092. H. M. Relles, *J. Org. Chem.*, 35, 4280 (1970). NMR: Associa-
       tion of substituted phenols with Cl⁻.

2093. J. Reuben, *J. Amer. Chem. Soc.*, 91, 5725 (1969). NMR: ^{17}O
       chemical shifts of H_2O and CH_3COOH.

2094. J. Reuben and D. Samuel, *Israel J. Chem.*, 1, 229 (1963). NMR:
       ^{17}O shifts, methanol.

2095. J. Reuben, A. Tzalmona, and D. Samuel, *Proc. Chem. Soc.*, 353
       (1962). NMR: ^{17}O shifts for water-acetone mixtures.

2095a. S. A. Rice and J. L. Wood, *J. Chem. Soc. Faraday Trans. 2*, 69,
       87, 91 (1973). IR: Gas-phase, trifluoroethanol-trimethylamine,
       effect of temperature of ν_s.

2095b. P. E. Rider and R. M. Hammaker, *Spectrochim. Acta*, 29A, 501
       (1973). Near-IR: K, di-*t*-butylcarbinol adducts.

2096. D. P. Ridge and J. L. Beauchamp, *J. Amer. Chem. Soc.*, 93, 5925
       (1971). Ion Cyclotron Resonance Spectroscopy: Organic ions in
       the gas phase.

2097. J. M. Rifkind and G. L. Eichhorn, *Biochemistry*, 9, 1753 (1970).
       BIOL: ORD, interaction of nucleotides with basic popypeptides.

2098. M. Riley, B. Maling, and M. J. Chamberlin, *J. Mol. Biol.*, 20,
       359 (1966). BIOL: Characterization of two- and three-stranded
       A—T and A—U homopolymer complexes.

2099. M. Riley and A. V. Paul, *J. Mol. Biol.*, 50, 439 (1970). BIOL:
       Formation and dissociation of two- and three-stranded complexes
       containing polyriboadenylic acid and polyribobromouridylic acid.

2100. J. W. Ring and P. A. Egelstaff, *J. Chem. Phys.*, 51, 762 (1969).
       Neutron Scattering: Liquid HF.

2101. C. D. Ritchie and R. E. Uschold, *Amer. Chem. Soc., Div. Petrol.
       Chem., Prepr.*, 13, A11-A14 (1968). Kinetics: Proton-transfer
       in dimethyl sulfoxides.

2102. H. A. Rizk and N. Youssef, *Z. Phys. Chem. (Leipzig)*, 239, 273
       (1968); 244, 413 (1969). Self-association constants for iso-,
       n-, *sec*-, *tert*-butanol at 30°C are 3.31, 2.99, 2.91, and 2.31,
       respectively.

2102a. J. D. Roberts, *Notes on Molecular Orbital Theory*, W. A. Benja-
       min, New York, 1961.

2103. E. A. Robinson, H. D. Schreiber, and J. N. Spencer, *Spectrochim.
       Acta*, 28A, 397 (1972). Theory: Potential function model of
       the O—H...halogen intramolecular bond.

2104. E. A. Robinson, H. D. Schreiber, and J. N. Spencer, *J. Phys.
       Chem.*, 75, 2219 (1971). IR: Temperature dependence of molar
       absorptivity of ν_{OH} of phenols in THF, C_6H_6, and C_6H_5Cl.

2105. R. E. Rodd, R. E. Miller, and W. F. K. Wynne-Jones, *J. Chem.
       Soc.*, 2790 (1961). IR: Intramolecular, maleic acid.

2105a. B. M. Rode, *J. Chem. Soc., Faraday Trans. 2*, 69, 1439 (1973).
       Theory: CNDO/2 on HCOOH...Cl⁻.

2105b. B. M. Rode, *Chem. Phys. Lett.*, 20, 366 (1973). Theory: CNDO/2
       on HCOOH...Cl⁻.

2105c. B. M. Rode, A. Engelbrecht, and W. Jakubetz, *Chem. Phys. Lett.*, 18, 285 (1973). Theory: CNDO/2 on acetic acid and halogenated acetic acid dimers.

2105d. N. Roesch, *Chem. Phys.*, 1, 220 (1973). Theory: Proton tunneling.

2106. M. T. Rogers and J. L. Burdett, *Can. J. Chem.*, 43, 1516 (1965). NMR: Intramolecular, β-diketones and β-keto esters.

2107. D. Rohrer and M. Sundaralingam, *Acta Crystallogr.*, *Sect. B*, 26, 546 (1970). BIOL, X-Ray: Structure and conformation of dihydrouracil.

2108. D. Rohrer and M. Sundaralingam, *J. Amer. Chem. Soc.*, 92, 4956 (1970). BIOL, X-Ray: Crystal and molecular structure of α-d-2'-amino-2-deoxyadenosine monohydrate.

2109. D. Rohrer and M. Sundaralingam, *J. Amer. Chem. Soc.*, 92, 4950 (1970). BIOL, X-Ray: Molecular structure and conformation of 2-pseudouridine monohydrate.

2110. G. Roland and G. Duyckaerts, *Spectrochim. Acta*, 22, 793 (1966). IR: Tributyl phosphine oxide-water mixtures.

2111. A. Ron and D. F. Hornig, *J. Chem. Phys.*, 39, 1129 (1963). Near-IR: Crystalline HCl at 77°K.

2111a. C. C. J. Roothaan, *Rev. Mod. Phys.*, 23, 69 (1951).

2112. M. C. Rose and J. Stuehr, *J. Amer. Chem. Soc.*, 90, 7205 (1968); 93, 4350 (1971); 94, 5532 (1972). Kinetics: Rates of internally hydrogen-bonded systems, temperature-jump technique.

2112a. N. J. Rose and R. S. Drago, *J. Amer. Chem. Soc.*, 84, 2037 (1962). Equation for determining K from spectroscopic data.

2113. P. D. Ross and R. L. Scruggs, *J. Mol. Biol.*, 45, 567 (1969). BIOL: Heat of reaction between polyribocytidylic acid and polyriboinosine acid.

2113a. J. Rossarie, J. C. Lavalley, and R. Romanet, *Compt. Rend. Acad. Sci.*, 273B, 185 (1971). Near-IR: Simultaneous excitation, $CH_3CN...HOR$.

2114. D. R. Rosseinsky and H. Kellawi, *J. Chem. Soc. A*, 1207 (1969). Error Analysis: Statistical assessment of association constants.

2115. G. P. Rossetti and K. E. van Holde, *Biochem. Biophys. Res. Commun.*, 26, 717 (1967). BIOL: Sedimentation equilibrium studies, association of adenosine-5'-phosphate in H_2O.

2116. H. S. Rossotti, *J. Inorg. Nucl. Chem.*, 33, 2037 (1971). Review: Anomalous water.

2116a. W. G. Rothschild, *J. Amer. Chem. Soc.*, 94, 8676 (1972). BIOL: Binding of hydrogen donors by peptide groups of lactams.

2117. J. Rouirere and J. Salvinien, *J. Chim. Phys.*, 66, 149 (1969). IR: K, adducts of Me_3SiOH and Me_3COH with bases and with phenols.

2118. M. M. Rousselot, *Compt. Rend. Acad. Sci.*, 262C, 26 (1966). NMR: Self-association, mercaptans.

2119. M. M. Rousselot and M. Martin, *Compt. Rend. Acad. Sci.*, 262C, 1445 (1966). NMR: Association chemical shifts for thiol adducts.

2120. R. S. Roy, *J. Indian Chem. Soc.*, 46, 739 (1969). Theory: Estimation of hydrogen bond length using an empirical length-frequency relation.

2121. M. Sh. Rozenberg and A. V. Iogansen, *Opt. Spektrosk.*, 31, 711 (1971); *Chem. Abstr.*, 76, 65721z (1972). IR: Torsional γ_{OH} vibrations of phenol adducts.

2121a. M. Sh. Rozenberg, A. V. Iogansen, and A. A. Mashkovskii, *Spectrosc. Lett.*, 5, 75 (1972). IR: Correlation of torsional frequency of AH groups with ΔH.

2121b. J. Roziere and J. Potier, *J. Mol. Struct.*, 13, 91 (1972). IR: $(H_5O_2^+)Cl^-$.

2121c. J. Roziere and J. Potier, *J. Inorg. Nucl. Chem.*, 35, 1179 (1973). IR: $(H\cdot(H_2O)_3)^+$.

2122. B. F. Rozsnyai and J. Ladik, *J. Chem. Phys.*, 52, 5711 (1970). BIOL: Calculation of the effects of hydration and divalent metal ions on nucleotide base pairs.

2123. H. Ruben, I. Olovsson, A. Zalkin, and D. H. Templeton, *Acta Crystallogr.*, 29B, 2963 (1973). X-Ray: Sodium chromate tetrahydrate.

2124. J. Rubin and G. S. Panson, *J. Phys. Chem.*, 69, 3089 (1965). Near-IR: Thermodynamic data, adducts of phenol with pyridine in various solvents.

2125. J. Rubin, B. Z. Senkowski, and G. S. Panson, *J. Phys. Chem.*, 68, 1601 (1964). Near-IR: K, phenol-pyridine adducts.

2127. R. Rusakowicz, G. W. Byers, and P. A. Leermakers, *J. Amer. Chem. Soc.*, 93, 3263 (1971). Electronic: Inversion of n, π^* and π,π^* states for aromatic carbonyl compounds in hydrogen bonding solvents.

2128. S. N. Rustamova et al., *Zh. Org. Khim.*, 7, 770 (1971); *Chem. Abstr.*, 75, 34833n (1971). IR: Intramolecular, 4-R-substituted-2-morpholinomethylphenols.

2130. Yu. S. Ryabokobylko and A. V. Chekunov, *Teor. Eksp. Khim.*, 5, 550 (1969); *Chem. Abstr.*, 72, 24980g (1970). IR: Correlation of $\Delta\nu_{OH}$ with Hammett constants for adducts of substituted phenols.

2131. Yu. S. Ryabokobylko and A. V. Chekunov, *Zh. Prikl. Spektrosk.*, 10, 603 (1969); *Chem. Abstr.*, 71, 90580g (1969). IR: Intramolecular, salicylaldehydes.

2132. Yu. S. Ryabokobylko and A. V. Chekunov, *Zh. Prikl. Spektrosk.*, 10, 304 (1969); *Chem. Abstr.*, 71, 43882x (1969). IR: Intramolecular, 4- and 5-substituted salicylaldehydes.

2133. V. Ya. Ryadov and N. I. Furashov, *Izv. Vyssh. Ucheb. Zaved. Radiofiz.*, 11, 1138 (1968); *Chem. Abstr.*, 70, 42418t (1969). Microwave: H_2O dimers.

2134. J. L. Ryan, *Inorg. Chem.*, $\underline{3}$, 211 (1964). UV-Visible: Hydrogen bonds to $UCl_6{}^{2-}$ and $UBr_6{}^{2-}$.

2135. Ya. I. Ryskin, G. P. Stavitskaya, and A. K. Shirvinskaya, *Izv. Akad. Nauk SSSR, Neorg. Mater.*, $\underline{8}$, 312 (1972); *Chem. Abstr.*, $\underline{76}$, 119418b (1972). IR: Hydrogermanate, α-2CaO—GeO_2—H_2O.

2135a. Ya. I. Ryskin and G. P. Stavitskaya, *Hydrogen Bonding and Structure of Hydrosilicates*, Nauka, Lenigrad. Otd., Leningrad, USSR, 1972; *Chem. Abstr.*, $\underline{77}$, 144070k (1972). Review.

2136. Ya. I. Ryskin, S. P. Zhdanov, and I. V. Gessan, *Teor. Eksp. Khim.*, $\underline{5}$, 422 (1969); *Chem. Abstr.*, $\underline{71}$, 65597k (1969). IR: Association of Y-type zeolites with organic bases.

2137. B. Ryutani, *Nippon Kagaku Zasshi*, $\underline{82}$, 520 (1961). IR: $\Delta\nu_{CH}$ deformation shifts for acetylenes in various solvents.

2137a. G. L. Ryzhova and A. M. Pogaleeva, *Zh. Fiz. Chim.*, $\underline{47}$, 1100, 1104 (1973); *Chem. Abstr.*, $\underline{79}$, 71765v (1973). UV, IR, NMR: Adducts of nitrophenols with secondary aliphatic amines.

2138. J. R. Sabin, *J. Chem. Phys.*, $\underline{56}$, 45 (1972). Theory: CNDO/2 study of $(HF)_6$ chains.

2138a. J. R. Sabin, *Theor. Chim. Acta*, $\underline{27}$, 69 (1972). Theory: CNDO/2 calculations on HF—N_2H_2 dimer, hydrogen bond to a π-system.

2139. J. R. Sabin, *J. Chem. Phys.*, $\underline{54}$, 4675 (1971). Theory: Gaussian SCF calculations on H_2S—HS^-.

2140. J. R. Sabin, *J. Amer. Chem. Soc.*, $\underline{93}$, 3613 (1971). Theory: $(H_2S)_2$ Gaussian SCF and CNDO/2.

2141. J. R. Sabin, *Int. J. Quant. Chem.*, $\underline{2}$, 31 (1968). Theory: Pariser-Parr-Pople calculations on pyridine-pyrrole.

2142. J. R. Sabin, *Int. J. Quant. Chem.*, $\underline{2}$, 23 (1968). Theory: Pariser-Parr-Pople calculations on pyridine-pyridinium ion.

2143. J. R. Sabin, S. F. Fischer, and G. L. Hofacker, *Int. J. Quant. Chem.*, $\underline{3S}$, 257 (1969). Theory: Proton-phonon coupling in a linear model of ice.

2144. J. R. Sabin, R. E. Harris, T. W. Archibald, P. A. Kollman, and L. C. Allen, *Theor. Chim. Acta*, $\underline{18}$, 235 (1970). Theory: Gaussian basis SCF calculations on cyclic H_2O hexamers.

2145. L. J. Sacks, R. S. Drago, and D. P. Eyman, *Inorg. Chem.*, $\underline{7}$, 1484 (1968). Gas-phase, $CHCl_3$ adduct of $(CH_3)_2NH$.

2146. I. D. Sadekov, V. I. Minkin, and A. E. Lutskii, *Usp. Khim.*, $\underline{39}$, 380 (1970); [English, 179]. Review: Effect of intramolecular hydrogen bonds on reactivity of organic compounds.

2147. A. J. Sadlej and J. Dabrowski, *J. Mol. Struct.*, $\underline{7}$, 1 (1971). Theory: Potential surface for proton motion in a 3-center hydrogen bond.

2148. W. Saenger and D. Suck, *Nature*, $\underline{227}$, 1046 (1970). BIOL, X-Ray: 1-Methyl-4-thiouracil: 9-methyladenine complex.

2149. W. Saffioti and A. Novak, *Compt. Rend. Acad. Sci., Ser. B*, $\underline{276}$, 709 (1973). IR: Imidazole-phenol adducts.

2150. G. J. Safford, P. C. Schaffer, P. S. Leung, G. F. Doebbler, G. W. Brady, and E. F. X. Lyden, *J. Chem. Phys.*, 50, 2140 (1969). Neutron Scattering, X-Ray: Aqueous solutions of $\overline{Me}_2SO$ and Me_2SO_2.

2150a. S. Saha, *Indian J. Chem.*, 11, 250 (1973). UV: Adducts of quinoline and 5,6-benzoquinoline with aliphatic alcohols.

2151. E. L. Saier, L. R. Cousins, and M. R. Basila, *J. Chem. Phys.*, 41, 40 (1964). IR: Asymmetry of OH band in alcohols.

2152. H. Saito and K. Nukada, *J. Amer. Chem. Soc.*, 93, 1072 (1971). NMR: ^{14}N hydrogen bond shifts of pyrroles and indole.

2153. H. Saito, K. Nukada, H. Kato, T. Yonezawa, and F. Fukui, *Tetrahedron Lett.*, 111 (1965). NMR: ^{14}N NMR spectrum of pyridine in methanol.

2154. H. Saito, K. Nukada, T. Kobayashi, and K. Morita, *J. Amer. Chem. Soc.*, 89, 6605 (1967). NMR: Thermodynamic data, $CDCl_3$ adduct with ethylenimine.

2154a. H. Saito, Y. Tanaka, and S. Nagata, *J. Amer. Chem. Soc.*, 95, 324 (1973). NMR: ^{14}N chemical shifts, 5- and 6-membered N-heterocycles.

2155. H. Saito, Y. Tanaka, and K. Nukada, *J. Amer. Chem. Soc.*, 93, 1077 (1971). NMR: ^{14}N, self-association, formamide, N-methylacetamide.

2156. T. K. Sakano, S. L. Rock, and R. M. Hammaker, *Spectrochim. Acta*, 25A, 1195 (1969). Near-IR: 2,6-di-*t*-Butylphenol in *n*-heptane.

2157. T. D. Sakore and H. M. Sobell, *J. Mol. Biol.*, 43, 77 (1969). BIOL, X-Ray: 9-Ethyl-8-bromoadenine and 9-ethyl-8-bromohypoxan-thine, 1:1 complexes.

2158. T. D. Sakore, H. M. Sobell, F. Mazza, and G. Kartha, *J. Mol. Biol.*, 43, 385 (1969). BIOL, X-Ray: 9-Ethyl-2,6-diaminopurine and 1-methylthymine, 9-ethyl-2,6-diaminopurine and 1-methyl-5-iodouracil, 1:2 complexes.

2159. T. D. Sakore, H. M. Sobell, and S. S. Tavale, *J. Mol. Biol.*, 43, 375 (1969). BIOL, X-Ray: 9-Ethyl-8-bromoadenine and 1-methyl-5-bromouracil, 1:1 complex.

2160. T. D. Sakore, S. S. Tavale, and H. M. Sobell, *J. Mol. Biol.*, 43, 361 (1969). BIOL, X-Ray: 9-Ethyladenine and 1-methyl-5-iodouracil, 1:2 complex.

2161. B. D. Saksena and P. H. Trevedi, *Trans. Faraday Soc.*, 58, 2082 (1962). IR: Thiophenol, $HSCH_2COOH$ in various solvents.

2162. N. Salaj, *Acta Chem. Scand.*, 24, 953 (1970). Theory: Point-charge model of N...H—O.

2163. N. Salaj, *Acta Chem. Scand.*, 23, 1534 (1969). Theory: Point-charge model of $H_9O_4^+$.

2164. J. A. Salthouse and T. C. Waddington, *J. Chem. Soc. A*, 28, (1966). IR: $HClNO_3^-$, $DClNO_3^-$, $DBrCl^-$, $DClI^-$, DCl_2^-, DBr_2^-, DI_2^-.

2165. T. E. Sampson and J. M. Carpenter, *Neutron Inelastic Scattering*,
       Proc. Symp., *4th*, 1, 491 (1968); *Chem. Abstr.*, 70, 73046z (1969).
       Thermal neutron inelastic scattering by CH₃OH and CH₃SH.

2166. C. Sander and P. O. P. Ts'o, *Biopolymers*, 9, 765 (1970). BIOL:
       Interaction of purine with DNA, tRNA, poly A, poly C, and poly
       A—poly U complex.

2167. C. Sandorfy, *Coll. Intern. du Centre Nat. de la Recherche
       Scient.*, 195, 237 (1971). A short review, mainly of theory.

2167a. C. Sandorfy, *Can. J. Spectrosc.*, 17, 24 (1972). IR: Anharmoni-
       city and hydrogen bonding, review.

2168. A. A. Sandoval, M. W. Sandoval, E. Lin, and Kuang Lu Cheng,
       J. Magn. Resonance, 3, 258 (1970). NMR: Intramolecular,
       diethylenetriaminepentaacetic acid and triethylenetetramine-
       hexaacetic acid.

2168a. G. V. Sandul *et al.*, *Zh. Prikl. Spektrosk.*, 16, 928 (1972);
       Chem. Abstr., 77, 47512y (1972). IR, NMR: 2,2'- and 4,4'-
       methylene- and thiobis(phenols).

2168b. G. V. Sandul, V. S. Kuts, and V. D. Pokhodenko, *Teor. Eksp.
       Khim.*, 8, 340 (1972); *Chem. Abstr.*, 78, 15123b (1973). NMR:
       Adducts of substituted phenols with substituted anilines,
       thermodynamic data.

2168c. G. V. Sandul, V. S. Kuts, and V. D. Pokhodenko, *Teor. Eksp.
       Khim.*, 9, 330 (1973); *Chem. Abstr.*, 79, 77883d (1973). NMR:
       Thermodynamic data, 2,4,6,-(Me₃C)₂RC₆H₂OH adducts with MeOC₆H₄R'.

2169. A. B. Sannigrahi and A. K. Chandra, *J. Phys. Chem.*, 67, 1106
       (1963). UV: K, adducts of diphenylamine, pyrrole, 2-methyl
       indole, and carbazole with bases.

2169a. T. Sano, N. Tatsumoto, T. Nuva, and T. Yasunaga, *Bull. Chem.
       Soc. Jap.*, 45, 2669, 2673 (1972). Kinetics: Ultrasonic absorp-
       tion, carboxylic acids.

2169b. V. Sara, J. Moravec, V. Horak, and M. Horak, *Collect. Czech.
       Chem. Commun.*, 34, 2390 (1969). IR: Thermodynamic data, phenol
       adducts of sulfides and sulfoxides.

2170. E. J. Sare, C. T. Moynihan, and C. A. Angell, *J. Phys. Chem.*,
       77, 1869 (1973). NMR: Hydrogen bonding in aqueous electrolyte
       solutions.

2170a. M. T. Saroechi, I. Courtois, and W. Guschlbauer, *Eur. J. Bio-
       chem.*, 14, 411 (1970). BIOL: Specific complex formation
       between poly C and guanosine or guanylic acid.

2171. G. Sarojini and A. N. Murty, *Indian J. Pure Appl. Phys.*, 6, 558
       (1968). Dielectric: Thermodynamic data, amine-phenol adducts.

2172. K. Sasaki and K. Arakawa, *Bull. Chem. Soc. Jap.*, 46, 2738 (1973).
       Kinetics: H₂O-tetramethylurea.

2173. I. Satake, M. Arita, H. Kimizuka, and R. Matsuura, *Bull. Chem.
       Soc. Jap.*, 39, 597 (1966). NMR: Theoretical dilution curves
       for EtOH—CCl₄, H₂O—acetone, and H₂O—dioxane.

2174.   D. P. N. Satchell and J. L. Wordell, *Trans. Faraday Soc.*, 61, 1199 (1965). IR: K, carboxylic acid dimers in *o*-dichlorobenzene.

2175.   K. Sato and A. Nishioka, *Bull. Chem. Soc. Jap.*, 44, 2042 (1971). Kinetics: Proton spin-lattice relaxation studies in acetic acid.

2176.   T. Sato, Y. Kyogoku, S. Higuchi, Y. Mitsui, Y. Iitaka, M. Tsuboi, and K. I. Niura, *J. Mol. Biol.*, 16, 180 (1966). BIOL: Double helix of rice dwarf virus RNA.

2177.   Y. Sato, *Nippon Kagaku Zasshi*, 79, 538 (1958). Near-IR: Thermodynamic data, adducts of methanol.

2178.   J. Saunders and J. R. K. May, *Chem. Ind. (London)*, 1355 (1963). UV: Intramolecular, malonaldehyde.

2179.   J. E. Saunders, F. F. Bentley, and J. E. Katon, *Appl. Spectrosc.*, 22, 286 (1968). Far-IR: Self-association, aliphatic monocarboxylic acids.

2180.   M. Saunders and J. B. Hyne, *J. Chem. Phys.*, 31, 270 (1959). NMR: Self-association, equilibrium constant determination.

2181.   M. Saunders and J. B. Hyne, *J. Chem. Phys.*, 29, 1319 (1958). NMR: Self-association, K, phenols.

2182.   M. Saunders and J. B. Hyne, *J. Chem. Phys.*, 29, 253 (1958). NMR: Self-association, K, alcohols.

2183.   P. Sauvageau and C. Sandorfy, *Can. J. Chem.*, 38, 1901 (1960). IR: Amine hydrohalides.

2184.   I. Savatinova, P. Simova, and M. Markov, *Zh. Prikl. Spektrosk.*, 10, 498 (1969); *Chem. Abstr.*, 70, 119870u (1969). Raman: 95 and 113 cm^{-1} of triglycine fluoroberyllate assigned to N—H...O and N—H...F.

2185.   V. A. Savel'ev and N. D. Sokolov, *Teor. Eksp. Khim.*, 6, 174 (1970); *Chem. Abstr.*, 73, 70048c (1970). IR: Isotopic stretching of hydrogen bonds O...O in carboxylic acid dimers.

2186.   J. P. Saxena, *J. Indian Chem. Soc.*, 45, 628 (1968). UV: Spectra of salts of 1-azacarbazole, 2-azacarbazole, and their 6-nitro derivatives.

2186a. L. J. Schaad, *Tetrahedron*, 26, 4115 (1970). Comparison of total and orbital energies.

2186b. H. F. Schaefer and C. F. Bender, *J. Chem. Phys.*, 55, 1720 (1971). CI calculations on H_2O monomer.

2187.   T. Schaefer and W. G. Schneider, *J. Chem. Phys.*, 32, 1224 (1960). NMR: Heterocyclic compounds in various solvents.

2188.   T. Schaefer and W. G. Schneider, *J. Chem. Phys.*, 32, 1218 (1960). NMR: *p*-Substituted toluenes in various solvents.

2188a. R. Schano, H. E. Affsprung, and F. Kohler, *Monatsh. Chem.*, 104, 389 (1973). IR, NMR: 1:1 adducts of H_2O with dioxane, Et_3N.

2189.   A. Schellenberger, W. Beer, and G. Oehne, *Spectrochim. Acta*, 21, 1345 (1965). IR: Intramolecular, gas-phase, α-keto acids.

2190.   A. Schellenberger and G. Oehne, *Z. Phys. Chem. (Leipzig)*, 227, 112 (1964). IR: Intramolecular, α-keto acids.

2190a. J. R. Scherer, M. K. Go, and S. Kint, *J. Phys. Chem.*, 77, 2108 (1973). Raman: H_2O-Me_2SO mixtures.

2191.   J. Schiffer and D. F. Hornig, *J. Chem. Phys.*, 49, 4150 (1968). Theory, IR: Liquid water.

2191a. D. Schioeberg and G. Zundel, *J. Chem. Soc., Faraday Trans. 2*, 69, 771 (1973). IR: $[H(H_2O)_3]^+$.

2192.   E. O. Schlemper, W. C. Hamilton, and S. J. La Placa, *J. Chem. Phys.*, 54, 3990 (1971). Neutron Diffraction: Intramolecular asymmetric hydrogen bond in *bis*(2-amino-2-methyl-3-butanone oximato)nickel(II) chloride monohydrate.

2193.   P. von R. Schleyer, *J. Amer. Chem. Soc.*, 83, 1368 (1961). IR: Intramolecular, propane-1,3-diols.

2194.   P. von R. Schleyer and R. West, *J. Amer. Chem. Soc.*, 81, 3164 (1959). IR: $\Delta\nu_{OH}$ for phenol and methanol adducts of organic halides.

2195.   H. Schmid, *Monatsh. Chem.*, 101, 194 (1970). Kinetics: Kinetic determination of entropy of hydrogen bonding of H_2O.

2196.   H. Schmidbaur, *Chem. Ber.*, 97, 830 (1964). NMR: Self-association, trimethyl silanol.

2197.   H. Schmidbaur, M. Schmidt, and W. Siebert, *Chem. Ber.*, 97, 3374 (1964). NMR: Sulfanes, H_2S_x.

2198.   V. H. Schmidt and E. A. Uehling, *Phys. Rev.*, 126, 447 (1962). NMR: KD_2PO_4.

2199.   K. S. Schmitz and J. M. Schurr, *Biopolymers*, 9, 697 (1970). BIOL: Cooperative binding of adenosine by poly U.

2200.   C. D. Schmulbach and D. M. Hart, *J. Org. Chem.*, 29, 3122 (1964). Thermodynamic data, phenol adducts of N-methyl lactams.

2201.   W. G. Schneider, *J. Phys. Chem.*, 66, 2653 (1962). NMR: Polar solutes associated with C_6H_6.

2202.   W. G. Schneider in *Hydrogen Bonding*, D. Hadži, ed., Pergamon Press, New York, 1957, p. 55. NMR: Gaseous and liquid ethanol.

2203.   W. G. Schneider, H. J. Bernstein, and J. A. Pople, *J. Chem. Phys.*, 28, 601 (1958). NMR: ^1H chemical shifts for hydrides in gaseous and liquid states.

2205.   E. E. Schrier, M. Pottle, and H. A. Scheraga, *J. Amer. Chem. Soc.*, 86, 3444 (1964). Influence of hydrogen and hydrophobic bonds on the stability of carboxylic acid dimers in H_2O.

2206.   R. Schroeder and E. R. Lippincott, *J. Phys. Chem.*, 61, 921 (1957). Theory: Potential function model for bent hydrogen bonds.

2207.   L. H. Schulman and R. W. Chambers, *Proc. Natl. Acad. Sci. U. S.*, 61, 308 (1968). BIOL: tRNA, structural basis for alanine acceptor activity.

2208. P. Schuster, *Theor. Chim. Acta*, 19, 212 (1970). Theory: CNDO and INDO studies of hydrogen-bonded dimers including some small organic proton acceptors.

2208a. P. Schuster, *Z. Chem.*, 13, 41 (1973). Theory: Review.

2209. P. Schuster, *Int. J. Quant. Chem.*, 3, 851 (1969). Theory: $>$C=O...H—O potential curves by CNDO/2.

2210. P. Schuster, *Chem. Phys. Lett.*, 3, 433 (1969). Theory: CNDO/2 calculations on hydrogen bond in acetyl acetone.

2211. P. Schuster and T. Funck, *Chem. Phys. Lett.*, 2, 587 (1968). Theory: CNDO/2 study of formic acid dimers.

2212. M. E. Schwartz and L. J. Schaad, *J. Chem. Phys.*, 47, 5325 (1967). Calculations on H_3^+.

2213. E. Schwarzmann, *Z. Naturforsch. B*, 24, 1104 (1969). IR: Review, crystalline hydrated oxides.

2214. E. Schwarzmann and H. Sparr, *Z. Naturforsch. B*, 24, 8 (1969). IR: Hydrogen bridge bonds in α-AlO$_2$H, α-ScO$_2$H, α-FeO$_2$H, and GaO$_2$H with diaspore structure.

2215. M. P. Schweizer, A. D. Broom, P. O. P. Ts'o, and D. P. Hollis, *J. Amer. Chem. Soc.*, 90, 1042 (1968). BIOL: PMR, stacking of 5'-, 3'-, and 2'-nucleoside monophosphates in H_2O.

2216. M. P. Schweizer, S. I. Chan, and P. O. P. Ts'o, *J. Amer. Chem. Soc.*, 87, 5241 (1965). BIOL: PMR, associations of pyrimidine nucleosides and their interactions with purine.

2217. J. F. Scott and P. Schofield, *Proc. Natl. Acad. Sci. U. S.*, 64, 931 (1969). BIOL: CD of tRNA.

2218. R. Scott, D. De Palma, and S. Vinogradov, *J. Phys. Chem.*, 72, 3192 (1968). UV: Proton transfer in complexes of *p*-nitrophenol with Et$_3$N, *n*-butylamine.

2219. R. Scott and S. Vinogradov, *J. Phys. Chem.*, 73, 1890 (1969). UV: Proton transfer in complexes of *p*-nitrophenol with Et$_3$N, *n*-butylamine.

2220. R. L. Scott, *Rec. Trav. Chem.*, 75, 787 (1956). Modification of Benesi-Hildebrand equation for calculating equilibrium constants.

2221. R. L. Scott and D. V. Fenby, *Ann. Rev. Phys. Chem.*, 20, 111 (1969). Hydrogen bonding, weak charge-transfer complexes.

2222. J. D. Schriebner and J. A. Miller, *J. Org. Chem.*, 32. 2348 (1967). IR: Intramolecular, *o*-methylmercapto derivatives of N-methylaniline and N-methyl-4-aminoazobenzene.

2223. R. L. Scruggs and P. D. Ross, *J. Mol. Biol.*, 47, 29 (1970). BIOL: Calorimetric study of monomer-polymer complexes formed by polyribouridylic acid and some adenine derivatives.

2224. A. V. Sechkarev, A. K. Petrov, and B. V. Makarow, *Opt. Spektrosk.*, supplement 3, 253 (1967); [English, 128 (1968)]. IR: Self-association, benzoic acid derivatives.

2225. R. M. Seel and N. Sheppard, *Spectrochim. Acta*, 25A, 1295 (1969). IR: Gas-phase, adducts of Me$_2$O with HCN and DCN.

2226. R. M. Seel and N. Sheppard, *Spectrochim. Acta*, 25A, 1287 (1969). IR: Solid phase spectra of Me_2O—HX adducts.

2227. I. Segal, *J. Phys. Chem.*, 65, 697 (1961). IR: $EtNH_2$ in $CHCl_3$.

2228. G. Sellier and B. Wojtkowiak, *Compt. Rend. Acad. Sci., Ser. C*, 271, 1341 (1970). ΔH, phenol adduct of 1-iodo-1-pentyne.

2229. G. Sellier and B. Wojtkowiak, *Compt. Rend. Acad. Sci.*, 265C, 75 (1967). IR: Interpretation of organic halide Δν-ΔH data.

2230. G. Sellier and B. Wojtkowiak, *J. Chim. Phys.*, 65, 936 (1968). Determination of hydrogen bond energy.

2231. K. Semba, *Bull. Chem. Soc. Jap.*, 34, 722 (1961). UV: K, *p*-nitrophenol-ether complex.

2232. K. Semba, *Bull. Chem. Soc. Jap.*, 33, 1640 (1960). UV: Intramolecular, *o*-substituted nitrobenzene derivatives.

2233. Yu. D. Semchikov, A. V. Ryabov, and V. N. Kashaeva, *Vysokomol. Soedin., Ser. B*, 12, 381 (1970); *Chem. Abstr.*, 73, 35882y (1970). IR: Copolymerization of N-vinylpyrrolidone in a carboxylic acid medium.

2234. T. Seno and S. Nishimura, *Biochim. Biophys. Acta*, 157, 97 (1968). BIOL: Specific hybridization of amino acid-specific tRNA and DNA.

2235. G. B. Sergeev, S. V. Zenin, V. A. Batyuk, L. P. Karunina, and T. D. Nekipelova, *Zh. Fiz. Khim.*, 43, 985 (1969); [English, 544]. NMR: Association of phosphoric acid esters with alcohols.

2236. M. Servanton, J. Biais, and B. Lemanceau, *J. Chim. Phys.*, 67, 800, 806 (1970). NMR: Self-association, acetic acid, alcohols.

2237. M. P. Serve and A. W. Bryant, *J. Org. Chem.*, 36, 3236 (1971). IR: Intramolecular, 1,2-diphenylethanol.

2237a. W. Seszler and G. Zundel, *Chem. Phys. Lett.*, 14, 356 (1972). IR: Aqueous solutions of purine.

2237b. W. Seszler and G. Zundel, *Z. Phys. Chem.*, 79, 180 (1972). IR: Double-minimum potential, proton-transfer, aqueous solutions of protonated amines.

2238. I. I. Shabalin and E. A. Kiva, *Zh. Prikl. Spektrosk.*, 13, 501 (1970); *Chem. Abstr.*, 74, 58867g (1971). IR: Intramolecular, O—H...π in hydroperoxide of *p*-cymene.

2239. R. R. Shagidullin, I. P. Lipatova, I. A. Nuretdinov, and S. A. Samartseva, *Dokl. Akad. Nauk SSSR*, 211, 1363 (1973); *Chem. Abstr.*, 79, 145816m (1973). IR: Phenol association with P—Se and P—Te groups.

2239a. R. R. Shagidullin *et al.*, *Izv. Akad. Nauk SSSR, Ser. Khim.*, 847 (1972); *Chem. Abstr.*, 77, 100303j (1972). IR: Self-association, phosphorus(V) dithio acids.

2239b. N. N. Shapet'ko, *Zh. Org. Khim.*, 8, 2226 (1972); *Chem. Abstr.*, 78, 7176h (1973). NMR: Intramolecular, β-dicarbonyls.

2239c. N. N. Shapet'ko, *Org. Magn. Resonance*, 5, 215 (1973). NMR: Intramolecular, β-dicarbonyl compounds.

2240. N. N. Shapetko and D. N. Shigorin, *Zh. Strukt. Khim.*, <u>8</u>, 538 (1967). NMR: Intramolecular, derivatives of hydroxynaphtho- and hydroxyanthroquinones, tropolones in $CDCl_3$.

2241. N. N. Shapetko, D. N. Shigorin, A. P. Skoldinov, T. S. Ryabchi- kova, and L. N. Reshetova, *Opt. Spektrosk.*, <u>17</u>, 459 (1964); [English, 247]. NMR: Intramolecular, alcohols.

2242. R. R. Sharp, H. T. Miles, and E. D. Becker, *Biochem. Biophys. Res. Commun.*, <u>23</u>, 194 (1966). BIOL: NMR evidence of specific base-pairing between purines and pyrimidines.

2243. G. Shaw, *Int. J. Quant. Chem.*, <u>3</u>, 219 (1969). Theory: Pertur- bation calculations on LiH—Li$^+$.

2244. Y. H. Shaw and N. C. Li, *Can. J. Chem.*, <u>48</u>, 2090 (1970). Near-IR, NMR: Adducts of N-methylacetamide.

2244a. Y. H. L. Shaw, S. M. Wang, and N. C. Li, *J. Phys. Chem.*, <u>77</u>, 236 (1973). NMR: $(H_2O)_2$ in H_2O—$CHCl_3$ solution.

2244b. A. S. Shawali, M. M. Naoum, and S. A. Ibrahim, *Bull. Chem. Soc. Jap.*, <u>45</u>, 2504 (1972). IR, UV: Intramolecular, keto-enol equilibrium, benzoylacetanilides.

2245. D. N. Shchepin and L. P. Belozerskaya, *Opt. Spektrosk.*, supple- ment 3, 204 (1967); [English, 101 (1968)]. IR: Gas-phase, configuration of HCl adducts using vibration-rotation spectra.

2246. D. N. Shchepin and E. V. Shuvalova, *Spektrosk. Vzaimodeistvuy- usch Ikh. Mol.*, M. O. Bulanin, ed., 98 (1970). Review: Spectroscopic analysis of hydrogen bond.

2247. E. Sheftner, M. N. G. James, and H. G. Mautner, *J. Pharm. Sci.*, <u>55</u>, 643 (1966). BIOL, X-Ray: Crystal structure of 2,4- diselenouracil.

2248. E. Sheftner and H. G. Mautner, *J. Amer. Chem. Soc.*, <u>89</u>, 1249 (1967). BIOL, X-Ray: Crystal structure of 2,4-dithiouracil.

2249. N. Sheppard in *Hydrogen Bonding*, D. Hadži, ed., Pergamon Press, New York, 1959, pp. 85-105. IR: Review.

2250. N. I. Shergina, T. V. Kashik, E. I. Kositsyna, Z. T. Dmitrieva, and B. A. Trofimov, *Izv. Akad. Nauk SSSR, Ser. Khim.*, 2703 (1969); *Chem. Abstr.*, <u>72</u>, 89421v (1970). IR: Intramolecular, (2-hydroxyethyl) propynylamines.

2251. J. P. Sheridan, D. E. Martire, and Y. B. Tewari, *J. Amer. Chem. Soc.*, <u>94</u>, 3294 (1972). GLC: Thermodynamic data for association of haloalkanes with di-*n*-octyl ether and di-*n*-octyl thioether.

2252. A. D. Sherry and K. F. Purcell, *J. Amer. Chem. Soc.*, <u>94</u>, 1853 (1972). CAL: Thermodynamic data, adducts of $(CF_3)_3COH$, ΔH-$\Delta \nu$ correlation.

2253. A. D. Sherry and K. F. Purcell, *J. Amer. Chem. Soc.*, <u>94</u>, 1848 (1972). CAL: Thermodynamic data, adducts of $(CF_3)_2CHOH$ and CF_3CH_2OH with sulfur donors.

2254. A. D. Sherry and K. F. Purcell, *J. Amer. Chem. Soc.*, <u>92</u>, 6386 (1970). CAL: ΔH, adducts of CF_3CH_2OH and $(CF_3)_2CHOH$ with pyridine and γ-collidine.

2255. A. D. Sherry and K. F. Purcell, *J. Phys. Chem.*, 74, 3535
       (1970). CAL: Thermodynamic data, adducts of CF_3CH_2OH, $\Delta\nu$-ΔH
       correlation.

2256. D. N. Shigorin, E. A. Gastilovich, T. S. Kopteva, and N. M.
       Viktorova, *Zh. Fiz. Khim.*, 42, 2328 (1968); *Chem. Abstr.*, 70,
       42438z (1969). IR: Intramolecular, dipivaloylmethane enolate.

2257. D. N. Shigorin, M. M. Shemyakin, M. N. Kosolov, and T. S.
       Ryabchikova, *Stroenie Ves. i Spektroskopiya Akad. Nauk SSSR*,
       36 (1960). IR: Spectra of phenylacetylene and deuterated
       forms in various solvents.

2258. T. Y. Shih and J. Bonner, *J. Mol. Biol.*, 50, 333 (1970). BIOL:
       Template properties of DNA-polypeptide complexes.

2259. H. Shimizu, *Nippon Kagaku Zasshi*, 81, 1025 (1960). NMR: Acetic
       acid in C_6H_6 and pyridine.

2259a. L. Shipman and R. E. Christoffersen, *J. Amer. Chem. Soc.*, 95,
       4733 (1973). Theory: *Ab initio* SCF calculations, polypeptides
       of glycine.

2260. H. N. Shrivastava and J. C. Speakman, *J. Chem. Soc.*, 1151 (1961).
       X-Ray: Acid salts of monocarboxylic acids.

2261. D. Shugar and W. Szer, *J. Mol. Biol.*, 5, 580 (1962). BIOL:
       Secondary structure in polyribothymidylic acid.

2262. R. G. Shulman, *Ann. N. Y. Acad. Sci.*, 158, 96 (1969). BIOL:
       Double-minimum of the adenine-thymine hydrogen bond.

2263. R. G. Shulman, H. Sternlicht, and B. J. Wyluda, *J. Chem. Phys.*,
       43, 3116 (1965). BIOL: ^{31}P NMR, aqueous solutions of RNA and
       $\overline{AMP}$ containing Mn^{2+} and Co^{2+}.

2264. E. V. Shuvalova, *Optics and Spectroscopy*, 13, 80 (1962). IR:
       Diacetylene adducts of bases.

2265. E. V. Shuvalova, *Opt. Spektrosk.*, 6, 696 (1959). IR: $\Delta\nu_{CH}$ for
       acetylene adducts of acetone, dioxane, and pyridine.

2266. T. H. Siddall, III, and W. E. Stewart, *J. Mol. Spectrosc.*, 24,
       290 (1967). NMR: Intramolecular, N-acylanilines.

2267. T. A. Sidorov, *Zh. Prikl. Spektrosk.*, 10, 562 (1969); *Chem.
       Abstr.*, 72, 7523t (1970). IR: Intensity changes of ν_{AH}.

2269. B. L. Silver, Z. Luz, S. Peller, and J. Reuben, *J. Phys. Chem.*,
       70, 1434 (1966). NMR: Intramolecular, anions of carboxylic
       acids.

2270. Z. Simon, *J. Theor. Biol.*, 9, 414 (1965). BIOL: Role of
       hydrogen bonding in mutual recognition by biological macromole-
       cules.

2271. H. Simpkins and E. G. Richards, *Biochemistry*, 6, 2513 (1967).
       BIOL: Spectrophotometric acid-base titrations of dinucleotides
       containing adenine and uracil derivatives, stacking association
       constants.

2272. O. Sinanoğlu in *Molecular Associations in Biology*, B. Pullman,
       ed., Academic Press, New York, 1968, p. 427. BIOL: Theoretical
       discussion, solvent effects on molecular associations.

2273. O. Sinanoğlu and S. Abdulnur, *Federation Proc.*, 24, S12 (1965). BIOL: Effect of water and other solvents on the structure of biopolymers.

2274. O. Sinanoğlu and S. Abdulnur, *Photochem. Photobiol.*, 3, 333 (1964). BIOL: Theoretical, hydrophobic stacking of bases and the solvent denaturation of DNA.

2275. O. Sinanoğlu, S. Abdulnur, and N. R. Kestner in *Electronic Aspects of Biochemistry*, B. Pullman, ed., Academic Press, 1964, p. 301. BIOL: Solvent effects on van der Waals dispersion attractions in DNA.

2276. S. S. Singh, *Indian J. Chem.*, 6, 393 (1968). IR: Self-association, monocarboxylic acids.

2277. S. Singh, Ph.D. Thesis, Indian Institute of Technology, Kampur, India, 1966. IR: Asymmetry of bonded peaks for adducts of alcohols and phenols with bases.

2278. S. Singh, A. S. N. Murthy, and C. N. R. Rao, *Trans. Faraday Soc.*, 62, 1056 (1966). IR: Thermodynamic data, adducts of phenol with bases, $\Delta\nu$-ΔH correlations.

2279. S. Singh and C. N. R. Rao, *J. Phys. Chem.*, 71, 1074 (1967). IR, NMR: Self-association, hindered alcohols.

2280. S. Singh and C. N. R. Rao, *Can. J. Chem.*, 44, 2611 (1966). IR: Phenol and phenol—OD as reference acids, thermodynamic data.

2281. S. Singh and C. N. R. Rao, *J. Amer. Chem. Soc.*, 88, 2142 (1966). IR: K, adducts of substituted phenols and alcohols with bases.

2282. S. Singh and C. N. R. Rao, *Trans. Faraday Soc.*, 62, 3310 (1966). IR: Thermodynamic data, alcohol adducts with tetra-n-heptylammonium iodide.

2283. T. R. Singh and J. L. Wood, *J. Chem. Phys.*, 50, 3572 (1969). Theory: Effect of deuterium substitution on hydrogen bond length.

2284. T. R. Singh and J. L. Wood, *J. Chem. Phys.*, 48, 4567 (1968). Theory: Calculation of vibrational levels in a symmetric double-minimum potential.

2285. L. Singurel, *Rev. Roum. Phys.*, 14, 465 (1969); *Chem. Abstr.*, 71, 86388h (1969). Raman, IR: Dioxane mixtures with Me_2CO, $POCl_3$, MeOH, CH_3COOH, HCOOH.

2286. L. Singurel, *Stud. cercet. Fiz.*, 24, 545 (1972); *Chem. Abstr.*, 78, 22068r (1973). Raman, IR: Effect of hydroxylic solvents on ν_{CO}, ν_{SO}, ν_{CN} of bases.

2286a. L. Singurel and G. Singurel, *Rev. Roum. Chim.*, 18, 789 (1973); *Chem. Abstr.*, 79, 71795e (1973). Raman, IR: Me_2SO associations in various solvents.

2286b. D. Sinka and J. E. Katon, *Appl. Spectrosc.*, 26, 599 (1972). IR: Self-association, cyanoacetic acid.

2287. L. Skulski, *Bull. Acad. Pol. Sci., Ser. Sci. Chim.*, 17, 253 (1969); *Chem. Abstr.*, 71, 60524z (1969). UV: Intramolecular acetyl- and benzoyl-substituted naphthols.

2288.   L. Skulski and W. Waclawek, *Bull. Acad. Pol. Sci., Ser. Sci.*
        Chim., 19, 277 (1971); *Chem. Abstr.*, 75, 55819c (1971). UV:
        Intramolecular, aromatic compounds.

2288a. I. M. Skvortsov and S. A. Kolesnikov, *Zh. Fiz. Khim.*, 47, 785
        (1973); [English, 447]. GLC: Intramolecular, thermodynamic
        data.

2289.   I. M. Skvortsov, V. M. Levin, and I. Ya. Evtushenko, *Khim.*
        Geterotsikl. Soedin., 7, 995 (1971); *Chem. Abstr.*, 76, 24282p
        (1972). IR: Intramolecular, π-hydrogen bond in 5-(β-
        hydroxyethyl)-1,2-dihydropyrrolizine.

2290.   F. L. Slejko and R. S. Drago, *Inorg. Chem.*, 12, 176 (1973).
        NMR, CAL: Adducts of quinuclidine, thermodynamic data.

2290a. F. L. Slejko and R. S. Drago, *J. Amer. Chem. Soc.*, 95, 6935
        (1973). NMR: ^{1}H chemical shifts, thermodynamic data.

2291.   F. L. Slejko, R. S. Drago, and D. G. Brown, *J. Amer. Chem. Soc.*,
        94, 9210 (1972). NMR: Thermodynamic data, adducts of $CHCl_3$,
        failure of Δν-ΔH correlations.

2292.   M. A. Slifkin, B. M. Smith, and R. W. Walmsley, *Spectrochim.*
        Acta, 25A, 1479 (1969). IR: Chloranilic acid complexes with
        amino acids.

2293.   D. W. Slocum and C. A. Jennings, *Tetrahedron Lett.*, 3543 (1972).
        NMR: Self-association, effect on spin-spin coupling for
        β-chloroethanols.

2293a. F. Smith and I. Brown, *Aust. J. Chem.*, 26, 691, 705 (1973).
        NMR, IR: Self-association, alcohol-alkane mixtures.

2294.   F. A. Smith and E. C. Creitz, *J. Res. Natl. Bur. Std.*, 46, 145
        (1951). IR: Hindered alcohols.

2295.   H. F. Smith and A. S. Rosenberg, *J. Chem. Soc.*, 5391 (1963).
        IR: Thermodynamic data for association of 1-octanol with hexyl
        hexanoate.

2296.   I. C. P. Smith, B. J. Blackburn, and T. Yamane, *Can. J. Chem.*,
        47, 513 (1969). BIOL, NMR: Ribonucleosides.

2297.   I. C. P. Smith, T. Yamane, and R. G. Shulman, *Can. J. Biochem.*,
        47, 480 (1969). BIOL, NMR: Alanine transfer, tRNA.

2298.   J. W. Smith, *Sci. Prog. (London)*, 52, 97 (1964). Review.

2299.   A. L. Smolyanskii, *Opt. Spektrosk.*, supplement 2, 254 (1963);
        [English, 133 (1966)]. IR: O—H...π, isobutyric acid-aromatic
        hydrocarbons.

2300.   A. L. Smolyanskii, *Opt. Spektrosk.*, 13, 475 (1962). IR: Self-
        association of acetic and butyric acids.

2301.   C. C. Snead, *J. Phys. Chem.*, 76, 774 (1972). IR: Dimerization
        of formic acid in anhydrous and wet CCl_4; K = 1.44 x 10^3 ℓ/mole.

2302.   L. C. Snyder, R. G. Shulman, and D. B. Neumann, *J. Chem. Phys.*,
        53, 256 (1970). Theory: Thymine, LCAO-SCF calculations.

2303.   R. G. Snyder and J. A. Ibers, *J. Chem. Phys.*, 36, 1356 (1962).
        IR: Double-minimum potential energy curves, $HCrO_2$, $DCrO_2$.

2303a. W. R. Snyder, H. D. Schreiber, and J. N. Spencer, *Spectrochim. Acta*, 29A, 1225 (1973). Theory: Schroeder-Lippincott potential function model of X—H...S systems.

2304. L. Sobczyk, ed., *Hydrogen Bonds*, PWN, Warsaw, Poland, 1969, 409 pp; *Chem. Abstr.*, 72, 115062a (1970).

2305. H. M. Sobell, *J. Mol. Biol.*, 18, 1 (1966). BIOL, X-Ray: 9-Ethyl-2-aminopurine and 1-methyl-5-fluorouracil complex.

2306. H. M. Sobell, K. Tomita, and A. Rich, *Proc. Natl. Acad. Sci. U. S.*, 49, 885 (1963). BIOL, X-Ray: 9-Ethylguanine: 1-methyl-5-bromocytosine complex.

2307. W. Sobotka, *Bull. Acad. Pol. Sci., Ser. Sci. Chim.*, 17, 85 (1969); *Chem. Abstr.*, 71, 60522x (1969). NMR, IR: Intramolecular, 1-(carbethoxymethylene) hydroisoquinolines.

2308. G. Socrates, *Trans. Faraday Soc.*, 66, 1052 (1970). NMR: Chemical shifts of fifty-five aromatic hydroxyl compounds in CCl_4 and Me_2SO.

2309. G. Socrates, *Trans. Faraday Soc.*, 64, 287 (1968). NMR: Correlation of chemical shifts and base pK_a for cresol adducts of bases.

2310. G. Socrates, *Trans. Faraday Soc.*, 63, 1083 (1967). NMR: Chemical shifts for phenol-base adducts in CCl_4 and CH_2Cl_2.

2311. E. Soczewinski and G. Matysik, *J. Chromatogr.*, 48, 57 (1970). Use of chromatography to study hydrogen bonding of phenols.

2312. G. Soda and T. Chiba, *J. Phys. Soc. Jap.*, 26, 723 (1969). NMR: ^2D, ^{23}Na, single crystals of $NaD_3(SeO_3)_2$.

2313. G. Soda and T. Chiba, *J. Phys. Soc. Jap.*, 26, 249 (1969). NMR: ^2D, single crystals of $Cu(DCO_3)_2 \cdot 4D_2O$.

2314. D. Soell, J. Cherayil, D. S. Jones, R. D. Faulkner, A. Hampel, R. M. Bock, and H. G. Khorana, *Cold Spring Harbor Symp. Quant. Biol.*, 31, 51 (1966). BIOL: sRNA specificity for codon recognition, ribosomal binding technique.

2315. N. D. Sokolov, *Ann. Chim.*, 10, 497 (1965). IR: Review, frequency shift correlations.

2316. N. D. Sokolov and V. M. Chulanovskii, eds., *Vodorodnaya Svyas*, Nauka, Moscow, 1964, 339 pp. Book on hydrogen bonding [also see *Chem. Abstr.*, 62, 2257a (1965)].

2316a. N. D. Sokolov, *Zh. Vses. Khim. Obshchest.*, 17, 299 (1972); *Chem. Abstr.*, 77, 118247d (1972). Review, geometry of hydrogen bond.

2317. E. Solcaniova and S. Kovac, *Chem. Zvesti*, 23, 687 (1969). IR: Intramolecular, alkyl derivatives of dihydroxydiphenylmethane, -sulfide, -disulfide, and -selenide.

2318. T. N. Solie and J. A. Schellman, *J. Mol. Biol.*, 33, 61 (1968). BIOL: Interactions of nucleosides in H_2O.

2319. B. G. Somers, Ph.D. Thesis, University of Illinois, Urbana, Ill., 1961. NMR: K_2 for self-association of hindered phenols.

2320. B. G. Somers and H. S. Gutowsky, *J. Amer. Chem. Soc.*, 85, 3065 (1963). NMR: K, adducts of phenols with bases.

2321.   R. L. Somorjai and D. F. Hornig, *J. Chem. Phys.*, 36, 1980
        (1962).   Theory: 1-Dimensional models for double-minimum hydro-
        gen bond potentials.

2322.   R. R. Soup, H. T. Miles, and E. D. Becker, *Biochem. Biophys.
        Res. Commun.*, 23, 194 (1966).   BIOL: NMR, 1-methylthymine,
        9-ethylguanine, 1-methylcytosine and 2-aminoadenosine associa-
        tions in DMSO and DMSO-DMF.

2323.   J. C. Speakman, *Chem. Commun.*, 32 (1967).   X-Ray: "Very short"
        hydrogen bonds.

2323a. J. C. Speakman, *Struct. Bonding*, 12, 141 (1972).   X-Ray:
        Review, "very short" hydrogen bonds.

2323b. J. C. Speakman, *MTP Int. Rev. Sci., Phys. Chem., Ser. One*, 11,
        1 (1972).   J. M. Robertson, ed., Butterworths, London.   Review:
        O...H...O symmetrical bonds.

2324.   J. N. Spencer, E. A. Robinson, and H. D. Schreiber, *J. Phys.
        Chem.*, 75, 2219 (1971).   IR: Temperature dependence of molar
        absorptivity of ν_{OH} of phenols in bases.

2325.   M. Spencer, *J. Chim. Phys. Physicochim. Biol.*, 65, 190 (1968).
        BIOL, X-Ray: Helical RNA fragments in the solid state and in
        solution.

2326.   E. Spinner and G. B. Yeoh, *Aust. J. Chem.*, 24, 2557 (1971).
        IR: 2,6-dihydroxypyridine, tautomerism.

2327.   T. M. Spotswood and C. I. Tanzer, *Tetrahedron Lett.*, 911 (1967).
        NMR: ^{1}H NMR spectra of nitrogen heterocycles in hydrogen bond-
        ing solvents.

2328.   C. S. Springer and D. W. Meek, *J. Phys. Chem.*, 70, 481 (1966).
        NMR: K_n, self-association of Et_2NH in cyclohexane.

2329.   R. A. Spurr and H. F. Byers, *J. Phys. Chem.*, 62, 425 (1958).
        IR: Self-association of mercaptans.

2330.   R. Srinivasan and K. K. Chacko, *Conformation of Biopolymers*,
        Vol. 2, G. N. Ramachandran, ed., Academic Press, New York, 1967,
        p. 607.   X-Ray: Survey, A—H...S, S—H...B structure data.

2331.   G. P. Srivastava and M. L. Goyal, *Proc. Phys. Soc., London, At.
        Mol. Phys.*, 1, 1212 (1968).   Microwave: CF_3COOH—HCOOH adduct.

2332.   G. P. Srivastava, M. L. Goyal, and H. O. Gautam, *J. Quant.
        Spectrosc. Radiat. Transfer*, 10, 659 (1970).   Microwave:
        CF_3COOH—HCOOH adduct.

2333.   G. P. Srivastava, M. L. Goyal, and H. O. Gautam, *J. Quant.
        Spectrosc. Radiat. Transfer*, 8, 1773 (1968).   Microwave:
        CF_3COOH—HCOOH adduct.

2334.   A. E. Stanevich, *Opt. Spektrosk.*, 16, 781 (1964); [English, 425].
        Far-IR: *m*- and *p*-nitrophenols, ν_σ.

2335.   A. E. Stanevich, *Opt. Spektrosk.*, 16, 446, 998 (1964); [English,
        243, 539].   IR: Self-association of CH_3COOD, CD_3COOD, trichloro-
        acetic acid, benzoic acid.

2336.   A. E. Stanevich, *Opt. Spektrosk.*, supplement 2, 205 (1963);
        [English, 104 (1966)].   Far-IR: Carboxylic acid dimers.

2337. A. E. Stanevich and N. G. Yavoslaosku, *Dokl. Akad. Nauk SSSR*, 137, 60 (1961). Far-IR: Self-association, H_2O.

2337a. S. Stankovsky, S. Kovac, M. Dandarova, and M. Livar, *Tetrahedron*, 29, 1825 (1973). IR: Thermodynamic data, phenol adducts of alkyl isothiocyanates.

2338. I. V. Stasyuk and R. R. Levitskii, *Ukr. Fiz. Zh. (Russ. Ed.)*, 14, 1100 (1969); *Chem. Abstr.*, 71, 117617p (1969). Theory: Elementary excitations in ferroelectrics with hydrogen bonding.

2339. G. Statz and E. Lippert, *Phys. Ice, Proc. Int. Symp.*, 3rd, N. Riehl, ed., 152 (1968). Theory: Review.

2340. G. Statz and E. Lippert, *Ber. Bunsenges, Physik. Chem.*, 71, 673 (1967). Far-IR: Self-association of twenty-seven monocarboxylic acids (deuterated and/or ^{18}O enriched).

2341. W. J. Stec, N. Goddard, and J. R. Van Wazer, *J. Phys. Chem.*, 75, 3547 (1971). NMR: $HP(O)(OR)_2$ and $DP(O)(OR)_2$ as HA acids.

2341a. J. Steigman and W. Cronkright, *Spectrochim. Acta*, 26A, 1805 (1970). IR: Self-association, chloroacetic acids, band overlap of monomer and dimer.

2342. A. I. Stekhanov and E. A. Popova, *Opt. Spektrosk.*, 11, 203 (1961). IR: Self-association, crystalline lithium hydroxide.

2344. H. Stenzenberger and H. Schindlbauer, *Monatsh. Chem.*, 99, 2474 (1968). IR: Association of Ph_3PO or Ph_3AsO with phenols.

2345. N. C. Stephenson, J. F. McConnell, and R. Warren, *Inorg. Nucl. Chem. Lett.*, 553 (1967). IR: (±)-Methioninepalladium(II) chloride.

2345a. J. Stepisnik and D. Hadži, *J. Mol. Struct.*, 13, 307 (1972). NMR: $KD(CF_3CO_2)_2$.

2346. H. Sterk, *Monatsh. Chem.*, 100, 916 (1969). IR, NMR: Intramolecular, anthrone, oxanthrone, and dihydroquinizarin.

2347. D. P. Stevenson, *J. Amer. Chem. Soc.*, 84, 2849 (1962). UV: K, alcohol adducts with Me_3N.

2348. G. R. Stevenson and H. Hidalgo, *J. Phys. Chem.*, 77, 1027 (1973). ESR: Thermodynamic data, nitrobenzene radical adduct with $[Me_2N]_3PO$.

2349. R. F. Stewart and L. H. Jensen, *Acta Crystallogr.*, 23, 1102 (1967). BIOL, X-Ray: Redetermination of the crystal structure of uracil.

2350. F. Stoenescu, Gr. Stanescu, and O. Radulescu, *Mater. Plast.*, 5, 125 (1968); *Chem. Abstr.*, 70, 4895p (1969). IR: Epoxide polymers.

2352. R. D. Stolow, *J. Amer. Chem. Soc.*, 86, 2170 (1964). IR: Intramolecular, p-menthane-2,5-diols.

2353. R. D. Stolow, *J. Amer. Chem. Soc.*, 83, 2592 (1961). IR: Intramolecular, 1,4-cyclohexanediol derivatives.

2354. R. D. Stolow, P. M. McDonagh, and M. M. Bonaventura, *J. Amer. Chem. Soc.*, 86, 2165 (1964). IR: Intramolecular, 1,4-cyclohexanediols.

2354a. W. Storek and H. Kriegsmann, *Ber. Bunsenges, Phys. Chem.*, 72, 706 (1968). NMR: Self-association, *t*-butanol, trimethyl-silanol.

2355. J. B. Stothers and P. C. Lauterbur, *Can. J. Chem.*, 42, 1563 (1964). NMR: ^{13}C chemical shifts in carbonyl groups.

2356. H. Strassmair, J. Engel, and G. Zundel, *Biopolymers*, 8, 237 (1969). IR: Alcohol adducts of polyproline.

2356a. M. Strat, D. S. Umreiko, and N. N. Khovratovich, *Zh. Prikl. Spektrosk.*, 19, 103 (1973); *Chem. Abstr.*, 79, 77607s (1973). IR: Intramolecular, 2-hydroxybenzophenone derivatives.

2357. H. Strehlow, *Ann. Rev. Phys. Chem.*, 16, 617 (1965). NMR: Review of proton-exchange reactions.

2357a. A. Streitwieser, Jr., *Molecular Orbital Theory for Organic Chemists*, John Wiley and Sons, New York, 1961.

2358. E. H. Strickland, M. Wilchek, J. Horwitz, and C. Billups, *J. Biol. Chem.*, 247, 572 (1972). UV: Tyrosine, *o*-methyltyrosine derivatives.

2358a. E. H. Strickland, C. Billups, and E. Kay, *Biochemistry*, 11, 3657 (1972). UV: 2,3-Dimethylindole, band shifts in hydrogen bonding solvents.

2359. F. Strohbusch and H. Zimmerman, *Ber. Bunsenges, Physik. Chem.*, 71, 679 (1967). NMR: K_2, self-association of 2-*t*-butylphenol.

2360. A. Sturis, I. Zuika, and J. Bankovskis, *Latv. PSR Zinat. Akad. Vestis, Kim. Ser.*, 750 (1968); *Chem. Abstr.*, 70, 77150b (1969). IR: Intramolecular, 1,2,3,4-tetrahydro-8-mercaptoquinoline.

2361. B. Stymne, H. Stymne, and G. Wettermark, *J. Amer. Chem. Soc.*, 95, 3490 (1973). IR: Thermodynamic data, adducts of substituted phenols with $MeC(O)NMe_2$.

2362. Y. S. Su and H-K. Hong, *Spectrochim. Acta*, 24A, 1461 (1968). IR: Hydrogen bonding between OH and NO_2 groups.

2363. E. Subramanian and D. J. Hunt, *Acta Crystallogr.*, Sect. B, 26, 303 (1970). X-Ray: Configuration of deoxycytidine hydrochloride.

2364. S. Subramanian and H. F. Fisher, *J. Phys. Chem.*, 76, 452 (1972). Near-IR: Association of water with acetone.

2365. S. Subramanian and H. F. Fisher, *J. Phys. Chem.*, 76, 84 (1972). Near-IR: Effect of ClO_4^- and BF_4^- on structure of water.

2366. T. Suga and T. Shishibori, *Chem. Ind.*, 733 (1971). Intramolecular, α-hydroxycyclohexanone.

2367. T. Suga and T. Shishibori, *Kagaku No Ryoiki*, 22, 995 (1968); *Chem. Abstr.*, 70, 67427h (1969). IR: Review, intramolecular hydrogen bonding and conformations of alicyclic alcohols.

2368. T. Suga, T. Shishibori, and T. Matsuura, *J. Chem. Soc., Perkin Trans.*, 1, 171 (1972). IR: Intramolecular, hydroxy keto steroids.

2369. T. Suga, S. Watanabe, and M. Kuniyoshi, *Yukagaku*, 19, 33 (1970);
       Chem. Abstr., 72, 90645c (1970). IR, NMR: Intramolecular,
       terpenic cyclic, unsaturated alcohols.

2370. T. Suga, S. Watanabe, T. Shishibori, and T. Matsuura, *Bull.
       Chem. Soc. Jap.*, 44, 204 (1971). IR: Intramolecular, OH with
       oxirane and cyclopropane rings.

2371. S. Suhai and J. Ladik, *Theor. Chim. Acta*, 28, 27 (1972). Theory:
       CNDO/2 and MINDO/2 on formamide polymers.

2372. H. Suhr, *Mol. Phys.*, 6, 153 (1963). NMR: p-Substituted ani-
       lines in various solvents.

2373. B. I. Sukhorukov and A. I. Finkelstein, *Opt. Spektrosk.*, 9, 330
       (1961). IR: Dimerization of guanylurea cation.

2374. B. I. Sukhorukov and L. A. Kozlova, *Biofizika*, 15, 539 (1970).
       BIOL: Structure of DNA containing various amounts of water
       studied by EPR.

2375. B. I. Sukhorukov, V. I. Poltev, U. R. Fed'kina, and Y. K. Knobel,
       Dokl. Acad. Nauk SSSR, 175, 945 (1967); *Chem. Abstr.*, 68, 279b
       (1968). BIOL: Tautomeric transformations and transitions of
       complementary pairs of DNA bases.

2376. J. Sulston, R. Lohrmann, L. E. Orgel, and H. T. Miles, *Proc.
       Natl. Acad. Sci. U. S.*, 60, 409 (1968). BIOL: Specificity of
       oligonucleotide synthesis directed by poly U.

2377. M. Sundaralingam, *Biopolymers*, 7, 821 (1969). BIOL: Allowed
       and preferred conformations of nucleosides, nucleoside mono-,
       di-, tri-, and tetraphosphates, nucleic acids and polynucleo-
       tides.

2378. J. Sunkel and H. Staude, *Ber. Bunsenges, Physik. Chem.*, 73, 203
       (1969). UV: Spectra of catechol, resorcinol, and hydroquinone
       derivatives.

2379. P. Suppan, *J. Chem. Soc. A*, 3125 (1968). UV: Effect of hydro-
       gen bonding on absorption bands.

2380. H. Susi, *J. Phys. Chem.*, 69, 2799 (1965). Near-IR: Self-
       association, δ-valerolactam in dioxane.

2380a. H. Susi, *Methods Enzymol.*, 26 (Pt.C), 381 (1972), C. H. W. Hirs,
       ed., Academic Press. IR: Review, amides.

2380b. H. Susi and J. S. Ard, *Arch. Biochem. Biophys.*, 117, 147 (1966).
       IR: Self-association, δ-valerolactam.

2380c. H. Susi, S. N. Timasheff, and J. S. Ard, *J. Biol. Chem.*, 239,
       3051 (1964). IR: Self-association, δ-valerolactam in H_2O.

2380d. R. Sustmann and F. Varenholt, *Theor. Chim. Acta*, 29, 305 (1973).
       Theory: Perturbation CNDO/2 treatment of $(H_2O)_2$.

2381. B. Sutherland, *Phys. Rev. Lett.*, 19, 103 (1967). Theory:
       2-Dimensional model of a hydrogen-bonded crystal.

2382. D. J. Sutor, *Nature*, 195, 68 (1962). Solids: C—H...O hydrogen
       bonds in crystals.

2383. I. Suzuki, *Bull. Chem. Soc. Jap.*, 35, 540 (1962). IR: Spectra
       of N-methyl formamides.

2384. I. Suzuki, M. Tsuboi, T. Shimanouchi, and S. Mizushima, *Spec-trochim. Acta*, <u>16</u>, 471 (1960). IR: Intramolecular, anilide derivatives.

2385. I. Suzuki, M. Tsuboi, and T. Shimanouchi, *Spectrochim. Acta*, <u>16</u>, 467 (1960). IR: Intramolecular, N-methyl-phenylacetamide.

2386. I. Suzuki, M. Tsuboi, and T. Shimanouchi, *J. Chem. Phys.*, <u>32</u>, 1263 (1960). IR: $\Delta\nu_{NH}$, N-methylacetamide in various solvents with phenyl groups.

2387. M. Suzuki and T. Shimanouchi, *J. Mol. Spectrosc.*, <u>29</u>, 415 (1969). IR, Raman: Adipic acid crystals.

2388. S. Suzuki and H. Baba, *J. Chem. Phys.*, <u>38</u>, 349 (1963). UV: Fluorescence spectra of anthrols.

2389. S. Suzuki and H. Baba, *Nippon Kagaku Zasshi*, <u>81</u>, 366 (1960). UV: Adducts of acetanilide, aceto-1-naphthylamide, and aceto-2-naphthylamide.

2390. C. G. Swain and E. C. Lupton, Jr., *J. Amer. Chem. Soc.*, <u>90</u>, 4328 (1968). Tabulation of σ_m, σ_p, σ_p^+.

2391. C. A. Swenson, *J. Phys. Chem.*, <u>71</u>, 3108 (1967). IR: Temperature dependence of the intensity of ν_{OH} for alcohols.

2392. A. Swinarski and J. Kornatowski, *Rocz. Chem.*, <u>43</u>, 1017 (1969). IR: Association of o-chlorophenol with 1,2-dimethoxyethane.

2392a. M. C. R. Symons, *Nature*, <u>239</u>, 257 (1972). Review, structure of water.

2392b. Yu. P. Syrnikov, *Strukt. Rol Vody Zhivom Organizme*, <u>3</u>, 50 (1970); *Chem. Abstr.*, <u>75</u>, 25634w (1971). Theory: Graph-theoretical description of hydrogen bonds in liquid water.

2393. M. Szafran, *Rocz. Chem.*, <u>42</u>, 1469 (1968). IR: Hydrogen bonding of quinoline N-oxide salts of dichloroacetic and benzoic acids.

2394. M. Szafran, *Bull. Acad. Pol. Sci.*, *Ser. Sci. Chim.*, <u>11</u>, 169 (1963). IR: 1:2 adducts of HI, $HClO_4$, $HSbCl_6$, $HSnCl_6$ with amine N-oxides.

2395. M. Szafran and Z. Dega-Szafran, *Rocz. Chem.*, <u>44</u>, 793 (1970); *Chem. Abstr.*, <u>74</u>, 3240s (1971). IR: Adducts of heterocyclic N-oxides with haloacetic acids.

2395a. M. Szafran and T. Dziembowska, *Rocz. Chem.*, <u>46</u>, 1531 (1972). IR: Intramolecular, 4-R-quinaldinic acids.

2396. M. Szafran and M. Rozwadowska, *Rocz. Chem.*, <u>44</u>, 1465 (1970); *Chem. Abstr.*, <u>74</u>, 8174h (1971). NMR: Adducts of heterocyclic N-oxides with haloacetic acids.

2397. K. Szczepaniak and M. Falk, *Spectrochim. Acta*, <u>26A</u>, 883 (1970). IR: Interaction of phenol and 2,6-diisopropyl phenol with weakly polar solvents.

2398. K. Szczepaniak, M. Golinska, and J. Mikolajczyk, *Acta Phys. Pol.*, <u>34</u>, 421 (1968). IR: Association constants for π-bases with phenols, naphthol, and CF_3CO_2H.

2399. K. Szczepaniak and A. Tramer, *J. Phys. Chem.*, 71, 3035 (1967). Theory: Charge-transfer model and IR spectra of hydrogen bonds with π-donors.

2400. H. H. Szmant and J. J. Rigau, *J. Org. Chem.*, 31, 2288 (1966). Intramolecular, *cis*-2-phenyl-mercaptoindanol.

2400a. M. Szostak, *Acta Phys. Pol. A*, 42, 279 (1972). Near-IR: *p*-Nitroaniline single crystals.

2401. W. Szybalski, H. Kubinski, and P. Sheldrick, *Cold Spring Harbor Symp. Quant. Biol.*, 31, 123 (1966). BIOL: Pyrimidine clusters, DNA.

2402. D. Tabuchi, *Mem. Inst. Sci. Ind. Res., Osaka Univ.; Chem. Abstr.*, 75, 53407t (1971). Ultrasonic: Self-association, acetic acid.

2403. D. Tabuchi, *Z. Electrochem.*, 64, 141 (1960). Kinetics: Ultrasonic relaxation, acetic acid.

2404. D. Tabuchi, *J. Chem. Phys.*, 26, 993 (1957). Kinetics: Equations for ultrasonic studies.

2405. I. Taesler and I. Olovsson, *J. Chem. Phys.*, 51, 4213 (1969). Crystal structure of $(H_3O^+)_2SO_4^{2-}$.

2406. R. W. Taft, *J. Phys. Chem.*, 64, 1805 (1960). σ° substituent constants.

2407. R. W. Taft, D. Gurka, L. Joris, P. von R. Schleyer, and J. W. Rakshys, *J. Amer. Chem. Soc.*, 91, 4801 (1969). Linear free energy relationships.

2408. A. A. Taha and S. D. Christian, *J. Phys. Chem.*, 73, 3430 (1969). Association of trifluoroacetic acid and benzophenone.

2409. F. Takahasi and N. C. Li, *J. Amer. Chem. Soc.*, 88, 1117 (1966). NMR: Thermodynamic data, 1:1 adducts of water with acetone, tetrahydrofuran, N,N-dimethylacetamide.

2410. F. Takahasi and N. C. Li, *J. Phys. Chem.*, 69, 2950 (1965). NMR: Thermodynamic data, adducts of *t*-butylamine, aniline with amides.

2411. F. Takahasi and N. C. Li, *J. Phys. Chem.*, 69, 1622 (1965). UV: Thermodynamic data, adducts of phenol with N-methylacetamide.

2412. F. Takahasi and N. C. Li, *J. Phys. Chem.*, 68, 2136 (1964). NMR: Thermodynamic data, 2-propanol with N-methylacetamide and DMA.

2413. H. Takahasi, K. Mamola, and E. K. Plyler, *J. Mol. Spectrosc.*, 21, 217 (1966). IR: Hydrogen bonding effects on skeletal vibrations of pyridine and pyrimidine.

2413a. T. Takahashi, M. Ota, and J. Azuma, *Kobe Daigaku Nogakubu Kenkyu Hokoku*, 10, 122, 128 (1971); *Chem. Abstr.*, 78, 75368v, 75369w (1973). IR: Hydrogen bonding in clays.

2414. S. Takashima, *Biopolymers*, 6, 1437 (1968). BIOL: Birefringence of poly-A, poly-U, and double-stranded poly-AU.

2414a. S. Takashima, *Biopolymers*, 11, 1903 (1972). BIOL: Nonempirical valence bond calculations, hydrogen bond energy in polypeptides.

2414b. Z. Talaikyte, V. Jasinskaite, and Z. Alaune, *Liet. TSR Mokalu Akad. Darb., Ser. B*, 75 (1971); *Chem. Abstr.*, 77, 4435h (1972). IR: Adducts of 1,2,3,4-tetrahydroquinolines.

2415. P. B. Talukdar, S. Banerjee, and S. K. Sengupta, *J. Indian Chem. Soc.*, 47, 267 (1970). IR, UV: Intramolecular, 2-(substituted aminomethyl)-4-acetamidophenol.

2416. M. Tamres and S. Searles, Jr., *J. Amer. Chem. Soc.*, 81, 2100 (1959). IR, CAL: Adducts of CHCl₃ with cyclic sulfoxides and cyclic ketones.

2417. M. Tamres, *J. Phys. Chem.*, 65, 654 (1961). Equations for calculating equilibrium constants, effects of solvent, concentration.

2418. T. Tanaka, *Yakugaku Zasshi*, 91, 324 (1971); *Chem. Abstr.*, 74, 140686f (1971). Electronic: Association of 1,2,3,4-substituted-3-pyrazoline-5-thiones with phenols and alcohols.

2419. T. Tanaka, T. Yokoyama, Y. Yamaguchi, S. Naganuma, and M. Furukawa, *Kogyo Kagaku Zasshi*, 74, 171 (1971). IR: Substituent effects, urethanes.

2420. S. Taniewska-Osinska, *Lodz. Tow. Nauk Wydz. III, Acta Chim.*, 12, 37 (1967); *Chem. Abstr.*, 71, 130375c (1969). IR, Raman: Self-association and hetero-association, benzyl alcohol-methanol.

2420a. V. Tatolis and E. Zurauskiene, *Liet. Fiz. Rinkinys*, 13, 149 (1973); *Chem. Abstr.*, 79, 31127g (1973). UV: Quinoline-ROH adducts.

2421. N. Tatsumoto, T. Sano, N. Matsunaga, E. Tochigi, and T. Yasunaga, *Bull. Chem. Soc. Jap.*, 45, 2083, 3096 (1972). Kinetics: Self-association, propionic acid.

2421a. B. N. Taylor, W. H. Parker, and D. N. Langenberg, *Rev. Mod. Phys.*, 41, 375 (1969). Revision of fundamental constants.

2422. C. A. Taylor, M. A. El-Bayoumi, and M. Kasha, *Proc. Natl. Acad. Sci., U. S.*, 63, 253 (1969). BIOL: Two-proton tautomerism in hydrogen-bonded N-heterocyclic base pairs.

2423. R. P. Taylor and I. D. Kuntz, Jr., *J. Phys. Chem.*, 74, 4573 (1970). IR: Association of phenol with anions.

2424. I. Tazawa, S. Tazawa, L. M. Stempel, and P. O. P. Ts'o, *Biochemistry*, 9, 3499 (1970). BIOL: Conformation and interaction of dinucleoside mono- and diphosphates.

2425. J. Tegenfeldt and I. Olovsson, *Acta Chem. Scand.*, 25, 101 (1971). NMR: Polycrystalline N₂H₅HC₂O₄.

2425a. R. Tellgren, D. Ahmad, and R. Liminga, *J. Chem. Soc., Faraday Trans. 2*, 68, 215 (1972). X-Ray: Ammonium trihydrogen selenite.

2426. R. Tellgren and I. Olovsson, *J. Chem. Phys.*, 54, 127 (1971). Crystal structures of NaHC₂O₄•H₂O and NaDC₂O₄•D₂O.

2426a. F. P. Temme, J. A. S. Smith, and T. C. Waddington, *J. Chem. Soc., Faraday Trans. 2*, 69, 1 (1973). Neutron Scattering: Ferroelectrics.

2426b. F. P. Temme and T. C. Waddington, *J. Chem. Phys.*, 59, 817 (1973). Neutron Scattering: CoO(OH), CrO(OH).

2427. A. N. Terenin, A. V. Shablya, G. J. Lashkov, and K. B. Demidov,
       Proc. Int. Conf. Lumin., 1, 137 (1966); *Chem. Abstr.*, 70, 33071x
       (1969). Fluorescence: 3-Aminoacridine and 3,6-diaminoacridine
       association with alcohols.

2427a. V. A. Terent'ev, *Zh. Fiz. Khim.*, 46, 1918, 2457, 2468 (1972);
       [English, 1103, 1413, 1418]. IR: $\Delta\nu$-ΔH correlations.

2427b. V. A. Terent'ev, *Zh. Obshch. Khim.*, 42, 2110 (1972); *Chem.
       Abstr.*, 78, 28659f (1973). IR: Hydrogen bonding by C—H in
       CCl_3CHO.

2428. V. A. Terent'ev, *Thermodynamics of Hydrogen Bonding*, Izd.
       Saratov. Univ., Saratov, USSR, 1973, 258 pp.

2429. V. A. Terent'ev and N. Kh. Shtivel, *Zh. Fiz. Khim.*, 43, 2929
       (1969); *Chem. Abstr.*, 72, 60866h (1970). IR: Determination of
       ΔH from temperature dependent absorbances or integrated inten-
       sities.

2429a. K. C. Tewari, F. K. Schweighardt, J. Lee, and N. C. Li, *J. Magn.
       Resonance*, 5, 238 (1971). NMR: Phenobarbital N—H bonding to
       Me_2SO, $MeC(\overline{O})NMe_2$.

2430. U. Thewalt and C. E. Bugg, *J. Amer. Chem. Soc.*, 94, 8892 (1972).
       X-Ray: Crystal structure of 6-thioguanosine monohydrate.

2431. U. Thewalt, C. E. Bugg, and R. E. Marsh, *Acta Crystallogr.*,
       Sect. B, 26, 1089 (1970). BIOL, X-Ray: Crystal structure of
       guanosine dihydrate and inosine dihydrate.

2432. L. Thil, J. J. Riehl, P. Rimmelin, and J. M. Sommer, *J. Chem.
       Soc. D*, 591 (1970). NMR: Intramolecular, protonated α-
       haloaldehydes.

2433. B. R. Thomas, *Biochem. Biophys. Res. Commun.*, 40, 1289 (1970).
       BIOL: Discussion of the origin of the genetic code.

2434. G. J. Thomas, Jr. in *Physical Techniques in Biological Research,
       Vol. 1, Part A, Optical Techniques*, G. Oster, ed., Academic
       Press, New York, 1971, Chapter 4. BIOL: Review, IR, Raman
       techniques in biological research.

2435. G. J. Thomas, Jr. and Y. Kyogoku, *J. Amer. Chem. Soc.*, 89, 4170
       (1967). BIOL: UV studies on binary mixtures of A, U, $\overline{G}$, C,
       and I in $CHCl_3$.

2436. G. J. Thomas, Jr. and M. Spencer, *Biochim. Biophys. Acta*, 179,
       360 (1969). BIOL: Ribosomal RNA structure.

2437. J. Thomas and D. F. Evans, *J. Phys. Chem.*, 74, 3812 (1970).
       Conductance: Electrolytes in formamide.

2437a. J. O. Thomas, *Acta Crystallogr.*, *Sect. B*, 28, 2037 (1972).
       X-Ray: $LiHC_2O_4 \cdot H_2O$ and $LiDC_2O_4 \cdot D_2O$.

2438. L. C. Thomas and R. A. Chittenden, *J. Opt. Soc. Am.*, 52, 829
       (1962). IR: Self-association, phosphorus acids.

2439. L. C. Thomas, R. A. Chittenden, and H. E. R. Hartley, *Nature*,
       192, 1283 (1961). IR: Self-association, alkyl phosphonic and
       alkyl phosphothionic acids.

2440. L. H. Thomas, *J. Chem. Soc.*, 1995 (1963). Viscosity and assoc-
       iation models for alcohols.

2441. L. H. Thomas and R. Meatyard, *J. Chem. Soc.*, 1986 (1963).
       Viscosity studies of self-association of alcohols.

2442. M. R. Thomas, H. A. Scheraga, and E. E. Schrier, *J. Phys. Chem.*,
       69, 3722 (1965). Near-IR: Self-association, H_2O, D_2O.

2443. R. K. Thomas, *Proc. Roy. Soc. (London)*, A325, 133 (1971). IR:
       Gas-phase, HCN—HF, DCN—HF, CH_3CN—HF.

2444. R. K. Thomas, *Proc. Roy. Soc. (London)*, A322, 137 (1971).
       Far-IR: Gas-phase, HF—ether adducts.

2444a. R. K. Thomas, *Proc. Roy. Soc. (London)*, A331, 249 (1972).
       Photoelectron: Self-association, carboxylic acids.

2445. R. K. Thomas and H. W. Thompson, *Proc. Roy. Soc. (London)*, A316,
       303 (1970). IR: Gas-phase, adducts of HCl and DCl with nitriles.

2446. H. W. Thompson, *Pure Appl. Chem.*, 7, 13 (1963). IR: Band
       shapes.

2447. W. E. Thompson and G. C. Pimentel, *Z. Elektrochem.*, 84, 748
       (1960). IR: Frozen mixtures of $CHCl_3$ and $(C_2H_5)_3N$ at 77°K.

2449. P. V. Tibanov, A. F. Vasil'ev, and Yu. A. Baskakov, *Khim.
       Geterotsikl. Soedin.*, 4, 746 (1968); *Chem. Abstr.*, 70, 52616v
       (1969). IR: Self-association, thermodynamic data for O-
       methyl-hydroxylamine derivatives of s-triazines.

2449a. P. V. Tibanov *et al.*, *Khim. Geterotskil. Soedin.*, 124 (1972);
       Chem. Abstr., 77, 47749f (1972). Self-association, O-methyl-
       hydroxylamine derivatives of s-triazines, thermodynamic data.

2450. P. A. Tice and D. B. Powell, *Spectrochim. Acta*, 21, 835 (1965).
       IR: Self-association, trithiocarbonic acid and related thio-
       acids.

2451. M. Tichy in *Advan. Org. Chem.*, R. A. Raphael, E. C. Taylor, and
       H. Wynberg, eds., 5, 115 (1965). IR: Review of intramolecular
       hydrogen bonding.

2452. M. Tichy and L. Kniezo, *Tetrahedron Lett.*, 1665 (1971). IR:
       Intramolecular, "twistane."

2453. S. F. Ting, S. M. Wang, and N. C. Li, *Can. J. Chem.*, 45, 425
       (1967). NMR: Thermodynamic data, H_2O adducts with DMF, DMSO.

2454. I. Tinoco, Jr., R. C. Davis, and S. R. Jaskunas in *Molecular
       Associations in Biology*, B. Pullman, ed., Academic Press, New
       York, 1968, p. 77. BIOL: Review, base pair interactions in
       nucleic acids.

2454a. E. V. Titov, V. M. Belobrov, and V. I. Shurpach, *Dokl. Akad.
       Nauk SSSR*, 212, 159 (1973); *Chem. Abstr.*, 79, 145825p (1973).
       Correlation of ΔH with ΔpK.

2455. E. V. Titov and S. I. Chekushin, *Zh. Prikl. Spektrosk.*, 16, 162
       (1972); *Chem. Abstr.*, 76, 119309s (1972). Theory: Calculation
       of HXH valence angle from X—H vibration frequencies in presence
       of hydrogen bonding.

2456.   M. C. Tobin, *Spectrochim. Acta, Part A*, 25, 1855 (1969). BIOL:
        Raman spectra of DNAs as solid and aqueous gels.

2457.   J. Toha, C. J. Edwards, A. Aguilera, and E. Colombara, *Z. Natur-
        forsch., B*, 22, 1014 (1967). BIOL: Analysis of an experimental
        model of DNA unwinding.

2457a. K. B. Tolpygo and I. P. Ol'khovskaya, *Chem. Phys. Lett.*, 16, 550
        (1972). BIOL: Theory, proton transitions in hydrogen bonds of
        DNA base pairs.

2458.   K. Tomita, L. Katz, and A. Rich, *J. Mol. Biol.*, 30, 545 (1967).
        BIOL, X-Ray: 9-Ethyladenine: 1-methyl-5-fluorouracil complex.

2459.   Y. Toshiyasu and R. Fujishiro, *Nippon Kagaku Kaishi*, 434 (1973);
        Chem. Abstr., 79, 4855n (1973). Intramolecular, dipole moments
        of 2-haloethanols.

2459a. Y. Toshiyasu and R. Fujishiro, *Nippon Kagaku Kaishi*, 429 (1973);
        Chem. Abstr., 78, 147163z (1973). Intramolecular, dipole mo-
        ments of α,ω-alkanediolethers.

2460.   H. Touhara, H. Shimoda, K. Nakanishi, and N. Watanabe, *J. Phys.
        Chem.*, 75, 2222 (1971). IR: $\Delta\nu_{HF}$ for HF solutions in organic
        solvents.

2461.   K. Toyoda, *Bull. Chem. Soc. Jap.*, 42, 1767 (1969). Application
        of Mulliken's charge-transfer complex theory.

2462.   A. Tramer and M. Zaborowska, *Acta Phys. Pol.*, 34, 821 (1968);
        Chem. Abstr., 71, 8185a (1969). UV: Solvent effects on absorp-
        tion, fluorescence, and phosphorescence spectra of 1-naphthol.

2463.   F. Travers, A. M. Michelson, and P. Douzou, *Biochim. Biophys.
        Acta*, 217, 1 (1970). BIOL: Conformational changes of nucleic
        acids in methanol-water solutions.

2464.   V. Trepadus, D. Bally, V. A. Parfenov, and V. G. Liforov,
        Neutron Inelastic Scattering, Proc. Symp., 4th, 1, 483 (1968);
        Chem. Abstr., 70, 119021t (1969). Neutron scattering in phenol,
        hydroquinone, and pyrogallol.

2465.   M. T. Tribble and J. G. Traynham, *J. Amer. Chem. Soc.*, 91, 379
        (1969). NMR: Intramolecular, intermolecular, *o*-substituted
        phenols in Me_2SO, Hammett relationship.

2466.   D. S. Trifan and R. Bacskai, *J. Amer. Chem. Soc.*, 82, 5010
        (1960). IR: Intramolecular, metal-hydrogen bonding in metallo-
        cenes.

2467.   D. S. Trifan and J. F. Ferenzi, *J. Polymer Sci.*, 28, 443 (1958).
        IR: Polyamides.

2468.   B. A. Trodimov *et al.*, *Reakts. Sposobnost Org. Soedin.*, 6, 902
        (1969); *Chem. Abstr.*, 73, 65691c (1970). IR: Association of
        phenol with vinyl ethers.

2469.   A. Trombetti, *J. Chem. Soc. B*, 1578 (1968). UV: Adducts of
        methanol with pyridine N-oxides.

2470.   P. J. Trotter and M. W. Hanna, *J. Amer. Chem. Soc.*, 88, 3724
        (1966). Analysis of the Benesi-Hildebrand method.

2470a. M. N. Tsarevskaya, *Zh. Fiz. Khim.*, 46, 2808 (1972). Thermodyna-
        mic data, acetic acid adducts in cyclohexane.

2471. P. O. P. Ts'o, *Ann. N. Y. Acad. Sci.*, <u>153</u>, 785 (1969). BIOL:
       Review, hydrophobic-stacking properties of the bases in nucleic
       acids.

2472. P. O. P. Ts'o in *Molecular Associations in Biology*, B. Pullman,
       ed., Academic Press, New York, 1968, p. 39. BIOL: Review,
       interactions of nucleic acids.

2473. P. O. P. Ts'o and S. I. Chan, *J. Amer. Chem. Soc.*, <u>86</u>, 4176
       (1964). BIOL: Association of 6-methylpurine and 5-bromouridine
       in H_2O.

2474. P. O. P. Ts'o, G. K. Helmkamp, and C. Sander, *Proc. Natl. Acad.
       Sci. U. S.*, <u>48</u>, 686 (1962). BIOL: Interaction of nucleosides
       with nucleic acids as indicated by the helix-coil transition
       temperature.

2475. P. O. P. Ts'o and W. M. Huang, *Biochemistry*, <u>7</u>, 2954 (1968).
       BIOL: Interactions of poly U and poly C with nucleoside mono-
       and triphosphates.

2476. P. O. P. Ts'o, N. S. Kondo, R. K. Robins, and A. D. Broom,
       J. Amer. Chem. Soc., <u>91</u>, 5625 (1969). BIOL: Association of
       7-methylinosine in H_2O, PMR, conductance studies.

2477. P. O. P. Ts'o, N. S. Kondo, M. P. Schweizer, and D. P. Hollis,
       Biochemistry, <u>8</u>, 997 (1969). BIOL: PMR data on fifteen nucleo-
       sides and nucleotides and on twenty-five dinucleoside mono- and
       diphosphates.

2478. P. O. P. Ts'o and P. Lu, *Proc. Natl. Acad. Sci. U. S.*, <u>51</u>, 17
       (1964). BIOL: Binding of thymine, adenine, steroids, and
       aromatic hydrocarbons to nucleic acids, equilibrium dialysis.

2480. P. O. P. Ts'o and M. P. Schweizer, *Biochemistry*, <u>7</u>, 2963 (1968).
       BIOL: PMR, interactions of poly U and poly C with nucleosides.

2481. M. Tsuboi, *Appl. Spectrosc. Rev.*, <u>3</u>, 45 (1969). BIOL: Review,
       IR, structure studies of nucleic acids.

2482. M. Tsuboi, *Bull. Chem. Soc. Jap.*, <u>25</u>, 60 (1952). IR: Intermo-
       lecular, intramolecular, substituted phenols.

2483. H. Tsubomura, *Bull. Chem. Soc. Jap.*, <u>27</u>, 445 (1954). Theory:
       Early approximate calculation on molecular fragments to investi-
       gate nature of hydrogen bond.

2484. S. V. Tsukerman, V. P. Izvekov, Yu. S. Rozym, and V. F. Lavrus-
       hin, *Khim. Geterotsikl. Soedin.*, 1011 (1968); *Chem. Abstr.*, <u>70</u>,
       72307y (1969). IR: Self-association, intramolecular, chalcones
       containing pyrrole nucleus.

2485. S. V. Tsukerman, L. A. Kutulya, Yu. N. Surov, N. S. Pivenko,
       and V. F. Lavrushin, *Zh. Obshch. Khim.*, <u>40</u>, 1337 (1970); *Chem.
       Abstr.*, <u>74</u>, 47579p (1971). NMR, IR: Association of 2,3,4,6-
       tetrachlorophenol with chalcones.

2486. R. Tubino and G. Zerbi, *J. Chem. Phys.*, <u>53</u>, 1428 (1970). Phonon
       curves and frequency distribution for solid HCOOH, HCOOD, DCOOH.

2486a. E. E. Tucker and E. D. Becker, *J. Phys. Chem.*, <u>77</u>, 1783 (1973).
       NMR, IR: Self-association, *t*-butylalcohol in hexadecane.

2486b. E. E. Tucker, S. B. Farnham, and S. D. Christian, *J. Phys. Chem.*, 73, 3820 (1969). Self-association, methanol.

2487. S. W. Tucker and S. Walker, *Trans. Faraday Soc.*, 62, 2690 (1966). Dielectric Studies: Pyrrole and indole complex with 1,4-diazo-bicyclo[2.2.2]octane.

2488. M. J. B. Tunis and J. E. Hearst, *Biopolymers*, 6, 1345 (1968). BIOL: Dependence of the net hydration of DNA on base composition.

2490. P. Tuomikoski and K. Blomster, *Suomen Kemistilehti B*, 44, 396 (1971). NMR: Use of coupling constants between ^{1}H and ^{13}C to measure K.

2491. A. A. Turovskii *et al.*, *Zh. Fiz. Khim.*, 47, 9 (1973); [English, 6]. IR: *n*-BuOH adducts of peroxides.

2492. A. J. Tursi and E. R. Nixon, *J. Chem. Phys.*, 52, 1521 (1970). IR: Water dimer in a solid nitrogen matrix.

2493. M. Uehara and J. Nakaya, *Bull. Chem. Soc. Jap.*, 43, 3136 (1970). Polarography: Intramolecular, effect on $E_{\frac{1}{2}}$ of azomethines.

2494. O. Uhlenbeck, R. Harrison, and P. Doty in *Molecular Associations in Biology*, B. Pullman, ed., Academic Press, New York, 1968, p. 107. BIOL: Noncomplementary bases, effects on helical complexes of polyribonucleotides.

2496. S. D. Ukeles, Ph.D. Thesis, Polytech. Inst. of Brooklyn, Brooklyn, N. Y., 1968; *Diss. Abstr. B*, 29, 1634 (1968). Hydrogen bonding in primary and secondary amines.

2497. H. E. Ungnade and L. Kissinger, *Tetrahedron Suppl.*, 19, 121 (1963). IR: Self-association, β-nitro-alcohols and diols in CHCl₃ and CCl₄.

2498. H. E. Ungnade, E. D. Loughran, and L. W. Kissinger, *J. Phys. Chem.*, 66, 2643 (1962). IR: Self-association, β-nitro-alcohols and diols in CHCl₃ and CCl₄.

2500. T. Urbanski, *Tetrahedron*, 6, 1 (1959). UV: Intramolecular, nitroparaffin derivatives.

2501. V. N. Ushkalova, V. D. Gol'tsev, and B. V. Tronov, *Spektrosk., Tr. Sib. Soveshch.*, *4th*, 1965, pp. 47-50, N. A. Prilezhaeva, ed.; *Chem. Abstr.*, 74, 17500p (1971). IR: Adducts of aldehydes with *p*-, *m*-, or *o*-cresol or α-, β-naphthol.

2501a. J. H. P. Utley, *J. Chem. Soc.*, 3252 (1963). Electronic: *p*-Nitroaniline interactions with solvents.

2502. L. I. Vachugova, I. P. Lipatova, S. A. Samartseva, and R. R. Shagidullin, *Mater. Nauck. Konf.*, *Inst. Org. Fiz. Khim.*, *Akad. Nauk SSSR*, 51 (1970); *Chem. Abstr.*, 76, 71628r (1972). IR: Adducts of phenol with P—S bases.

2503. C. Valdemoro and S. Fraga, *Afinidad*, 25, 221 (1968); *Chem. Abstr.*, 69, 49159x (1968). Theory: LCAO SCF calculations, DNA bases.

2504. F. B. van Duijneveldt, *J. Chem. Phys.*, 49, 1424 (1969). Theory: Perturbation studies of the effect of lone-pair hybridization.

2505.   F. B. van Duijneveldt and J. N. Murrell, *J. Chem. Phys.*, 46, 1759 (1967). Theory: Double perturbation calculations on a rough model of a hydrogen bond, many approximations.

2506.   J. G. C. M. van Duijneveldt-van de Rijdt and F. B. van Duijneveldt, *J. Amer. Chem. Soc.*, 93, 5644 (1971). Theory: Perturbation calculations on small first-row hydrogen-bonded systems.

2507.   J. G. C. M. van Duijneveldt-van de Rijdt and F. B. van Duijneveldt, *Theor. Chim. Acta*, 19, 83 (1970). Theory: Perturbation studies of hydrogen bond geometry.

2507a. J. G. C. M. van Duijneveldt-van de Rijdt and F. B. van Duijneveldt, *Chem. Weekbl.*, 68, 21 (1972). Theory: Comparison of *ab initio* SCF MO and electrostatic methods.

2508.   J. G. C. M. van Duijneveldt-van de Rijdt and F. B. van Duijneveldt, *Chem. Phys. Lett.*, 2, 565 (1968). Theory: Perturbation studies of hydrogen bond length.

2509.   B. L. Van Duuren, *J. Org. Chem.*, 26, 2954 (1961). UV: Solvent effects on fluorescence emission spectra of indole derivatives.

2510.   C. H. Van Dyke and A. G. MacDiarmid, *J. Phys. Chem.*, 67, 1930 (1963). IR: Relative basicities of Me_2O and Et_2O toward phenol and MeOH.

2511.   R. Van Loon, J. P. Dauchot, and A. Bellemans, *Bull. Soc. Chim. Belges*, 77, 397 (1968). Self-association, *t*-butanol, dielectric constant measurements.

2512.   H. C. Van Ness, J. Van Winkle, H. H. Richtol, and H. B. Hollinger, *J. Phys. Chem.*, 71, 1483 (1967). IR: Self-association, ethanol.

2513.   M. Van Thiel, E. D. Becker, and G. C. Pimentel, *J. Chem. Phys.*, 27, 486 (1957). IR study of water polymers in a solid nitrogen matrix.

2514.   M. Van Thiel, E. D. Becker, and G. C. Pimentel, *J. Chem. Phys.*, 27, 95 (1957). IR: Self-association, MeOH, matrix isolation technique.

2516.   L. K. Vasyanina *et al.*, *Zh. Obshch. Khim.*, 42, 447 (1972); [English, 440]. NMR: Thermodynamic data, *t*-butanol adducts.

2517.   A. A. Vedenor and A. M. Dykhre, *Zh. Eksp. Teor. Fiz.*, 55, 357 (1968). BIOL: Theory of helix-coil transitions in DNA.

2518.   K. Venkataramiah and P. G. Puranik, *Proc. Indian Acad. Sci.*, A56, 96, 155 (1962). Theory: Calculated IR shifts, ν_{NH}, amides.

2519.   K. Venkataramiah and P. G. Puranik, *J. Mol. Spectrosc.*, 7, 89 (1961). IR: Aminopyridines.

2521.   F. Vernon, *J. Chromatogr.*, 63, 249 (1971). GLC: ΔH, adducts of aromatic ketones.

2522.   R. E. Verrall and W. A. Senior, *J. Chem. Phys.*, 50, 2746 (1969). Vacuum UV: Liquid D_2O and H_2O.

2523.   A. A. Viktorova and S. A. Zhevakin, *Dokl. Akad. Nauk SSSR*, 171, 1061 (1966); *Chem. Abstr.*, 66, 60532x (1967). Experimental: Spectroscopic study of $(H_2O)_2$ vapor.

2524.   R. Vilcu and E. Lucinescu, *Rev. Roum. Chim.*, 14, 283 (1969);
        Chem. Abstr., 71, 25198z (1969). Raman: $(CH_3COOH)_2$.

2525.   L. I. Vinogradov, Yu. Yu. Samitov, E. G. Yarkova, and A. A.
        Muratova, *Opt. Spektrosk.*, 26, 959 (1969); *Chem. Abstr.*, 71,
        75983x (1969). NMR: Solvent effects on $^1J_{PH}$ for organophos-
        phorus compounds.

2526.   S. N. Vinogradov, *Can. J. Chem.*, 41, 2719 (1963). IR: Self-
        association, equilibrium constants.

2527.   S. N. Vinogradov and M. Kilpatrick, *J. Phys. Chem.*, 68, 181
        (1964). IR: Self-association, 3,5-dimethylpyrazole.

2528.   S. N. Vinogradov and R. H. Linnell, *Hydrogen Bonding*, Van
        Nostrand-Reinhold Co., New York, 1971.

2529.   P. O. I. Virtanen, L. Pajari, and E. Rahkamaa, *Suomen Kemistil-
        ehti*, 44B, 146 (1971). NMR: Association in N-methylmethane-
        sulfonamide-water mixtures.

2530.   R. Vivilecchia and B. L. Karger, *J. Amer. Chem. Soc.*, 93, 6598
        (1971). Gas Chromatography: Adducts of bicyclic alcohols with
        tris(p-t-butylphenyl) phosphate.

2531.   D. Voet and A. Rich, *Prog. Nucl. Acid Res. Mol. Biol.*, 10, 183
        (1970). BIOL: Review, crystal structures of purines, pyrimi-
        dines and their intermolecular complexes.

2532.   D. Voet and A. Rich, *J. Amer. Chem. Soc.*, 91, 3069 (1969).
        BIOL, X-Ray: 1:1 intermolecular complex containing cytosine
        and 5-fluorouracil.

2532a. D. Voet and A. Rich, *J. Amer. Chem. Soc.*, 94, 5888 (1972).
        BIOL, X-Ray: Ethyladenine adduct of 5-isopropyl-5-bromoalkyl-
        barbituric acid.

2533.   G. C. Vogel and R. S. Drago, *J. Amer. Chem. Soc.*, 92, 5347
        (1970). CAL: Thermodynamic data, adducts of substituted phen-
        ols with sulfur bases.

2534.   V. E. Volkov, V. F. Shabanov, and A. V. Korshunov, *Izv. Vyssh.
        Ucheb. Zaved. Fiz.*, 14, 134 (1971); *Chem. Abstr.*, 75, 42807x
        (1971). Raman: Solid solutions of 2,4-Cl-6-Br- and 2,4-Br-6-
        Cl-phenol.

2535.   R. C. Von Borstel, O. L. Miller, Jr., and F. J. Bollum, *Genetics*,
        61, 401 (1969). BIOL: Unlikelihood of extensive single strands
        of DNA during replication.

2536.   P. H. Von Hippel, *J. Cell. Physiol.*, 74, 235 (1969). BIOL:
        Discussion, conformational aspects of the nucleic acid-protein
        recognition problem.

2537.   P. H. Von Hippel and K. Y. Wong, *J. Biol. Chem.*, 240, 3909
        (1965). BIOL: Conformational stability of globular proteins.

2538.   P. H. Von Hippel and K. Y. Wong, *Science*, 145, 577 (1964).
        BIOL: Generality of the effects of neutral salts on macromolecu-
        lar conformations.

2540.   J. N. Vournakis, H. A. Scheraga, G. W. Rushizky, and H. A.
        Sober, *Biopolymers*, 4, 33 (1966). BIOL: Neighbor-neighbor
        interactions in single-strand polynucleotides.

2541. S. Wada, *Bull. Chem. Soc. Jap.*, 35, 710 (1962). IR: ν_{OH} for catechol, vanillin, benzoic acid, and cinnamic acid complexes with squalene.

2542. S. Wada, *Bull. Chem. Soc. Jap.*, 35, 707 (1962). IR: Thermodynamic data, phenol π-base complexes.

2542a. T. C. Waddington, *Trans. Faraday Soc.*, 54, 25 (1958). Bond strength of HF_2^-.

2542b. D. D. Wagman, W. H. Evans, V. B. Parker, I. Halow, S. M. Bailey, and R. H. Schamm, *Selected Values of Chemical Thermodynamic Properties*, National Bureau of Standards Technical Note, 270-3, Washington, D. C., 1968.

2543. K. G. Wagner and K. Wulff, *Biochem. Biophys. Res. Commun.*, 41 (4), 813 (1970). BIOL: Specific circular dichroism spectra of poly L-lysine complexes with guanosine monophosphates.

2544. R. G. Wake and R. L. Baldwin, *J. Mol. Biol.*, 5, 201 (1962). BIOL: Physical studies on the replication of DNA *in vitro*.

2545. P. Waldstein and L. A. Blatz, *J. Phys. Chem.*, 71, 2271 (1967). Raman: Self-association, formic and acetic acids in aqueous and hydrocarbon solutions.

2546. P. M. B. Walker, *Prog. Nucl. Acid Res. Mol. Biol.*, 9, 301 (1969). BIOL: Review, specificity of molecular hybridization of nucleic acids.

2547. C. Walling and L. Heaton, *J. Amer. Chem. Soc.*, 87, 48 (1965). IR: Self-association, *t*-butyl hydroperoxide.

2547a. S. C. Wallwork, *Acta Crystallogr.*, 15, 758 (1962). X-Ray: Hydrogen bond radii.

2548. G. E. Walrafen, *J. Chem. Phys.*, 50, 560 (1969). Raman: H_2O in D_2O.

2549. G. E. Walrafen, *J. Chem. Phys.*, 40, 3249 (1964). Raman: Effects of temperature and electrolytes on H_2O spectrum.

2549a. G. E. Walrafen and L. A. Blatz, *J. Chem. Phys.*, 59, 2646 (1973). Raman: H_2O, D_2O.

2550. W. Walter and U. Sewekow, *Justus Liebigs Ann. Chem.*, 761, 104 (1972). NMR, IR: Intramolecular *o*-thioacetanilides.

2551. S. M. Wang and N. C. Li, *J. Amer. Chem. Soc.*, 90, 5069 (1968). NMR: Self-association, nucleosides.

2552. M. J. Waring, *Ann. Rep. Prog. Chem.*, 64, 493 (1967). BIOL: Review of structural aspects of nucleic acids.

2553. I. D. Warren and E. B. Wilson, *J. Chem. Phys.*, 56, 2137 (1972). Microwave: Intramolecular, trifluoroethylamine.

2554. M. M. Warshaw and C. R. Cantor, *Biopolymers*, 9, 1079 (1970). BIOL: CD spectra of all sixteen ribonucleoside phosphates containing A, U, C, and G compared with corresponding deoxy compounds, stacking.

2555. M. M. Warshaw and I. Tinoco, Jr., *J. Mol. Biol.*, 13, 54 (1965). BIOL: ORD of six dinucleoside phosphates, stacking.

2556. M. M. Warshaw and I. Tinoco, Jr., *J. Mol. Biol.*, $\underline{20}$, 29 (1966).
       BIOL: ORD, structures of dinucleoside phosphates in H_2O.

2557. R. Wasylishen and T. Schaefer, *Can. J. Chem.*, $\underline{48}$, 1263 (1970).
       NMR: Intramolecular, *o*-disubstituted aniline derivatives.

2558. M. Watabe, *Nippon Kagaku Kaishi*, 225 (1972); *Chem. Abstr.*, $\underline{76}$,
       105973s (1972). $\Delta\nu_{OH}$ for adducts of ring-substituted phenols
       with hexamethylbenzene.

2558a. N. Watanabe, K. Nakanishi, H. Touhara, and H. Shimoda, *Asahi
       Garasu Kogyo Gijutsu Shoreikai Kenkyu Hokoku*, $\underline{20}$, 181 (1972);
       Chem. Abstr., $\underline{78}$, 104003e (1973). IR: $\Delta\nu_{HF}$ for adducts of
       high-purity HF.

2559. D. G. Watson, D. J. Sutor, and P. Tollin, *Acta Crystallogr.*,
       $\underline{19}$, 111 (1965). BIOL, X-Ray: Crystal structure of deoxyadeno-
       sine monohydrate.

2560. J. D. Watson, *Molecular Biology of the Gene*, 2nd edition,
       Benjamin, New York, 1970. BIOL: General reference, molecular
       biology.

2561. J. D. Watson and F. H. C. Crick, *Cold Spring Harbor Symp. Quant.
       Biol.*, $\underline{18}$, 123 (1953). BIOL: Analysis of X-Ray data, secondary
       structure of DNA.

2562. J. D. Watson and F. H. C. Crick, *Nature*, $\underline{171}$, 964 (1953).
       BIOL: Genetic implications of the structure of DNA.

2563. J. D. Watson and F. H. C. Crick, *Nature*, $\underline{171}$, 737 (1953).
       BIOL: Analysis of X-Ray data, secondary structure of DNA.

2563a. J. S. Waugh, F. B. Humphrey, and B. M. Yost, *J. Phys. Chem.*,
       $\underline{57}$, 486 (1953). NMR: Symmetrical single-minimum in HF_2^-.

2564. B. B. Wayland and R. S. Drago, *J. Amer. Chem. Soc.*, $\underline{86}$, 5240
       (1964). IR: $\Delta\nu_{OH}$, phenol adducts of anisole, thioanisole.

2565. E. L. Wehry, *Fluorescence News*, $\underline{6}$, 1 (1971); *Chem. Abstr.*, $\underline{75}$,
       62671b (1971). Effect of hydrogen bonding on fluorescence of
       5-hydroxyquinoline and 8-hydroxyquinoline.

2566. E. G. Weidemann and G. Zundel, *Z. Naturforsch. A*, $\underline{25}$, 627
       (1970). $H_5O_2^+$, polarizability of hydrogen bond with the tunnel-
       ing proton.

2567. I. Weinberg and J. R. Zimmerman, *J. Chem. Phys.*, $\underline{23}$, 748 (1955).
       NMR: Concentration dependence of chemical shifts, H_2O, EtOH.

2568. J. H. Weiner and A. Askar, *Nature*, $\underline{226}$, 842 (1970). Theory:
       Model for hydrogen-bonded chains.

2569. M. A. Weiner and P. Schwartz, *J. Organometal. Chem.*, $\underline{35}$, 285
       (1972). IR, NMR: $\Delta\nu_{OH}$ and chemical shifts, organosilicon- and
       organogermanium substituted pyridine N-oxides as bases with
       MeOH and phenol.

2570. M. Weissman, L. Blum, and N. V. Cohan, *Chem. Phys. Lett.*, $\underline{1}$, 95
       (1967). Theory: Hydrogen bond in ice-like structures, SCF
       with approximate integrals.

2570a. M. Weissmann and N. V. Cohan, *J. Chem. Phys.*, $\underline{59}$, 1385 (1973).
       Theory: CNDO/2 on hydrated electron.

2571. M. Weissmann and N. V. Cohan, *J. Chem. Phys.*, 43, 119 (1965).
       Theory: Hydrogen bond in ice, Slater basis SCF calculation
       with approximate multicenter integrals.

2571a. H. A. Wells, *Nature (London)*, *Phys. Sci.*, 244, 95 (1973).
       Raman: Intramolecular, alcohols.

2572. W. E. Wentworth, W. Hirsch, and E. Chen, *J. Phys. Chem.*, 71,
       218 (1967). Association constant determinations, least-squares
       treatment of spectrophotometric data.

2573. D. Weres and S. A. Rice, *J. Amer. Chem. Soc.*, 94, 8983 (1972).
       Theory: Liquid water structure.

2574. R. West, reported at 140th ACS Meeting, St. Louis, Mo., March,
       1961. Near-IR: Thermodynamic data, phenol adducts with bases
       in CCl_4.

2575. R. West, *J. Amer. Chem. Soc.*, 81, 1614 (1959). IR: $\Delta\nu_{OH}$,
       adducts of phenols with olefins.

2576. R. West, R. H. Baney, and D. L. Powell, *J. Amer. Chem. Soc.*, 82,
       6269 (1960). IR: $\Delta\nu_{OH}$, acid-base studies of Ph_3XOH where X =
       Group IV atoms.

2577. R. West and C. S. Kraihanzel, *J. Amer. Chem. Soc.*, 83, 765
       (1961). IR: $\Delta\nu_{CH}$, adducts of terminal acetylenes with bases.

2578. R. West, D. L. Powell, M. K. T. Lee, and L. S. Whatley, *J. Amer.
       Chem. Soc.*, 86, 3227 (1964). Near-IR: Thermodynamic data for
       adducts of phenol with ethers.

2579. R. West, D. L. Powell, L. S. Whatley, M. K. T. Lee, and P. von
       R. Schleyer, *J. Amer. Chem. Soc.*, 84, 3221 (1962). Near-IR:
       Thermodynamic data, adducts of phenol with alkyl halides, lack
       of $\Delta\nu$-ΔH correlation.

2580. J. G. Wetmur and N. Davidson, *J. Mol. Biol.*, 31, 349 (1968).
       BIOL: Kinetics of renaturation of DNA.

2581. E. Whalley, *Can. J. Chem.*, 50, 310 (1972). Theory: Dipole
       moment derivative of hydrogen bond in ice.

2582. K. B. Whetsel, *Applied Spectroscopy Reviews*, 2, 1 (1968).
       Near-IR: Review, near-IR region.

2583. K. B. Whetsel, *Spectrochim. Acta*, 17, 614 (1961). IR: Associa-
       tion of amines in $CHCl_3$.

2584. K. B. Whetsel and R. E. Kagarise, *Spectrochim. Acta*, 18, 329
       (1962). UV: K, adducts of $CHCl_3$ with cyclohexanone.

2585. K. B. Whetsel and R. E. Kagarise, *Spectrochim. Acta*, 18, 315
       (1962). IR: K, *p*-cresol-ketone complexes.

2586. K. B. Whetsel and J. H. Lady, *Amer. Chem. Soc.*, *Fuel Chem.
       Preprints*, 11, 210 (1967). Near-IR, IR: Self-association,
       phenol.

2587. K. B. Whetsel and J. H. Lady, *J. Phys. Chem.*, 69, 1596 (1965).
       Near-IR: Thermodynamic data, adducts of aniline with N-
       methylaniline.

2588. K. B. Whetsel and J. H. Lady, *J. Phys. Chem.*, 68, 1010 (1964).
       Near-IR: Thermodynamic data, adducts of $CHCl_3$.

2589. N. E. White and M. Kilpatrick, *J. Phys. Chem.*, 59, 1044 (1955). Self-association of N—H compounds.

2590. S. C. White and H. W. Thompson, *Proc. Roy. Soc. (London)*, A291, 460 (1966). IR: K, phenol-nitriles in CCl_4.

2591. D. C. Whitney and R. M. Diamond, *J. Phys. Chem.*, 67, 209 (1963). Adduct of tributyl phosphate with H_2O.

2592. R. R. Wiederkehr and H. G. Drickamer, *J. Chem. Phys.*, 28, 311 (1958). IR: Model for solvent effects on stretching frequencies.

2593. G. R. Wiley and S. I. Miller, *J. Amer. Chem. Soc.*, 94, 3287 (1972). NMR: Thermodynamic data, $CHCl_3$ adducts.

2594. I. J. Wilk and R. T. Lundquist, *J. Mol. Struct.*, 3, 247 (1969). IR: Intramolecular, ν_{NH} and ν_{PO} of tris(butylamine)phosphine oxides.

2595. M. H. F. Wilkins, S. Arnott, D. A. Marvin, and L. D. Hamilton, *Science*, 167, 1693 (1970). BIOL: Structure of DNA, misconceptions on Fourier analysis and Watson-Crick base pairing.

2596. M. H. F. Wilkins, R. G. Gosling, and W. E. Seeds, *Nature*, 167, 759 (1951). BIOL: X-Ray diffraction photographs of moist DNA.

2597. M. H. F. Wilkins and J. T. Randall, *Biochem. Biophys. Acta*, 10, 192 (1953). BIOL: B form of DNA exists in cells.

2598. M. H. F. Wilkins, A. R. Stokes, and H. R. Wilson, *Nature*, 171, 737 (1953). BIOL: X-Ray, B form of DNA from widely differing sources.

2599. M. H. F. Wilkins, H. R. Wilson, and L. D. Hamilton, *Proc. Natl. Acad. Sci. U. S.*, 65, 761 (1970). BIOL: X-Ray diffraction patterns of DNA do not support a four-stranded helical structure.

2600. G. Will, *Angew. Chem., Int. Ed. Engl.*, 8, 356 (1969). Neutron Diffraction: A review of various applications including hydrogen bonding.

2600a. D. R. Williams, L. J. Schaad, and J. N. Murrell, *J. Chem. Phys.*, 47, 4916 (1967). Double perturbation theory.

2601. J. G. Williams and O. Delatycki, *J. Polymer Sci.*, Part A-2, 8, 295 (1970). IR: Epoxy-diamines.

2602. J. M. Williams and S. W. Peterson, *J. Amer. Chem. Soc.*, 91, 776 (1969). Neutron Diffraction: $[H_5O_2]^+$ in $HAuCl_4 \cdot 4H_2O$.

2602a. J. M. Williams and L. F. Schneemeyer, *J. Amer. Chem. Soc.*, 95, 5780 (1973). Neutron Diffraction: p-Toluidinium bifluoride.

2603. R. L. Williams, *Ann. Reports (Chem. Soc. London)*, 58, 34 (1961). IR: Review of solvent effects on stretching frequencies.

2604. H. R. Wilson, *Nature*, 225, 545 (1970). BIOL: Changes in nucleoside conformation.

2605. H. R. Wilson, A. Rahmon, and P. Tollin, *J. Mol. Biol.*, 46, 585 (1969). BIOL: Furanose ring conformation in nucleosides and nucleic acids.

2606. H. J. Wimette and R. H. Linnell, *J. Phys. Chem.*, 66, 546 (1962).
       IR: Thermodynamic data, pyrrole-pyridines in CCl₄.

2607. H. Winde, *Z. Chem.*, 10, 101 (1970). Theory: Short review in
       German on the quantum chemistry of intermolecular forces,
       including the hydrogen bond.

2608. A. Witkowski, *J. Chem. Phys.*, 52, 4403 (1970). Far-IR: Acetic
       acid dimer.

2609. A. Witkowski, *J. Chem. Phys.*, 47, 3645 (1967). Theory, IR:
       Coupling of ν_{OH} with hydrogen bond vibrations in carboxylic
       acid dimer.

2610. A. Witkowski and M. Wojcik, *Bull. Acad. Pol. Sci.*, *Ser. Sci.
       Chim.*, 19, 577 (1971); *Chem. Abstr.*, 76, 19654g (1972). IR:
       Derivation of vibrational Hamiltonian.

2610a. A. Witkowski and M. Wojcik, *Chem. Phys. Lett.*, 20, 615 (1973).
       IR: Crystalline 1-methylthymine.

2611. T. A. Wittstruck and J. F. Cronan, *J. Phys. Chem.*, 72, 4243
       (1968). NMR: Competition between inter- and intramolecular
       hydrogen bonding in alcohols.

2612. J. Witz and V. Luzzati, *J. Mol. Biol.*, 11, 620 (1965). BIOL:
       Small angle X-Ray scattering, structure of poly A.

2613. B. Wladislaw, R. Rittner, and H. Viertler, *J. Chem. Soc. B*,
       1859 (1971). IR, NMR: $\Delta\nu_{OH}$, chemical shifts, adducts of α-
       ethylthio-substituted carbonyl compounds with $C_6H_5C\equiv CH$ and
       C_6H_5OH.

2614. C. Woese, *Nature*, 226, 817 (1970). BIOL: Reciprocating ratchet
       mechanism for translation, RNA.

2615. C. Woese, *Prog. Nucl. Acid Res. Mol. Biol.*, 7, 107 (1967).
       BIOL: Discussion of the genetic code.

2616. B. Wojtkowiak, *Compt. Rend. Acad. Sci.*, 268C, 24 (1969). IR:
       Method for determining K.

2617. R. Wolf, D. Houalla, and F. Mathis, *Spectrochim. Acta*, A23, 1641
       (1967). IR: Self-association $R_2PH(O)$ compounds.

2618. S. Wolf, *Mutation Res.*, 10, 405 (1970). BIOL: Tertiary struc-
       ture of chromosomes.

2619. F. H. Wolfe, K. Oikawa, and C. M. Kay, *Can. J. Biochem.*, 47, 977
       (1967). BIOL: UV, CD, complexes of synthetic polynucleotides.

2620. F. H. Wolfe, K. Oikawa, and C. M. Kay, *Can. J. Biochem.*, 47, 637
       (1969). BIOL: UV, CD, polynucleotides.

2621. R. V. Wolfenden, *J. Mol. Biol.*, 40, 307 (1969). BIOL: UV,
       tautomeric equilibria, inosine and adenosine.

2622. H. Wolff, *Z. Elektrochem.*, 66, 529 (1962). IR: Self-
       association, $MeNH_2$.

2623. H. Wolff and G. Gamer, *J. Phys. Chem.*, 76, 871 (1972). IR:
       Self-association, Me_2NH.

2623a. H. Wolff and G. Gamer, *Spectrochim. Acta*, 28A, 2121 (1972).
       IR: Hydrogen bonding and rotational isomerism of $(C_2H_5)_2NH$.

2623b. H. Wolff and D. Horn, *Ber. Bunsenges, Physik. Chem.*, 71, 467
(1967). IR: Self-association, 2,2,2-trifluoroethylamine.

2624. H. Wolff and D. Mathias, *J. Phys. Chem.*, 77, 2081 (1973). IR:
Adducts of aniline, Fermi resonance.

2624a. E. M. Wolley and L. G. Hepler, *J. Phys. Chem.*, 76, 3058 (1972).
CAL: Self-association, phenol in CCl_4, C_6H_6, and C_6H_{12}.

2624b. K. F. Wong, T. S. Pang, and Ng Soon, *J. Chem. Soc., Chem.
Commun.*, 55 (1974). NMR: Correlation of ΔH with hydrogen bond
chemical shifts.

2624c. P. T. T. Wong and E. Whalley, *J. Chem. Phys.*, 55, 1830 (1971).
Far-IR: Crystalline α-methanol.

2625. A. B. Wood, F. V. Robinson, and R. C. Araujo Lago, *Chem. Ind.
(London)*, 1738 (1969). IR, UV: Intramolecular, gossypol.

2625a. J. L. Wood, *J. Mol. Struct.*, 12, 283 (1972). IR, Theory:
Slightly asymmetric double minimum potential functions.

2625b. J. L. Wood, *J. Mol. Struct.*, 13, 141 (1972). Far-IR:
[pyHpy]$^+$, proton potential.

2625c. J. L. Wood, *J. Mol. Struct.*, 16, 349 (1973). IR: Proton
tunneling in asymmetric complexes.

2625d. J. L. Wood, *J. Mol. Struct.*, 17, 307 (1973). IR: Effects of
proton potential on vibrational spectra of hydrogen-bonded
systems.

2625e. R. A. Work, III and R. L. McDonald, *J. Phys. Chem.*, 77, 1148
(1973). Far-IR: Benzoic acid-halide association.

2626. J. D. Worley and I. M. Klotz, *J. Chem. Phys.*, 45, 2868 (1966).
Near-IR: ΔH, adducts of HOD, self-association of H_2O, D_2O.

2627. W. Woycicki and B. Trebicka-Mojska, *Przem. Chem.*, 48, 280 (1969);
Chem. Abstr., 71, 60545g (1969). Thermodynamic data for phenol-
nitrile systems.

2628. R. W. Wright and R. H. Marchessault, *Can. J. Chem.*, 46, 2567
(1968). IR: Intramolecular, cyclopentane- and cyclohexane-
1,2-diol monoacetates.

2629. F. Y. Wu, *Phys. Rev. Lett.*, 24, 1476 (1970). Model for
hydrogen-bonded ferroelectrics.

2630. T. T. Wu, *Bull. Math. Biophys.*, 30, 681 (1968). BIOL: X-Ray,
strandedness of DNA at 92% relative humidity, possibility of a
four-stranded helical model.

2631. T. T. Wu, *Proc. Natl. Acad. Sci. U. S.*, 63, 400 (1969). BIOL:
Secondary structure of DNA.

2632. O. R. Wulf, U. Liddel, and S. B. Hendricks, *J. Amer. Chem. Soc.*,
58, 2287 (1936). IR: Intramolecular, *o*-substituted phenols.

2632a. O. P. Yablonskii, V. A. Belyaev, and A. N. Vinogradov, *Usp.
Khim.*, 41, 1260 (1972); [English, 565]. Review of association
of hydrocarbon hydroperoxides.

2632b. O. P. Yablonskii, N. M. Rodionova, L. F. Lapuka, and S. Yu.
Pavlov, *Zh. Fiz. Khim.*, 47, 799, 1922 (1973); [English, 454].
IR, NMR: ΔH, adducts of 2-methyl-but-1-en-3-yne, 1-pentyne.

2633.   V. I. Yakutin, S. S. Dubov, A. V. Fokin, Yu. M. Kosyrev, and
        N. P. Novoselov, *Zh. Obshch. Khim.*, 38, 2174 (1968); *Chem.
        Abstr.*, 70, 52641z (1969). IR: NH and NF frequency shifts of
        HNF_2 mixtures with bases.

2634.   I. Yamaguchi, *Bull. Chem. Soc. Jap.*, 34, 1606 (1961). NMR:
        Amino proton associated with acetone.

2635.   I. Yamaguchi, *Bull. Chem. Soc. Jap.*, 34, 1602 (1961). NMR:
        p-Cresol adducts of dioxane, acetone, pyridine.

2636.   I. Yamaguchi, *Bull. Chem. Soc. Jap.*, 34, 744 (1961). NMR:
        Dimethyl phenols in various solvents.

2637.   I. Yamaguchi, *Bull. Chem. Soc. Jap.*, 34, 451 (1961). NMR:
        Substituted phenols in various bases.

2638.   I. Yamaguchi, *Bull. Chem. Soc. Jap.*, 34, 353 (1961). NMR:
        Intramolecular, 2-hydroxybenzophenones in CCl_4.

2638a. T. Yamana, A. Tsuji, and Y. Mizukami, *Chem. Pharm. Bull.*, 20,
        922 (1972). IR: Intramolecular, 4-hydroxyvaleramide.

2639.   H. Yamatera, B. Fitzpatrick, and G. Gordon, *J. Mol. Spectrosc.*,
        14, 268 (1964). Near-IR: Spectra of H_2O, D_2O mixtures.

2639a. S. Yambe, S. Kato, H. Fujimoto, and K. Fukui, *Theor. Chim. Acta*,
        30, 327 (1973). Theory: LCAO SCF study of $HF + NH_3 \rightarrow NH_4F$.

2640.   R. Yamdagni and P. Kebarle, *J. Amer. Chem. Soc.*, 93, 7139 (1971).
        Gas-phase: Thermodynamic data, $(BHR)^- \rightarrow B^- + HR$.

2640a. R. Yamdagni and P. Kebarle, *J. Amer. Chem. Soc.*, 95, 3504 (1973).
        Gas-phase: Hydrogen bonding in proton-bound amine dimers.

2641.   H. B. Yang and R. W. Taft, *J. Amer. Chem. Soc.*, 93, 1310 (1971).
        NMR: Hydrogen-bonded ion pairs.

2642.   J. T. Yang and T. Samejima, *Prog. Nucl. Acid Res. Mol. Biol.*,
        9, 223 (1969). BIOL: Review, ORD and CD of nucleic acids.

2643.   J. T. Yang and T. Samejima, *Biochem. Biophys. Res. Commun.*, 33,
        739 (1968). BIOL: CD, ORD, effect of base tilting in the
        optical activity of nucleic acids.

2644.   T. Yano, T. Miyata, and M. Hasegawa, *J. Phys. Soc. Jap.*, 31,
        1290 (1970). Theory: Effect of electron correlation on the
        hydrogen bond.

2645.   T. Yasunaga, S. Nishikawa, and N. Tatsumoto, *Bull. Chem. Soc.
        Jap.*, 44, 2308 (1971). Kinetics: Ultrasonic, self-association,
        benzoic acid derivatives.

2646.   T. Yasunaga, N. Tatsumoto, H. Inoue, and M. Muira, *J. Phys.
        Chem.*, 73, 477 (1969). Ultrasonic: Kinetic, intramolecular,
        salicylates.

2647.   Y. Yeh, *J. Chem. Phys.*, 52, 6218 (1970). BIOL: Helix-coil
        transition of the co-polymer deoxadenylate-deoxythymidylate.

2648.   S. Yomosa, *Biopolymers*, Symp. No. 1, 527 (1964). Theory:
        Energy levels in hydrogen-bonded polypeptides.

2649.   T. Yonezawa, H. Saito, S. Matsuoka, and K. Fukui, *Bull. Chem.
        Soc. Jap.*, 38, 1431 (1965). NMR: Self-association, diols.

2650. S. Yoshida and M. Asai, *Chem. Pharm. Bull. (Tokyo)*, 7, 162
       (1959). IR: Self-association, 3- and 4-pyridine carboxylic
       acids, intramolecular bonding in pyrazine- and pyridine-2-
       carboxylic acids.

2651. Z. Yoshida and M. Haruta, *Tetrahedron Lett.*, 3745 (1965). IR:
       Intramolecular, 2-hydroxy-4-substituted acetophenones.

2652. Z. Yoshida and M. Haruta, *Tetrahedron Lett.*, 2631 (1964). IR:
       Intramolecular, 2-hydroxy-2,5-substituted acetophenones.

2653. Z. Yoshida and N. Ishibe, *Bull. Chem. Soc. Jap.*, 42, 3263
       (1969). IR: Adducts of phenol with monoolefins.

2654. Z. Yoshida and N. Ishibe, *Bull. Chem. Soc. Jap.*, 42, 3259
       (1969). IR: Adducts of 2,6-xylenols with mesitylene.

2655. Z. Yoshida and N. Ishibe, *Bull. Chem. Soc. Jap.*, 42, 3254
       (1969). IR: Adducts of phenols and naphthols with aromatic
       hydrocarbons.

2656. Z. Yoshida and N. Ishibe, *Spectrochim. Acta*, 24A, 893 (1968).
       IR: Thermodynamic values, phenol adducts of π-bases, nitriles,
       ketones, esters.

2657. Z. Yoshida, N. Ishibe, and H. Kusumoto, *J. Amer. Chem. Soc.*, 91,
       2279 (1969). IR: Adducts of phenol with cyclopropane ring.

2658. Z. Yoshida, N. Ishibe, and H. Ozoe, *J. Amer. Chem. Soc.*, 94,
       4948 (1972). IR: Thermodynamic data, adducts of phenol with
       acetylenes and allenes.

2659. Z. Yoshida, H. Ogoshi, and T. Tokumitsu, *Tetrahedron*, 26, 5691
       (1970). IR, NMR: Intramolecular, 3-substituted-2,4-pentane-
       dione.

2660. Z. Yoshida and E. Ōsawa, *J. Amer. Chem. Soc.*, 88, 4019 (1966).
       IR: K, phenol—π-base complexes in CCl_4.

2661. Z. Yoshida and E. Ōsawa, *J. Amer. Chem. Soc.*, 87, 1467 (1965).
       IR: Intermolecular, thermodynamic data for phenol adducts of
       π-bases.

2661a. M. Yoshimine and A. D. McLean, *Int. J. Quantum Chem.*, 1S, 313
       (1967). Theory: SCF calculations on HF_2^-.

2662. D. W. Young, P. Tollin, and H. R. Wilson, *Acta Crystallogr.*,
       Sect. B, 25, 1423 (1969). BIOL, X-Ray: Thymidine.

2663. R. Young, E. Fishman, and P. Luner, *Tappi*, 53, 1126 (1970).
       IR: Association of methanol with tetrahydropyran derivatives.

2664. E. E. Yudovich and V. V. Pal'chevskii, *Zh. Obshch. Khim.*, 39,
       62 (1969); *Chem. Abstr.*, 70, 101325k (1969). UV: Spectra of
       aqueous solutions of nitrophenols and nitrophenolates.

2665. G. V. Yukhnevich, *Opt. Spektrosk.*, supplement 2, 223 (1963);
       [English, 114 (1966)]. IR: Vibrational spectrum of H_2O per-
       turbed by hydrogen bond.

2666. G. V. Yukhnevich and A. V. Karyakin, *Dokl. Akad. Nauk SSSR*, 153,
       681 (1964). Theory: Relation between vibration frequencies of
       water and the hydrogen bond energy.

2667. G. V. Yukhnevich, A. A. Vetrov, and B. P. Shelyukhaev, *Aust. J. Chem.*, 23, 1507 (1970). IR: Self-association models for H$_2$0.

2668. S. Yun-bo and V. M. Chulanovskii, *Opt. Spektrosk.*, supplement 2, 214 (1963); [English, 109 (1966)]. IR: Self-association, methanol.

2669. S. Yun-bo and V. M. Chulanovskii, *Opt. Spektrosk.*, supplement 2, 218 (1963); [English, 111 (1966)]. IR: Δv_{CO}, alcohols.

2670. H. G. Zachau, *Angew. Chem., Int. Ed., Engl.*, 8, 711 (1969). BIOL: Review on tRNA.

2671. B. A. Zadorozhnyi, *Zh. Fiz. Khim.*, 39, 1944 (1965); [English, 1031]. IR: Intramolecular, *o*-hydroxy-carbonyl derivatives of naphthalene.

2672. B. A. Zadorozhnyi and I. K. Ishchenko, *Opt. Spektrosk.*, 19, 551 (1965); [English, 306]. IR: Correlation of Δv_{CO} for adducts of phenol and methanol with carbonyl donors.

2673. B. E. Zaitsev, L. B. Preobrazhenskaya, B. N. Kolokolov, and V. M. Allenov, *Zh. Fiz. Khim.*, 44, 2129 (1970); *Chem. Abstr.*, 74, 3142m (1971). IR: Aminoanthraquinones.

2674. M. Zanger, W. W. Simons, and A. R. Gennaro, *J. Org. Chem.*, 33, 3672 (1968). NMR: Intramolecular, N-acylanilines.

2675. V. Zanker, H. H. Mantsch, and E. Erhardt, *An. Quim.*, 64, 659 (1968). IR: NH shifts in 9,10-dihydroacridines.

2676. N. G. Zarakhani and M. I. Vinnik, *Zh. Fiz. Khim.*, 38, 632 (1964); [English, 332]. Raman: Self-association, aqueous carboxylic acids.

2677. N. G. Zarakhani and M. I. Vinnik, *Zh. Fiz. Khim.*, 37, 2550 (1963); [English, 1376]. Raman: Self-association, aqueous formic acid.

2678. O. A. Zasyadko, Yu. L. Frolov, and R. G. Mirskov, *Zh. Prikl. Spektrosk.*, 15, 939 (1971); *Chem. Abstr.*, 76, 65904m (1972). IR: Association of ethynylstannanes with solvents.

2679. M. Zaucer, J. Koller, and A. Ažman, *Z. Naturforsch. A*, 25, 1769 (1970). NMR: Proton-shielding constants for HF, HF$_2^-$, H$_2$F$_3^-$.

2680. N. S. Zaugg, S. P. Steed, and E. M. Woolley, *Thermochimica Acta*, 3, 349 (1972). CAL: Self-association, acetic acid, dimer model.

2680a. N. S. Zaugg, L. E. Trejo, and E. M. Woolley, *Thermochimica Acta*, 6, 293 (1973). CAL: Self-association, chloro-substituted acetic and propionic acids in CCl$_4$.

2681. A. Zecchina, E. Guglielminotti, S. Coluccia, and E. Borello, *J. Chem. Soc., A*, 2196 (1969). IR: Nitriles adsorbed on chromia-silica.

2682. Th. Zeegers-Huyskens, *Ind. Chim. Belge*, 33, 4018 (1968). IR: Review.

2683. Th. Zeegers-Huyskens, *Spectrochim. Acta*, 23A, 855 (1967). IR: Proton-transfer, adducts of aniline with alcohols and phenols.

2684. Th. Zeegers-Huyskens, *Spectrochim. Acta*, 21, 221 (1965). IR: Degree of proton transfer for adducts of alcohols with *n*-propylamine.

2685. S. V. Zenin, M. Orban, and G. B. Sergeev, *Zh. Fiz. Khim.*, 44, 2914 (1970); *Chem. Abstr.*, 74, 41729r (1971). NMR: K, association of spatially hindered phenols with tri-p-tolyl phosphate.

2686. V. V. Zharkov, A. V. Zhitinkina, and F. A. Zhokhova, *Zh. Fiz. Khim.*, 44, 223 (1970); [English, 120]. IR: Thermodynamic properties for adducts of alcohols and tertiary amines.

2687. D. A. Zhogolev and I. V. Matyash, *Chem. Phys. Lett.*, 10, 444 (1971). Theory: Extended Hückel studies on small water polymers.

2688. E. L. Zhukova and I. I. Shman'ko, *Opt. Spektrosk.*, 26, 532 (1969); *Chem. Abstr.*, 71, 43971a (1969). IR: Intensities of NH_2 bands of amines.

2688a. E. L. Zhukova and I. I. Shman'ko, *Opt. Spektrosk.*, 32, 514 (1972); *Chem. Abstr.*, 77, 11763s (1972). IR: Shift in $-NH_2$ deformation vibration, amine adducts.

2689. E. L. Zhukova and I. I. Shman'ko, *Opt. Spektrosk.*, 25, 500 (1968); *Chem. Abstr.*, 70, 24379b (1969). IR: Effect of hydrogen bonding on NH_2 vibrations of amines.

2690. D. Ziessow, U. Jentschura, and E. Lippert, *Ber. Bunsenges, Physik. Chem.*, 75, 901 (1971). NMR: ^{17}O, CH_3COOH.

2691. D. Ziessow and E. Lippert, *Ber. Bunsenges, Physik. Chem.*, 74, 13 (1970). NMR: ^1H, ^{13}C, $PhC{\equiv}CH$ association with hexamethylphosphoramide.

2691a. F. Zigan, *Ber. Bunsenges, Physik. Chem.*, 76, 686 (1972). Theory: Bent hydrogen bonds.

2692. R. Zilversmit, *Mol. Pharmacol.*, 7, 674 (1971). IR: Intramolecular, lucanthone.

2694. C. Zimmer, G. Luck, H. Venner, and J. Fric, *Biopolymers*, 6, 563 (1968). BIOL: Predominant protonation effects, structural changes of DNA, spectrophotometric titrations.

2695. C. Zimmer and H. Triebel, *Biopolymers*, 8, 573 (1969). BIOL: Reversible and irreversible conformational changes in acid-induced denaturation of DNA.

2696. H. Zimmermann, *Z. Elektrochem.*, 65, 821 (1961). IR: Self-association, imidazole, proton tunneling.

2698. B. Zmudzka and D. Shugar, *FEBS Lett.*, 8, 52 (1970). BIOL: Role of the 2'-hydroxyl in polynucleotide conformation.

2699. I. Zuika, J. Bankovskis, A. Sturis, D. Zaruma, J. Cviule, and M. Cviule, *Latv. PSR Zinat. Akad. Vestis, Kim. Ser.*, 650 (1971); *Chem. Abstr.*, 76, 85018d (1972). IR: Intramolecular, 8-mercaptoquinoline.

2699a. I. Zuika and J. Bankovskis, *Usp. Khim.*, 42, 39 (1973); [English, 22]. Review of -SH and >S interactions, 306 references.

2700. G. Zundel, *Allg. Prakt. Chem.*, 21, 329 (1970). Theory: Hydrogen bonds with tunnel protons.

2700a. G. Zundel, *Hydration and Intermolecular Interaction*, Academic Press, New York, 1969, Chapter 5. IR: Potential well, $H_5O_2^+$ and $H_9O_4^+$.

2700b. G. Zundel and H. Metzger, *Spectrochim. Acta*, 23A, 759 (1967).
        IR: Association of acid groups in polyelectrolytes, asymmetric
        potential minimum.

2701.   G. Zundel and J. Muehlinghaus, *Z. Naturforsch. B*, 26, 546 (1971).
        IR: Poly-L-histidine films, aqueous imidazole, pyrazole.

2702.   G. Zundel and E. G. Weidemann, *Trans. Faraday Soc.*, 66, 1941
        (1970). IR: Tunnel frequency dependence.

2703.   G. Zundel and E. G. Weidemann, *Z. Phys. Chem.*, 83, 327 (1973).
        IR: Polarization of hydrogen bonds of hydration water molecules.

AUTHOR INDEX

This index refers both to the text and
to the Bibliography. If an author has more
than one bibliography number and his work
is referred to in the text by bibliography
number only, the page number is followed by
the bibliography number in parentheses. In
addition, a complete listing of bibliography
numbers is given in brackets for each author.

A

Abboud, J. L., [195]
Abdulaev, N. D., [1150]
Abdulnur, S., [2273-2275]
Abel, E. W., [1]
Abello, L., 371-372(2), 375(2),
   [2-4, 1871]
Abraham, R. J., [5]
Abramovich, M. A., [6, 864]
Acharya, P. K., [6a,b,c]
Ackermann, T. [1057]
Adam, W., 129, 137, [7]
Adamek, P., 161, 321(8), 345(8),
   347(8), 348(8), 351(8),
   [8-10]
Adreichikov, Yu. S., [1374b]
Afanas'eva, A. M., 363, 364
   [1899a]
Afans'ev, M. L., [1287]
Affsprung, H. E., 181(452a), [11,
   12, 451, 452a, 1466, 2188a]
Agaeva, E. A., [1367a]
Agami, C., [13]
Agarwal, R. P., [747]
Agashkin, O. V., [1853]
Ageno, M., [14-18]
Aguilera, A., [2457]
Ahlf, J., [20]
Ahmad, D., [2425a]
Aida, K., 333, [1119]
Aihara, A., [20a]
Akashi, K., [1829]
Akazawa, H., [1355]
Akermark, B., [784]
Akhunov, T. F., [1899b-1901b]
Aknurina, I. S., [830]
Aksel'rod, V. D., [455]
Aksnes, G., 247(25), 322(22, 25),
   323(22, 25), 324(22), [21-25]
Akutsu, H., [26]

Al-Adhami, L., 5, 24, [27]
Alagona, G., 128, 129, 139, [27a,b]
Alaune, Z., [2414b]
Albers, R. J., [1132a]
Albery, W. J., [28]
Albriktsen, P., 322-324, [22, 23]
Aldo, M., [1473]
Aldridge, M. H., 28, [690]
Alei, M., Jr., 47, 49, 50, 52,
   [29, 773, 1488, 1489]
Alemagna, A., [1500, 1501]
Alexeeva, T. L., [270]
Alford, A. L., [827]
Alford, D. O., [30]
Alger, T. D., [30a]
Allan, E. A., 241(2073), 255(32),
   262, [31, 32, 2073]
Allen, G., 287, [34]
Allen, G. A., [33]
Allen, L. C., 53, 68(36), 92-101,
   105, 110, 112, 119-121, 124-
   126, 129-133, 149-151, [35, 36,
   1292-1298, 1300, 2144]
Allenov, V. M., [2673]
Allerhand, A., 199(40), 201, 202,
   203, 206, 207, 220(40), 274,
   [37-41]
Alley, S. K., Jr., 370, 371, [42]
Allred, A. L., 366(536), 368(536),
   [43, 535-537]
Almlöf, J., 110, 114, 131, 134,
   140-143, [44-48]
Altardi, G., [49]
Alves, A. C. P., [50]
Amaldi, F., [49]
Amis, E. S., [51, 1056]
Amosov, A. V., [52]
Ananthanarayanan, V., [53]
Anderegg, J. W., [1392]
Anderson, C. M. W., 290, [54]
Anderson, G. R., 129, 131, [54a,b,

555

Mehl, J., 108, 141, 145, [475]
Meilahn, M. K., [1636]
Meister, R., [763a]
Meister, T. G., [1637]
Melchers, F., [679]
Melli, M., [250]
Mellier, A., [1638]
Melnick, A. M., [11]
Mely, B., [2005]
Menefee, A., [30]
Menon, B. C., [1263]
Men'shova, I. I., [1537]
Meos, A. I., [1982]
Mercer, E. E., 289, [675]
Merlet, P., 120, [1638a]
Merlin, J. C., [792]
Merrill, J. R., 241, [1639]
Merrill, S. H., [1070]
Meshitsuka, S., 22, [1639a]
Mesley, R. J., [1640]
Messmer, R. P., [1641]
Mettee, H. D., 68(1642), 272
 (1346), 277(1346), 293(1350),
 296(1350), 297(1350), [1345-
 1350, 1642]
Metzger, H., 12, [2700b]
Meunier, A., 68(1638b), 93, 121,
 [1638b]
Meyer, W., 129, [1642a]
Meyer, W. C., [1643]
Michell, A. J., 204, [485, 487,
 1644]
Michelson, A. M., [303, 304, 1442,
 1568, 1645-1648, 1956, 1957,
 2463]
Middaugh, R. L., [1794]
Middleton, W. J., [1649]
Mierzecki, R., [1649a]
Mikawa, Y., 15(1158, 1159), 17
 (1158, 1159), 19(1158, 1159),
 20, 21, 24(311), [311, 312,
 1157-1161, 1650-1653]
Mikolajczyk, J., 326, 327, 345,
 347, 350, 351, 362, [2398]
Miksa-Spiric, A., 367, [1590a]
Miles, D. W., [1654-1657]
Miles, H. T., [1090, 1091, 1658-
 1660, 2242, 2322, 2376]
Miles, M. H., [1024, 1661, 1662]
Milesch, G., [1287d]
Miligy, F., 294, 295, 300, 301,
 [1906]
Millar, D. B., [1664]
Millen, D. J., 5(27, 73, 75, 1666,
 1667), 6(220), 24, [27, 73-75,
 220-222, 1665-1668]

Miller, C., [1669]
Miller, D., [1863]
Miller, J. A., [2222]
Miller, Jeffrey H., [1670]
Miller, John H., 85-86, [1670a]
Miller, O. L., Jr., [2535]
Miller, P. J., [1670b]
Miller, R., [1669]
Miller, R. E., [630, 2105]
Miller, S. I., 173(1096a, 2593),
 175, 176, 279, 280, 364(1096a),
 365(1096a), 366-369(2593),
 [1096a, 1097, 1597, 2593]
Miller, W. R., Jr., [839]
Minakata, K., [1670c]
Minesinger, R. R., 28, 29, 236,
 [1227, 1227a]
Minkin, V. I., [1671, 2146]
Minton, A. P., 88, 101, 123, 148,
 149, [1672]
Mirek, J., [1673]
Mirskov, R. G., [2678]
Mirza, H. N., [424]
Mislow, K., [111]
Mitchell, E. J. (see also Arnett,
 E. M.), 247(1674), 366-343
 (1674), [70, 71, 1674]
Mitra, S. S., 296, 321, 322, 376,
 [1675]
Mitsky, J., 236, 244, 245, 377-379
 (1676), 335-343(1201), [1201,
 1676]
Mitsui, Y., [2176]
Miyata, T., 68(2644), [1677, 2644]
Miyazaki, H., [1354]
Miyazawa, T., [1678]
Mizukami, Y., [2638a]
Mizushima, S., [2384]
Mocek, M. M., 44, [1563]
Moffitt, W., 72, [1678a]
Mohr, S. C., [1679]
Moliton, C., [1679a]
Moll, F., 15, [1680, 1681]
Mollendal, H., [1604a]
Moller, K. D., [1681a]
Momany, F. A., 140, 143, 152,
 [1564, 1682]
Monaselidze, D. R., [1987]
Moniz, W. B., [1683]
Monny, C., [1646, 1647]
Montanari, F., 232, 297, 298, 346,
 [847a]
Monti, L., [1571]
Moore, C. E., 85, [1683a]
Moore, L. F., [1114, 1115]
Moore, L. S., 123, [1684]

Slavik, V., [66]
Slejko, F. L., 42(2290), 173(2290, 2291), 177(2291), 218(2291), 220(2291), 222-224, 241(2290), 244, 302(2290a), 315(2290), 345(2290a), 348-349(2290), 349(2290a), 366-369(2291), 370(2290a), 371(2290, 2290a), [2290-2291]
Sletaniak, L., [557]
Slifkin, M. A., [2292]
Slocum, D. W., 44(2293), 45, [1177a, 2293]
Slodkova, I. A., [1907]
Smerkolj, R., 234, [615]
Smirnov, L. D., [1453]
Smissman, E., 272, [1065]
Smith, B. M., [2292]
Smith, C. P., [778]
Smith, D. A., [1037]
Smith, D. W., [1352]
Smith, F., [2293a]
Smith, F. A., [2294]
Smith, G. P., [347]
Smith, H. F., 307, [2295]
Smith, I. C. P., [1096, 2296, 2297]
Smith, J. A. S., [2462a]
Smith, J. W., [154, 2298]
Smith, P. A., [339]
Smolikova, J., [1085]
Smolyanskii, A. L., 14(610a), [610a, 957, 958, 1487, 2299, 2300]
Smyth, C. P., [548]
Snatzke, G., [1515]
Snead, C. C., [2301]
Snyder, L. C., [2302]
Snyder, R. G., 10, [2303]
Snyder, W. R., [2303a]
Sobczyk, L., 234, 357(1031a), 373 (1033), 374(1032, 1033), [615, 977, 1031a-1033, 2304]
Sobell, H. M., [1016-1019, 1391, 1632, 1670, 2157-2160, 2305, 2306]
Sober, H. A., [2540]
Sobotka, W., [694, 2307]
Sobue, H., [1824a]
Socha, J., [173a, 174]
Socrates, G., 241, 241(2309, 2310), 242(2309), 284, 285, [937, 2308-2310]
Soczewinski, E., [2311]
Soda, G., [2312, 2313]
Soell, D., [2314]

Sokolov, M. P., [1474a]
Sokolov, N. D., [1990-1992, 2185, 2315, 2316]
Solcaniova, E., [1322, 1322a, 2317]
Solie, T. N., [2318]
Solmajer, T., 10, [975a]
Someno, K., [1799]
Somers, B. G., 44(2320), 352-353 (2320), 356-357(2320), [2319, 2320]
Sommer, J. M., [1210, 2432]
Somorjai, R. L., 9, [1561b, 2321]
Sondheimer, F., [317]
Soon, Ng, 244, [2624b]
Sorarrain, O. M., [616]
Sorm, F., [940]
Sostman, H. D., [1200a]
Soup, R. R., [2322]
Souter, R. W., [1492a]
Souty, N., [1438, 1439, 1530]
Sparr, H., [2214]
Speakman, J. C., 10, 35, 37, 38 (551a, 1571a), [551a, 885, 1571a, 2260, 2323-2323b]
Spencer, J. H., [413]
Spencer, J. N., 254(2103), 256, [2103, 2104, 2303a, 2324]
Spencer, M., [2325, 2436]
Sperber, G., 107, 146, [1528]
Spicer, C. K., 272, [1545]
Spinner, E., [2326]
Spivey, H. O., 186, 187, [998]
Spotswood, T. M., [2327]
Springer, C. S., Jr., 289, [1635, 2328]
Sprowls, J., [1784]
Spunta, G., 23, [845a,b]
Spurr, R. A., [2329]
Srinivasan, R., [2330]
Srivatsava, G. P., 27(510), [510, 511, 2331-2333]
Stacey, M., [788]
Stachura, R., [554]
Stafford, F. E., [1887]
Stanescu, Gr., [2350]
Stanevich, A. E., 16(2334), [2334-2337]
Stanford, S. C., 198, [891, 892]
Stankovsky, S., 333, [2337a]
Starosta, J., [610b]
Stasyuk, I. V., [2338]
Statz, G., 19(2340), 24(2340), [2339, 2340]
Staude, H., [2378]
Stavitskaya, G. P., [2135, 2135a]
Stec, W. J., 290, [2341]

Watabe, M., [2558]
Watanabe, N., [2460, 2558a]
Watanabe, S., [2369, 2370]
Watanabe, T., 309, [1219]
Watari, F., 333, [1119]
Watkinson, J. G., 182(1190), 208
    (1190), 255(1189), 287, 325
    (1190), 335(1190), 343(1190),
    345(1190), 346(1190), [34,
    1189-1191]
Watson, D. G., [2559]
Watson, I. D., 366, 367, [123]
Watson, J. D., [2560-2563]
Watson, R. E., 56, [265a]
Waugh, J. S., 104, [2563a]
Wayland, B. B., 171(2564), 221,
    222, 230(2564), 332(655),
    [654, 655, 2564]
Webb, H. M., [90a]
Webb, J. G. K., [1547]
Webb, K. H., 287, [34]
Wehry, E. L., [2565]
Weidemann, E. G., 13(1164, 1164a,
    2703), 129, [1164, 1164a,
    2566, 2703]
Weil, J. A., [1045]
Weiler-Feilchenfeld, H., [207,
    2011]
Weinberg, I., [2567]
Weiner, J. H., [2568]
Weiner, M. A., [2569]
Weir, C. E., [1483]
Weisleder, D., [405]
Weissman, S. I., [1332]
Weissmann, M., 150, [482b, 2570-
    2571]
Wells, H. A., [2571a]
Wells, R. D., [1686]
Wen, W.-Y., [796]
Wendricks, R. N., [43]
Wenkert, E., [605, 632]
Wentworth, W. E., 161, [2572]
Weres, O., [2573]
Werner, R. L., 229-231, 373-377
    (1996), 379(1996), 380(1996),
    [349, 1996, 1997, 2069]
Wertz, D. W., [676]
West, R., 171, 208(1981, 2579),
    272(1065), 284, 310(1981),
    317-319(2578), 319(1981), 326
    (1981), 334(2574, 2578),
    [1065, 1586, 1981, 2194,
    2573-2579]
West, W., [1273]
Westrum, E. F., Jr., [423]
Wetmur, J. G., [2580]

Wettermark, G., 217, 310, 336, 344-
    351, [2361]
Whalley, E., [737a, 1809, 2581,
    2624b]
Whatley, L. S., 208(2579), 317-319
    (2578), 331(2578), 334(2578),
    [2578, 2579]
Whetsel, K. B., 25, 26, 172, 176
    (2588), 275(1402), 276, 289,
    347(2585), 366(2588), 367
    (2584), 371(1401, 2587), 375
    (1401, 2587), [1216a, 1401-
    1403, 2582-2588]
White, N. E., [2589]
White, S. C., 321, 322, [2590]
Whitney, D. C., [2591]
Wiebe, R., [1333]
Wieder, M. J., 276, [96]
Wiederkehr, R. R., [2592]
Wigfield, Y. Y., [1546]
Wilchek, M., [2358]
Wiley, G. R., 173, 175, 176, 366-
    369, [2593]
Wilezok, T., [1513]
Wilford, L. D. R., [1560]
Wilk, I. J., [2594]
Wilk, W. D., [1679]
Wilkins, M. H. F., [1411, 1413,
    2595-2599]
Wilks, A. D., 164, 165, [1046]
Will, G., [2600]
William, A. G., [1414]
Williams, D. A., [1978]
Williams, D. L., [1037]
Williams, D. R., 70(1742, 2600a),
    99, [1742, 2600a]
Williams, E. M., [1198]
Williams, H. C. W. L., [1684a]
Williams, J. G., [2601]
Williams, J. M., 104, [1527a,
    2602, 2602a]
Williams, P. P., 266, [1034]
Williams, R. J. P., [144]
Williams, R. L., 195, 196(2603),
    198, 199, [178, 181, 191-193,
    2603]
Williamson, A. G., 366(123), 367
    (123), [123, 161]
Willis, H. A., 173, 176, [883]
Willmot, F. W., 178(1491), [1491,
    1492]
Willson, W., [695]
Wilson, D. L., [834]
Wilson, E. B., [99, 2553]
Wilson, H. R., [1411, 1413, 2598,
    2599, 2604, 2605, 2662]

SUBJECT INDEX

This index refers both to the text and to the Bibliography. Text page numbers are followed by Bibliography reference numbers in brackets.

Table 4 of Chapter 2 and the Appendix are not indexed here since they are already in a suitable form for data retrieval.

A

Ab initio methods, 64-70
Acetamide, N-methyl
  adducts of, 229-231 [1996]
  far-IR, [1146]
  IR, [300, 2386]
  near-IR, [233]
  NMR, 280-282 [900, 901, 1414a]
    ^{14}N, 46 [2155]
  self-association, 280-282
    [900, 901, 1132a, 1147,
    1414a]
Acetamide, N,N-dimethyl-
  as reference base, 216, 217,
    225-227 [652, 2252, 2255]
Acetamide, N-methyl-phenyl-
  intramolecular, [2385]
Acetamidines
  IR, [1198]
Acetanilides
  adducts of, [2389]
  far-IR, [1146]
Acetanilides, *ortho-*
  intramolecular, 277 [56, 58,
    1536, 1546, 1982a, 2034,
    2244b, 2550]
  IR, [2244b, 2550]
  NMR, [56, 1546, 1982a, 2034,
    2550]
  UV, [1536]
Acetate salts
  bis (trichloro-), [975a]
  bis (trifluoro-),
    IR, [1670b, 1808a]
    NMR, [1191b]
    neutron diffraction, [1571a]
    X-Ray diffraction, [885]
  methylhalo-, [869a]
Acetic acid (see also Appendix,
  Carboxylic acids)

adducts of, [2470a]
  acetonitrile, [1899a]
  aniline, [1191a]
  dimethylsulfoxide, [1470]
  halosubstituted acetic acids,
    [11, 610, 610a, 612a, 869a,
    2341a]
  phosphoric acid, [1195a]
  sulfuric acid, [1272, 1319]
  triethylamine, [1287d]
bromo-, [1239]
chloro-, [1157a]
cyano-, [2286b]
electron diffraction, 20 [1233]
far-IR, 20, 21, 25 [404, 857,
  1161, 1276, 1449, 1765,
  2608]
halo-, 9 [967, 969, 973, 976a]
IR, [1030, 1031, 1319, 2300]
kinetics, [507, 2175, 2402]
o-methoxy-phenoxy-, [1062a]
near-IR, 27 [1704a]
normal coordinate analysis,
  [1276]
NMR, [1578, 1877, 1909, 1914,
  2070, 2074, 2236, 2259,
  2690]
^{17}O, 48-50 [449, 2093]
neutron diffraction, [1195]
phenylthio-, [1062a]
Raman, [2524, 2545]
relaxation, [30a]
self-association, 287 [182,
  364a, 460, 507, 667, 754b,
  882a, 1102, 1276, 1478,
  1727, 1849, 1877, 1909,
  2072, 2074, 2236, 2300,
  2402, 2524, 2545, 2680]
thio-, [868]
trichloro-, 9, 10 [965, 969,
  973, 1196]

thiocyanates, [1119]
collision complexes, 199 [1083]
correlations (see also Correlations)
  enthalpy, 246-250 [70, 167, 847, 1136, 1139, 1937, 2055]
  frequency shift, 210-215, 217, 219-222, 246 [719, 918, 1182, 1778]
  far-IR, 16, 23 [310, 845a, 858, 859, 978, 978a, 1109, 1156]
  IR, 198-206 [40, 181, 252, 524, 1083, 1110, 1140, 1141, 1418, 1602a]
  near-IR, 171, 172 [1981, 2586]
  normal coordinate analyses, 23 [859]
  NMR, [731, 911, 1573, 2044, 2586]
  reference acid, 210-222, 227, 230, 235, 237, 242, 243 [653, 654, 719, 917a, 1182, 1778, 2255, 2310]
  relaxation measurements, [1451]
  self-association, 183 [20, 188, 270, 295, 930, 2044, 2586]
  thermodynamic measurements
    calorimetry, 167-171 [70, 666, 719, 1811a]
    IR, 172 [70]
    near-IR, 171, 172 [1981]
    NMR, 174, 175, 177 [1772]
    UV, 172 [100, 1181, 2347]
  torsional vibrations, [2121]
Phenols
  adducts of (see also Appendix), 23 [70, 100, 101, 156, 176, 378, 379, 442, 978a, 1031a, 1032, 1033, 1100, 1100a, 1227a, 1234, 1303, 1304, 1455, 1456, 1456a, 1724a, 1900, 1923-1924a, 2043, 2092, 2137a, 2168c, 2392]
  chromatography, [2311]
  2,6-diisopropyl-
    NMR, 43, 44 [2320]
  2,6-dimethyl-, 18, 19 [1156]
  far-IR, [1112, 1156]
  m-fluoro-, 164, 181, 182, 219, 222 [651, 653, 1811a]
  frequency shift correlations, 210, 212, 215, 217, 230 [650, 1421, 1779, 2533, 2564]
  IR, [1901a, 1901b, 2042, 2397]
  near-IR, [1586, 2156]

NMR, [1606, 1607, 2636, 2637]
ortho-
  electronic, 262-264 [317, 590, 591]
  IR, 264, 265 [555]
  intramolecular, 260, 262-266 [31, 1062, 1085, 1123, 1339, 1375, 1534, 1539, 1582, 1582a, 1759, 1815, 1867, 1901, 2028, 2128, 2415, 2465, 2482, 2632]
  NMR, 262 [1975, 2073]
  self-association, 283, 284 [2029]
ortho-t-butyl-, [2359]
ortho-halo-
  gas phase, 182 [1467]
  IR, 254, 255 [115, 402b, 1189, 1191, 1884, 1920]
  NMR, 255, 256 [31, 632, 1287b]
  NQR, [1331b]
  potential function, 256, 257 [2103, 2206]
  thermodynamic data, 254-256 [32, 118, 402a, 1189, 1467, 2103]
ortho-nitro-
  intramolecular, 263-265 [555, 590]
  microwave, [1425b]
  UV, [1541, 2664]
ortho-trifluoromethyl-, [1081, 1314, 1315]
para-t-butyl-, 222 [653]
para-chloro-, 219
para-fluoro-, see p-Fluorophenol
para-methyl-, 222 [653]
para-nitro-, see p-Nitrophenol
Raman, [2534]
reference acids, 210, 211 [650]
self-association, 284, 285 [20, 59, 188, 270, 411, 561, 778, 930, 937, 1112, 1142, 1156, 1606, 1607, 1881, 1901a, 2051, 2181, 2319]
sterically hindered, [1795]
UV, [1142]
Phenylacetylene
  adducts, 176 [720, 828, 883, 1766, 2613]
  IR, [13, 2257, 2613]
  NMR, [13, 766, 2613]
Phenylenediamines, [1255, 1341]
Phosphorus compounds (see also Organophosphorus compounds)
  phosphates, 10 [425, 514, 618a,